Abiotic Stresses in Crop Plants

Abiotic Stresses in Crop Plants

Edited by

Usha Chakraborty

Plant Biochemistry Laboratory, Department of Botany,
University of North Bengal, India

and

Bishwanath Chakraborty

Immuno-Phytopathology Laboratory, Department of Botany,
University of North Bengal, India

CABI is a trading name of CAB International

CABI	CABI
Nosworthy Way	38 Chauncy Street
Wallingford	Suite 1002
Oxfordshire OX10 8DE	Boston, MA 02111
UK	USA
Tel: +44 (0)1491 832111	Tel: +1 800 552 3083 (toll free)
Fax: +44 (0)1491 833508	Tel: +1 (0)617 395 4051
E-mail: info@cabi.org	E-mail: cabi-nao@cabi.org
Website: www.cabi.org	

© CAB International 2015. All rights reserved. No part of this publication may be reproduced in any form or by any means, electronically, mechanically, by photocopying, recording or otherwise, without the prior permission of the copyright owners.

A catalogue record for this book is available from the British Library, London, UK.

Library of Congress Cataloging-in-Publication Data

Abiotic stresses in crop plants / edited by Usha Chakraborty and Bishwanath Chakraborty.
   pages cm
  ISBN 978-1-78064-373-1 (hbk : alk. paper) 1. Crops--Effect of stress on. 2. Crops--Physiology. I. Chakraborty, Usha. II. Chakraborty, Bishwanath. III. C.A.B. International.

 SB112.5.A24 2014
 631.5'82--dc23

                          2014011777

ISBN-13: 978 1 78064 373 1

Commissioning editor: Sreepat Jain
Editorial assistant: Emma McCann
Production editor: Shankari Wilford

Typeset by SPi, Pondicherry, India.
Printed and bound in the UK by CPI Group (UK) Ltd, Croydon, CR0 4YY.

Contents

Contributors	vii
Introduction	ix
About the Editors	xiii

PART 1: TEMPERATURE, WATER AND SALINITY STRESS

1. Heat-Shock Proteins and Molecular Chaperones: Role in Regulation of Cellular Proteostasis and Stress Management — 1
   M. Kapoor and S.S. Roy

2. Heat Response, Senescence and Reproductive Development in Plants — 23
   Renu Khanna-Chopra and Vimal Kumar Semwal

3. Ethylene, Nitric Oxide and Haemoglobins in Plant Tolerance to Flooding — 43
   Luis A.J. Mur, Kapuganti J. Gupta, Usha Chakraborty, Bishwanath Chakraborty and Kim H. Hebelstrup

4. Monitoring the Activation of Jasmonate Biosynthesis Genes for Selection of Chickpea Hybrids Tolerant to Drought Stress — 54
   Palmiro Poltronieri, Marco Taurino, Stefania Bonsegna, Stefania De Domenico and Angelo Santino

5. Genetic Engineering of Crop Plants to Sustain Drought Tolerance — 71
   J. Amudha and G. Balasubramani

6. Physiology and Biochemistry of Salt Stress Tolerance in Plants — 81
   André Dias de Azevedo Neto and Elizamar Ciríaco da Silva

7. Sugarcane (*Saccharum* sp.) Salt Tolerance at Various Developmental Levels — 102
   Ch.B. Gandonou and N. Skali-Senhaji

PART 2: HEAVY METALS AND OZONE

8   The Impact of Ozone Pollution on Plant Defence Metabolism: Detrimental Effects on Yield and Quality of Agricultural Crops   112
Fernanda Freitas Caregnato, Rafael Calixto Bortolin, Armando Molina Divan Junior and José Cláudio Fonseca Moreira

9   Potentiality of Ethylene in Sulfur-Mediated Counteracting Adverse Effects of Cadmium in Plants   136
Mohd Asgher, Mehar Fatma and Nafees A. Khan

10   Heavy Metal and Metalloid Stress in Plants: The Genomics Perspective   164
Piyalee Panda, Lingaraj Sahoo and Sanjib Kumar Panda

11   Influence of Arsenic and Phosphate on the Growth and Metabolism of Cultivated Plants   178
Asok K. Biswas

PART 3: GENERAL ABIOTIC STRESSES AND THEIR ALLEVIATION BY MICROBES

12   Plant Responses to Abiotic Stresses in Sustainable Agriculture   196
Zlatko Stoyanov Zlatev

13   Interactive Role of Polyamines and Reactive Oxygen Species in Stress Tolerance of Plants   212
Rup Kumar Kar

14   Indirect and Direct Benefits of the Use of *Trichoderma harzianum* Strain T-22 in Agronomic Plants Subjected to Abiotic and Biotic Stresses   222
Antonella Vitti, Maria Nuzzaci, Antonio Scopa and Adriano Sofo

15   Role of Microorganisms in Alleviation of Abiotic Stresses for Sustainable Agriculture   232
Usha Chakraborty, Bishwanath Chakraborty, Pannalal Dey and Arka Pratim Chakraborty

Index   255

Contributors

J. Amudha, Central Institute for Cotton Research, Post Box 2, Shankar Nagar (PO), Nagpur, 440 010 Maharashtra, India.
Mohd Asgher, Department of Botany, Aligarh Muslim University, Aligarh 202 002, India.
André Dias de Azevedo Neto, Laboratory of Biochemistry, Centre of Exact and Technological Sciences, Federal University of Recôncavo of Bahia, Cruz das Almas, 44380-000, BA, Brazil.
G. Balasubramani, Central Institute for Cotton Research, Post Box 2, Shankar Nagar (PO), Nagpur 440 010 Maharashtra, India.
Asok K. Biswas, Plant Physiology and Biochemistry Laboratory, Centre for Advanced Study, Department of Botany, University of Calcutta, 35 Ballygunge Circular Road, Kolkata 700 019, India.
Stefania Bonsegna, CNR-ISPA, Institute of Sciences of Food Productions, via Monteroni, 73100 Lecce, Italy.
Rafael Calixto Bortolin, Department of Biochemistry, Center for Oxidative Stress Research, Federal University of Rio Grande do Sul (UFRGS), Porto Alegre, RS, Brazil.
Fernanda Freitas Caregnato, Department of Biochemistry, Center for Oxidative Stress Research, Federal University of Rio Grande do Sul (UFRGS), Porto Alegre, RS, Brazil.
Arka Pratim Chakraborty, Immuno-Phytopathology Laboratory, Department of Botany, University of North Bengal, Siliguri 734013, West Bengal, India.
Bishwanath Chakraborty, Immuno-Phytopathology Laboratory, Department of Botany, University of North Bengal, Siliguri 734013, West Bengal, India.
Usha Chakraborty, Plant Biochemistry Laboratory, Department of Botany, University of North Bengal, Siliguri 734013, West Bengal, India.
Elizamar Ciríaco da Silva, Laboratory of Applied Botany, Department of Biology, Federal University of Sergipe, Aracaju, 49100-000, SE, Brazil.
Stefania De Domenico, CNR-ISPA, Institute of Sciences of Food Productions, via Monteroni, 73100 Lecce, Italy.
Pannalal Dey, Plant Biochemistry Laboratory, Department of Botany, University of North Bengal, Siliguri 734013, West Bengal, India.
Armando Molina Divan Junior, Laboratory of Plant Bioindication, Center of Ecology, Federal University of Rio Grande do Sul (UFRGS), Porto Alegre, RS, Brazil.
Mehar Fatma, Department of Botany, Aligarh Muslim University, Aligarh 202 002, India.
Ch.B. Gandonou, Laboratoire de Biologie et Santé, Faculté des Sciences de Tétouan, Université Abdelmalek Essaâdi, BP 2121 Tétouan, Morocco.

Kapuganti J. Gupta, Department of Plant Sciences, University of Oxford, South Parks Road, Oxford OX1 3RB, UK.

Kim H. Hebelstrup, Department of Molecular Biology and Genetics, Aarhus University, Forsøgsvej 1, DK-4200 Slagelse, Denmark.

M. Kapoor, Department of Biological Sciences, The University of Calgary, Calgary, Alberta, Canada T2N 1N4.

Rup Kumar Kar, Plant Physiology & Biochemistry Laboratory, Department of Botany, Visva-Bharati University, Santiniketan 731 235, West Bengal, India.

M. Iqbal R. Khan, Department of Botany, Aligarh Muslim University, Aligarh 202 002, India.

Nafees A. Khan, Department of Botany, Aligarh Muslim University, Aligarh 202 002, India.

Renu Khanna-Chopra, Stress Physiology Lab, Water Technology Centre, Indian Agricultural Research Institute, New Delhi 110012, India.

José Cláudio Fonseca Moreira, Department of Biochemistry, Center for Oxidative Stress Research, Federal University of Rio Grande do Sul (UFRGS), Porto Alegre, RS, Brazil.

Luis A.J. Mur, Institute of Environmental and Rural Science, Aberystwyth University, Edward Llwyd Building, Aberystwyth SY23 3DA, UK.

Maria Nuzzaci, School of Agricultural, Forestry, Food and Environmental Sciences, University of Basilicata, Viale dell'Ateneo Lucano, 10-85100 Potenza, Italy.

Piyalee Panda, Plant Molecular Biotechnology Laboratory, Department of Life Science and Bioinformatics, Assam University, Silchar 788 011, India.

Sanjib Kumar Panda, Plant Molecular Biotechnology Laboratory, Department of Life Science and Bioinformatics, Assam University, Silchar 788 011, India.

Palmiro Poltronieri, CNR-ISPA, Institute of Sciences of Food Productions, via Monteroni, 73100 Lecce, Italy.

S.S. Roy, Department of Biological Sciences, The University of Calgary, Calgary, Alberta, Canada T2N 1N4.

Lingaraj Sahoo, Department of Biotechnology, Indian Institute of Technology, Guwahati, India.

Angelo Santino, CNR-ISPA, Institute of Sciences of Food Productions, via Monteroni, 73100 Lecce, Italy.

Antonio Scopa, School of Agricultural, Forestry, Food and Environmental Sciences, University of Basilicata, Viale dell'Ateneo Lucano, 10-85100 Potenza, Italy.

Vimal Kumar Semwal, Stress Physiology Lab, Water Technology Centre, Indian Agricultural Research Institute, New Delhi 110012, India.

N. Skali-Senhaji, Laboratoire de Physiologie Végétale et d'Etude des Stress Environnementaux, Faculté des Sciences et Techniques, Université d'Abomey-Calavi, 01 BP 526 Cotonou, République du Bénin.

Adriano Sofo, School of Agricultural, Forestry, Food and Environmental Sciences, University of Basilicata, Viale dell'Ateneo Lucano, 10-85100 Potenza, Italy.

Marco Taurino, CNR-ISPA, Institute of Sciences of Food Productions, via Monteroni, 73100 Lecce, Italy.

Antonella Vitti, School of Agricultural, Forestry, Food and Environmental Sciences, University of Basilicata, Viale dell'Ateneo Lucano, 10-85100 Potenza, Italy.

Zlatko Stoyanov Zlatev, Department of Plant Physiology and Biochemistry, Agricultural University-Plovdiv, 4000 Plovdiv, Bulgaria.

Introduction

The existence of life on earth depends on the interaction of environment and living organisms and unless this is maintained at a steady balance this whole existence is at risk. Fast changing environment, increasing population, urbanization and a multitude of related factors affect food productivity by plants – the main food factory of the earth. What is of major concern is the ever-increasing population, projected to be around 9.2 billion by 2050, making demands on food production on the one hand, coupled with decreasing crop productivity on the other. In this scenario, looking for ways and means of maintaining sustainable food production seems a daunting task. Abiotic stresses, mainly due to changing climatic conditions, provide the main challenge to sustainable agriculture. Plant abiotic stressors include: fluctuating temperatures, from very low to extremely high; water level shifts, ranging from water scarcity to flooding; excessive soil salinity, caused in part by prolonged use of irrigation water and low quantities of rainfall, combined with rising saline groundwater levels; heavy metals and other pollutants in soil and air etc. Fortunately for life on earth, many plants are resilient and have developed degrees of tolerance to such stresses. The major thrust for increasing food productivity would be to accelerate such tolerance or resistance mechanisms in the plant at physiological, cellular or molecular levels, leading to improved crop health. However, sufficient caution has to be exercised while dealing with the intricate molecular mechanisms in the plant, as interference in nature's mechanisms may sometimes be counter-productive. To this end, scientists across the globe are working on developing tools for engineering enhanced resistance of plants against abiotic stresses with subsequent increase in productivity.

This book is a compilation of articles that focus on the above problem and will give an overall perspective on the current progress being made in the area of abiotic stresses and their management for sustainable agriculture. The 15 chapters in the book are divided into three sections: Temperature, water and salinity stress; Heavy metals and ozone; and General abiotic stresses and their alleviation by microbes.

The section on temperature, water and salinity stress covers seven chapters and occupies the bulk of the book. All the three major abiotic stresses have been clubbed together as there is a definite interrelationship among all three. Elevated temperatures can lead to rapid water loss, which in turn leads to drought conditions. Similarly, excess salinity also reduces available water to the plant, leading to symptoms related to water deficiency. The first chapter in this section deals with heat-shock proteins (Hsps), which show accelerated synthesis and accumulation in eukaryotes immediately following hyperthermia and confers thermotolerance as well as the capacity to withstand subsequent exposure to lethal temperature and other metabolic insults.

Many of the Hsps on the other hand, are molecular chaperones with vital functions in metabolic pathways, signal transduction, cell proliferation, differentiation and apoptosis under permissive growth conditions. Understanding the role of Hsps in thermotolerance can lead to development of strategies for induction of heat tolerance in plants. The reproductive phase in crops is particularly vulnerable to heat and drought stress and their combination, and in the second chapter the authors discuss how the interplay between leaf senescence, oxidative stress and sugar signalling in reproductive tissues contributes towards reduction in growth and yield in heat-stressed plants.

Both an excess and a deficit of water are abiotic stressors. Three chapters are devoted to this specifically. Chapter 3 will detail our understanding of the roles of nitric oxide, ethylene and haemoglobin in flooding stress and consider how this can be exploited in breeding programmes and sustainable agricultural practice. Nitric oxide (NO) has been shown to trigger the biosynthesis of ethylene during stress and also play key roles in programmed cell death and the hyponastic response. It is discussed as to how the expression of non-symbiotic haemoglobins which oxidize NO to NO_3 play an important role in controlling NO production and thus ethylene-mediated responses to submergence. In Chapter 4, the authors focus on the defence mechanisms against stresses at the molecular level, with special reference to oxylipin metabolism, which according to the authors, represents one of the main defence mechanisms employed by plants. One of the members of this family, jasmonic acid, is well known to be involved in resistance to both abiotic and biotic stresses. Authors have taken the specific example of chickpea hybrids to illustrate the roles. In Chapter 5, the authors discuss how, in contrast to conventional breeding techniques, genetic engineering offers a fast and efficient tool to produce drought-resistant and drought-tolerant plants and thus improved water uptake, use and retention by plants. In order to genetically manipulate plants to be drought tolerant or resistant, genes from the plants that are tolerant or even from other organisms can be selected, which can be grouped into three drought-tolerance engineering strategies: the engineering of functional proteins, manipulating the expression of transcription factors and the regulation of signalling pathways involved in drought tolerance. Chapters 6 and 7 deal with salinity. In Chapter 6, the authors provide an overview of the physiological, biochemical and molecular mechanisms underlying salt tolerance, combining knowledge from classic physiology with information from recent findings. Special emphasis has been given on salt signal perception and transduction and mechanisms related to maintenance of osmotic, ionic, biochemical and redox homoeostasis in salt-stressed plants. A fundamental biological knowledge in conjunction with the understanding of the salt-stress effects on plants is necessary to provide additional information for the dissection of the plant response to salinity and in trying to find future applications for reducing the deleterious effects of salinity on plants, improving the productivity of species important to agricultural sustainability. In Chapter 7, based on results from sugarcane, the authors discuss the results that indicate that the salt tolerance of a variety depends on the stage of development and the level considered. Consequently, salt tolerance of a given cultivar at whole-plant level does not guarantee salt tolerance of tissue or cell cultures issued from this cultivar.

The section on heavy metals, ethylene and ozone consists of four chapters, which deal with the negative effects of heavy metals and air pollutants. Chapter 8 deals with ozone phytotoxicity caused mainly because of its high oxidation potential to generate reactive oxygen species in exposed plant tissue. The balance between the production and the scavenging of activated oxygen is crucial to plant growth maintenance and overall environmental stress tolerance. While increased accumulation of plant secondary metabolites in leaves in response to ozone exposure has been reported, the changes on crop plants' composition and nutritional quality needs to be further studied and discussed to guide our efforts to select ozone-tolerant crops in an attempt to provide a secure food supply for a developing world. Chapters 9 and 10 deal with heavy metal toxicity including cadmium and arsenic among others. In Chapter 9, the authors have mainly focused on the interactive role of ethylene, sulfur, antioxidant system and tolerance of cadmium in plants. Ethylene is the gaseous plant hormone and is now considered to regulate many plant developmental processes throughout the plant's life from germination to senescence and also mediate the plant's responses to abiotic and biotic stress. The basic mechanisms and functional genomics perspective underlying heavy metal toxicity in plants, knowledge of which is essential for development of sustainable agriculture, are dealt with in Chapter 10. Several genetic studies have revealed major signalling pathways that are interconnected and lead to multiple responses in plants under heavy-metal stress. Functional genomics is now considered as an important dissecting tool to understand heavy-metal toxicity as well as tolerance in plants. In Chapter 11, the author has dealt with the negative effects of arsenic, which is a naturally occurring highly toxic metalloid to all forms of life, taking the example of the growth and metabolism of cereals and pulses. Combined application of phosphate with arsenate can ameliorate the damaging effects caused by arsenate treatment alone in cereal and legume seedlings. Hence, the use of phosphate-enriched fertilizers in arsenic-contaminated soil may help normal growth of cereals and legumes.

In the final section, which deals with abiotic stresses in general and their alleviation by microbes, four chapters have been included. In Chapter 12, the authors have focused mainly on recent information about the effects of abiotic stress on plant growth, water relations and photosynthesis, as well as mechanisms of adaptation. The higher acclimation capacity, and hence greater resistance to a given stress factor, is determined by the plant's capacity to maintain its physiological processes within the reaction norm, at a greater variation of this factor. Chapter 13 deals with small molecules such as polyamines, which may play a definitive role in protective or adaptive mechanisms that combat the potential stress-induced injuries in plants encountering abiotic stresses regularly under natural conditions apart from abrupt natural calamities for which the plant may not be prepared. Moreover, it is apprehended that PA-ROS-mediated signalling under stress may have a cross-talk with the phytohormones, figuring a further complex network of signalling for stress tolerance, analysis of which will be a challenging task in near future. The last two chapters deal with a recent, ecofriendly, cost-effective mechanism for stress alleviation through the use of beneficial soil microbes. Chapter 14 deals with the potential of *Trichoderma harzianum* to directly increase plant tolerance against abiotic stresses, such as drought, salinity and soils with low fertility, though traditionally it has been successfully used for the biological control of many plant pathogens through chemiotropic mycoparasitic interactions with the target fungal or bacterial organism. This could promote a rational and non-empirical inclusion of this important fungal species in modern agricultural sustainable practices. The possibility that soil microorganisms could play a significant role in evolving efficient low-cost technologies for abiotic stress management has been dealt with in Chapter 15. Their unique properties of tolerance to extremities, their ubiquity, genetic diversity and their interaction with crop plants can be exploited in order to develop methods for their successful deployment in agricultural production. Soil microorganisms can help crops withstand abiotic stresses, such as drought, chilling injury, salinity, metal toxicity and high temperature, through different mechanisms such as the induction of osmo-protectants and Hsps etc. in plant cells more efficiently. This ability in alleviating abiotic stress conditions in different crop systems can be used for cost-effective sustainable agriculture.

We have endeavoured to compile this book taking a holistic approach from basics to advanced technologies, with the main objective being to put together sufficient information on how to take forward sustainable agriculture in the face of mild to extreme environmental changes occurring in nature. The whole book is well focused and offers insights into the various factors reducing crop productivity and highlights different mechanisms of resistance and approaches that could be used in sustainable agriculture. The editors and authors hope that this book will be of use to agricultural scientists, the agro-industry, academicians and researchers working in the area of abiotic stress and its management.

We would like to thank all the authors who responded in time, which made it possible to bring out this book within the prescribed time. Finally, it is our pleasure to thank CABI for making this possible. Special thanks are also due to Dr Sreepat Jain, Commissioning Editor, CABI and Emma McCann, Editorial Assistant, CABI for their involvement at various stages of publication.

Usha Chakraborty
Bishwanath Chakraborty

About the Editors

U. Chakraborty

Dr Usha Chakraborty, an MSc Gold Medallist of Calcutta University, is Professor of Plant Biochemistry in the Department of Botany, University of North Bengal, Siliguri. She joined the department as a lecturer in 1986 and has since been Head of the Department of Botany twice, been member of University Court, Executive Council and has generally been involved in university administration at various levels. Currently, she is the Programme Co-ordinator of UGC SAP DRS-III level. Her main research focus is on the elucidation of mechanisms of abiotic and biotic stress responses in plants and development of markers for tolerance. She is also engaged in research on crop improvement through biological means and determination of mechanisms of action of such plant growth promoting rhizobacteria. To date, 22 students have received PhD under her guidance and many more are registered. She has published more than 120 research papers in national and international journals, 14 chapters in books and three edited books. She has travelled widely in Canada, USA, Malaysia, UK, Japan, Russia, Italy, China and made representation in International forums. Dr Chakraborty has been elected as Fellow of IPS, ISMPP and Royal Society of Chemistry, London.

B. Chakraborty

Dr Bishwanath Chakraborty is Professor of Plant Pathology in the Department of Botany, University of North Bengal, Siliguri. He has served the institution in various capacities, namely Dean, Faculty of Science; Director, Centre for Development Studies, Programme Coordinator, SAP of the UGC on the subject of Microbiology including mycology and plant pathology and plant diversity. He has to his credit over 140 research papers published in national and international journals, ten review articles, 15 chapters in books, four edited books which have opened up a new line of research on molecular plant pathology and fungal biotechnology. Twenty six students obtained PhD under his guidance. He has travelled widely in Japan, Canada, USA, UK, Germany, Australia, Italy, Russia, New Zealand and made representation in International forums and established International Collaboration in UK and Canada. Dr Chakraborty has been elected as Fellow of IPS, ISMPP, West Bengal Academy of Science and Royal Society of Chemistry, London.

1 Heat-Shock Proteins and Molecular Chaperones: Role in Regulation of Cellular Proteostasis and Stress Management

M. Kapoor* and S.S. Roy
Department of Biological Sciences, The University of Calgary, Alberta, Canada

Abstract

The multiplicity of environmental and physiological stresses experienced by all organisms presents a formidable challenge to survival. Encounters with near-lethal temperature, extreme cold, pathogens and parasites, metabolic toxins, heavy metals, nutrient deficit, hypoxia and desiccation comprise some of the more common forms of stress that negatively impact all three domains of life. Many of these agents lead to protein unfolding and structural damage to intracellular organelles and cell membranes, and genome replication, transcriptional and translational machinery. The prime strategy to ameliorate the effect of adverse conditions relies upon the evolutionarily conserved stress response: the rapid and transient production of numerous defence-capable proteins, the molecular chaperones. The most prominent and extensively investigated amongst this group are the heat shock proteins (Hsps). Their accelerated synthesis and accumulation, immediately following hyperthermia, confers thermotolerance: the capacity to withstand subsequent exposure to lethal temperature and other metabolic insults. The appearance of aberrant, unfolded or mis-folded aggregation-prone proteins is a signal for mounting the heat shock-stress response. Many of the Hsps are molecular chaperones with vital functions in metabolic pathways, signal transduction, cell proliferation, differentiation and apoptosis even under permissive growth conditions. The accumulation of molecular chaperones under adverse conditions provides the basic strategy for stress management. Molecular chaperone families are classified into two general categories. The first comprises the 'foldases', including the ATP-dependent chaperonins, Hsp70, Hsp90 and Hsp110 families, involved in folding nascent polypeptides and refolding proteins unfolded as a result of stress. The second group, the 'holdases', sequester unfolded or partially folded proteins, which are subsequently processed by the foldases. The ubiquitous set of small Hsps (sHsps) represents the ATP-independent holdases that play a major role in protection against hyperthermia, oxidative stress and a variety of other abiotic stresses. In plants, sHsps have an important role in development of thermotolerance and adaptation to osmotic and high salinity stress. In addition, some subfamilies of plant sHsps are not heat shock-inducible but are expressed constitutively during specific developmental stages.

1.1 Introduction

Numerous factors in the life of an organism elicit moderate to severe physiological/environmental stress. The most prevalent forms of stress experienced on a regular basis include hyperthermia, exposure to ultraviolet light, nutrient deficit, dehydration/drought and metabolic

---

*E-mail: mkapoor@ucalgary.ca

© CAB International 2015. *Abiotic Stresses in Crop Plants*
(eds U. Chakraborty and B.N. Chakraborty)

poisons, heavy metals, microbial pathogens and other toxic substances in the milieu. Consequently, during the course of evolution, a wide variety of strategies for managing potentially lethal effects of stress have been perfected in different organisms to counteract specific threats to survival. The best understood and the most extensively investigated strategy for protection against hyperthermal conditions, in prokaryotes and eukaryotes alike, is the evolutionarily conserved phenomenon referred to as the heat shock (stress) response. In addition to high temperature, a surfeit of reactive oxygen species (ROS) is a universal elicitor of stress response, while persistence of herbicides and toxins in the soil, salinity and bacterial and fungal infections also induce a powerful stress response in plants.

Hyperthermia and other stresses present a serious threat to survival by causing unfolding and mis-folding of proteins, resulting in disturbance of intracellular protein homoeostasis. As partially or completely unfolded/mis-folded proteins are intrinsically aggregation-prone, reversal of the process by refolding or removal of the offending proteins constitutes the prime strategy for stress management. Appearance of unfolded proteins in the cytosol acts as the principal signal for immediate deployment of the heat shock-stress response: elevated expression of a plethora of stress-inducible genes and the rapid synthesis of defence-capable proteins (Hsps) fortifies the target organism against adverse environmental conditions. Such a defence mechanism can react swiftly to a wide range of physiological and chemical challenges leading to protein unfolding and is encountered universally in all three biological kingdoms: the Eubacteria, Archaea and Eukarya. The Hsps, also known as molecular chaperones, are exquisitely designed for shielding the cellular machinery from damaged macromolecules. Although most Hsps are required at low levels during normal growth and metabolism, a dramatic up-regulation of their synthesis is necessitated under stress conditions, as molecular chaperones are required in stoichiometric amounts relative to the population of unfolded/mis-folded or aggregated polypeptides.

In the eukaryotes, pathogen attack and several genetic/physiological factors also cause a substantial build-up of unfolded proteins in the endoplasmic reticulum (ER), the lumen of ER being the locale of synthesis of secretory and membrane-specific proteins. Perturbations in the redox status, calcium homoeostasis or post-translational modifications of secretory proteins, can result in substandard local folding capacity culminating in the accumulation of unfolded or mis-folded ER macromolecules. To counter this cytotoxic hazard, a robust surveillance system – designated the unfolded protein response (UPR) – conserved in plant, fungal and mammalian species, is launched. UPR is critical for adjustment of ER homoeostasis under stress elicited by mis-folded proteins (Walter and Ron, 2011). This system implements an immediate cessation of normal protein synthesis and activation of a preferred set of genes encoding chaperone and co-chaperone proteins, affording protection by induction of the ER-specific degradation system, or apoptosis as a last resort. Fortuitously these chaperones also promote resumption of proper folding (Lai et al., 2006). Irreversibly damaged ER proteins are moved to the cytoplasm and subjected to degradation by the ERAD system (ER-associated degradation).

Persisting ER stress is linked to several metabolic disorders, such as obesity, diabetes, diseases of the liver and atherosclerosis. Avenues are being explored to develop therapeutic approaches targeting specific components of the UPR for treatment of human diseases (reviewed in Lee and Ozcan, 2014). Insightful analyses of the heat shock response, protein folding, aggregation, macromolecular assemblies, and structure and function of molecular chaperones are available in recent reviews (Pearl and Prodromou, 2006; Hartl and Hayer-Hartl, 2009; Richter et al., 2010; Tyedmers et al., 2010; Waters, 2013). The following is a brief overview of commonly encountered environmental stresses and properties and structural features of selected, typical molecular chaperones that respond to them.

1.2 Molecular Chaperones: Functions and Properties

During the last two decades a large number of molecular chaperone families (exceeding 100) have been recorded, found in virtually every compartment/organelle of the cell – the endoplasmic reticulum, the cytosol, mitochondria, chloroplasts and nucleus. Molecular chaperones span an ever-expanding, wide-ranging class of proteins, with a majority of the members being present at a basal level throughout the life cycle; their rate of synthesis is accelerated manifold, immediately and transiently under stress-inducing stimuli. As stated in the Introduction, with an increase in the population of unfolded and aggregated proteins during growth at super-optimal temperatures or elevated ROS levels, a bank of defensive proteins (Hsps and other stress proteins) is vital for survival. In the immediate term, Hsps aid survival by conducting repair/reversal of damage and subsequently protect the organism from potential lethality by conferring tolerance/adaptation towards other potent abiotic and biotic stresses.

In eukaryotic cells some of the major consequences of hyperthermia include defects in the cytoskeleton by disruption of intermediate actin and tubulin networks, erroneous localization of organelles, fragmentation of ER and the Golgi, changes in membrane fluidity, deficit of mitochondria, aggregation of ribosomal proteins and defective processing of ribosomal RNA (Nover et al., 1989; Richter et al., 2010; Toivola et al., 2010). As stated in the foregoing, Hsps and their constitutively expressed cognates are essential, even under stress-free conditions, for the assembly of macromolecular complexes, trans-membrane trafficking of proteins, development and differentiation, cell cycle signalling pathways, regulation of gene expression and apoptosis.

Molecular chaperones also play vital roles in normal metabolism by regulating crucial steps in DNA replication and repair, maintenance of genome integrity, chromatin architecture, membrane stability, ribosome biogenesis, metabolic pathways, signal transduction, control of cell proliferation, differentiation and apoptosis (Frydman, 2001; Calderwood et al., 2006). They intercede at every step of protein biogenesis and maturation, from the emergence of the nascent polypeptide chain to the conclusion of its synthesis and the final design of a stable three-dimensional native structure. Stress-inducible proteins sustain the conformational integrity of structural proteins and enzymes by enabling requisite folding of nascent and partially unfolded polypeptides as well as by directing the timely degradation of irreparably impaired proteins.

Prior to the completion of polypeptide synthesis, hydrophobic segments of nascent chains, released from the translational apparatus, display a strong tendency to associate with each other. The intracellular environment in the cytosol is extremely crowded; the high local concentration of macromolecules – estimated at ~300–400 mg ml^{-1} – provides congenial conditions for self- and cross-aggregation of proteins. In such a milieu, the primary target of molecular chaperones is the linear polypeptide chain exiting from the ribosome in the process of synthesis. Molecular chaperones sequester the newly synthesized hydrophobic patches – which would normally be buried in the interior of the mature folded protein – thereby precluding the intra- or inter-molecular aggregation of exposed surfaces while guiding the proper folding of the polypeptide and subsequent assembly into biologically functional oligomers or macromolecular complexes. Physico-chemical studies, in conjunction with Cryo-electron microscopy (Cryo-EM) and other imaging techniques, demonstrate protein aggregates to be either amorphous – typically seen in bacterial inclusion bodies – or amyloid-like in nature (Tyedmers et al., 2010). Aggregates of endogenous proteins form in bacteria under heat or oxidative stress conditions, as well as in host cells over-expressing recombinant proteins. In the latter situation, preferential high level synthesis and build-up of heterologous polypeptides mimics an internal stress condition, when the host cell may lack the capacity to garner adequate quantities of chaperones. It is noteworthy that the abundance of bacterial proteins predisposed to thermal unfolding – with structural features amenable to aggregate formation – may form up to 1.5 to 3% of total protein contingent in *Escherichia coli* (Winkler et al., 2010).

Under heat stress or otherwise unfavourable conditions, chaperones overcome the devastating effects of damage either directly by refolding of mis-folded, non-native proteins or by channelling the malformed proteins towards degradation by energy-dependent protease systems (Young *et al.*, 2004). Some molecular chaperones mediate macromolecular trafficking across membranes while others assist in re-mediation of damaged multi-component assemblages. In all encounters, the chaperones engage only transiently with their substrates and readily dissociate upon conclusion of the folding/refolding/disaggregation reactions. Many molecular chaperones are endowed with the facility to differentiate between the native and non-native states of proteins. In coordination with the protease system, they provide a robust scheme of 'quality control' to cleanse the cell of dysfunctional proteins (Bukau *et al.*, 2006). Molecular chaperones often form complex interacting networks in cooperation with some of the other major chaperones and co-chaperones (reviewed in Söti *et al.*, 2005). Co-chaperones perform critical functions in facilitating substrate recognition and selectivity, and assist its productive binding to the chaperone protein. In the case of chaperones that utilize ATP binding and hydrolysis (e.g. Hsp70 and 90 families), their co-chaperones partners often act directly by positive or negative modulation of the ATPase activity of the nucleotide binding domain (NBD). Furthermore, co-chaperone impart malleability to the chaperone protein enabling it to engage in precise interactions with diverse substrates. The structure–function relations of the major molecular chaperones and co-chaperones have been thoroughly discussed in insightful reviews (Frydman, 2001; Tsigelny and Nigam, 2004; Bukau *et al.*, 2006).

1.3 Factors Promoting Protein Mis-Folding and Aggregation

It is abundantly clear that both the intracellular environment and exogenous stimuli contribute significantly to protein mis-folding and aggregation. Internal factors that elicit aggregation include products of normal metabolism – such as ROS – and spontaneous mutations imparting a tendency on affected proteins to mis-fold and aggregate. Examples of the latter are seen in devastating human conformational diseases such as Huntington's disease, familial forms of Parkinson's disease and Alzheimer's and type II diabetes. Defective steps in protein biogenesis during translation or post-translational modifications can give rise to anomalous products. During ageing, too, progressive deterioration of the quality control system leads to an increased propensity for protein aggregation (Münch and Bertolotti, 2010). In addition to heat, external stimuli for generation of anomalous proteins comprise oxidants, prolonged exposure to UV and toxic chemical agents. If the level of mis-folded proteins were to surpass the total folding or degradation capacity of a cell, accumulation of aggregates would follow. A scenario of this type is likely under hyperthermia and oxidative stress when unfolding of cellular proteins is witnessed on a global scale. Therefore, the immediate action of the heat-stress response is to raise the relative level of molecular chaperones and stress-linked proteases to cope with the overload of non-native proteins. Given appropriate conditions, heat-induced unfolding may be completely reversible by the action of molecular chaperones, however, proteins damaged by oxidative stress, marked with ubiquitin for degradation or amyloidogenic proteins, are not revertible to their native state.

1.4 Reactive Oxygen Species: Positive and Negative Impacts

ROS are produced endogenously during crucial steps in normal metabolic pathways and are also necessary components of several important reactions *in vivo*. ROS are generated continuously at a basal rate by the electron transport chain during the operation of TCA cycle in the mitochondria and are germane to the control of normal cell proliferation, differentiation and signalling, regulation of immunity as well as ageing (Sena and Chandel, 2012). In addition to respiratory chain proteins, enzymes such as glycerol-3-phosphate dehydrogenase produce superoxide radicals in the mitochondria, thereby contributing to the overall increase in ROS (Murphy, 2009; Collins *et al.*, 2012). Oxidative stress, engendered by intermediary metabolic pathways, is perceived as an inevitable, potentially damaging agent for cellular macromolecules. Nevertheless, there are numerous essential functions requiring ROS activity: mitochondrial ROS are important activators of the protective response against hypoxia and, in plants, ROS play a central role in defence against bacterial and fungal pathogens (Shetty et al., 2008). Incomplete reduction of oxygen in oxidative phosphorylation reactions produces superoxide free radicals ($O_2\cdot$), convertible by the ubiquitous enzyme superoxide dismutase (SOD) to a relatively benign product, H_2O_2. The latter serves as a regulator of some redox-sensitive proteins and also as an activator of transcriptional and translational pathways in human cells. Superoxide free radicals and H_2O_2 are implicated in the development of numerous human pathological conditions including inflammatory diseases, hypertension and atherosclerosis (Lyle et al., 2014 and references therein).

H_2O_2 can also be transformed into a highly reactive, deleterious derivative, the hydroxyl radical. Over the course of evolution, a variety of systems for protection from/avoidance of oxidative damage have been refined in all organisms. Crucial to this quest is a battery of powerful antioxidant enzymes – catalases, glutathione peroxidases and peroxiredoxins – that offers a sturdy line of defence by catalysing reduction of H_2O_2 to water, thereby neutralizing a source of subsequent damage. Evidently, under normal circumstances, intracellular ROS level is stringently proscribed and confined within a tolerable range by the concerted action of antioxidant enzymes.

In plants, ROS are also generated by chloroplasts and peroxisomes; other sources are enzymes, including glycolate oxidase, oxalate oxidase, cell wall NADP oxidase, peroxidases and amino acid oxidase. At permissive levels ROS serve as signalling conduits and regulators of metabolism, as well as in defence against pathogens. And as with mammalian cells, higher levels of intracellular ROS are damaging to plant macromolecules.

Persistent oxidative stress can cause additional irreversible modifications of proteins: free radical-induced fragmentation of the polypeptide backbone and replacement of side chains of proline (Pro), arginine (Arg), lysine (Lys) and threonine (Thr) residues by carbonyl groups. Carbonyl derivatives can accrue from direct oxidative modification or from reactions of Lys, cysteine (Cys) and histidine (His) residues with reactive carbonyl compounds or glycoxidation end-products. Furthermore, carbonyl groups can react with α-amino group of Lys residues giving rise to cross-linked products that are recalcitrant to degradation by the proteasome system (Stadtman and Levine, 2000; Nystrom, 2005). Such irreparable modifications lead to mis-folding and aggregation of the target protein. Paradoxically, while ROS are required for crucial functions in cellular metabolism, aerobic organisms remain fundamentally vulnerable to oxidative stress.

As alluded to in the preceding, exposure of animal cells to low oxygen (hypoxia) also augments production of ROS by the mitochondria. Hypoxic conditions promulgate adaptive metabolic changes designed to reduce oxygen uptake and energy consumption. Interestingly, H_2O_2 liberated from mitochondria during hypoxia is the chief factor in deployment of the adaptive response to hypoxia, which is dependent on specific transcriptional activators, HIFs (hypoxia-inducible transcriptional factors). HIFs are dimeric proteins, maintained in a quiescent state during normoxic conditions by virtue of the presence of one unstable subunit; the latter is converted to a stable form under hypoxia, upon release of H_2O_2. The stabilized HIF dimer acts as an inducer of expression of an array of genes regulating the cell cycle and signal transduction in various pathways (Chandel et al., 1998; Weidemann and Johnson, 2008). While activation of HIFs is dependent on mitochondrial ROS production, an unremitting increase in the ROS level is definitely detrimental. The efficacy of control of mitochondrial ROS by antioxidant enzymes in mammalian cells is reinforced by the localization of isoforms, SOD1 and SOD2, of superoxide dismutase, in the inter-membrane space and mitochondrial matrix, respectively, along with some isoforms of peroxiredoxins. Increased levels of mitochondrial ROS, decline in antioxidant enzymes, enhanced oxidation of DNA and the resulting higher mutation rates, constitute some of the contributing factors associated with a variety of cancers and neurodegenerative diseases (Wellen and Thompson, 2010).

1.5 Principal Molecular Chaperones: Heat-Shock Proteins

Sub-lethal temperature, metabolic poisons, heavy metals, anoxia and ROS, among other physiological stresses, stimulate the rapid and transitory over-production of a set of evolutionarily conserved molecular chaperones, the Hsps (Ellis et al., 1989). Proteins in this group are the principal protagonists in development of resistance to subsequent encounters with lethal stress. A variety of investigations of heat-shock response by bioinformatics analyses, transcriptional arrays and tools of proteomics methodology have uncovered the induction of >200 genes in various model organisms among the Eubacteria, Archaea, fungi, plants and mammalian cells. Stress-inducible proteins can be considered to fall into several functionally distinguishable categories (Richter et al., 2010). The best characterized and historically the most prominent is, of course, represented by the Hsps. The other significant groupings include: proteolysis systems, for cleansing the cells of residual debris of misfolded and irreversibly aggregated proteins left after refolding/de-aggregation by Hsps; catalytic proteins implicated in repair of genome wide damage; enzymes in metabolic pathways directly linked to cellular energy generation; transcriptional and translational regulators; proteins involved in the preservation of cytoskeleton structure, the macromolecular transport machinery, and membrane integrity and stability (Richter et al., 2010).

Hsps are categorized into the following groupings on the basis of their approximate molecular mass: small Hsp family (sHsps) including Hsp26, Hsp10, Hsp12 and many other members (Hsp20 family); Hsp40; Hsp60 (chaperonins); Hsp70, Hsp90 and the Hsp110 families, distant relatives of Hsp70 (Lindquist and Craig, 1988). The Hsp90 family members in lower eukaryotes, including yeasts, filamentous fungi and protists, have molecular masses between 80 and 83 kDa. Although the defining features of the heat-stress response are preserved universally, notable variations in the distribution and disposition of individual Hsps are well documented. For instance, bacterial and lower eukaryotic genomes encode all of the Hsp families, while Hsp70 is apparently not encoded in some extremophilic genomes; Hsp90 and Hsp100 families have not been reported in the Archaea, and cytosolic orthologues of Hsp100 are apparently absent in the nematodes, arthropods and vertebrates. It has been noted that compared to prokaryotes and the Archaea, higher eukaryote genomes harbour far more numerous loci for chaperones and regulatory factors (Richter et al., 2010). This is not surprising as genome size and complexity must dictate the nature, diversity and abundance of regulators.

As stated in the preceding, Hsps oversee vital functions during normal growth, differentiation and cell-cycle progression. The 'foldases' – eukaryotic ATP-dependent 70-kDa, 90-kDa, the bacterial equivalents DnaK and HtpG, respectively, and the 60-kDa ubiquitous chaperonin families (prokaryotic GroEL/ES) – capable of folding nascent polypeptides or re-folding unfolded proteins, are an extensively investigated group. While some of their members can interact with a wide spectrum of substrates, others are more specialized recognizing only a narrow structurally defined range of client proteins. The widespread super-family of ATP-independent small Hsps, designated the 'holdases', though incapable of folding polypeptides per se, forms a vital force in prevention of aggregation by binding to unfolded intermediates that are subsequently delivered to the 'foldase' complexes. For optimal management of stress, the requisite balance between intracellular levels of holdases versus foldases is achieved by restricting the production of the former mainly to stress conditions, while the expression of foldases has to be, of necessity, constitutive as well as stress-induced. Some Hsps may be able to perform both the holding and folding functions. In the following brief overview we examine the properties, structures and mechanistic underpinnings of some of the well characterized, typical representatives of the holdase and foldase class.

1.6 The Ubiquitous Holdases: Small Heat-Shock Proteins

The best known holdases, the foot soldiers of the anti-stress contingent, are the sHsps, encompassing a ubiquitous, diverse superfamily of ATP-independent molecular chaperones, indispensable to all three life domains. Elevated expression of sHsps vastly enhances tolerance toward heat and oxidants. Contrasted with the ATP-dependent chaperone families, sHsps do not interact with nascent polypeptides and native or near-native proteins. Although functionally analogous, sHsps encompass an assortment of multi-subunit members that are heterogeneous in size and sequence, displaying an overall low level of conservation (Kriehuber et al., 2010; Basha et al., 2012; Delbecq and Klevit, 2013). They can interact with a wide variety of structurally unrelated, unfolded or partially folded substrates, keeping them in a folding-amenable state until the requisite equipment, namely the ATP-dependent Hsp70 or Hsp110 cohorts, becomes available. A dual objective is thus accomplished: preempting the aggregation and irreversible denaturation of target proteins and blocking their inopportune conveyance into degradation pathways.

At non-permissive temperatures and in conjunction with unfolding client proteins, the monomers of sHsps undergo conformational changes exposing hydrophobic surfaces, which associate to form higher-order oligomers. Thus, contrasted with the 'foldase' class, the activity of sHsps is directly responsive to modulation by temperature. Moreover, it is an energy conserving system as it does not utilize ATP hydrolysis-linked cycles of substrate binding and release (Stengel et al., 2010). The structure, disposition and properties of some of the sHsps, however, diverge significantly from the standard paradigm.

Currently sHsps are also the focus of intense research in the biomedical field as they are implicated in several human diseases associated with protein aggregation, such as cataract, myopathies and neurodegenerative conditions – Alzheimer's and Alexander's disease (Kato et al., 2001; Goldstein et al., 2003). Their importance in cellular functions can be appreciated by the fact that higher eukaryote genomes often encode multi-gene families of sHsps. The human and other mammalian genomes encode 10–11 sHsps while plants have many more representatives compared to other eukaryotes. For instance, the model organism *Arabidopsis thaliana* has 19 sHsps while *Oryza sativa* has 23. Furthermore, in angiosperms, several subfamilies of sHsps are distinguishable according to their localization in various cellular sites and compartments. Distinct subfamilies are expressed in the endoplasmic reticulum, peroxisomes, chloroplasts, mitochondria, the nucleus and the cytosol, indicating a remarkable degree of functional specialization. In addition to stress-induced synthesis, many sHsps are constitutively expressed in tissues, such as leaves and apical meristems, and during specific developmental stages in plants: embryogenesis, maturation of pollen, seeds and fruits. Some sHsp subfamilies are also expressed in response to oxidative stress and UV exposure in plants. Expression of some *A. thaliana* sHsps is thought to be associated with thermotolerance and adaptive response to osmotic stress and salinity (reviewed in Waters, 2013).

1.7 Structural Characteristics of Small Heat-Shock Proteins

In striking contrast with the ATP-dependent molecular chaperones, the biologically mature configuration of different sHsps is concocted by a bewildering array of permutations and combinations of subunits. The monomers of individual sHsps range from 12 to 40 kDa, the majority being between 15 and 22 kDa in size; hence the sHsp family is also referred to as the Hsp20 family. Although the products of sHsps genes are relatively small in size, they can associate to form gigantic oligomeric assemblages of 12–48 monomers in their native state: dynamic homo- or hetero-oligomeric complexes. In spite of vast differences in quaternary structure and overall dimensions, all small Hsps are structurally related with respect to one conserved central domain of 90–100 amino acids – the α-crystallin domain (α-CD) – characteristic of the eye lens α-crystallin. In individual sHsps, the α-CDs are flanked by highly variable in length, unrelated sequences (averaging 55 residues) on the N-terminal side and moderately variable sequences (10–20 residues) at the C-terminal end; in the latter, an iconic small conserved motif of hydrophobic residues, I/L-X-I/L, is found in virtually all sHsps. The variable regions are thought to be involved in determination of specificity of client recognition and the nature of dynamic oligomeric assemblages (Stengel et al., 2010; Basha et al., 2012).

Structural parameters of sHsps have been unravelled by X-ray crystallographic analyses of Hsp16.5, a 24-unit oligomer from the archaeal species *Methanocaldococcus jannaschii*, and the *Triticum aestivum* (wheat) Hsp16.9, a dodecameric member of the cytosolic sHsp family (van Moncroft et al., 2001; Lyle et al., 2012; Mchaourab et al., 2012). The quaternary structure of these two sHsps is based on the use of dimers as the foundation stone, but the final products are very different in shape and size. The quaternary structure of wheat Hsp16.9 consists of two circles, each with six α-CDs arranged as a trimer-of-dimers. The stacked hexameric rings and the dimeric form are shown in Plate 1 (A, B). In the monomer, the α-CD is folded into two anti-parallel β-sheets with a flexible loop emanating from it and an N-terminal segment of 42 residues folded into small α-helices, separated by random coil regions; the C-terminal segment is mostly unstructured (Plate 1C). Residues in the flexible loops from the β-sandwich participate in strand swap with the partner monomer. The hydrophobic motif I/L-X-I/L (conserved in most sHsps) in the C-terminal sequence binds to a hydrophobic pocket formed by β strands of the α-CD in the opposite (*trans*) monomer. The N-terminal domains of the dimers interact with each other to generate the hexameric ring. Inter-ring interaction between monomers via the N-terminal sequences of apposed rings stabilizes the dodecameric structure; α-helices from adjacent dimers project into the cavity in the interior of the double circle. The assembly of *M. jannaschii* Hsp16.5, on the other hand, involves 24 monomers associated to form octahedrally symmetrical, hollow spherical oligomers (Plate 1, D–F). As for Hsp16.9, the dimeric form of Hsp16.5 is also considered to be the building block for larger oligomeric assemblies. The monomer contains the conserved α-crystallin domain with the standard β sandwich motif.

Wheat Hsp16.9 dodecamer, like other sHsps, is subject to reversible temperature-dependent dissociation into smaller oligomers *in vitro*, while forming complexes with substrates at higher temperatures. Higher temperatures lead to increased exposure of hydrophobic surfaces of subunits, some of which furnish the binding sites for substrates whereas the others mediate higher-order oligomer formation. It appears that the active substrate-capturing entity may not be uniform in all sHsps. However, both the dimer as well as higher-order-oligomers are likely to bind substrates with exposed hydrophobic patches with equal facility. Studies with a diverse range of sHsps from different organisms have led to proposals of mechanistic models of oligomeric assemblies and regulation of substrate binding. So far, a single comprehensive model elucidating the mechanism of action of the highly diverse sHsp collective has not been formulated (Basha et al., 2012).

1.8 The Yeast Hsp12: an Unusual Small Heat-Shock Protein

In addition to the typical sHsps, described above, that form multi-subunit oligomeric assemblies, a unique sHsp in yeast (Hsp12) is monomeric and completely unstructured in its naturally occurring native state. Furthermore, in sharp contrast with the conventional sHsps, it appears to be devoid of any appreciable chaperoning activity *in vitro*. Conventional wisdom has it that an unstructured Hsp should be intrinsically incapable of binding unfolded or aggregated proteins, which is indeed the case with Hsp12. Strikingly, Hsp12 does adopt an α-helical structure, but only upon interaction with membrane lipids and, in turn, stabilizes the membranes by modulating their fluidity (Welker et al., 2010). In functional terms, it acts as a 'membrane lipid chaperone'. Recent studies suggest that it is also one of the proteins implicated in the progression of ageing and extension of lifespan witnessed in several model organisms subjected to calorie-restricted diet (Herbert et al., 2012). Studies based on NMR analysis show that on addition of dodecyl-phospho-choline (DPC) the residues in the disordered Hsp12 monomer fold into a single α-helix near the

C-terminal end (Plate 2A) while the structure of SDS micelle-bound monomer shows folding of Hsp12 into four dynamic amphipathic α-helices (Plate 2B), with polar residues and hydrophobic regions on opposite surfaces (Herbert et al., 2012). The chaperoning activity of Hsp12 is highly specialized; it acts primarily as a re-modeller of cellular membranes undergoing deleterious alterations during environmental stress. In yeast, Hsp12 – along with Hsp26, Hsp31 and a number of other proteins – has been shown to be induced by osmotic stress as well.

1.9 Hsp60 Family of Chaperones: the Chaperonins

1.9.1 The *Escherichia coli* GroEL-ES complex

The Hsp60 family molecular chaperones – the chaperonins – are found universally in all organisms and are classified into Groups I (in eubacteria and eukaryotic organelles of endosymbiotic origin) and II (in the archaea and eukaryotic cytosol). Higher-than-normal bacterial growth temperature, oxidative stress and other detrimental treatments trigger protein unfolding on a comprehensive scale. Management of this situation is achieved by an enormous up-regulation of chaperonin genes to trap the burgeoning population of unfolded proteins. The *E. coli* GroE chaperonin system was the first to be described – close to half a century ago – and is mechanistically the best characterized protein unfolding–refolding equipment. This system comprises a set of exquisitely designed ATP-dependent, ring-shaped multi-subunit, allosteric protein assemblies that encapsulate non-native proteins that are unfolded and refolded in cooperation with the DnaK (Hsp70) system.

The bacterial GroE ensemble (Group I chaperonin) consists of 14 identical GroEL protomers of 57 kDa, organized as a multimeric structure of two heptameric back-to-back circles that interacts with one heptameric ring of the co-chaperonin GroES (subunit M_r 10 kDa) to form the biologically functional unit (Plate 3). The tertiary structure of GroEL protomers is defined by a predominantly α-helical equatorial domain harbouring a single ATP-binding site and a so-called apical domain with a mix of α-helices and β-strands; these two domains are linked by a short middle domain. The GroES monomer, on the other hand, is folded almost exclusively into β-strands with flexible unstructured regions; the monomers are in contact with each other, arranged in a heptameric circle with smaller loops projecting into the interior (Plate 3). The following is a brief summary of the current understanding of events involved in restoration of the native structure of damaged proteins by the GroE super-assembly (reviewed in Saibil et al., 2013).

In each heptameric GroEL ring, the equatorial domains of subunits are in physical contact with their nearest neighbours. The stack of the two rings forms a central substrate-binding cavity with the hydrophobic residues of the apical domains lining its periphery; the physical dimensions of the cavity are commensurate with accommodation of non-native proteins/polypeptides of up to ~60 kDa. In the unliganded apo state, the GroEL double ring is in the 'open' configuration where ATP binds to subunits of one ring with positive cooperativity and to the other ring with negative cooperativity, such that only one ring is in the 'active' functional configuration at any given time. The hydrophobic surface formed by the seven apical domains will then bind, in short order, multiple exposed hydrophobic segments of the non-native substrate protein. Concomitantly with admittance of the substrate, the GroES heptameric ring binds to the apical domain (Plate 3) and closes the receptor cavity, acting like the proverbial 'lid' or a 'hat' (Horwich et al., 2006; Saibil et al., 2013). In this state, the GroES subunits forge contacts with the hydrophobic surface of the apical domains of ATP-bound GroEL subunits, via a small hydrophobic loop. As a direct consequence, a series of events is set in motion: the apical domains undergo a powerful rotatory movement forcing switching of the predominantly hydrophobic surface with a negatively charged, hydrophilic stretch. This action disengages the bound polypeptide and firmly places it into the latter, newly fabricated cavity, topped by the GroES ring. The microenvironment of the cavity provides the necessary conditions, conducive for proper folding of the polypeptide, as dictated by its primary structure. Hydrolysis of ATP occurs after completion of one cycle of substrate binding and release from the folding chamber; the second cycle is initiated by ATP binding to the second ring; the two rings of GroE thus function alternately in the process. Thus we see that the unfolding/folding by the bacterial chaperonin machine is a complex process, dependent upon highly sophisticated inter- and intra-ring allosteric interactions.

1.10 The Eukaryotic Chaperonin Complex

The eukaryotic cytosol contains a functionally similar chaperonin (Group II) known as CCT or TriC (TCP1 ring complex), a huge oligomer (~1 MDa) composed of eight heterogeneous subunits, arranged in two stacked rings. Other Group II chaperonins, exemplified by the archaeal 'thermosomes', are also multi-subunit complexes of either octameric or nonameric rings of identical subunits, or alternatively, a blend of different gene products. The CCT oligomer – like GroEL – is a structure with two back-to-back rings forming a central chamber where binding and folding of the substrates is accomplished. Interestingly, CCT is not a conventional Hsp – it is actually down-regulated during hyperthermal stress. The major substrates of CCT are actins and tubulins, but it is also known to recognize some newly synthesized proteins. It interacts with other chaperones, including Hsp70; however, no GroES-like co-factors are involved in its activity.

The individual subunits of CCT, like all chaperonins, are constructed of an archetypical three-domain structure: a central equatorial domain housing the ATP-binding site; an apical domain with the substrate-binding site and the middle domain, bridging the two. A recently published crystal structure of the bovine (*Bos taurus*) CCT shows that a helical segment at the tip of the flexible apical domain forms a lid on the chaperonin folding chamber (Muñoz *et al.*, 2011). In TriC, as in GroEL, binding and hydrolysis of ATP are governed by intricate allosteric interactions, underscored by positive cooperativity between subunits of each ring and negative cooperativity between the two rings (Horovitz and Williams, 2005; Yèbenes *et al.*, 2011). Remarkably, the mechanism underlying cooperativity in TriC has recently been shown to be distinct from that documented for GroEL. Using confocal microscopic analyses of single molecules of TriC in solution, bound to fluorescently labelled ATP(ADP), Jiang *et al.* (2011) unravelled a unique scenario showing that of the total of 16 ATP-binding sites in the double ring structure only four subunits per ring – high-affinity protomers – are occupied by nucleotides at any given time. The resulting conformational change promotes positive cooperativity of ATP hydrolysis at these sites, while binding of ATP to the remaining, low-affinity subunits inhibits ATPase activity, thus contributing to negative cooperativity. An outstanding feature of molecular chaperones, in general, is the exquisite use of allosteric interactions for perfecting and fine-tuning the system.

1.11 Dynamic Hexameric ATPases: Clp/Hsp100 Disaggregases

Protein aggregates, formed on encounters with severe hyperthermal and other varieties of damaging stress, are catastrophic for cells. Hence efficient systems for their removal or dissolution have evolved in all three biological kingdoms. Molecular chaperones of the Hsp100 family are a conserved group of AAA+ ATPases (AAA+, ATPases Associated with diverse cellular Activities) proficient in reversing the denaturation of proteins enmeshed in aggregates (Barends *et al.*, 2010). While not necessary under permissive conditions, their presence under stress conditions is reported to improve cellular survival by as much as three orders of magnitude. Two major classes are recognized within this family: Class I with dynamic, hexameric structures containing two nucleotide-binding (NBD) sites per monomer, exemplified by the bacterial ClpA, B, C and E proteins, and the yeast

Hsp104; Class II monomers harbour a single NBD as seen in bacterial ClpX and Y. The ATPase domains of AAA+ family enzymes, homologous to Hsp100, are found in mammalian cells and in various ATPases associated with proteasome assemblies (Schrader *et al.*, 2009).

The yeast Hsp104 and bacterial ClpB (caseinolytic peptidase B) are two prominent, well characterized members of the AAA+ super-family that function as hexamers, with orthologues in filamentous fungi, plants and the mitochondria. The protomeric structure of these chaperones entails an N-terminal domain, connected to the first NBD (NBD1) by a flexible linker, the middle domain (MD), followed by the C-terminal NBD2 (Lee *et al.*, 2007; Plate 4). The C-terminal end of yeast Hsp104 has a 38-residue extension – not seen in ClpB – that is involved in the assembly of protomers to form the hexamer and presumably has a role in acquisition of thermotolerance (Tkach and Glover, 2004). The two NBDs show highly conserved secondary structures, folded into a five-stranded β-sheet, flanked by α-helices containing the signature motifs, Walker A and B – catalytic sites for ATP binding and hydrolysis (Plate 4C, D). Both NBDs have conserved, strategically located Arg residues that are involved in ATP hydrolysis and oligomeric assembly. The M domain in Hsp104 is folded into a long α-helical (~200 amino acids) construct with two 'wings' each formed of anti-parallel α-helices joined together by hydrophobic surfaces (reviewed in Liberek *et al.*, 2008; Doyle and Wickner, 2009; Barends *et al.*, 2010). In the ATP-bound state, the oligomeric structure is stable, capable of engaging in productive interactions with protein aggregates; the energy of ATP hydrolysis fuels the dissagregase reaction.

According to the currently available models, based on X-ray crystallography and Cryo-EM imaging, the quaternary structures of Hsp104 and ClpB appear to be similar in external morphology: two hexameric rings forming a central lumen within which the aggregated proteins are enclosed and processed (Plate 4A, B). The substrates are threaded through a central lumen between the two rings – seen in both ClpB and Hsp104 – that is lined by sub-structures referred to as 'pore loops' with tyrosine residues linked to binding of the substrate and its movement along the hexameric rings. However, there are significant differences in the substrate binding and nucleotide binding modules of Hsp104 and ClpB. For instance, the central chamber of Hsp104 is more capacious than that of ClpB (diameter of 78 Å versus 25 Å), perhaps signifying an evolutionary adaptation to facilitate processing of bulkier substrates. In addition, the MD of ClpB is a coil located on the outer surface of the hexamer (Doyle and Wickner, 2009; Barends *et al.*, 2010).

Information on the mechanism of the disaggregation and the threading of the substrate has emerged from X-ray structure of isolated NBD2 domain of *Thermus thermophilus* ClpB (Biter *et al.*, 2012). This study shows that in the hexamer formed by the NBD2 domains, a critical Arg residue in the Arg-finger motif is required for ATPase activity of the ring as its mutation (Arg to Ala) abolishes this activity. The model proposed by Biter *et al.* (2012) suggests a sequence of events whereby initially an ATP-bound monomer of ClpB interacts with the substrate with high affinity. The energy of ATP hydrolysis drives the translocation of the substrate on the hexamer, ADP is released next and simultaneously the substrate binds to the next ATP-bound monomer. The same sequence of events is repeated until the substrate has traversed through the monomers of the entire hexamer. It has been reported that mutations of Arg residues in the M domain (Arg to Ala) as well as those in the NBDs compromise development of thermotolerance. Hsp100 family proteins are known to act in coordination with Hsp70 (and DnaK) to conclude the dissolution of aggregates (Doyle and Wickner, 2009; Barends *et al.*, 2010).

The disaggregation and ultimate retrieval of proteins trapped in aggregates is dependent upon the cooperative action of Hsp100s and Hsp70 family chaperones. The details of the mechanism of the interaction between these two chaperoning machines have been revealed in a recent investigation by nuclear magnetic resonance (NMR) spectroscopy analysis of ClpB interaction with specifically labelled DnaK (Rosenzweig *et al.*, 2013). It was shown in this study that the ClpB hexamer binds to the cleft between subdomains, SBD-α and SBD-β, of the substrate binding site of DnaK, overlapping with the GrpE binding site (see section on Hsp70 for details). In this system, binding of GrpE to DnaK was observed to inhibit the disaggregation reaction. Both ClpB and DnaK are required for the disaggregation reaction to commence. According to the proposed model of the mechanism of dissolution of the aggregate and recovery of properly folded client protein, the aggregated protein (pAg) is first bound by the DnaK-DnaJ binary complex – which had been pre-primed by binding and hydrolysis of ATP – forming a ternary complex (DnaK-J-pAg). ClpB, also in its active form, is postulated to interact with the latter complex, the disaggregation process being initiated by threading of the denatured protein through the double hexameric ring lumen (Plate 4). At this stage the DnaJ-DnaK is ejected, concomitantly with the release of bound ADP. The newly liberated polypeptide emerging from the ClpB ring structure is finally refolded by re-association with the DnaK-DnaJ-GrpE machinery. Thus, the DnaK-J complex is hypothesized to engage in the reaction at two stages: first, the initial capture of the aggregated substrate and second, the final refolding to generate the native configuration (Rosenzweig *et al.*, 2013).

1.12 The Hsp70 (DnaK) Family: Highly Conserved Allosteric Foldases

It is well established that the Hsp70 and Hsp90 families embody the most important molecular chaperones for regulation of proteostasis in both prokaryotes and eukaryotes. While heat-induced Hsp70s are critical for folding during stress conditions, their constitutively expressed isoforms are essential for folding of nascent polypeptides during stress-free growth conditions. The Hsp70 family is one of the most structurally and functionally conserved of molecular chaperones exhibiting a striking sequence similarity/identity between orthologues and preservation of the three-dimensional structure, across biological kingdoms. The invariant structural features of the Hsp70 family include the ~40 kDa N-terminal nucleotide (ATP)-binding/ATPase domain (NBD) and the C-terminal ~25 kDa peptide binding domain (SBD) linked by a short, flexible hydrophobic connector first demonstrated in crystal structures of *E. coli* DnaK, published more than two decades ago (Flaherty *et al.*, 1990; Wang *et al.*, 1993). Since then, insightful information has been derived from numerous biochemical and structural studies of isolated, ligand-free and nucleotide-bound NBD and peptide-bound SBD domains of the DnaK. An understanding of the properties and functional features gained from studies of DnaK has proved invaluable in deciphering the molecular details of the mechanism of action of the Hsp70 family in general. While NBD and SBD, in isolation, are proficient at binding their respective ligands, the chaperoning activity of the protein is strictly dependent on elaborate allosteric interactions between the two domains. In the nucleotide-free Apo state or ADP-bound state, substrates bind to SBD with high affinity. Binding of ATP (to NBD) allosterically regulates the conformation of SBD reducing its substrate-binding affinity by a few orders of magnitude, thus promoting the release of the bound peptide (Schmid *et al.*, 1994; Qi *et al.*, 2013). The following summarizes the structural parameters of DnaK, gleaned from crystal structures and mutational analyses.

The substrate binding domain of DnaK is further divisible into two subdomains, SBD-α (12 kDa) and SBD-β (15 kDa), connected by a short linker. The subdomain SBD-β is composed of eight β-strands separated by seven loops and SBD-α is connected to the SBD-β by five α-helices. Unfolded/mis-folded substrate proteins with exposed hydrophobic surfaces are enclosed within a cavity formed by the β-strands. The α-helical region of SBD-α forms a lid-like cover on the substrate binding cleft and interacts with and stabilizes the loop regions. The lid region is believed to be capable of acquiring different conformational states, one of which is an 'open' state in the ATP-bound DnaK, wherein the substrate can gain access to the cavity (Plate 5). Thus DnaK and eukaryotic homologues alternate between ATP-bound state – with a slow rate of substrate association/dissociation – and the ADP-bound state with a rapid rate of substrate binding/dissociation. The DnaK substrate binding site has been shown to be capable of admitting a stretch of from five to seven hydrophobic residues *in vitro*. However, *in vivo* it can also bind to extended stretches of polypeptides as well as to regions containing some tertiary structure (Rüdinger et al., 1997; Schlecht et al., 2011).

Substrates are bound by Hsp70s in an ATP-dependent manner and ATP hydrolysis is essential for performance of the folding reaction. However, most Hsp70s have a very low intrinsic ATPase activity and assistance by co-chaperones is essential to raise it to a workable level. The Hsp40 family proteins and bacterial DnaJ proteins, described in the following section, are co-chaperones that bind to Hsp70s and alter their conformation by promoting a rearrangement of residues in the catalytic site to optimize its hydrolytic activity. Eukaryotic Hsp70s differ from DnaK in the presence of a ~35-residue extension at the C-terminal end and their cytosolic isoforms have a characteristic motif, M/VEEVD, that is missing in bacterial orthologues. This motif is involved in direct interaction with co-chaperones containing TPR repeats.

1.13 Bacterial J-Proteins and Eukaryotic Hsp40: Co-Chaperones of Hsp70s

It is well established that the bacterial Hsp70 chaperone, DnaK, works in collaboration with the co-chaperone DnaJ (eukaryotic homologue Hsp40) and the nucleotide exchange factor GrpE. The latter is a critical component of the bacterial Hsp70 core machinery; it interacts with the ADP-bound protein to dislodge the resident ADP and release the client protein/polypeptide after a cycle of substrate folding. ATP then rebinds to the vacated, ligand-free site, the cycle of ATP binding and hydrolysis being repeated multiple times until conclusion of the folding process. The DnaK-DnaJ-GrpE cohort forms the functional unit responsible for folding of bacterial polypeptides emerging from the GroEL-ES chaperonin complex. The eukaryotic Hsp40 (J protein) families encompass a diverse collage of proteins exhibiting marked sequence diversity and molecular mass that may deviate considerably from the prototypic 40 kDa. Some J proteins can even act as molecular chaperones independently of the Hsp70 machine (Kampinga and Craig, 2010). There are six human nucleotide exchange factors – functionally equivalent to bacterial GrpE – designated the BAG (Bcl2-associated anthanogene) family proteins, containing common BAG domains. They interact with Hsp70 and distinct J family members and catalyse release of ADP and substrate from the SBD (Rauch and Gestwicki, 2014).

Three classes of DnaJ proteins have been documented in the bacteria. Class I is represented by the *E. coli* DnaJ, where the monomer is folded into the following domains: an N-terminal highly conserved J domain containing the signature HPD motif; a flexible glycine-phenylalanine (GF) rich region; four repeats of a zinc-finger motif and a C-terminal region with two similar domains (CTD I and CTD II) mainly composed of β-strands. A hydrophobic pocket in CTD I constitutes the critical binding site for client proteins, and a zinc-finger emanating from it may assist in substrate binding. Terminal sequences in CTD II participate in fabrication of a dimerization domain (Plate 6). The latter may be involved in controlling the affinity for substrates. However, Class II DnaJ proteins, exemplified by the *Thermus thermophilus* orthologue, are devoid of the Zn-finger domain. This motif has been shown to be dispensable for biological function as the *T. thermophilus* DnaJ can replace the *E. coli* DnaJ for protein folding *in vitro* by the DnaK-J-GrpE system. In Class III, on the other hand, are housed diverse entities, including the human Hsp40 and other mammalian orthologues, with structural features that are inconsistent with the criteria listed for classes I and II.

Apart from lacking the Zn-finger domains, DnaJ of *T. thermophilus* is similar to Class I entities with respect to the overall structural parameters. It has the classical J domain in the N-terminus with the conserved HPD motif, but next to it there is a polyproline motif of six Pro residues, followed by the GF domain which is connected to the C-terminal region by a short flexible linker. The J domain molecular structure, determined by X-ray crystallography of the spin-labelled recombinant protein, shows it to be composed of four α-helices, the HPD motif, necessary for binding to Hsp70, being located in a loop between the second and the third helix (Plate 6B). The C-terminal domain is folded into two subdomains defined by β barrel-type secondary structure. Next to the HPD motif the GF domain makes hydrophobic contacts with J domain by its phenylalanine residues. Residues at the extreme end of the C-terminus form a helical segment that participates in domain swapping with another monomer resulting in a homodimeric unit (Barends et al., 2013). Plate 6A shows the relative positions of the three domains in each monomer forming the dimeric complex. The yeast Hsp40 J domain structure is similar to that of a typical Class I J protein.

However, the structure of the mammalian type III DnaJs is entirely different from Class I and II. X-ray crystallographic analysis of the human type III J protein, P58^{IPK}, which is a monomer, unlike the Class I and II dimeric proteins, shows it to be elongated in shape with a completely α-helical secondary structure (Svärd et al., 2011). This J protein is a co-chaperone of BiP (binding immunoglobulin protein, also known as glucose regulated protein 78 (GRP78)), the ER equivalent of Hsp70 with the terminal sequence KDEL. BiP is a molecular chaperone responsible for folding of ER proteins under stress and it plays a key role in maintenance of calcium homoeostasis. P58^{IPK} is localized in the ER along with BiP; both proteins are induced by ER stress and form a major component of the UPR system. It has the standard J domain, consistent with corresponding proteins in other organisms, with the conserved HPD motif. But, contrary to the class I and II proteins, the J domain herein is situated in the C-terminal region and linked by a flexible linker to the N-terminal segment, which is folded into 19 α-helices arranged into three subdomains, each containing three TPR repeats (Plate 6C, D). The J domains in different Hsp40s are similar in that they are all composed of four α-helices – two short ones flanking two longer ones. The canonical HPD motif in P58^{IPK}, for interaction with Hsp70 (BiP in this case), is presumed to be located in a loop between the second and third helix. The segments adjacent to the HPD-containing loop comprise a mix of positively charged polar and hydrophobic surface residues, involved in the putative BiP binding site. As for other eukaryotic Hsp70s, interaction of BiP with its partner Hsp40 elevates its ATPase activity (Svärd et al., 2011).

1.14 Hsp90, the Multifaceted Chaperone: Myriad Functions and Varied Clientele

The Hsp90 class chaperones regulate a variety of critical cellular processes – cell proliferation, signal transduction, biogenesis of microRNAs and damage repair – through their extensive repertoire of client proteins. They recognize hydrophobic stretches on non-native, unfolded or partially folded polypeptides as well as proteins in a near-native state. The spectrum of their substrates or client proteins is enormous, straddling across transcription factors, cell cycle/checkpoint kinases, steroid hormone receptors, chromatin remodelling factors, nitric oxide synthase, telomerase, plant pathogen resistance gene products, mammalian innate immunity response proteins and numerous others – the list is endless – with no apparent commonality of structural or functional attributes (see http://www.picard.ch/downloads/hsp90interactors.pdf). The only universal feature of the well over 300 client proteins of Hsp90 is their inherent instability. It has been demonstrated repeatedly that mammalian Hsp90, in concert with other chaperones and co-chaperones, is a major factor in cytoprotection and maintenance of cellular integrity.

In the fungi and higher eukaryotes the presence of at least one active cytosolic Hsp90 conformer is required, believed to be indispensable for survival. As documented with the Hsp70 family, interaction of Hsp90s with substrates is dependent upon cycles of ATP binding and hydrolysis (Young and Hartl, 2000; Frydman, 2001). Although ATPase activity is imperative for function of Hsp90s, ATP hydrolysis is a very slow process: turnover numbers for yeast and human Hsp90 ATPase are 1 min^{-1} and 0.04 min^{-1}, respectively. Eukaryotic Hsp90s form complexes with co-chaperones of which Aha1 (activation of Hsp90 ATPase) is an activator of its ATPase activity, as the name implies. In addition, several other co-chaperones are known to be inhibitors of ATPase activity and/or subsequent steps in binding and processing of substrates (Pearl and Prodromou, 2006; Richter *et al.*, 2006).

Details of Hsp90 structure have been unravelled by the use of a variety of biochemical, biophysical and imaging techniques including limited proteolysis, x-ray crystal structures of isolated domains, NMR spectroscopy, Cryo-EM, rotary shadowing EM and small-angle x-ray scattering (SAXS). All Hsp90s are dimers in their active state; the monomer is composed of three major structural domains, with secondary and tertiary structure that is virtually unchanged in diverse species: a conserved ~25 kDa N-terminal domain (NTD) containing the ATP-binding pocket; a highly conserved ~40 kDa middle domain (MD) important for substrate binding; and a partially conserved ~12 kDa C-terminal domain (CTD). An assortment of crystal structures of isolated individual domains of Hsp90s of various species, along with a few of the complete protein, are currently listed in the PDB database (Pearl and Prodromou, 2006; Picard, 2013). In the eukaryotic Hsp90s, NTD and MD are connected by a charged linker varying in length from 30 to 70 residues in different species, with alternating segments of acidic and basic charged residues – poly-glutamate and poly-arginine stretches interspersed with a few aliphatic residues. In contrast, the linker segment is absent in HtpG, the bacterial Hsp90, and the mitochondrial orthologue. Another significant divergence between the bacterial and eukaryotic cytosolic Hsp90s is the absence in the bacterial and mitochondrial Hsp90s of a ~35 residue C-terminal sequence seen in the eukaryotic Hsp90s – terminating in the motif MEEVD. This motif is also present in the C-terminal ends of eukaryotic Hsp70s and is the binding site for some co-chaperones with tetratricopeptide repeats (TPR) – tandem repeats of degenerate 34-amino acid motifs – known for mediating protein–protein interactions. TPR domains of co-chaperones with subdomains folded into anti-parallel α-helices can bind to the C-terminal tails of both Hsp70 and Hsp90, linking them together in chaperoning complexes active in maturation of mammalian hormone receptors.

In the ligand-free apo state, the residues in the C-terminal region of two monomers of Hsp90 are intertwined to form an 'open' dimerized state, resembling a V-shaped molecule. Structural studies of *E. coli* HtpG, human Hsp90α, yeast Hsc82 and the native form of pig brain Hsp90 – using SAXS, single-particle Cryo-EM and 3-D reconstructions – demonstrate that the apo-dimers, in solution, exist in a state of dynamic equilibrium between two or more 'open' conformations (Shiau *et al.*, 2006; Bron *et al.*, 2008; Southworth and Agard, 2008). Upon binding of ATP to the NTD, large scale conformational changes and re-arrangements lead to dimerization at the N-terminal end yielding a structure with the characteristic twisted appearance (Harris *et al.*, 2004; Ali *et al.*, 2006).

An identical sequence of events leading to the formation of the biologically active closed ATP-bound Hsp90 dimer is witnessed in all members of the Hsp90 family. The NTD domain is fairly rigid with two mobile segments that are involved in ATP binding and dimerization: the initial ~22 residues at the N-terminal end and the 'lid' region comprising residues 93–124 in *Neurospora* Hsp80 and 91–122 in *Saccharomyes cerevisiae* Hsp82. In the ligand-free, inactive state the lid is in a predominantly α-helical conformation and curled back, the ATP binding pocket being partly exposed. Subsequent to binding of the nucleotide and dimerization, conformational changes result in the mobilization of the 'lid' motif enabling it to interact with residues lining the catalytic site thereby enclosing the bound nucleotide. Release of the lid segment is closely linked to a marked change in the position and structure of the first ~20 residues of the NTD from a β-strand structure in an intra-domain interaction to the extended coil conformation. The latter is stabilized by forming multiple contacts with the corresponding sequence in the NTD of the opposite ATP-bound monomer chain in the dimer. Concomitantly, a catalytically active ATPase is generated following the movement into the ATP binding pocket of a key arginine residue (R380 in yeast) from the MD. Thus NTDs of the two arms of the V-shaped structure are joined together to form a 'closed' unit, which then adopts the twisted shape visible in EM images.

The structural representations of the extended Apo-dimer of the bacterial HtpG

(Krukenberg et al., 2008) and *Neurospora crassa* Hsp80(90) active 'closed' dimer (Plate 7A, B; Plate 7C, D, respectively) are shown with surface views, and details of arrangement of the three domains – NTD, MD and CTD. In the *N. crassa* dimer (Plate 7C, D) considerable inter- and intra-monomer interactions occur between residues in the N-terminal region and those in the C-terminal end (Roy et al., 2013). Furthermore, in the dimer structure movement of MD relative to NTD and that of CTD away from MD is evident. Such large scale conformational shifts result in establishment of contact with the corresponding domains in the opposite monomer.

The human orthologue of Hsp90 is of particular interest as it is known to be the major contributor to conformational maturation of some oncogenic signalling proteins and its inhibitors have been observed to selectively destroy cancer cells. Over-expression of Hsps in tumour cells is closely associated with the increased expression of oncogenes and development of resistance to anticancer therapy and inhibition of apoptosis (Sreedhar and Csermely, 2004; Calderwood et al., 2006). Compared to the isoforms in normal cells, Hsp90 isolated from tumour cells exhibits higher ATPase activity *in vitro*. In view of the role of Hsp90 in promotion of unregulated cell proliferation, it is recognized as a significant target for anticancer drugs, such as Geldanamycin, that block the ATP binding site in the N-terminal domain. The compound 17-allylaminogeldanamycin (17-AAG), an analogue of Geldanamycin, is reported to bind to Hsp90 from tumour cells with a 100-fold higher affinity compared to that from normal cells (Kamal et al., 2003). However, it has been known for several years that the C-terminal domain of Hsp90 also contains a nucleotide binding site (Garnier et al., 2002; Söti et al., 2005; Sgobba et al., 2010). This second NBD is the preferred target of antibiotics and anticancer drugs such as Novobiocin and a multitude of its more potent, chemically engineered derivatives. Consequently, unravelling the mode of binding of nucleotides and pharmaceuticals at this site is the subject of intense efforts at knowledge-based design of the ideal site for optimized, productive binding of these and other promising anticancer drugs (Zhao et al., 2011; Zhao and Blagg, 2013).

1.15 Co-Chaperones of Hsp90 Influence its Catalytic Activity

Interaction with specific co-chaperones directly modulates both binding and hydrolysis of ATP at the nucleotide binding site in the NTD. Several co-chaperones – known to associate with the Hsp90 homodimer and acting on its ATPase activity – have been characterized. This group includes both positive and negative regulators of ATPase activity as well as cofactors that impact substrate binding and downstream processing, without interference with ATP hydrolysis (reviewed in Röhl et al., 2013). The best known activator of Hsp90 ATPase activity is the co-chaperone Aha1. The N-terminal domain of the latter binds to the MD of the Hsp90 dimer in its closed state and alters its conformation to empower the catalytic site, resulting in ~10-fold stimulation of ATPase activity (Pearl and Prodromou, 2006; Wandinger et al., 2008; Prodromou, 2012). Binding of co-chaperones Cdc37, Sba1 and Sti1 diminishes ATPase activity of yeast Hsp90. Apparently Sti1 can bind to the MEEVD motif of Hsp90 by one of its TPR domains and it can also bind to a second site in the NTD/MD region. In addition, it can block the N-terminal dimerization of Hsp90 by interacting with the first 24 residues of the NTD. Human Cdc37 is reported to bind to the catalytic pocket 'lids' of the two chains, thereby preventing dimerization. Crystal structure of a complex between yeast Hsp82 and Sba1 shows that this co-chaperone binds to the N-terminal domain and stabilizes its closed (ATP-bound) dimerized state (Ali et al., 2006). As discussed in the following, Sgt1 and Rar1 are additional co-chaperones of Hsp90, with vital roles in innate immunity response of plants to fungal and bacterial pathogens.

1.16 Hsp90 and Co-Chaperones: a Vital Role in Plant and Animal Pathology

In addition to its multifarious activities in promoting maturation and assembly of numerous metabolic regulators and oncogenic proteins, Hsp90 is a key component of the system involved in recognition of potential pathogens and induction of the disease-defence response in higher eukaryotes. In plants, specific resistance to pathogens is conferred by the so-called resistance (R) genes. A large number of R genes have been isolated from a variety of plant species and most of them encode cytosolic sensors of products produced by pathogens. As these proteins contain nucleotide binding sites and leucine-rich repeats (LRR) they are referred to as NLR proteins (NLR: acronym for nucleotide-binding site (NB) and leucine-rich repeats (LRR)). Plant NLR proteins have N-terminal coiled coil domains, followed by a middle nucleotide binding domain and a C-terminal, leucine-rich domain. NLR proteins, in complexes with Hsp90, act as regulators of innate immunity in both plants and animals. Upwards of 150 NLR protein-encoding genes have been documented in the *Arabidopsis thaliana* genome and up to 600 or so in the rice genome. In humans, 21 NLR proteins (also referred to as NOD-like proteins) are widely believed to be implicated in sensing pathogen products and regulation of the innate immune responses (Shirasu, 2009).

In higher plants NLR sensors recognize specific pathogen-encoded 'effector' molecules in the host cells that enhance the virulence of the invading organism upon infection. Various genetic screens have identified key innate immunity-regulating molecules in plants, the most well characterized are the R gene products, Rar1 (required for MLA12 resistance 1) and Sgt1 (suppressor of the G2 allele of Skp1) that bind to NTD of Hsp90 and act as co-chaperones. Sgt1 is highly conserved in eukaryotes and it contains three distinct structural domains: (i) the N-terminal TPR domain, similar to TPR domains of other proteins that bind to the MEEVD motif of Hsp90; (ii) a middle CS domain (CHORD-containing protein and SGT1) – CS of Sgt1 is also defined as the crystallin and small heat-shock protein-like domain; it is structurally related to the α-CD domain of small Hsps and to Hsp90 co-chaperone, Sba1/p23; and (iii) a C-terminal SGS (Sgt1-specific) domain. However, it appears that Sgt1 does not use its TPR domain to bind to Hsp90, but instead, it serves to interact with another host protein, Skp1 (reviewed in Kadota *et al.*, 2010). Interestingly, although the site of Sgt1 interaction with Hsp90 lies in its CS (Plate 8A–C) domain with homology to Sba1, unlike Sba1 it does not bind to the ATP pocket lids and, therefore, does not inhibit ATPase activity. Its mode of action is reported to involve binding to another distinct site in the Hsp90 NTD, where it functions by attracting Rar1, a low-level activator of ATPase, to the complex. Crystal structure of the Hsp90-Sgt1 complex shows interaction of the CS domain β-sandwich domain with different residues of Hsp90 NTDs in partner monomeric chains (Zhang *et al.*, 2008). The role of Sgt1 in stable [Hsp90-Sgt1-CHORD] is twofold: its CS domain furnishes distinct binding sites for Hsp90 and Rar1 (Plate 8), while its SGS domain is involved in interaction with the LRR domain of NLR proteins in plants and animals.

Rar1 is an important conserved eukaryotic regulator, containing two so-called CHORD domains, 1 and 2 (structurally similar to cysteine and histidine-rich, zinc-containing domains), one of which, CHORD1, interacts with NTD of Hsp90 while CHORD2 binds to the CS of Sgt1, forming the [Hsp90-Sgt1-Rar1] ternary complex (Plate 8D–F). The importance of this complex in development of resistance to various pathogens was confirmed by experiments showing that silencing or deletions of Rar1, Sgt1 or Hsp90 – or treatment with Geldanamycin that inhibits ATPase activity – led to a marked diminution of resistance to pathogens (Hubert *et al.*, 2009). More direct and convincing biochemical evidence for association between the N-terminal region of Hsp90 and Sgt1 proteins *in vivo* was obtained by the use of yeast 2-hybrid protein screen and by co-immunoprecipitation analyses *in vitro* (Takahashi *et al.*, 2003; Kadota *et al.*, 2010). Structural studies of R protein complexes with Hsp90, coupled with information from biochemical and mutational analyses of the components parts, have provided an insight into the basic mechanism, but many questions remain unanswered. Elucidation of the web of interactions between sensors, molecular chaperones, co-chaperones, nucleotides and other participating molecules in regulation of the innate immune response of plants, is beset with further complexity by the presence of multiple cytosolic isoforms of Hsp90 with affinity for diverse substrates.

1.17 Concluding Comments

The pace of progress in unravelling the mechanism(s) of defence against environmental and endogenously generated stress conditions has accelerated rapidly in the last two decades. With the availability of more in-depth information, newer and at times unexpected ambiguities arise within an ever-expanding network of interactions involving molecular chaperones and assisting entities. Considering the activities of the molecular chaperones, both of the holdase and foldase category, the following picture emerges.

For passive binding and sequestration of unfolded/mis-folded proteins, the major attribute in a chaperone, with simple or complex external morphology is, for the most part, the presence or unmasking of hydrophobic surfaces, as seen in the sHsps. For the process of repair/unfolding/refolding, energy for propelling the chaperone machines and design of an energy generator is incorporated into the chaperone(s). In all of the foldase class molecular chaperones – chaperonins, Hsp70 and Hsp90 families – similarity in design of the nucleotide binding pockets is witnessed. The multimeric/oligomeric chaperonins have similar, especially constructed chambers or cavities, formed by rings or circles of monomers for capturing the substrate protein/polypeptide.

The energy-generating sites, ATP-binding/hydrolysis domains, located in individual subunits are allosterically regulated. Inter-domain conformational changes within monomers create the appropriate design of the catalytic sites. Inter-protomer and inter-subassembly allosteric interactions, prevalent throughout the gamut of molecular chaperones, underlie the fine tuning, efficiency and the sustainability of the system. Diverse components, the enormous array of co-chaperones and related factors respond to specific features, unique to the client protein. The ubiquitous folding-capable Hsp70 and Hsp90 families function in collaboration with partner co-chaperones to optimize the efficiency of their respective machines in combating proteotoxic stress. Thus a common basic theme in design and operation of different molecular chaperone systems is clearly discernible.

On encounter with mild stress conditions one or more of the chaperoning systems would suffice, but under more severe or extreme forms of stress concerted action of virtually all of the chaperoning systems would be imperative. With mild forms of stress, one can envisage batteries of defence-related enzymes engaged in development of tolerance towards or avoidance of stress (e.g. endogenous ROS). This constitutes the first line of defence, not the second, as accepted by conventional wisdom. In the face of conditions leading to massive unfolding/aggregation of proteins, the overexpression and concerted action of various chaperoning machines operating cooperatively to process the substrates in precisely coordinated steps, is the ultimate stress-managing strategy in all organisms.

The mechanism underlying thermotolerance is not completely understood. Genetic/mutational studies suggest the involvement of the sHsp in plants and Hsp104 family members in some other organisms, based on compromise of thermotolerance in mutants. For a comprehensive view, the contribution of enzymes, such as glutathione peroxidases and small molecules including carbohydrates and membrane components in adaptation against various stresses, needs to be evaluated vis-à-vis that of stress-inducible molecular chaperone proteins.

References

Ali, M.M., Roe, S.M., Vaughan, C.K., Meyer, P., Panaretou, B., Prodromou, C. and Pear, L.H. (2006) Crystal structure of an Hsp90-nucleotide-p23/Sba1 closed chaperone complex. *Nature* 440, 1013–1017.

Barends, T.R., Werbeck, N.D. and Reinstein, J. (2010) Disaggregases in 4 dimensions. *Current Opinion in Structural Biology* 20, 46–53.

Barends, T.R.M., Brosi, R.W.W., Steinmetz, A., Scherer, A., Hartmann, E., Eschenbach, J., Thorsten, L., Seidel, R., Shoeman, R.L., Zimmermann, S., Bittl, R., Schlinchting, I. and Reinstein, J. (2013) Combining crystallography and EPR: crystal and solution structure of the multidomain cochaperone DnaJ. *Acta Crystallographica* D69, 1540–1552.

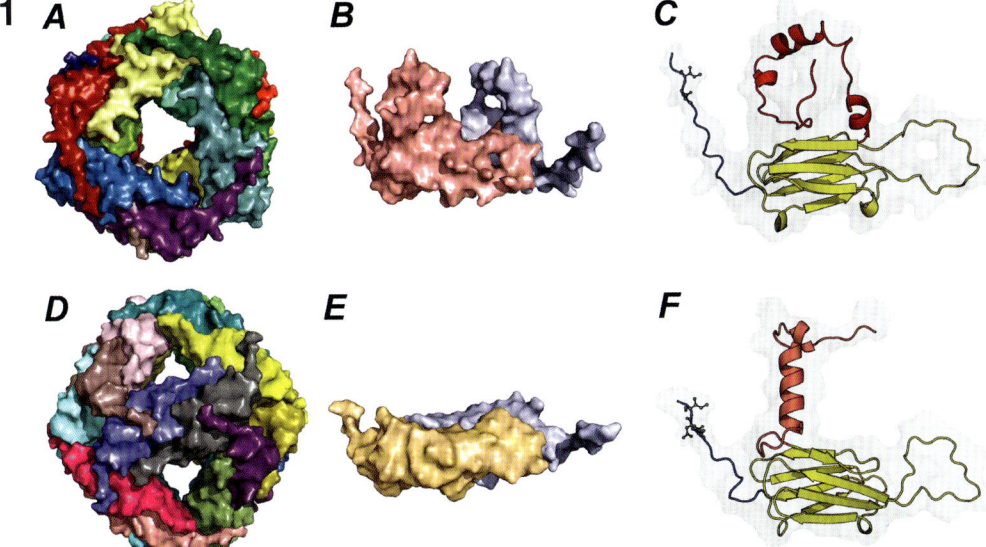

Plate 1. Structural features of small Heat Shock Proteins (sHSPs).

A, The stable oligomer of Hsp16.9 from *Triticum aestivum* (pdb id = 1gme) is a dodecamer made up of two interlocking hexameric discs with a central hole. B, The dimeric form of Hsp16.9 is widely considered to be the building block for larger assemblies of that protein. C, The monomer of Hsp16.9 contains the structurally conserved α-crystallin domain (yellow) flanked by an N-terminal region (red) and a C-terminal extension (blue). The position of a conserved motif necessary for oligomer formation is highlighted in the C-terminal extension. D, The stable oligomer of Hsp16.5 from *Methanocaldococcus jannaschii* (pdb id = 1shs) is a spherical and hollow 24-meric complex with octahedral symmetry. It has eight triangular and six square windows large enough to allow entry of small molecules and extended peptides. E, The dimeric form of Hsp16.5 is also considered to the building block for larger assemblies of that protein. F, The monomer of Hsp16.5 contains the structurally conserved α-crystallin domain (yellow) flanked by a reconstructed N-terminal region (pink) and a C-terminal extension (blue). The position of a conserved motif necessary for oligomer formation, similar to the one seen in C, is highlighted in the C-terminal extension.

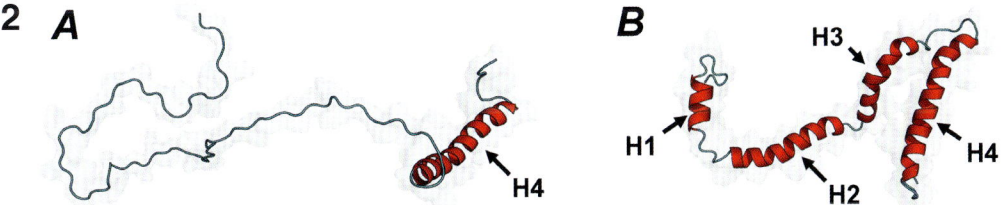

Plate 2. Representative solution structures of Hsp12 in the presence of membrane-mimetic micelles.

A, NMR analysis of the structure for Hsp12 in aqueous solution suggests that it is dynamically disordered. The addition of dodecyl-phospho-choline (DPC) results in the gain of a single helical region (H4) near the C-terminal end of the protein as seen in the structure (pdb id = 2ljl). B, The addition of sodium-dodecyl-sulfate (SDS) results in the gain of three more helical regions (H1, H2 and H3) in the structure (pdb id = 4axp).

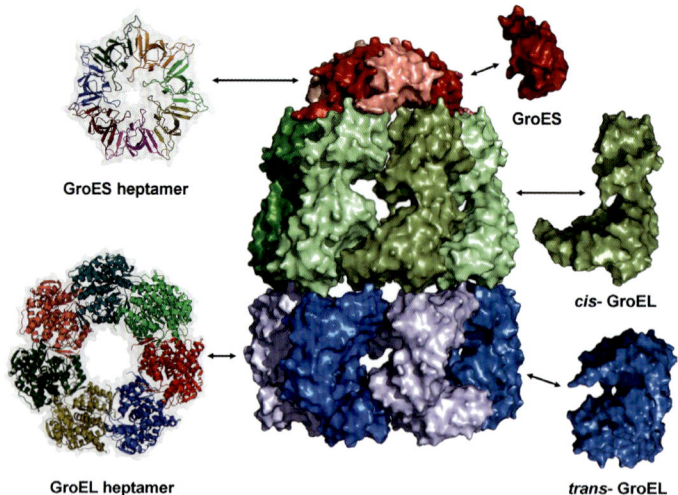

Plate 3. Structural representation of the bacterial chaperonin complex.

The bacterial GroE ensemble (Group I chaperonin), consists of 14 identical GroEL protomers, organized as a multimeric structure of two heptameric back-to-back rings that interacts with one heptameric ring of the co-chaperonin GroES to form the biologically functional unit.

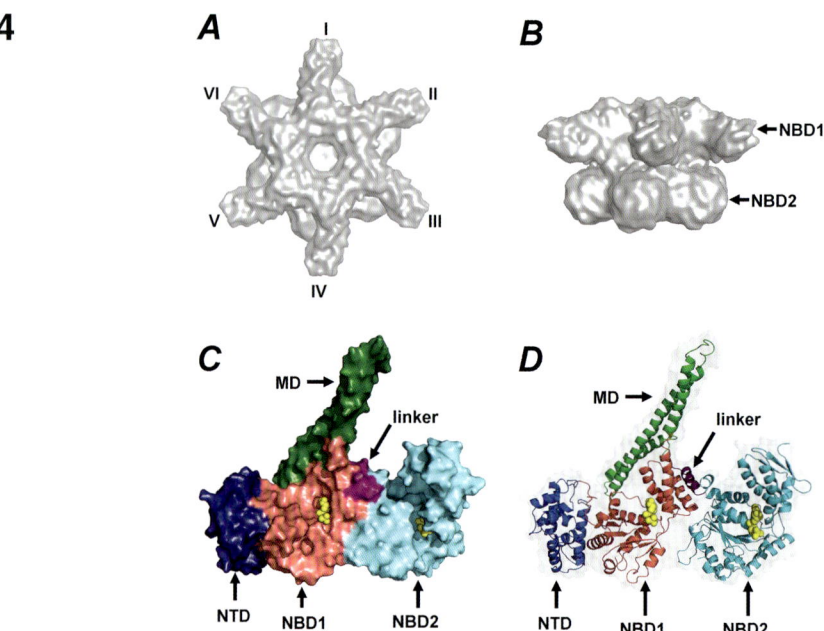

Plate 4. Structure of the hexameric and monomeric forms of ClpB.

A, The locations of the six monomers constituting the discoidal hexamer of ClpB are marked in this top-down view of a cryo-EM reconstruction (12.1 Å) of the full-length *Thermus thermophilus* ClpB construct (E271A/E668A) as described by Lee *et al.* (2007). B, A side view of the same cryo-EM reconstruction reveals the relative positions of NBD1 and NBD2 in a monomer of the ClpB hexamer. C, The shape and relative sizes of the four major domains in a ClpB monomer are highlighted (NTD, blue; NBD1, pink; MD, green; linker, violet; NBD2, cyan; ligand, yellow) (pdb id =1qvr). D, Secondary structural characteristics of the four major domains in a ClpB monomer are shown (NTD, blue; NBD1, pink; MD, green; linker violet; NBD2, cyan; ligand, yellow).

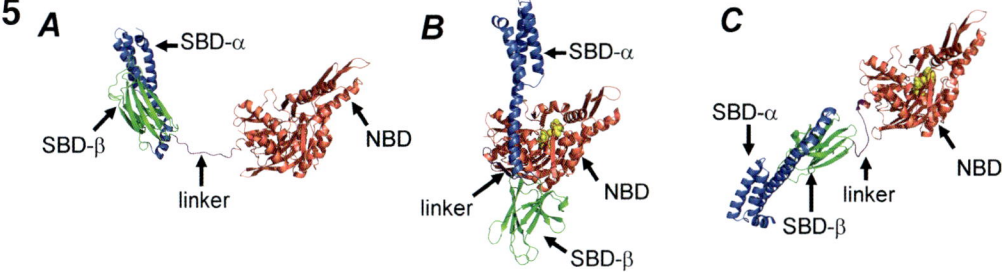

Plate 5. Conformational changes induced by ATP binding and hydrolysis in DnaK.

A, In its apo-state, the nucleotide binding domain (NBD) of DnaK and the substrate binding domains (SBD-α and SBD-β) are not in contact with each other, except via linker. This structural form exhibits a high affinity for substrate proteins. It was constructed by using *1dkg* as a structural scaffold for the residue sequence of *E. coli* DnaK. *B,* ATP binding to the NBD causes a significant conformational shift in the relative positions of SBD-α and SBD-β resulting in a more compact structure (pdb id = 4b9q). This change in conformation also releases bound substrate proteins. *C,* The post-ATP hydrolysis, or ADP-bound state of DnaK (pdb id = 2kho) has considerable similarity to its apo-state and exhibits a high affinity for substrate proteins.

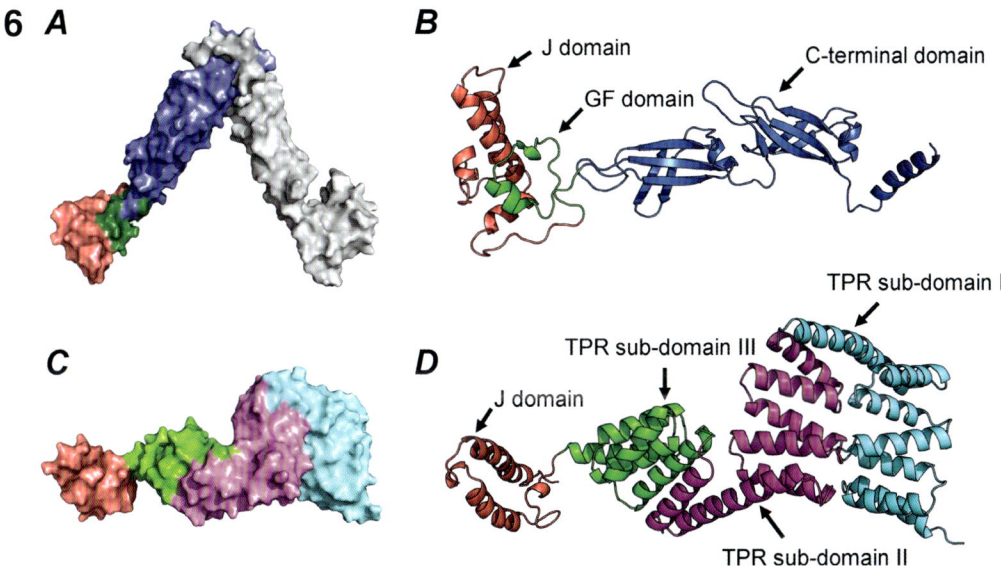

Plate 6. Structural features of members of the DnaJ family.

A, The dimeric form of a type II DnaJ protein from *Thermus thermophilus* (pdb id = 4j80) was used to show the relative positions of the three domains in each monomer. The J-domain (pink) is located near the N-terminus and is followed by the GF domain (green) and a large C-terminal domain (blue). *B,* The J domain (pink) is composed of four α-helices and contains the HPD motif, necessary for binding to Hsp70. The GF domain (green) lacks significant defined secondary structural features, yet is necessary for normal chaperoning activity of DnaJ. The C-terminal domain (blue) is folded into two sub-domains defined by β barrel-type secondary structure and terminates into a helical segment involved in dimerization. *C,* The crystal structure of the human type III J protein, P58[IPK] (pdb id = 2y4u), shows an elongated monomer in which the J domain (pink) is located at the C-terminal end of the protein. The rest of the protein is made up of a TPR-type domain containing three sub-domains: I (cyan), II (magenta) and III (green). *D,* The monomer has a completely α-helical secondary structure, in contrast to most type I and type II DnaJ proteins. The J domain (pink) in P58[IPK] is structurally similar to other J domains. It is widely believed that the three TPR sub-domains (cyan, magenta, green) in the protein are involved in coordinating interactions between the DnaJ-Hsp70 complex and other chaperones/co-chaperones.

7

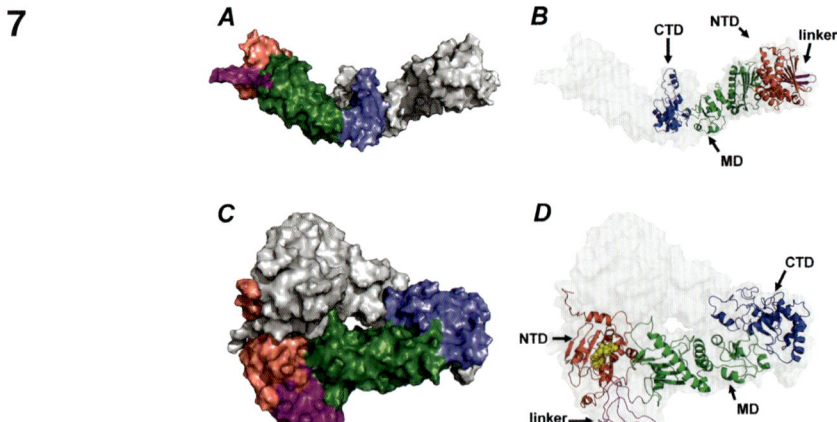

Plate 7. Apo- and ATP-bound dimers of Hsp90.

A and *B*, The extended form and domain structure of bacterial Hsp90 (HtpG) are presented (NTD, pink; linker, violet; MD, green; CTD, blue)(Krukenberg *et al.*, 2008). *C* and *D*, The closed dimer of *Neurospora crassa* Hsp80 (90) shows the conformational changes and inter-domain interactions caused by ATP binding (Roy *et al.*, 2013). The colour scheme is identical to that used in *A* and *B*.

8

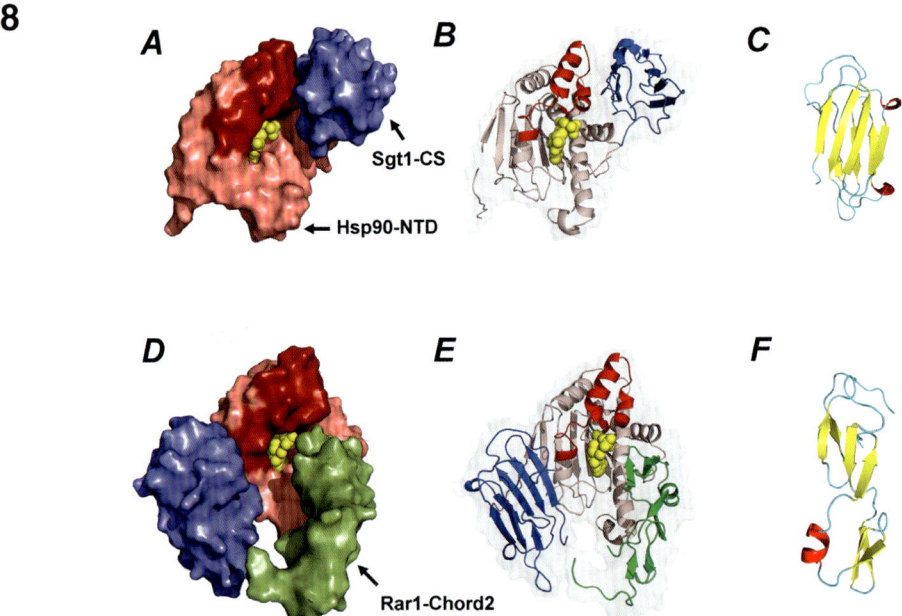

Plate 8. Interactions of Sgt1 and Rar1 with Hsp90.

A, The CS domain of Sgt1 interacts with the NTD of Hsp90 near the nucleotide-binding site (pdb id = 2jki). *B*, Secondary structure of the interacting domains as shown in *A*. *C*, Secondary structure of the CS domain of Sgt1. *D*, The CHORD2 domain of Rar1 interacts with the NTD of Hsp90 partially displacing the CS domain of Sgt1 (pdb id = 2xcm). *E*, Structure of the three interacting domains shown in *D*. *F*, Secondary structure of the CHORD2 domain of Rar1. NTD domain of Hsp90, pink; lid in NTD domain, red; CS domain in Sgt1, blue; CHORD2 domain in Rar1, green; ligand, yellow.

Basha, E., O'Neill, H. and Vierling, E. (2012) Small heat shock proteins and α-crystallins: dynamic proteins with flexible functions. *Trends in Biochemical Sciences* 37, 106–117.

Biter, A.B., Lee, S., Sung, N. and Tsai, F.T.F. (2012) Structural basis for intersubunit signaling in a protein disaggregating machine. *Proceedings National Academy of Sciences USA* 109, 12515–12520.

Bron, P., Giudice, E., Rolland, J.-P., Buey, R.M., Barbier, P., Diaz, F., Peyrott, V., Thomas, D. and Garnier, C. (2008) Apo-Hsp90 coexists in two open conformational states in solution. *Biological Cell* 100, 413–425.

Bukau, B., Weissman, J. and Horwich, A. (2006) Molecular chaperones and protein quality control. *Cell* 125, 443–451.

Calderwood, S.K., Khaleque, M.A., Sawyer, D.B. and Ciocca, D.R. (2006) Heat shock proteins in cancer: chaperones of tumorigenesis. *Trends in Biochemical Sciences* 31, 64–172.

Chandel, N.S., Maltepe, E., Goldwasser, E., Mathieu, C.E., Simon, M.C. and Schumacker, P.T. (1998) Mitochondrial reactive oxygen species trigger hypoxia-induced transcription. *Proceedings of the National Academy of Sciences USA* 95, 11715–11720.

Collins, Y., Chouchani, E.T., James, A.M., Menger, K.E., Cocheme, H.M. and Murphy, M.P. (2012) Mitochondrial redox signaling at a glance. *Journal of Cell Science* 125, 801–806.

Delbecq, S.P. and Klevit, R.E. (2013) One size does not fit all: the oligomeric states of αB crystallin. *FEBS Letters* 587, 1073–1080.

Doyle, S.M. and Wickner, S. (2009) Hsp104 and ClpB: protein disaggregating machines. *Trends in Biochemical Sciences* 34, 40–48.

Ellis, R.J., van der Vies, S.M. and Hemmingsen, S.M. (1989) The molecular chaperone concept. *Biochemical Society Symposium* 55, 145–153.

Flaherty, K.M., DeLuca-Flaherty, C. and McKay, D.B. (1990) Three dimensional structure of the ATPase fragment of a 70-K heat shock cognate protein. *Nature* 346, 623–628.

Frydman, J. (2001) Folding of newly translated proteins *in vivo*: the role of molecular chaperones. *Annual Review of Biochemistry* 70, 603–647.

Garnier, C., Lafitte, D., Tsvetkov, P.O., Barbier, P., Leclerc-Devin, J., Millot, J.-M., Briand, C., Makarov, A.A., Catelli, M.G. and Peyrot, V. (2002) Binding of ATP to heat shock protein 90: evidence for an ATP-binding site in the C-terminal domain. *Journal of Biological Chemistry* 277, 12208–12214.

Goldstein, L.E., Muffat, J.A., Cherny, R.A., Moir, R.D., Ericsson, M.H., Huang, X., Mavros, C., Coccia, J.A., Faget, K.Y., Fitch, K.A., Masters, C.L., Tanzi, R.E., Chylack, Jr, L.T. and Bush, A.I. (2003) Cytosolic β-amyloid deposition and supranuclear cataract in lenses from people with Alzheimer's disease. *Lancet* 361, 1258–1265.

Harris, S.F., Shiau, A.K. and Agard, D.A. (2004) The crystal structure of the carboxy-terminal dimerization domain of htpG, the *Escherichia coli* Hsp90, reveals a potential substrate binding site. *Structure* 12, 1087–1097.

Hartl, F.-U. and Hayer-Hartl, M. (2009) Converging concepts of protein folding *in vitro* and *in vivo*. *Nature Structural and Molecular Biology* 16, 574–581.

Herbert, A.P., Riesen, M., Bloxam, L., Kosmidou, E., Wareing, B.M., Johnson, J.R., Phelan, M.M., Pennington, S.R., Lian, L.-Y. and Morgan, A. (2012) NMR structure of Hsp12, a protein induced by and required for dietary restriction-induced lifespan extension in yeast. *PLoS One* 7(e41975), 1–12.

Horovitz, A. and Williams, K.R. (2005) Allosteric regulation of chaperonins. *Current Opinion in Structural Biology* 15, 646–651.

Horwich, A.L., Farr, G.W. and Fenton, W.A. (2006) GroEL-GroES-mediated protein folding. *Chemical Reviews* 106, 1917–1930.

Hubert, D.A., He, Y., Tonero, P. and Dangl, J.L. (2009) Specific Arabidopsis Hsp90.2 alleles recapitulate RAR1 cochaperone function in plant NB-LRR disease resistance protein regulation. *Proceedings of the National Academy of Sciences USA* 106, 9556–9563.

Jiang, Y., Douglas, N.R., Conley, N.R., Miller, E., Frydman, J. and Moerner, W.E. (2011) Sensing cooperativity of ATP hydrolysis for single multisubunit enzymes in solution. *Proceedings of the National Academy of Sciences USA* 108, 16962–19967.

Kadota, Y., Shirasu, K. and Guerois, R. (2010) NLR sensors meet at the SGT1-Hsp90 crossroad. *Trends in Biochemical Sciences* 36, 199–207.

Kamal, A., Thao, L., Sensintaffar, J., Zhang, L., Boehm, M.F., Fritz, L.C. and Burrows, F.J. (2003) A high-affinity conformation of Hsp90 confers tumour selectivity on Hsp90 inhibitors. *Nature* 425, 407–410.

Kampinga, H.H. and Craig, E.A. (2010) The Hsp70 chaperone machinery: J proteins as drivers of functional specificity. *Nature Reviews in Molecular and Cell Biology* 11, 579–592.

Kato, K., Inaguma, Y., Ito, H., Iwamoto, I., Kamei, K., Ochi, N., Ohta, H. and Kishikawa, M. (2001) Ser-59 is the major phosphorylation site in αB-crystallin accumulated in the brains of patients with Alexander's disease. *Journal of Neurochemistry* 7, 730–736.

Kriehuber, T., Rattel, T., Weinmaier, T., Bepperling, A., Haslbeck, M. and Buchner, J. (2010) Independent evolution of the core domain and its flanking sequences in small heat shock proteins. *FASEB Journal* 24, 3633–3642.

Krukenberg, K.A., Förster, F., Rice, L.H., Sali, A. and Agard, D.A. (2008) Multiple conformations of *E. coli* Hsp90 in solution: insights into conformational dynamics of Hsp90. *Structure* 16, 755–765.

Lai, E., Teodoro, T. and Volchuk, A. (2006) Endoplasmic reticulum stress: signaling the unfolded protein response. *Physiology* 22, 193–201.

Lee, J. and Ozcan, U. (2014) Unfolded protein response signaling and metabolic diseases. *Journal of Biological Chemistry* 289, 1203–1211.

Lee, S., Choi, J.M. and Tsai, F.T. (2007) Visualizing the ATPase cycle in a protein disaggregating machine: structural basis for substrate binding by ClpB. *Molecular Cell* 25, 261–271.

Liberek, K., Lewandowska, A. and Zietkięwicz, S. (2008) Chaperones in control of protein disaggregation. *EMBO Journal* 27, 328–335.

Lindquist, S. and Craig, E.A. (1988) The heat shock proteins. *Annual Review of Genetics* 22, 15–29.

Lyle, A.N., Remus, E.W., Fan, A.E., Lassègue, B., Walter, G.A., Kiyosue, A., Mchaourab, H.S., Lin, Y.-L. and Spiller, B.W. (2012) Crystal structure of an activated variant of small heat shock protein Hsp16.5. *Biochemistry* 51, 5105–5112.

Lyle, A.N., Remus, E.W., Fan, A.E., Lassègue, B., Walter, G.A., Kiyosue, A., Griendling, K.K. and Taylor, R. (2014) Hydrogen peroxide regulates osteopontin expression through activation of transcription and translational pathways. *Journal of Biological Chemistry* 289, 275–285.

Mchaourab, H.S., Lin, Y.-L. and Spiller, B.W. (2012) Crystal structure of an activated variant of small heat shock protein Hsp16.5. *Biochemistry* 51, 5105–5112.

Münch, C. and Bertolotti, A. (2010) Exposure of hydrophobic surface initiates aggregation of diverse ALS causing Superoxide dismutase1 mutations. *Journal of Molecular Biology* 399, 512–523.

Muñoz, I.G., Yébenes, H., Zhou, M., Mesa, P., Serna, M., Park, A.Y., Bragado-Nilsson, E., Beloso, A., de Cárcer, G., Malumbres, M., Robinson, C.V., Valpuesta, J.M. and Montoya, G. (2011) Crystal structure of the open conformation of the mammalian chaperonin CCT in complex with tubulin. *Nature Structural and Molecular Biology* 18, 14–20.

Murphy, M.P. (2009) How mitochondria produce reactive oxygen species. *Biochemical Journal* 417, 1–13.

Nover, L., Scharf, K.D. and Neumann, D. (1989) Cytoplasmic heat shock granules are formed from precursor particles and are associated with a specific set of mRNAs. *Molecular and Cellular Biology* 9, 1298–1302.

Nystrom, T. (2005) Role of oxidative carbonylation in protein quality control and senescence. *EMBO Journal* 24, 1311–1317.

Pearl, L.H. and Prodromou, C. (2006) Structure and mechanism of the Hsp90 molecular chaperone machinery. *Annual Review of Biochemistry* 7, 271–294.

Picard, D. (2013) Hsp90: facts and literature. Available at: http://www.picard.ch (accessed December 2013).

Prodromou, C. (2012) The 'active' life of Hsp90 complexes. *Biochimica Biophysica Acta* 1823, 614–623.

Qi, R., Sarbeng, E.B., Zhou, L., Liu, Q., Le, K.Q., Xu, X., Xu, H., Yang, J., Wong, J.L., Vorvis, C., Hendrickson, W.A., Zhou, L. and Liu, Q. (2013) Allosteric opening of the polypeptide-binding site when an Hsp70 binds ATP. *Nature Structural and Molecular Biology* 20, 900–910.

Rauch, J.N. and Gestwicki, J.E. (2014) Binding of human nucleotide exchange factors to heat shock protein 70 (Hsp70) generates functionally distinct complexes. *Journal of Biological Chemistry* 289, 1402–1414.

Richter, K., Moser, S., Hagn, F., Friedrich, R., Hainzl, O., Heller, M., Schlee, S., Kessler, H., Reinstein, J. and Buchner, J. (2006) Intrinsic inhibition of the Hsp90 ATPase activity. *Journal of Biological Chemistry* 281, 11301–11311.

Richter, K., Haslbeck, M. and Buchner, J. (2010) The heat shock response: life on the verge of death. *Molecular Cell* 40, 253–266.

Röhl, A., Rohrberg, J. and Buchner, J. (2013) The chaperone Hsp90: changing partners for demanding clients. *Trends in Biochemical Sciences* 38, 253–262.

Rosenzweig, R., Moradi, S., Zarrine-Afsar, A., Glover, J.R. and Kay, L.S. (2013) Unraveling the mechanism of protein disaagregation through a ClpB-DnaK interaction. *Science* 339, 1080–1083.

Roy, S.S., Wheatley, R.W. and Kapoor, M. (2013) Homology modeling, ligand docking and in silico mutagenesis of neurospora Hsp80(90): insight into intrinsic ATPase activity. *Journal of Molecular Graphics and Modeling* 44, 54–69.

Rüdinger, S., Buchberger, A. and Bukau, B. (1997) Interactions of Hsp70 chaperones with substrates. *Nature Structural Biology* 4, 342–349.

Saibil, H.R., Fenton, W.A., Clare, D.K. and Horwich, A.L. (2013) Structure and allostery of the chaperonin GroEL. *Journal of Molecular Biology* 425, 1476–1487.

Schlecht, R., Erbse, A.H., Bukau, B. and Mayer, M.P. (2011) Mechanics of Hsp70 chaperone enables differential interaction with client proteins. *Nature Structural and Molecular Biology* 18, 345–351.

Schmid, D., Baici, A., Gehring, H. and Christen, P. (1994) Kinetics of molecular chaperone action. *Science* 263, 971–973.

Schrader, E.K., Harstad, K.G. and Matouschek, A. (2009) Targeting proteins for degradation. *Nature Chemical Biology* 5, 815–822.

Sena, L.A. and Chandel, N.S. (2012) Physiological roles of mitochondrial reactive oxygen species. *Molecular Cell* 48, 158–167.

Sgobba, M., Forestiero, R., Degliesposti, G. and Rastelli, G. (2010) Exploring the binding site of C-terminal Hsp90 inhibitors. *Journal of Chemical Information and Modeling* 50, 1522–1528.

Shetty, N.P., Jorgensen, H.J.L., Jensen, J.D., Collinge, D.B. and Shetty, H.S. (2008) Roles of reactive oxygen species in interaction between plants and pathogens. *European Journal of Plant Pathology* 121, 267–280.

Shiau, A.K., Harris, S.F., Southworth, D.R. and Agard, D.A. (2006) Structural analysis of *E. coli* hsp90 reveals dramatic nucleotide-dependent conformational rearrangements. *Cell* 127, 329–340.

Shirasu, K. (2009) The Hsp90-SGT1 chaperone complex for NLR immune sensors. *Annual Review of Plant Biology* 60, 139–164.

Söti, C., Pál, C., Papp, B. and Csermely, P. (2005) Molecular chaperones as regulatory elements of cellular networks. *Current Opinion in Cell Biology* 17, 210–215.

Southworth, D.R. and Agard, D.A. (2008) Species-dependent ensembles of conserved conformational states define the Hsp90 chaperone ATPase cycle. *Molecular Cell* 32, 631–640.

Sreedhar, A.S. and Csermely, P. (2004) Heat shock proteins in the regulation of apoptosis: new strategies in tumour therapy. A comprehensive review. *Pharmacolology & Therapeutics* 101, 227–257.

Stadtman, E.R. and Levine, R.L. (2000) Protein oxidation. *Annals of New York Academy of Science* 899, 191–208.

Stengel, F., Baldwin, A.J., Painter, A.J., Jaya, N., Basha, E., Kay, L.E., Vierling, E., Robinson, C.V. and Benesch, J.L.P. (2010) Quaternary dynamics and plasticity underlie small heat-shock protein chaperones function. *Proceedings of the National Academy of Sciences USA* 107, 2007–2012.

Svärd, M., Biterova, E., Bourhis, J.-M. and Guy, J. (2011) The crystal structure of the human co-chaperone p58^{IPK}. *PloS One* 6, e22337.

Takahashi, A., Casais, C., Ichimura, K. and Shirasu, K. (2003) Hsp90 interacts with RAR1 and SGT1 and is essential for RPS2-mediated disease resistance in *Arabidopsis*. *Proceedings of the National Academy of Sciences USA* 100, 11777–11782.

Tkach, J.M. and Glover, J.R. (2004) Amino acid substitutions in the C-terminal AAA+ module of Hsp104 prevent substrate recognition by disrupting oligomerization and cause high temperature inactivation. *Journal of Biological Chemistry* 279, 35692–35701.

Toivola, D.M., Strnad, P., Habtezion, A. and Omary, M.B. (2010) Intermediate filaments take the heat as stress proteins. *Trends in Cellular Biology* 20, 79–91.

Tsigelny, I.F. and Nigam, S.K. (2004) Complex dynamics of chaperone-protein interactions under cellular stress. *Cellular Biochemistry and Biophysics* 40, 263–276.

Tyedmers, J., Mogk, A. and Bukau, B. (2010) Cellular strategies for controlling protein aggregation. *Nature Review of Molecular and Cell Biology* 11, 777–788.

Van Moncroft, R.L.M., Basha, E., Friedrich, K.L., Slingsby, C. and Vierling, E. (2001) Crystal structure and assembly of a eukaryotic small heat shock protein. *Nature Structural Biology* 8, 1025–1030.

Walter, P. and Ron, D. (2011) The unfolded protein response: from stress pathway to homeostatic regulation. *Science* 334, 1081–1086.

Wandinger, S.K., Richter, K. and Buchner, J. (2008) The Hsp90 chaperone machinery. *Journal of Biological Chemistry* 289, 18473–18477.

Wang, T.-F., Chang, J. and Wang, C. (1993) Identification of the peptide binding domain of hsc70. 18-kilodalton fragment located immediately after ATPase domain is sufficient for high affinity binding. *Journal of Biological Chemistry* 268, 26049–26051.

Waters, E.R. (2013) The evolution, function, structure, and expression of the plant sHsps. *Journal of Experimental Botany* 64, 391–403.

Weidemann, A. and Johnson, R.S. (2008) Biology of HIF-1alpha. *Cell Death and Differentiation* 15, 621–627.

Welker, S., Rudolph, B., Frenzel, E., Hagn, F., Liebisch, G., Schmitz, G., Scheuring, J., Kerth, A., Blume, A., Weinkauf, S., Haslbeck, M., Kessler, H. and Buchner, J. (2010) Hsp12 is an intrinsically unstructured stress protein that folds upon membrane association and modulates membrane function. *Molecular Cell* 39, 507–520.

Wellen, K.E. and Thompson, C.B. (2010) Cellular metabolic stress: considering how cells respond to nutrient excess. *Molecular Cell* 40, 323–332.

Winkler, J., Seybert, A., König, L., Pruggnaller, S., Haselmann, U., Sourjik, V., Weiss, M., Frangakis, A.S., Mogk, A. and Bukau, B. (2010) Quantitative and spatio-temporal features of protein aggregation in *Escherichia coli* and consequences on protein quality control and cellular ageing. *EMBO Journal* 29, 910–923.

Yébenes, H., Mesa, P., Munoz, I.G., Montaya, G. and Valpuesta, J.M. (2011) Chaperonins: two rings for folding. *Trends in Biochemical Sciences* 36, 424–432.

Young, J.C. and Hartl, F.U. (2000) Polypeptide release by Hsp90 involves ATP hydrolysis and is enhanced by the co-chaperone p23. *EMBO Journal* 19, 5930–5940.

Young, J.C., Agashe, V.R., Siegers, K. and Hartl, F.U. (2004) Pathways of chaperone-mediated protein folding in the cytosol. *Nature Review of Molecular and Cell Biology* 5, 781–791.

Zhang, M., Botër, M., Li, K., Kadota, Y., Panaretou, B., Prodromou, C., Shirasu, K. and Pearl, L.H. (2008) Structural and functional coupling of Hsp90- and Sgt1-centred multi-protein complexes. *EMBO Journal* 27, 2789–2798.

Zhao, H. and Blagg, B.J.S. (2013) Novobiocin analogues with second-generation noviose surrogates. *Bioorganic & Medicinal Chemistry Letters* 23, 552–557.

Zhao, H., Donnelly, A.C., Kusuma, B.R., Brandt, G.E.L., Brown, D., Rajewski, R.A., Vielhauer, G., Holzbeierlein, J., Cohen, M.S. and Blagg, B.S.J. (2011) Engineering an antibiotic to fight cancer: optimization of the novobiocin scaffold to produce anti-proliferative agents. *Journal of Medicinal Chemistry* 54, 3839–3853.

2 Heat Response, Senescence and Reproductive Development in Plants

Renu Khanna-Chopra* and Vimal Kumar Semwal

Stress Physiology Lab, Water Technology Centre, Indian Agricultural Research Institute, New Delhi

---

Abstract

Heat stress (HS)-induced reduction in crop productivity is likely to enhance because of the impending climate change. HS affects both vegetative and reproductive phase in plants leading to loss in yield. HS at the vegetative stage results in less biomass accumulation and at reproductive stage anther development is particularly susceptible to heat episodes in crops. Heat acclimation is a major protective mechanism in plants enabling survival under HS in the field. Development of thermotolerance is associated with reactive oxygen species (ROS), sugar and hormone signalling, expression of heat-shock transcription factors (HSFs) genes and activation of heat-shock proteins (Hsps), enhancement in antioxidant capacity, accumulation of compatible solutes etc. HS can induce/accelerate monocarpic senescence in crops, and hasten the process of grain development by shortening the duration and enhancing the growth rates of developing sinks leading to shrivelled grains. More studies are needed to address species-specific physiological criteria for HS tolerance to define climate change vulnerability of crops and breeding of HS-tolerant genotypes. There is a need to understand the mechanism/s enabling pollen fertility under HS in crops in order to reduce the yield loss. In this chapter we discuss current understanding of HS effect on leaf senescence, reproductive development and mechanisms enabling heat-stress tolerance with special emphasis on ROS, sugar metabolism, Hsps and transcription factors (TFs).

2.1 Introduction

Abiotic stresses are a major cause for limiting crop productivity throughout the world (Long and Ort, 2010). Global climate change due to continuously increasing temperatures has been suggested to be one of the most critical factors for agricultural productivity in almost every part of the world. Exposure to heat stress (HS) can cause a series of changes in morphology, anatomy and physiology of plants, leading to metabolic rearrangements in cells, which in turn negatively affects growth and development of plants and ultimately, loss in yield (Wahid *et al.*, 2007).

Heat stress response starts with sensing of stress and results in signal transduction and gene expression. Plasma membrane is the first to face and sense heat and is the major heat-sensing part of the cell. High temperature-induced changes are perceived and then transduced to the nucleus where the transcriptome

---

*E-mail: renu_wtc@rediffmail.com

is altered (Ruelland and Zachowski, 2010). This affects fluidity of membranes, changes to protein conformation, disassembly of cytoskeleton, rearrangements in metabolic processes and finally results in expression of an array of target genes.

Many organisms have an innate ability to withstand exposure to elevated temperatures and this has been termed as thermotolerance. This ability is generally acquired by plants on sudden exposure to high temperatures and is a rather rapid process, enabling them to survive under otherwise lethal high temperatures. Elevation of protective gene expression just before exposure to high temperatures may be one of the mechanisms for acquiring thermotolerance (Larkindale and Vierling, 2008). Acclimation to elevated temperatures triggered by pre-exposure to temperatures which are high but non-lethal is under genetic control (Wang and Li, 2006). Adaptation to high temperatures is due to change in gene expression and adjustment in metabolic processes, which enables plants to minimize heat injury. Some of the processes associated with tolerance to elevated temperatures include chaperone synthesis and enhancement of antioxidant activity as well as compatible solutes accumulation (Wahid et al., 2007). Each plant is able to acclimate to changing environmental conditions within its given genetic potential. In addition, the adaptation ability of plants has been widened during evolution by genetic variation.

During oxidative stress, certain toxic molecules referred to as reactive oxygen species (ROS) are formed and these can cause damage to cellular components such as DNA, proteins and lipids (Vacca et al., 2004). In order to detoxify these ROS, plants have a battery of antioxidative mechanisms in their cells that may be either enzymatic or nonenzymatic. This is an important mechanism for survival during various stresses (Mittler et al., 2004). In plants, temperature stress is generally associated with synthesis of heat shock proteins (Hsps) (Kotak et al., 2007). Integrated response of ROS detoxification systems and Hsps protects cells from damage caused by HS.

Soluble sugars are highly sensitive to HS, which act on the supply of carbohydrates from source to sink. Impairment of carbon metabolism and utilization is one of the factors causing yield loss under HS (Ruan et al., 2010). Both vegetative and reproductive tissues are sensitive to HS leading to less biomass accumulation, reduced reproductive potential and thus final yield. Even after successful fertilization, grain growth and filling is also sensitive to HS in many crops (Barnabás et al., 2008). Therefore, understanding plant responses to HS and breeding for heat tolerance is of the highest importance in order to stabilize agricultural productivity under adverse environments. Periods of destructively high temperature, which have occurred in past perhaps once every century, are predicted to become much more frequent by the end of this century (Semenov and Shewry, 2010).

The present review discusses current understanding of HS with special emphasis on ROS, sugar metabolism, Hsps and different TFs involved in HS tolerance. Leaf senescence in response to HS will be discussed briefly as this process is an important determinant of grain yield. HS response of major grain crops during reproductive development will also be highlighted.

2.2 Leaf Senescence Under High Temperature Stress

Heat stress during grain development phase enhances leaf senescence coupled with an increase in chlorophyllase activity, decline in photosynthesis and photosynthetic pigments (Harding et al., 1990). The detrimental effect of heat on chlorophyll and the photosystems are also associated with the production of injurious ROS (Camejo et al., 2006), which cause oxidative damage to cell membranes. Loss of leaf viability during senescence results in a close link between the duration of photosynthetically active leaf area and grain yield in wheat (Chauhan et al., 2009).

Senescence, a programmed cell death process (PCD), is associated with the overproduction of ROS, which results in oxidative stress. ROS contribute to progression of leaf senescence as the antioxidant capacity of the leaf declines (Khanna-Chopra, 2012). There have been several reports on the oxidative stress and the response of the antioxidant defence mechanism under HS in plants (Larkindale and Knight, 2002; Wahid *et al.*, 2007). A heat-tolerant wheat cultivar exhibited a slower rate of monocarpic senescence coupled with higher Rubisco activity and Rubisco content than the susceptible cultivar during HS (Chauhan *et al.*, 2009). The heat-tolerant cultivar exhibited lower ROS and lipid peroxidation coupled with better and less altered superoxide dismutase (SOD), ascorbate peroxidase (APX) activities and less decline in reduced glutathione/oxidized glutathione (GSH/GSSG) ratio under HS compared to control during monocarpic senescence (Fig. 2.1). High diurnal temperatures hastened flag leaf senescence in wheat coupled with decline in SOD, APX and catalase

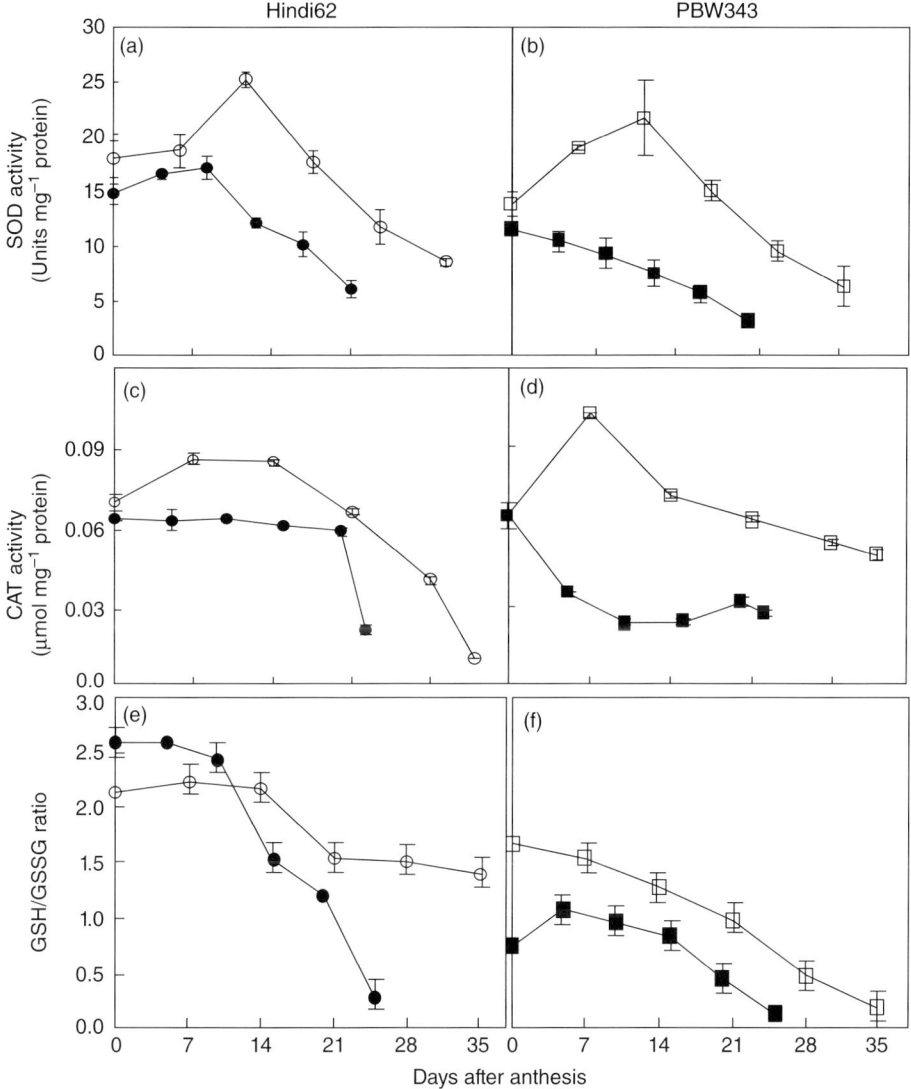

Fig. 2.1. High temperature-induced changes in SOD and CAT activity and GSH/GSSG ratio in flag leaf of wheat cvs Hindi62 (heat tolerant) and PBW343 (heat susceptible) during monocarpic senescence. Control (open symbol) and HS (closed symbol). Vertical bars indicate SE (n = 3).

(CAT) activity (Zhao et al., 2007). While scavenging of ROS during stress is important, reducing the rate of its production is equally so (Mittler, 2006). During senescence increase in levels of ROS was followed by induction of senescence associated genes (SAG) transcripts such as senescence-related transcription factor *PeWRKY6-1*, *WRKY53* and cysteine protease homologue *SAG12* (Miao et al., 2004; Rosenvasser et al., 2006).

Both leaf senescence and environmental stresses have similar responses in terms of phytohormones such as abscisic acid (ABA), jasmonic acid, ethylene and salicylic acid (SA; Khanna-Chopra, 2012). ABA induces H_2O_2 accumulation and expression of antioxidant genes and enhances the activities of antioxidative enzymes such as SOD, APX and CAT, which in turn protect the cellular functions needed for senescence reactions (Hung and Kao, 2004). High SA level in senescing leaves is also involved in up-regulation of several *SAGs* and the change of transcriptome mediated by the SA pathway was shown to have great similarity to that related to age-dependent senescence (Lim et al., 2007).

Sugars are increasingly implicated as a signalling molecule during leaf senescence (Thomas, 2013). Separate sensors are present in plant cells for sensing sucrose and hexoses such as glucose and fructose. These sensory systems can detect changes in the sucrose to hexose ratio signalling different transduction pathways as well as changed gene transcription, which may be induced or repressed (Smeekens et al., 2010). Protein kinase SnRK1 is actively involved in an important sugar-sensing pathway during senescence. It acts as a post-translational inhibitor and also induces transcription, and thus has a significant influence on developmental processes and HS. Low glucose and high sucrose concentrations, and/or nutrient starvation, which are generally found to be associated with heat stress and senescence induction, can activate this sensor (Baena-Gonzalez et al., 2007). Premature senescence occurs in plants where SnRK1 expression is down-regulated (Thelander et al., 2004). The enzyme hexokinase (HXK) is another senescence regulating pathway and it interacts with the SnRK1 network during regulation. While the mitochondrion is the major region where the glucose-sensing property of HXK resides, a fraction of it also exists in the nucleus as high-molecular weight complexes, which function towards repression of photosynthetic gene expression as well as activation of transcription factor degradation mediated by proteasome which functions in signalling pathways induced by plant hormones (Smeekens et al., 2010). Glucose-insensitive mutants of *Arabidopsis*, i.e. *Arabidopsis hxk1* show delayed flowering as well as senescence, indicating that sugars regulate these processes in an antagonistic manner (Moore et al., 2003).

ROS and sugars are central regulators of senescence in plants. The literature reveals a strong correlation between ROS, sugars and phytohormone signalling during HS and developmental senescence (Love et al., 2008). However, there could be additional levels of cell-death control that extend beyond those of hormones, sugars and ROS.

2.3 Reproductive Development Under Heat Stress In Crop Plants

Heat stress due to increasing air temperatures is a worldwide threat for sustainable agricultural productivity. HS and its impact on crops depend on growth stages and stress intensity, which may vary from mild to severe, and stress duration. Different stages of reproductive development have different optimum temperatures in major crop plants. During reproductive development, exposure to high temperatures either just preceding or during anthesis, causes greater damage leading to reduced seed-set in several crops including wheat (Saini and Aspinall, 1982), rice (Mackill et al., 1982; Sakata et al., 2010) and maize (Wilhelm et al., 1999). HS is more detrimental during anther and pollen development in major crop plants although both male and female gametophytes are sensitive to HS (Giorno et al., 2013). High temperatures also affect dehiscence of anther, quantity, morphology, chemical composition and metabolism of pollen, as well as architecture of the pollen wall (Zinn et al., 2010).

2.3.1 Wheat

Development of reproductive structures starts at the four to five leaf stage in spring wheat followed by tillering. Head initiation takes place from shoot apex and every ridge on the head develops in to spikelets after head elongation. Spikelet formation is a temperature-dependent process and the time taken for spikelet initiation is about 9°days per spikelet. Longer duration at this stage of development results in higher spikelet formation. The final stage of spikelet initiation, i.e. terminal spikelet, is followed by floret formation in every spikelet. Optimum temperature at this stage is 10–13°C and temperatures above 24°C can cause severe reduction in spikelets. Terminal spikelet formation to heading and heading to anthesis phase in wheat crop is highly sensitive to HS. Temperatures above 30°C during floret development and above 34°C at anthesis cause complete sterility (Saini and Aspinall, 1982). If temperatures increase between 30 days before anthesis to anthesis potential grain yield can decrease 4–5% for each degree increase above optimum temperature, i.e. 15°C. The final stage of crop development, i.e. grain filling, is also sensitive to HS episodes. Temperatures between 20 and 24°C are optimum for grain filling in different spring wheat cultivars while temperatures above 35°C can limit proper grain filling leading to shrivelling of grains hence reducing the grain weight (Saini and Aspinall, 1982; Blum, 1988). HS at post-anthesis phase accelerates the rate of canopy senescence reducing grain growth and filling leading to improper nutrient relocation to developing grains.

2.3.2 Maize

Optimum temperatures for maize reproductive development during initial reproductive stages, i.e. tasselling and silking, are 26–30°C, but it can tolerate higher temperatures when irrigation is sufficient. However, above 38°C it is difficult for the plant to maintain adequate water movement even at full irrigation. Pollen grains can lose viability within a few minutes at and above 40°C at this stage. The relative water content (RWC) of the pollen should be 80% for successful germination while at 40% RWC pollen can die. There is an acceleration of stigma and ovule development induced by elevated temperatures, thereby reducing the time span during which they are receptive to pollen and pollen tubes. However, pollen grains may not be ready by this time, resulting in decrease of successful mating (Zinn et al., 2010). Silks can also dry rapidly under HS and may not contain sufficient moisture to support pollen germination and growth of pollen tube to ovary. After successful fertilization and kernel growth, filling and physiological maturity comes in the next 50 to 60 days. The number of ears and kernels per ear is set soon after fertilization. HS can reduce photosynthesis resulting in inadequate supply of sugars needed for kernel growth leading to loss in yield. HS during kernel growth affects kernel size and weight, and determines whether kernels at the tip will fill even after successful pollination. HS during grain filling above 40°C affects kernel growth and filling and hastens grain maturation leading to yield loss (Wilhelm et al., 1999).

2.3.3 Rice

Reproductive development phase in rice begins with panicle initiation and differentiation, which is sensitive to HS. The important yield component, i.e. number of potential grains per panicle, is set at this stage. HS at booting also results in yield loss by interfering with meiosis at this stage. Temperatures above 35°C for two consecutive nights 1 to 2 weeks prior to flowering can cause excessive sterility or blanks. During flowering, day temperatures above 35°C cause increased sterility resulting in a higher number of blanks in panicles (Rang et al., 2010). During grain filling HS results in poor nutrient relocation, shortened grain growth and filling period and thus poor assimilate mobilization and loss of yield.

In rice, it has been shown that the reproductive processes that are most sensitive to elevated temperatures are those that occur within 1 h of anthesis such as anther dehiscence, pollen shedding, germination of pollen grains on stigma, as well as elongation of pollen tubes. In fact, they could be disrupted at day temperatures above 33°C (Satake and Yoshida, 1978). Night temperatures higher than 29°C have been reported to enhance susceptibility of rice to sterility along with reduction in seed-set leading to loss in grain yield (Satake and Yoshida, 1978). It is known generally that in rice, male reproductive development is more sensitive to HS and it is reported that HS during flowering in rice led to a reduction in pollen production as well as pollen shed (Wassmann et al., 2009). This could be due to an inhibition of pollen grain swelling and indehiscence of anthers combined with poor release of pollen grains (Shah et al., 2011). Extremely high temperatures (>35°C) during ripening can lead to reduced grain filling by inhibiting the deposition of storage materials such as starch and protein resulting in reduced final yield (Yamakawa and Hakata, 2010).

Taken together, expression profiling, proteomic and physiological studies on multiple plant species supports a perspective that the genetic programme for the male gametophyte life cycle is highly complex, and sensitive to changes in environment. All these environmental factors can result in reduced numbers of viable pollen and indirectly limit female fitness by reducing the number of seed sired, hence limiting crop yield (Barnabas et al., 2008).

2.4 Mechanisms of Heat Tolerance

Heat stress results in imbalance in cellular metabolic processes as different enzymes involved in these processes have different temperature optima. This imbalance may result in damage to different cellular components due to accelerated accumulation of ROS capable of damaging cellular macromolecules, i.e. proteins, lipids and nucleic acids (Mittler, 2006). Under HS accelerated ROS can be produced in different parts of cell, importantly in membranes by NADPH oxidases and subcellular locations, i.e. chloroplasts, mitochondria and peroxisomes, during photosynthesis, respiration and other processes (Miller and Mittler, 2006). The primary targets of HS damage in chloroplasts are the oxygen evolving complex, factors associated in PSII and deactivation of Rubisco (Allakhverdiev et al., 2008). Decline in photosynthesis under HS results in reduced sugar pool in the cells leading to decline in supply of carbohydrates from source organs to sinks (Rosa et al., 2010).

The ability of the plant to grow and reproduce under HS conditions is defined as heat tolerance. This trait is highly specific, and differences in this trait have been observed not only between closely related species, but also between different tissues and organs of a plant. During the course of evolution, several mechanisms have developed in plants for growing under high temperatures, which may be adaptive, avoidance or acclimative in nature. When exposed to high temperatures, there is an alteration of physiological and biochemical processes and in order to overcome these detrimental changes, several other mechanisms related to tolerance are activated. These activated mechanisms generally involve chaperones and other proteins, osmoprotectants and ion transporters, antioxidants, as well as factors involved in transcriptional control and signalling cascades (Wahid et al., 2007). Thus, it is quite obvious that survival of plants under HS depends on their ability to perceive, generate and transmit the stress-induced signal, with subsequent alterations in their metabolism.

The major components of cellular defence during HS-induced oxidative stress in plants are antioxidants and enzymes that can scavenge ROS formed under such conditions. Hsps are also synthesized under temperature stress and these act as molecular chaperones for quality control of proteins. Though Hsps are highly conserved, both qualitative and quantitative variations are known to occur among and within species in plants and hence may be involved in differences in tolerance to stress (Barua et al., 2003; Kotak et al., 2007). HS transcription factors (HSFs) are the central regulons of HS transcriptome in plants. Previous studies have shown that HSFs promote thermotolerance by regulating Hsps and other defence genes (Kotak et al., 2007). Major HS signalling molecules include ROS, NO, sugars and phytohormones, which transduce signals mainly through Ca-dependent protein kinases (CDPKs) and mitogen-activated protein kinase (MAPK/MPKs).

2.4.1 Reactive oxygen species defence under heat stress

Heat stress induced oxidative damage is manifested in lipid and damage to DNA and proteins (Vacca *et al.*, 2004). Expression of genes encoding ROS-scavenging enzymes was found to be elevated by heat acclimation and heat shock (Vacca *et al.*, 2004). Previous studies indicate that the changes in antioxidant enzymes and metabolites contribute to plants' resistance to HS and heat acclimation is achieved through enhancing antioxidant response (Wahid *et al.*, 2007). Exposure to high temperature for longer periods resulted in increased transcript level and net activities of both CAT and APX in *Brassica juncea* plants. A correlation between activities of antioxidant enzymes and heat tolerance was demonstrated by comparing tolerant and sensitive cultivars of wheat (Sairam *et al.*, 2000). Thermotolerance in terms of plant survival and related antioxidant enzyme activities was induced by heat acclimation in grapevine (Wang and Li, 2006).

In the past several years different groups of researchers have studied the response of *Arabidopsis thaliana* transcriptome to HS as a model (Wahid *et al.*, 2007; Larkindale and Vierling, 2008). It was shown that large numbers of transcripts altered their expression in response to HS, and up-regulation of transcripts for well-characterized Hsps were obtained. In addition, transcripts which increased expression significantly included members of the dehydration responsive element binding protein (DREB2) family of transcription factors, HSFs and ascorbate peroxidase2 (APX2). APX2, which has been documented as HS induced in several studies, seems to be regulated by HSFA2 (Panchuk *et al.*, 2002). However, these studies could not determine how these transcriptional changes correlated with HS damage or thermotolerance.

The role of different antioxidant enzymes in heat acclimation and HS tolerance has been evaluated by several groups by using over-expressing transgenics or knockout mutants (Table 2.1). Over-expression of single antioxidant protein gene or two genes simultaneously resulted in enhanced thermotolerance under highly defined conditions. The majority of the work was performed using young seedlings and the natural HS conditions were not properly simulated. A few studies conducted on mature plants did not show any impact on yield or other related traits. By using transcript expression profiles it has already been reported that acquired thermotolerance is a highly complex and multigenic trait involving multiple gene regulation processes (Larkindale and Vierling, 2008). Hence, over-expression of one or two genes related to the defence process will probably lead to only limited success in terms of enhanced thermotolerance.

The stability of enzymes during extremes of environmental conditions such as very high temperatures can be considered as an adaptive mechanism in plants. The *in vitro* heat stability is also reported for some antioxidant defence proteins in the temperature-tolerant weed plants *Chenopodium album* and *Chenopodium murale*. Further, it was observed that in both species, isozymes of SOD and APX retained their activity even after boiling treatment (Khanna-Chopra and Sundaram, 2004; Khanna-Chopra and Semwal, 2011; Khanna-Chopra *et al.*, 2011). In both vegetative and reproductive tissues of *C. album* a higher number of heat-tolerant isozymes of SOD and APX were present in chloroplasts in comparison to mitochondria (Khanna-Chopra *et al.*, 2011). The chloroplastic heat-stable isoforms of SOD and APX may be playing an important role in basal and acquired thermotolerance and survival of these weeds in very high temperatures as chloroplasts are the major centre of damage under HS.

Antioxidant metabolites such as reduced ascorbic acid (AsA) and reduced glutathione (GSH) are very important and involved in both direct and indirect control of ROS concentration during abiotic stresses (Foyer and Noctor, 2009). The importance of GSH was shown in case of HS, which resulted in higher total glutathione content in maize (Dash and Mohanty, 2002). Maintenance of reduced/oxidized (GSH/GSSG) ratio coupled with higher GR activity during HS was correlated

Table 2.1. Genetic engineering for heat tolerance in plants by using genes involved in ROS metabolism, heat shock proteins and transcription factors.

Gene/Source	Protein	Transgenic plant/ Growth stage	Effect on heat tolerance of transgenic plants[a]	References
Antioxidant defence				
APX 1/*Hordeum vulgare*	APX	*Arabidopsis thaliana* seedling stage	Maintained significantly higher green fresh mass after heat stress (35°C for 5 days) and recovery.	Shi et al., 2001
SOR/*Pyrococcus furiosus*	SOD	*Arabidopsis* seedling stage	Higher seedling survival than WT plants after direct heat stress (45°C for 2 h) or acclimation (38°C for 1.5 h) followed by heat stress (45°C for 2 h).	Im et al., 2009
SOD/wheat	SOD	Potato mature plants	Increased seedling survival after heat stress/recovery cycles (44°C/22°C, 16 h/8 h for 2 days).	Waterer et al., 2010
Cu/Zn SOD/*Manihot esculenta* APX/pea	SOD/APX	*Solanum tuberosum* seedling stage	Enhanced tolerance to oxidative stress, better PS-II efficiency and growth after heat stress (42°C for 20 h) and recovery.	Tang et al., 2006
Heat shock proteins and transcription factors				
Hsp21/tomato	Hsp 21	Tomato mature plants	Less PS-II damage during high light with heat shock (47°C or 50°C for 2 h) in detached leaves and enhanced carotenoid accumulation in fruits.	Neta-Sharir et al., 2005
Hsp26/*Saccharomyces cerevisiae*	Hsp26	*Arabidopsis thaliana* seedling stage	Maintained higher chlorophyll content, PS-II activity, proline accumulation, seedling survival, less membrane damage than WT plants after heat stress (45°C for 16 h or 36 h) and recovery.	Xue et al., 2010
CaHsp26/sweet pepper	Hsp26	Tobacco seedlings stage	Enhanced tolerance of PS-II during heat stress (42°C for 2 h).	Guo et al., 2007
AtHsp101/*Arabidopsis*	Hsp101	Rice seedling stage	Enhanced growth during recovery period after heat stress (47°C for 3 h).	Katiyar-Agarwal et al., 2003
ZmHsp16.9/maize	Hsp16.9	Tobacco seedling stage	Increased root length and higher activities of antioxidant defence enzymes than WT plants under heat stress (40°C for 9h).	Sun et al., 2012
HsfA1d/*Thellungiella salsuginea*	HsfA1d	*Arabidopsis* seedling stage	Greater heat tolerance than WT plants due to higher induction of HSP 17.6 and DREB2A genes than WT plants under heat stress (42°C for 1.5 h).	Higashi et al., 2013
Hsf3/*Arabidopsis*	Hsf3	*Arabidopsis* seedling stage	Total soluble APX activity enhanced and a new heat induced isoform of APX was visible after acclimation (28°C for 3 days) followed by heat stress (44°C for 1–4 h).	Panchuk et al., 2002

Gene/species	Gene	Plant/stage	Response	Reference
HsfA2/Arabidopsis	HsfA2	Arabidopsis seedling stage	CS plants were highly sensitive to HS (44°C for 30 min) or acclimation (37°C for 1 h/24°C 2 day) followed by heat stress (44°C for 45 min), showed reduction in transcript levels of many heat stress-inducible genes and less capacity of acquired thermotolerance.	Charng et al., 2007
BhHsf1/Boea hygrometrica	Hsf1	Tobacco seedling stage	Enhanced basal and acquired thermotolerance as measured by seedling survival rate after HS (48°C for 2.5 h) or acclimation (40°C for 3 h) followed by heat stress (50°C for 2 h).	Zhu et al., 2009
DREB2A CA/Arabidopsis	DERB2A	Arabidopsis seedling stage	Increased accumulation of many stress related genes and higher seedling survival rate after heat stress (45°C for 1 h) or acclimation (37°C for 1 h) followed by heat stress (49°C for 1 h). Survival was decreased in CS plants.	Sakuma et al., 2006
WRKY25/Arabidopsis	WRKY25	Arabidopsis seedling stage	Increased germination, hypocotyl, root growth, less membrane damage and increased expression of defence genes than WT and CS plants after heat stress (45°C for 4 h). CS plants were highly sensitive to heat stress.	Li et al., 2009
bZIP28/Arabidopsis	bZIP28	Arabidopsis seedling stage	CS plants exhibited severe chlorosis and decrease in expression of HSPs after heat stress (45°C for 2 h). Thermotolerance was regained in CS line after transformation with bZIP28.	Gao et al., 2008
ROB 5/bromegrass	ROB5 LEA group	Potato mature plants	Less membrane damage and increased plant survival after heat stress/recovery cycles (44°C /22°C, 16 h/8 h for 2 days).	Waterer et al., 2010

[a]WT, wild-type plants; CS, gene suppression lines

with heat tolerance in wheat and maize (Szalai et al., 2009). HS induced a greater increase in GSH synthesis in tolerant wheat and maize genotypes than in susceptible ones. *Arabidopsis* mutants defective in GSH biosynthesis *cad2-1* (mutants of γ-glutamylcysteine synthetase gene) have shown strong decrease in basal thermotolerance, reduced heat acclimation capacity and increased TBARS content (Larkindale et al., 2005). AsA content increased in grape plants acclimated to high temperatures (Li et al., 2010). Mutants defective in AsA biosynthesis *vtc-1*, *vtc-2* (mutants of GDP-mannose phosphorylase gene) and *vtc-4* (mutants of L-galactose 1-P phosphatase gene) have shown reduced capacity to heat acclimation and basal thermotolerance than the wild-type plants (Larkindale et al., 2005). *vtc-2* has greater thermo-induced photon emission and increased lipid peroxidation at high temperatures (Havaux et al., 2003). Transgenic potato plants over-expressing L-gulono-γ-lactone oxidase (GLOase) gene, accumulated elevated AsA levels and showed increased tolerance to oxidative stress (Hemavathi et al., 2010).

Tocopherols and carotenoids are lipid soluble antioxidants, which have multi-functional roles in plants including oxidative stress tolerance (Kruk et al., 2005). There are four isomers of tocopherols in plants, of which α-tocopherol has the highest antioxidant capacity. The ability of tocopherols to scavenge free radicals is because it can prevent chain elongation during lipid auto-oxidation. Bergmüller et al. (2003) reported that during oxidative stress induced by high temperature α-tocopherol and γ-tocopherol increased in wild-type plants and γ-tocopherol in *vte4-1* mutant (mutant of tocopherol cyclase gene which lacks α-tocopherol), thus suggesting that α-tocopherol can be replaced by γ-tocopherol in *vte4-1* for achieving protection of photosynthetic apparatus against HS (Giacomelli et al., 2007). Mutants of *Arabidopsis npq1-2* (mutants of violaxanthin deepoxidase gene) with reduced capacity of violaxanthin synthesis showed decreased acquired and basal thermotolerance (Larkindale et al., 2005). Similarly, *Arabidopsis* plants over-expressing the *chyB* gene that encodes β-carotene hydroxylase show greater tolerance to HS, and it has been suggested that the protection from stress may most probably be due to the action of zeaxanthin in preventing oxidative damage to membranes (Meiri et al., 2010).

In summary, enzymes that scavenge ROS and antioxidant metabolites are major components of cellular defence during HS-induced oxidative stress in plants. Under HS up-regulation of enzyme activities, enhanced transcript levels of these enzymes and enhanced antioxidant level is reported by many groups and they appear to be essential for the process of heat acclimation.

2.4.2 Heat shock proteins and transcription factors during heat stress

The synthesis of Hsps when plants are exposed to elevated temperatures is a universal phenomenon in higher plants. The Hsp gene network in plants is very complex and many of the Hsps are necessary for growth and development of plants under normal conditions. Although almost all Hsps show induction after exposure to high temperatures, specific information on how these contribute to survival under such conditions in plants is still lacking. The major Hsp groups are Hsp100, Hsp90, Hsp70, Hsp60 and small Hsps (sHsps) (Kotak et al., 2007).

It is now an established fact that induction of Hsps is an integral part of acquired thermotolerance in plants. In particular, Hsp101 has been shown to be necessary for heat acclimation in plants (Hong and Vierling, 2000; Lee et al., 2005). The role of Hsp101 in acquisition of thermotolerance is reported in *Arabidopsis hot1* mutants (mutation in *Hsp 101* gene) and *Hsp101* T-DNA insertion mutant (Hong and Vierling, 2000). The *hot1* plants were thermosensitive while transformation of *hot1* plants with Hsp101 genomic DNA restored thermotolerance of *hot1* plants similar to wild-type phenotype (Hong and Vierling, 2000). Similar to *Arabidopsis*, knockouts of Hsp101 in maize were also defective in both basal and acquired thermotolerance (Nieto-Sotelo et al., 2002).

Heat stress results in heat shock granule (HSG) formation in cytoplasm. It has been shown that the presence of HSGs is also necessary for the survival of cells subjected to HS. Smykal et al. (2000) showed that HSGs are predominantly composed of low molecular weight (LMW) Hsps, Hsp40 and Hsp70. A role for Hsp90, in addition to Hsp17-CII and Hsp70, in the restoration of HSFs is also suggested (Kotak *et al.*, 2007). LMW Hsps (16–30 kDa) belong to a super family of chaperones having a conserved C-terminal α-crystallin domain. sHsps are mainly targeted to the nucleus, chloroplasts, mitochondria, peroxisomes, endoplasmic reticulum and cytosol, implicating their importance in protection of different cellular compartments. The chloroplastic sHsps can probably protect photosynthetic electron transport in the chloroplast. Quantitative variation in csHsps could be correlated to thermotolerance of net photosynthesis and photosystem-II in different *C. album* ecotypes (Barua *et al.*, 2003). Thermotolerance was enhanced in chloroplastic sHsp or mitochondrial sHsp over-expressing tomato and tobacco plants (Table 2.1). sHsps are important in the protection of thermo-sensitive tissues such as anthers in tomato and embryos in cork oak (Puigderrajols *et al.*, 2002; Giorno *et al.*, 2010). Higher expression of sHsps was also observed in anthers of HS-tolerant rice compared to the susceptible variety (Jagadish *et al.*, 2010).

Heat stress transcription factors (HSFs) are the central regulons of HS transcriptome expression in plants. HSFs promote thermotolerance by regulating Hsps (Kotak *et al.*, 2007). The N-terminal domain of all HSFs recognizes HS promoter elements (HSE). HSEs are located upstream of TATA box of HS-inducible genes. Among those which have been studied most extensively, *LpHsfA1* has been shown to be a master regulator for induced thermotolerance in tomato, while *AtHsfA2* is the dominant HSF involved in acquired thermotolerance in *Arabidopsis*. Over-expression of both of these genes enhances thermotolerance while thermotolerance is inhibited when the genes are knocked-out or interfered with (Table 2.1). In tomato, HSFA1 is constitutively expressed and regulates the HS-induced expression of HSFA2 and HSFB1 as well as synthesis of Hsp genes including sHsps and Hsp101 (Mishra *et al.*, 2002).

Over-expression of Hsps or HSFs in different plant species resulted in enhanced thermotolerance of different physiological processes including photosynthesis, antioxidant defence and increased seedling survival after imposition of high-temperature stress (Table 2.1). Due to different initial growth temperatures (18–25°C for *Arabidopsis*), thermotolerance assay temperatures and time periods of HS (35 to 49°C for 1 h to 5 days in *Arabidopsis*) it is difficult to compare across studies and draw definite inferences (Table 2.1). In a number of studies the thermotolerance generated was not dependent on the amount of the transgene expression and in some cases the over-expression did not result in imparting thermotolerance. Many reports have shown altered expression of many other Hsps and/or stress-related genes in plants over-expressing HSFs and Hsps or knockout plants (Panchuk *et al.*, 2002; Zhu *et al.*, 2009). Whole genome microarray studies during HS also showed altered expression of more than 4000 genes (mainly Hsps) during heat acclimation in *Arabidopsis* (Larkindale and Vierling, 2008).

Expression of mRNAs and activities of ROS scavengers, such as APX, increased under HS conditions and was controlled by HSFs (Panchuk *et al.*, 2002). ROS may also function as signal transduction molecules and cause HSF activation. In *Arabidopsis* HS-induced H_2O_2 is required for effective expression of Hsp genes (Volkov *et al.*, 2006). *HsfA4a* and *HsfA8* are now being considered as prominent candidates to function as H_2O_2 sensors in *Arabidopsis* and rice (Miller and Mittler, 2006).

Besides HSFs many other potential TFs are now being increasingly implicated in acquired thermotolerance including DREB family, bZIP and WRKY. In *Arabidopsis* HSFA3 was shown to be regulated by DREB2A (Sakuma *et al.*, 2006). These findings suggest HSF-mediated cross-talk between HS and other abiotic stress signalling, notably, the high HS up-regulation of *LeDREB1* in tomato (Frank *et al.*, 2009). Over-expression of bZIP-type TF *ABF3* in *Arabidopsis* resulted in induced HS tolerance and mutants of *bZIP28 Arabidopsis* showed HS sensitivity (Table 2.1).

Hence, in plants, response to HS is a highly complex process involving more than 20 different HSFs and other TFs which are responsible for regulation of gene expression of Hsps and other defence proteins. Hsps are also important for cellular maintenance during stress and act as molecular chaperones during protein folding and play multitudes of other functions in the heat acclimation process.

2.4.3 Sugar metabolism under high temperature stress

Soluble sugars are highly sensitive to HS, which act on the supply of carbohydrates from source to sink and hence reduce yield under HS (Ruan et al., 2010). Reproductive development of many crops including cowpea are more sensitive to high night temperature than day temperature indicating that the lack of photoassimilate supply at night aggravates heat-induced damage. Hydrolyses of sucrose into glucose and fructose is dependent on invertases (INVs). Down-regulation of different INVs is reported in many plants under HS (Ruan et al., 2010). Hexokinase (HSK) is the major sensor of glucose level in cells (Thomas, 2013). Expression of HSK gene was down-regulated in rice flag leaves during HS at the reproductive stage (Zhang et al., 2013).

Carbohydrate availability (e.g. glucose and sucrose) during HS represents an important physiological trait associated with HS tolerance (Liu and Huang, 2000). Sucrose is the principal end-product of photosynthesis, which translocates from source leaves to sink organs through the phloem. Sucrose and its cleavage products regulate plant development and response to stresses through carbon allocation and sugar signalling (Roitsch and González, 2004). At low concentrations sucrose acts as a signalling molecule while in high concentrations it may become a ROS scavenger (Sugio et al., 2009). HS down-regulates sucrose synthase (SS) and several cell wall and vacuolar INVs in the developing pollen grains; as a consequence, sucrose and starch turnover are disrupted and thus soluble carbohydrates accumulate at reduced levels (Sato et al., 2006). Enhanced expression of SS under HS in maturing tomato pollen resulted in higher sucrose synthesis, which may play a role as osmoprotectant in maintaining cell membrane integrity and cellular function under HS (Frank et al., 2009).

In tomato, the reduction of sink- and source-strength even under moderately elevated temperatures leads to a depletion in available carbohydrates at critical stages of plant development, leading to reduced fruit set and other yield-related parameters (Sato et al., 2006). Studies on a tomato genotype that was tolerant to HS demonstrated that it is the enhanced activities of invertases of cell walls and vacuoles along with an increase of sucrose import into young tomato fruit that contributes to heat tolerance by increasing sink strength as well as sugar signalling activities (Li et al., 2012). Similarly, the carbohydrate content of developing and mature pollen grains may be an important factor in determining pollen quality, as heat-tolerant tomato cultivars appear to maintain an appropriate carbohydrate content under HS through specific mechanisms (Firon et al., 2006). In sorghum, HS reduced the accumulation of carbohydrates in pollen grains and ATP in the stigmatic tissue (Jain et al., 2007). Chickpea plants under HS showed lower SS and INV activity coupled with a drop in the concentration of sugars in the anther walls, pollen grains and in the locular fluid, resulting in decreased sugar concentration in the mature pollen grains and decreased pollen viability (Kaushal et al., 2013).

In summary, impairment of carbon metabolism and utilization appears to be among central factors causing abnormal development and yield loss under HS. Reduction of INV and HSK activity are the major features under HS, which leads to disrupted source–sink relationship. Do INVs and HSK play a role in seed and fruit abortion observed under HS in many crop plants? There is need for more studies.

2.4.4 Epigenetic regulation of heat-stress response in plants

Acclimation responses of plants to HS reveal the retention of HS memory for short durations (Mittler et al., 2004). Heat stress causes greater genetic instability as well as higher rates of somatic homologous recombination (Pecinka et al., 2010). It has been shown that if reprogramming in phenology and morphology of plants is involved in this response, the memory can last longer. Stress memory can be made to be retained for longer duration through epigenetic processes such as stable or heritable DNA methylation and histone modifications. It has further been suggested that within-generation and trans-generational stress memory may be achieved through heritable, epigenetic modifications (Chinnusamy and Zhu, 2009). Prolonged exposure to high temperatures has been shown to activate several repetitive elements of *A. thaliana* that are under epigenetic regulation by transcriptional gene silencing at ambient temperatures and also upon short-term heat exposure. While such activation can occur without DNA methylation loss or/and with only minor changes to histone modifications, it involves a loss of nucleosomes as well as decondensation of heterochromatin (Pecinka *et al.*, 2010). In *Arabidopsis*, a variant of histone H2A which acts as a thermo-sensor and regulates temperature-dependent gene expression has been identified (Kumar and Wigge, 2010). It is also possible to transmit heat acclimation to subsequent generations through an epigenetic mechanism. Transient shift to a higher temperature (37 or 42°C) destabilized *Arabidopsis* loci that had already been shown to be subject to epigenetic regulation, i.e. suppressed by transcriptional gene silencing (Pecinka *et al.*, 2010; Popova *et al.*, 2013).

Small RNAs (smRNAs), including small interfering RNAs (siRNAs), that are able to alter DNA methylation in sequence-specific manner and micro-RNAs that promote sequence specific mRNA degradation are also important in HS response in plants. Chromatin states can be modulated by double-stranded RNA and derived 24 nucleotide short RNAs in a process called RNA-directed DNA methylation (RdDM). The RdDM-mediated pathway is essential for basal heat tolerance in *Arabidopsis* (Popova *et al.*, 2013).

It was demonstrated that when *Arabidopsis* seedlings were subjected to HS, a *copia*-type retrotransposon *ONSEN* became transcriptionally active and also synthesized copies of extra-chromosomal DNA. In mutants where biogenesis of siRNAs were impaired, heat-induced *ONSEN* accumulation was stimulated. The progeny of stressed plants deficient in siRNAs showed a high frequency of new *ONSEN* insertions, which were found to occur during flower development and before gametogenesis. Interestingly, in those plants with impaired siRNA biogenesis, stress memory was maintained throughout the course of development, with subsequent priming of *ONSEN* to transpose during differentiation of generative organs. Genes close to *ONSEN* insertions are also conferred with heat responsiveness (Ito *et al.*, 2011). *Arabidopsis* mutants defective in DNA methylation, histone modifications, chromatin-remodelling, or siRNA-based silencing pathways were found to be highly sensitive to temperature stress. Hence, it has been suggested that the transcriptional response to temperature stress may be dependent, at least to some extent, on the integrity of the RNA-dependent DNA methylation pathway (Popova *et al.*, 2013). Exposure to HS in three previous generations accelerated flowering in the fourth generation even under controlled conditions in *Arabidopsis* (Suter and Widmer, 2013). Early flowering in fourth generation plants was due to stress memory from previous generations. Early induced flowering is a common phenomenon in many plants when subjected to HS and is considered important for survival under HS (Barnabas *et al.*, 2008).

The antioxidant defence system is also known to interact with smRNAs, for example, microRNA-dependent down-regulation of antioxidant proteins such as SOD and post-transcriptional regulation or inactivation of APX (Foyer and Noctor, 2009). *Arabidopsis* miR398 regulates two SOD isozymes and a cytochrome-c oxidase subunit. miR398 acts as translational repressor at post-transcriptional level (Dugas and Bartel, 2008). An overexpressing mutant version of SOD that is resistant to miR398 regulation increases tolerance to oxidative stress in transgenic *Arabidopsis* (Foyer and Noctor, 2009). Hence, HS leads to epigenetic stress memory in plants.

2.5 Conclusions and Future Research

The increasing threat of global warming and limited water availability is one of the major concerns for sustainable agricultural productivity to satisfy increasing human demands in future. Rising global temperatures and more frequent droughts will act to drive down yields. However, grain production per unit of land will need to more than double over this century to address rising population and demand (Long and Ort, 2010).

In this scenario, under field conditions, one of the vital strategies of crops would be acquisition of thermotolerance through a gradual exposure to episodes of heat or by autonomous synthesis of pertinent compounds. It is also possible to induce thermotolerance by exposing plants to gradually increasing temperature up to lethal highs, as would be experienced under natural conditions. Such induction would no doubt involve a number of processes. Studies with *Arabidopsis* mutants have revealed that other than Hsps (sHsp and Hsp101), absiscic acid, ROS and salicylate pathways are also involved in the development and maintenance of acquired thermotolerance (Fig. 2.2) (Larkindale *et al.*, 2005; Charng *et al.*, 2007).

The reproductive process is the most vulnerable phase of crops to high temperature. HS during this phase may result in serious loss

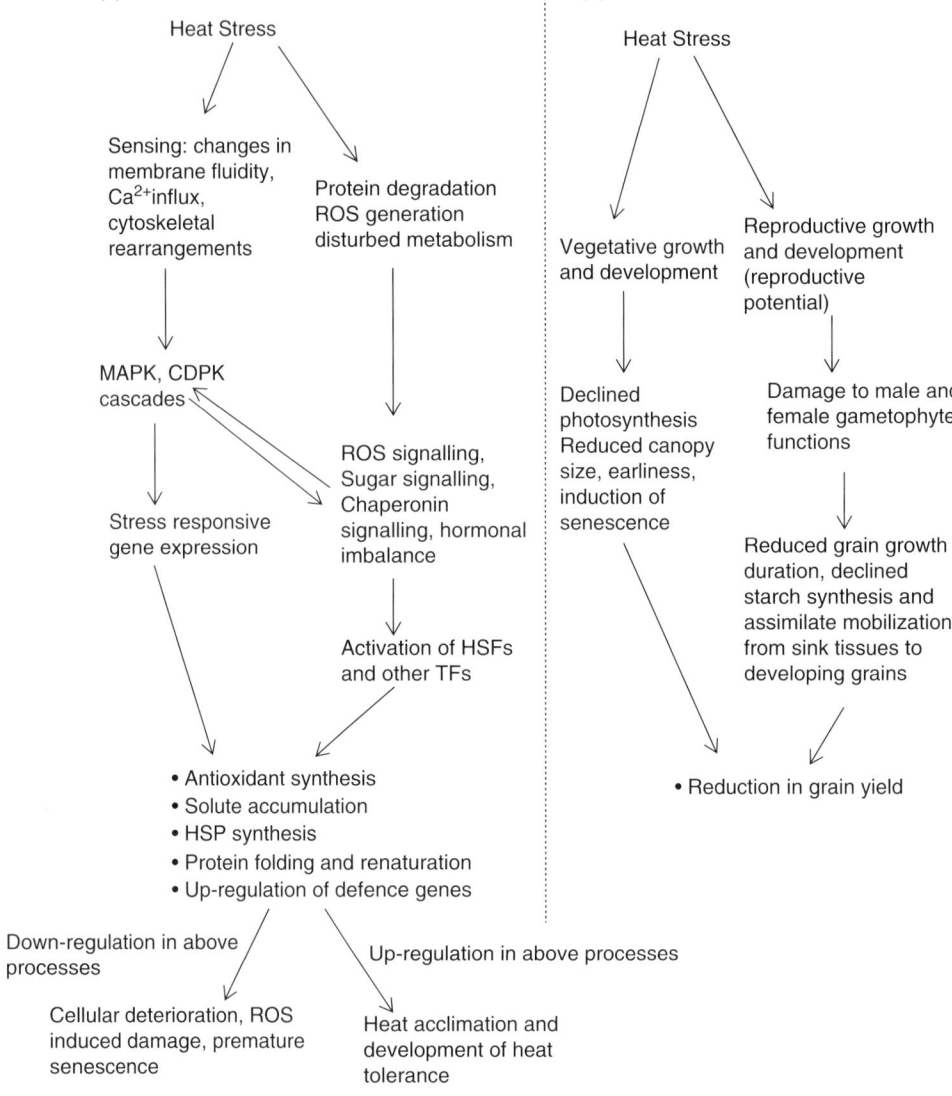

Fig. 2.2. (a) Effect of heat stress on different metabolic processes in plants and generation of thermotolerance and (b) effect of heat stress in crop plants during vegetative and reproductive stages culminating in reduced grain yield.

of sink potential and thus to economic yield. Crops are susceptible to HS during the post-anthesis phase, resulting in shrivelled grains. Some heat-tolerant varieties have better protection of reproductive parts by Hsps than the susceptible varieties. In heat-tolerant rice cv. N22, higher pollen fertility in stressed plants was due to higher accumulation of sHsps in the anthers compared to the susceptible cultivars (Jagadish et al., 2010). HS results in accelerated rate of leaf senescence culminating in early maturation of plants (Fig. 2.2). Stay-green character is a desirable trait in crops under abiotic stresses including HS (Mondal et al., 2013). Crops with delayed onset of senescence during HS generally maintain higher yield potential (Chauhan et al., 2009).

ROS are the common outputs of a number of abiotic stresses. For plants to survive under HS, it is important that antioxidants work in cooperation. ROS act as central signalling molecules in response to HS (Fig. 2.2). There is a dearth of information on ROS regulation during multiple stresses and/or stress combinations. More studies on response of mutants having deficient or altered ability to produce or scavenge ROS may provide a clearer picture on oxidative stress and HS tolerance in plants. Hsps and HSFs are the major regulators of HS response in plants. Detailed studies on HSFs are still mainly restricted to *Arabidopsis* and tomato. The role of HSFs in HS tolerance in grain crops needs to be elucidated.

Stresses when combined under field conditions may result in synergistic damage caused by different stresses. HS is often accompanied by drought stress in field conditions. There is a need to understand the physiological aspects of abiotic stress combinations in different crops. Crops with resistance to these stress combinations will be needed to fulfil the future grain demands. Area- and crop-specific programmes need to be developed according to the changes expected due to global climate change.

Many HS-tolerant plants have been developed by transgenic tools in the past decade (Table 2.1). However, the majority of the work done was limited to model systems. Few studies reported in crop plants were performed in greenhouse or growth chamber-based experiments. The biggest concern is the difference in the conditions of laboratories and crop fields. It is difficult to simulate the natural conditions in greenhouses and growth chambers. Transgenic plants developed with the objective of HS tolerance are needed to be evaluated and observed under natural field environments. Moreover, in future studies on transgenic crops we need to focus on the growth and yield-related parameters to gain sufficient knowledge and economic output from these strategies.

Identification of high temperature tolerant germplasm of different crops and its further transfer to high yielding varieties with the help of conventional breeding programmes and marker assisted selection (MAS) will help in better understanding of the genetic and physiological basis of heat tolerance in crops. For use of MAS, genetic markers that are associated with genes or quantitative trait loci (QTLs) which affect stress tolerance in whole plants or individual components contributing to it will need to be identified. Though QTLs are promising approaches to dissect the genetic basis of thermotolerance, till now only limited work has been done in crops to identify genetic markers.

Acknowledgement

This research was supported by the financial grants of CSIR Emeritus Scientist Scheme awarded to Dr (Mrs) R.K. Chopra.

References

Allakhverdiev, S.I., Kreslavski, V.D., Klimov, V.V., Los, D.A., Carpentier, R. and Mohanty, P. (2008) Heat stress: an overview of molecular responses in photosynthesis. *Photosynthesis Research* 98, 541–550.

Baena-Gonzalez, E., Rolland, F., Thevelein, J.M. and Sheen, J. (2007) A central integrator of transcription networks in plant stress and energy signalling. *Nature* 448, 938–942.

Barnabás, B., Jäger, K. and Fehér, A. (2008) The effect of drought and heat stress on reproductive process in cereals. *Plant Cell and Environment* 31, 11–38.

Barua, D., Downs, C.A. and Heckathorn, S.A. (2003) Variation in chloroplast small heat shock protein function is a major determinant of variation in thermo-tolerance of photosynthetic electron transport among ecotypes of *Chenopodium album*. *Functional Plant Biology* 30, 1071–1079.

Bergmüller, E., Porfirova, E. and Dőrmann, P. (2003) Characterization of an *Arabidopsis* mutant deficient in γ-tocopherol methyltransferase. *Plant Molecular Biology* 52, 1181–1190.

Blum, A. (1988) *Plant Breeding for Stress Environments*. CRC Press, Florida.

Camejo, D., Jiménez, A., Alarcón, J.J., Torres, W., Gómez, J.M. and Sevilla, F. (2006) Changes in photosynthetic parameters and antioxidant activities following heat-shock treatment in tomato plants. *Functional Plant Biology* 33, 177–187.

Charng, Y.Y., Liu, H.C., Liu, N.Y., Chi, W.T., Wang, C.N., Chang, S.H. and Wang, T.T. (2007) A heat inducible transcription factor, HsfA2, is required for extension of acquired thermo-tolerance in *Arabidopsis*. *Plant Physiology* 143, 251–262.

Chauhan, S., Srivalli, S., Nautiyal, A.R. and Khanna-Chopra, R. (2009) Wheat cultivars differing in heat tolerance show a differential response to monocarpic senescence under high-temperature stress and the involvement of serine proteases. *Photosynthetica* 47, 536–547.

Chinnusamy, V. and Zhu, J.K. (2009) Epigenetic regulation of stress responses in plants. *Current Opinion in Plant Biology* 12, 133–139.

Dash, S. and Mohanty, N. (2002) Response of seedlings to heat stress in cultivars of wheat: growth temperature-dependent differential modulation of photosystem 1 and 2 activity, and foliar antioxidant defense capacity. *Journal of Plant Physiology* 159, 49–59.

Dugas, D.V. and Bartel, B. (2008) Sucrose induction of *Arabidopsis* miR398 represses two Cu/Zn superoxide dismutases. *Plant Molecular Biology* 67, 403–417.

Firon, N., Shaked, R., Peet, M.M., Phari, D.M., Zamskı, E., Rosenfeld, K., Althan, L. and Pressman, N.E. (2006) Pollen grains of heat tolerant tomato cultivars retain higher carbohydrate concentration under heat stress conditions. *Sciencia Horticulture* 109, 212–217.

Foyer, C.H. and Noctor, G. (2009) Redox regulation in photosynthetic organisms: signaling, acclimation and practical implications. *Antioxidants and Redox Signalling* 11, 861–905.

Frank, G., Pressman, E., Ophir, R., Altman, L., Shaked, R., Freedman, M., Shen, S. and Firon, M. (2009) Transcriptional profiling of maturing tomato (*Solanum lycopersicum* L.) microspores reveals the involvement of heat shock proteins, ROS scavengers, hormones and sugars in the heat stress response. *Journal of Experimental Botany* 60, 3891–3908.

Gao, H., Brandizzi, F., Benning, C. and Larkin, R.M. (2008) A membrane-tethered transcription factor defines a branch of the heat stress response in *Arabidopsis thaliana*. *Proceedings of National Academy of Sciences USA* 105, 16398–16403.

Giacomelli, L., Masi, A., Ripoll, D.R., Lee, M.J. and Van Wijk, K.J. (2007) *Arabidopsis thaliana* deficient in two chloroplast ascorbate peroxidases shows accelerated light induced necrosis when levels of cellular ascorbate are low. *Plant Molecular Biology* 65, 627–644.

Giorno, F., Wolters-Arts, M., Grillo, S., Scharf, K.D., Vriezen, W.H. and Mariani, C. (2010) Developmental and heat stress-regulated expression of *HsfA2* and small heat shock proteins in tomato anthers. *Journal of Experimental Botany* 61, 453–462.

Giorno, F., Wolters-Arts, M., Mariani, C. and Rieu, I. (2013) Ensuring reproduction at high temperatures: the heat stress response during anther and pollen development. *Plants* 2, 489–506.

Guo, S.J., Zhou, H.Y., Zhang, X.S., Li, X.G. and Meng, Q.W. (2007) Overexpression of *CaHSP26* in transgenic tobacco alleviates photoinhibition of PS-II and PS-I during chilling stress and low irradiance. *Journal of Plant Physiology* 164, 126–136.

Harding, S.A., Guikema, J.A. and Paulsen, G.M. (1990) Photosynthetic decline from high temperature stress during maturation of wheat. I. Interaction with senescence processes. *Plant Physiology* 92, 648–653.

Havaux, M., Lütz, C. and Grimm, B. (2003) Chloroplast membrane photostability in chlP transgenic tobacco plants deficient in tocopherols. *Plant Physiology* 132, 300–310.

Hemavathi, Upadhyaya, C.P., Akula, N., Young, K.E., Chun, S.C., Kim, D.H. and Park, S.W. (2010) Enhanced ascorbic acid accumulation in transgenic potato confers tolerance to various abiotic stresses. *Biotechnology Letters* 32, 321–330.

Higashi, Y., Ohama, N., Ishikawa, T., Katori, T., Shimura, A., Kusakabe, K. and Taji, T. (2013) HsfA1d, a protein identified via FOX hunting using *Thellungiella salsuginea* cDNAs improves heat tolerance by regulating heat stress responsive gene expression. *Molecular Plant* 6, 411–422.

Hong, S.W. and Vierling, E. (2000) Mutants of *Arabidopsis thaliana* defective in the acquisition of tolerance to high temperature stress. *Proceedings of National Academy of Sciences USA* 97, 4392–4397.

Hung, K.T. and Kao, C.H. (2004) Hydrogen peroxide is necessary for abscisic acid-induced senescence of rice leaves. *Journal of Plant Physiology* 161, 1347–1357.

Im, Y.J., Ji, M., Lee, A., Killens, R., Grunden, A.M. and Boss, W.F. (2009) Expression of *Pyrococcus furiosus* superoxide reductase in *Arabidopsis* enhances heat tolerance. *Plant Physiology* 151, 893–904.

Ito, H., Gaubert, H., Bucher, E., Mirouze, M., Vaillant, I. and Paszkowski, J. (2011) An siRNA pathway prevents transgenerational retrotransposition in plants subjected to stress. *Nature* 472, 115–119.

Jagadish, S.V.K., Muthurajan, R., Oane, R., Wheeler, T.R., Heuer, S., Bennett, J. and Craufurd, P.Q. (2010) Physiological and proteomic approaches to address heat tolerance during anthesis in rice (*Oryza sativa* L.). *Journal of Experimental Botany* 61, 143–156.

Jain, M., Prasad, P.V.V., Boote, K.J., Hartwell, A.L. and Chourey, P.S. (2007) Effects of season-long high temperature growth conditions on sugar-to-starch metabolism in developing microspores of grain sorghum (*Sorghum bicolor* L. Moench). *Planta* 227, 67–79.

Katiyar-Agarwal, S., Agarwal, M. and Grover, A. (2003) Heat tolerant basmati rice engineered by overexpression of hsp101 gene. *Plant Molecular Biology* 51, 677–686.

Kaushal, N., Awasthi, R., Gupta, K., Gaur, P., Siddique, K. and Nayyar, H. (2013) Heat-stress-induced reproductive failures in chickpea (*Cicer arietinum*) are associated with impaired sucrose metabolism in leaves and anthers. *Functional Plant Biology* 40, 1334–1349.

Khanna-Chopra, R. (2012) Leaf senescence and abiotic stresses share reactive oxygen species-mediated chloroplast degradation. *Protoplasma* 249, 469–481.

Khanna-Chopra, R. and Semwal, V.K. (2011) Superoxide dismutase and ascorbate peroxidase are constitutively more thermotolerant than other antioxidant enzymes in *Chenopodium album*. *Physiology and Molecular Biology of Plants* 17, 339–346.

Khanna-Chopra, R. and Sundaram, S. (2004) Heat-stable chloroplastic Cu/Zn superoxide dismutase in *Chenopodium murale*. *Biochemical and Biophysical Research Communications* 320, 1187–1192.

Khanna-Chopra, R., Jajoo, A. and Semwal, V.K. (2011) Chloroplast and mitochondria contain multiple heat stable isozymes of SOD and APX in leaves and inflorescence in *Chenopodium album*. *Biochemical and Biophysical Research Communications* 412, 522–525.

Kotak, S., Larkindale, J., Lee, U., von Koskull-Doring, P., Vierling, E. and Scharf, K.D. (2007) Complexity of the heat stress response in plants. *Current Opinion in Plant Biology* 10, 310–316.

Kruk, J., Hollander-Czytko, H., Oettmeier, W. and Trebst, A. (2005) Tocopherol as singlet oxygen scavenger in photosystem-II. *Journal of Plant Physiology* 162, 749–757.

Kumar, S.V. and Wigge, P.A. (2010) H2A.Z-containing nucleosomes mediate the thermosensory response in *Arabidopsis*. *Cell* 140, 136–147.

Larkindale, J. and Knight, M.R. (2002) Protection against heat stress-induced oxidative damage in *Arabidopsis* involves calcium, abscisic acid, ethylene and salicylic acid. *Plant Physiology* 128, 682–695.

Larkindale, J. and Vierling, E. (2008) Core genome response involved in acclimation to high temperature. *Plant Physiology* 146, 748–761.

Larkindale, J., Hall, J.D., Knight, M.R. and Vierling, E. (2005) Heat stress phenotypes of *Arabidopsis* mutants implicate multiple signaling pathways in the acquisition of thermotolerance. *Plant Physiology* 138, 882–897.

Lee, U., Wie, C., Escobar, M., Williams, B., Hong, S.W. and Vierling, E. (2005) Genetic analysis reveals domain interactions of *Arabidopsis* Hsp100/ClpB and cooperation with the small heat shock protein chaperone system. *The Plant Cell* 17, 559–571.

Li, S., Fu, Q., Huang, W. and Yu, D. (2009) Functional analysis of an *Arabidopsis* transcription factor WRKY25 in heat stress. *Plant Cell Reports* 28, 683–693.

Li, Y., Liu, Y. and Zhang, J. (2010) Advances in the research on the AsA-GSH cycle in horticultural crops. *Frontiers in Agricultural Sciences China* 4, 84–90.

Li, Z.M., Palmer, W.M., Martin, A.P., Wang, R.Q., Rainsford, F., Jin, Y., Patrick, J.W., Yang, Y.J. and Ruan, Y.L. (2012) High invertase activity in tomato reproductive organs correlates with enhanced sucrose import into, and heat tolerance of, young fruit. *Journal of Experimental Botany* 63, 1155–1166.

Lim, P.O., Kim, H.J. and Nam, H.G. (2007) Leaf senescence. *Annual Review of Plant Biology* 58, 115–136.

Liu, X. and Huang, B. (2000) Carbohydrate accumulation in relation to stress tolerance in two creeping bentgrass cultivars. *Journal of American Society of Horticultural Sciences* 125, 442–447.

Long, S.P. and Ort, D.R. (2010) More than taking heat: Crops and global climate change. *Current Opinion in Plant Biology* 13, 241–248.

Love, A.J., Milner, J.J. and Sadanandom, A. (2008) Timing is everything: regulatory overlap in plant cell death. *Trends in Plant Science* 13, 589–595.

Mackill, D.J., Coffman, W.R. and Rutger, J.N. (1982) Pollen shedding and combining ability for high temperature tolerance in rice. *Crop Science* 22, 730–733.

Meiri, D., Tazat, K., Cohen-Peer, R., Farchi-Pisanty, O., Aviezer-Hagai, K., Avni, A., et al. (2010) Involvement of *Arabidopsis* ROF2 (FKBP65) in thermotolerance. *Physiology and Molecular Biology of Plants* 72, 191–203.

Miao, Y., Laun, T.M., Zimmermann, P. and Zentgraf, U. (2004) Targets of WRKY53 transcription factor and its role during leaf senescence in *Arabidopsis*. *Plant Molecular Biology* 55, 853–867.

Miller, G. and Mittler, R. (2006) Could heat shock transcription factors function as hydrogen peroxide sensors in plants? *Annals of Botany* 98, 279–288.

Mishra, S.K., Tripp, J., Winkelhaus, S., Tschiersch, B., Theres, K., Nover, L. and Scharf, K.D. (2002) In the complex family of heat stress transcription factors, *HsfA1* has a unique role as master regulator of thermotolerance in tomato. *Genes and Development* 16, 1555–1567.

Mittler, R. (2006) Abiotic stress, the field environment and stress combination. *Trends in Plant Science* 11, 15–19.

Mittler, R., Vanderauwera, S., Gollery, M. and Van Breusegem, F. (2004) Reactive oxygen gene network of plants. *Trends in Plant Science* 9, 490–498.

Mondal, S., Singh, R., Crossa, J., Huerta-Espino, J., Sharma, I., Chatrath, R., et al. (2013) Earliness in wheat: a key to adaptation under terminal and continual high temperature stress in South Asia. *Field Crop Research* 151, 19–26.

Moore, B., Zhou, L., Rolland, F., Hall, Q., Cheng, W.H., Liu, Y.X., Hwang, I., Jones, T. and Sheen, J. (2003) Role of the *Arabidopsis* glucose sensor HXK1 in nutrient, light, and hormonal signaling. *Science* 300, 332–336.

Neta-Sharir, I., Isaacson, T., Lurie, S. and Weiss, D. (2005) Dual role for tomato heat shock protein 21: protecting photosystem ii from oxidative stress and promoting color changes during fruit maturation. *The Plant Cell* 17, 1829–1838.

Nieto-Sotelo, J., Martínez, L.M., Ponce, G., Cassab, G.I., Alagón, A., Meeley, R.B., Ribaut, J.M. and Yang, R. (2002) Maize *HSP101* plays important roles in both induced and basal thermotolerance and primary root growth. *The Plant Cell* 14, 1621–1633.

Panchuk, I.I., Volkov, R.A. and Schöffl, F. (2002) Heat stress- and heat shock transcription factor-dependent expression and activity of ascorbate peroxidase in *Arabidopsis*. *Plant Physiology* 129, 838–853.

Pecinka, A., Dinh, H.Q., Baubec, T., Rosa, M., Lettner, N. and Mittelsten Scheid, O. (2010) Epigenetic regulation of repetitive elements is attenuated by prolonged heat stress in *Arabidopsis*. *The Plant Cell* 22, 3118–3129.

Popova, O.V., Dinh, H.Q., Aufsatz, W. and Jonak, C. (2013) The RdDM pathway is required for basal heat tolerance in Arabidopsis. *Molecular Plant* 6, 396–410.

Puigderrajols, P., Jofre, A., Mir, G., Pla, M., Verdaguer, D., Huguet, G. and Molinas, M. (2002) Developmentally and stress-induced small heat shock proteins in cork oak somatic embryos. *Journal of Experimental Botany* 53, 1445–1452.

Rang, Z.W., Jagadish, S.V.K., Zhou, Q.M., Craufurd, P.Q. and Heuer, S. (2010) Effect of high temperature and water stress on pollen germination and spikelet fertility in rice. *Environmental and Experimental Botany* 70, 58–65.

Roitsch, T. and González, M.C. (2004) Function and regulation of plant invertases: sweet sensations. *Trends in Plant Sciences* 9, 606–613.

Rosa, S.B., Caverzan, A., Teixeira, F.K., Lazzarotto, F., Silveira, J.A.G., Ferreira-Silva, S.L., Abreu-Neto, J., Margis, R. and Margis-Pinheiro, M. (2010) Cytosolic APX knockdown indicates an ambiguous redox response in rice. *Phytochemistry* 71, 548–558.

Rosenvasser, S., Mayak, S. and Friedman, H. (2006) Increase in reactive oxygen species (ROS) and in senescence-associated gene transcript (SAG) levels during dark-induced senescence of *Pelargonium* cuttings, and the effect of gibberellic acid. *Plant Science* 170, 873–879.

Ruan, Y.L., Jin, Y., Yang, Y.J., Li, G.J. and Boyer, J.S. (2010) Sugar input, metabolism, and signaling mediated by invertase: roles in development, yield potential, and response to drought and heat. *Molecular Plant* 3, 942–955.

Ruelland, E. and Zachowski, A. (2010) How plants sense temperature. *Environmental and Experimental Botany* 69, 225–232.

Saini, H.S. and Aspinall, D. (1982) Abnormal sporogenesis in wheat (*Triticum aestivum* L.) induced by short episodes of high temperature. *Annals of Botany* 49, 835–846.

Sairam, R.K., Srivastava, G.C. and Saxena, D.C. (2000) Increased antioxidant activity under elevated temperatures: a mechanism of heat stress tolerance in wheat genotypes. *Biologia Plantarum* 43, 245–251.

Sakata, T., Oshino, T., Miura, S., Tomabechi, M., Tsunaga, Y., Higashitani, N., Miyazawa, Y., Takahashi, H., Watanabe, M. and Higashitani, A. (2010) Auxins reverse plant male sterility caused by high temperature. *Proceedings of the National Academy of Sciences USA* 107, 8569–8574.

Sakuma, Y., Maruyama, K., Osakabe, Y., Qin, F., Seki, M., Shinozaki, K. and Yamaguchi-Shinozaki, K. (2006) Functional analysis of an *Arabidopsis* transcription factor, DREB2A, involved in drought-responsive gene expression. *The Plant Cell* 18, 1292–1309.

Satake, T. and Yoshida, S. (1978) High temperature-induced sterility in indica rices at flowering. *Japanese Journal of Crop Science* 4, 6–17.

Sato, S., Kamiyama, M., Iwata, T., Makita, N., Furukawa, H. and Ikeda, H. (2006) Moderate increase of mean daily temperature adversely affects fruit set of *Lycopersicon esculentum* by disrupting specific physiological processes in male reproductive development. *Annals of Botany* 97, 731–738.

Semenov, M.A. and Shewry, P.R. (2010) Modelling predicts that heat stress and not drought will limit wheat yield in Europe. Nature Proceedings. Available at: http:/hdl.handle.net/10101/npre.2010.4335.1 (accessed 20 December 2013).

Shah, F., Huang, J., Cui, K., Nie, L., Shah, T., Chen, C. and Wang, K. (2011) Impact of high-temperature stress on rice plant and its traits related to tolerance. *Journal of Agricultural Sciences* 149, 545–556.

Shi, W.M., Muramoto, Y., Ueda, A. and Takabe, T. (2001) Cloning of peroxisomal ascorbate peroxidase gene from barley and enhanced thermotolerance by overexpressing in *Arabidopsis thaliana*. *Gene* 273, 23–27.

Smeekens, S., Ma, J., Hanson, J. and Rolland, F. (2010) Sugar signals and molecular networks controlling plant growth. *Current Opinion in Plant Biology* 13, 274–279.

Smykal, P., Hrdy, I. and Pechan, P.M. (2000) High-molecular-mass complexes formed *in vivo* contain smHSPs and HSP70 and display chaperone-like activity. *European Journal of Biochemistry* 267, 2195–2207.

Sugio, A., Dreos, R., Aparicio, F. and Maule, A.J. (2009) The cytosolic protein response as a subcomponent of the wider heat shock response in *Arabidopsis*. *The Plant Cell* 21, 642–654.

Sun, L., Liu, Y., Kong, X., Zhang, D., Pan, J., Zhou, Y. and Yang, X. (2012) ZmHSP16.9, a cytosolic class I small heat shock protein in maize (*Zea mays*), confers heat tolerance in transgenic tobacco. *Plant Cell Reports* 31, 1473–1484.

Suter, L. and Widmer, A. (2013) Environmental heat and salt stress induce transgenerational phenotypic changes in *Arabidopsis thaliana*. *PLoS One* 8, e60364.

Szalai, G., Kellős, T., Galiba, G. and Kocsy, G. (2009) Glutathione as an antioxidant and regulatory molecule in plants under abiotic stress conditions. *Plant Growth Regulation* 28, 66–80.

Tang, L., Kwon, S.Y., Kim, S.H., Kim, J.S., Choi, J.S., Cho, K.Y., Sung, C.K., Kwak, S.S. and Lee, H.S. (2006) Enhanced tolerance of transgenic potato plants expressing both superoxide dismutase and ascorbate peroxidase in chloroplasts against oxidative stress and high temperature. *Plant Cell Reports* 25, 1380–1386.

Thelander, M., Olsson, T. and Ronne, H. (2004) Snf1-related protein kinase 1 is needed for growth in a normal day–night light cycle. *EMBO Journal* 23, 1900–1910.

Thomas, H. (2013) Senescence, ageing and death of the whole plant. *New Phytologist* 197, 696–711.

Vacca, R.A., de Pinto, M.C., Valenti, D., Passarella, S., Marra, E. and De Gara, L. (2004) Production of reactive oxygen species, alteration of cytosolic ascorbate peroxidase and impairment of mitochondrial metabolism are early events in heat shock induced programmed cell death in tobacco Bright-Yellow 2 cells. *Plant Physiology* 134, 1100–1112.

Volkov, R.A., Panchuk, I.I., Mullineaux, P.M. and Schöffl, F. (2006) Heat stress-induced H_2O_2 is required for effective expression of heat shock genes in *Arabidopsis*. *Plant Molecular Biology* 61, 733–746.

Wahid, A., Gelani, S., Ashraf, M. and Foolad, M.R. (2007) Heat tolerance in plants: an overview. *Environmental Experimental Botany* 61, 199–223.

Wang, L.J. and Li, S.H. (2006) Thermotolerance and related antioxidant enzyme activities induced by heat acclimation and salicylic acid in grape (*Vitis vinifera* L.). *Plant Growth Regulation* 48, 137–144.

Wassmann, R., Jagadish, S.V.K., Heuer, S., Ismail, A., Redona, E., Serraj, R. and Sumfleth, K. (2009) Climate change affecting rice production: the physiological and agronomic basis for possible adaptation strategies. *Advances in Agronomy* 101, 59–122.

Waterer, D., Benning, N.T., Wu, G., Luo, X., Liu, X., Gusta, M., McHughen, A. and Gusta, L.V. (2010) Evaluation of abiotic stress tolerance of genetically modified potatoes (*Solanum tuberosum* cv. Desiree). *Molecular Breeding* 25, 527–540.

Wilhelm, E.P., Mullen, R.E., Keeling, P.L. and Singletary, G.W. (1999) Heat stress during grain filling in maize: effects on kernel growth and metabolism. *Crop Science* 39, 1733–1741.

Xue, Y., Peng, R., Xiong, A., Li, X., Zha, D. and Yao, Q. (2010) Over-expression of heat shock protein gene hsp26 in *Arabidopsis thaliana* enhances heat tolerance. *Biologia Plantarum* 54, 105–111.

Yamakawa, H. and Hakata, M. (2010) Atlas of rice grain filling-related metabolism under high temperature: joint analysis of metabolome and transcriptome demonstrated inhibition of starch accumulation and induction of amino acid accumulation. *Plant and Cell Physiology* 51, 795–809.

Zhang, X., Rerksiri, W., Liu, A., Zhou, X., Xiong, H., Xiang, J. and Xiong, X. (2013) Transcriptome profile reveals heat response mechanism at molecular and metabolic levels in rice flag leaf. *Gene* 530, 185–192.

Zhao, H., Dai, T., Jing, Q., Jiang, D. and Cao, W. (2007) Leaf senescence and grain filling affected by post anthesis high temperatures in two different wheat cultivars. *Plant Growth Regulation* 51, 149–158.

Zhu, Y., Wang, Z., Jing, Y., Wang, L., Liu, X., Liu, Y. and Deng, X. (2009) Ectopic over-expression of *BhHsf1*, a heat shock factor from the resurrection plant *Boea hygrometrica*, leads to increased thermotolerance and retarded growth in transgenic *Arabidopsis* and tobacco. *Plant Molecular Biology* 71, 451–467.

Zinn, K.E., Tunc-Ozdemir, M. and Harper, J.F. (2010) Temperature stress and plant sexual reproduction: uncovering the weakest links. *Journal of Experimental Botany* 61, 1959–1968.

3 Ethylene, Nitric Oxide and Haemoglobins in Plant Tolerance to Flooding

Luis A.J. Mur,[1]* Kapuganti J. Gupta,[2] Usha Chakraborty,[3] Bishwanath Chakraborty[3] and Kim H. Hebelstrup[4]

[1]*Institute of Biological, Environmental and Rural Sciences, Aberystwyth University, UK;* [2]*Department of Plant Sciences, University of Oxford, UK;* [3]*Department of Botany, University of North Bengal, Siliguri, India;* [4]*Department of Molecular Biology and Genetics, Aarhus University, Slagelse, Denmark*

Abstract

As much as 12% of the world's soils may suffer excess water so that flooding is a major limiting factor on crop production in many areas. Plants attempt to deal with submergence by forming root aerenchyma to facilitate oxygen diffusion from the shoot to the root, initiating a hyponastic response where petiole elongation facilitates access to atmospheric oxygen or initiating a bio-energetically conserving quiescence phase. Ethylene has well established roles in the initiation of programmed cell death (PCD) to form air-spaces in aerenchyma and in the hyponastic responses in petioles. The flooding-tolerant species *Rumex palustris* and the model plant *Arabidopsis thaliana* have been extensively exploited to reveal some key molecular events. Our groups have recently demonstrated that nitric oxide (NO) triggers the biosynthesis of ethylene during stress and that NO plays key roles in PCD and the hyponastic response. NO is formed from the reduction of NO_3/NO_2 via several pathways, which are differentially utilized depending on the availability of O_2. In fact, NO production and responses to flooding can be directly dependent on the nitrogen status of soil, which reflects local agricultural practice. This chapter will detail our understanding of the roles of ethylene, NO and haemoglobin in flooding stress.

3.1 Introduction

The human population has been estimated to reach 9 billion by 2050 and one of the grand challenges for plant scientists is to increase crop production (Godfray *et al.*, 2010). This will involve exploitation and expansion of our knowledge of tolerance mechanisms to drought (Setter, 2012), excessive salt (Zhang and Shi, 2013), micronutrient deficiency (Mayer *et al.*, 2008) and also resistance to pathogens and pests (Gregory *et al.*, 2009). Equally, the effects of crop flooding need to be considered in the food security question. Recent floods in Europe, Australia, the USA and Pakistan have all impacted on crop production with economically significant consequences (review by Bailey-Serres *et al.*, 2012).

It should be noted that waterlogging refers to the inundation only of the soil and rhizosphere. If plants are totally submerged this effectively filters sunlight as well as dramatically reducing available oxygen and CO_2 levels by ~10^4-fold (Colmer and Pedersen, 2008).

---

*E-mail: lum@aber.ac.uk

As a result both respiration and photosynthesis may be severely disrupted and could be maintained only through dissolved CO_2 in the floodwater (Pedersen *et al.*, 2010). In leaves with a sufficiently hydrophobic cuticle, the relative availability of CO_2 and O_2 is likely to be influenced by gas film formation (Pedersen *et al.*, 2009).

The effect of reduced O_2 for ATP production via mitochondrial oxidative phosphorylation has been extensively characterized (Lasanthi-Kudahettige *et al.*, 2007; Narsai *et al.*, 2009). These studies demonstrated the switch to anaerobic respiratory pathways based on glycolysis and fermentation – the so-called Pasture effect – as shown by increased sucrose catabolism and the expression of alcohol dehydrogenase (ADH) (Bailey-Serres and Voesenek, 2008) and increases in some organic acids, e.g. succinate and lactate. These latter changes suggested decreased carbon flow into the tricarboxylic acid cycle (TCA) linked to lower levels of reducing equivalents (NADH) in the mitochondrial matrix, causing inhibition of TCA enzymes such as pyruvate dehydrogenase. These and increases in lactate dehydrogenase are indicative of anaerobic metabolism (Menegus *et al.*, 1988, 1989). Accompanying these were changes in nitrogen metabolism with individual amino acids either increasing or decreasing in low O_2 and γ-aminobutyric acid (GABA) increasing (Reggiani *et al.*, 1988; Narsai *et al.*, 2009). Narsai *et al.* (2009) also noted gene-expression changes indicating changes in lipid metabolism. Anoxic plants are less able to synthesize desaturated lipids due to lipid desaturases requiring oxygen. It may be that a resulting bias toward saturated lipids could – as with temperature stress – be part of an O_2 sensing mechanism (Penfield, 2008).

Given the lack of O_2 it may be surprising that submergence also leads to the generation of reactive oxygen species (ROS) and heat shock proteins (Hsps) that may be a direct consequence of oxidative damage (Mustroph *et al.*, 2010). The insensitivity to prolonged submergence exhibited by the rice cultivar cv. M202 may be a consequence of its inability to reduce ROS (Fukao *et al.*, 2006). A major source of ROS during anoxia is a nicotinamide adenine dinucleotide phosphate (NADPH) oxidase (Baxter-Burrell *et al.*, 2002). The activity of low-O_2 responsive NADPH oxidase is regulated by a Rop small GTP-binding (G) protein, which is in turn suppressed by the GTPase activating protein ROPGAP4. Baxter-Burrell *et al.* (2002) have shown that NADPH oxidase-derived ROS were required for ADH expression but this needs to be modulated by ROPGAP4 in order to survive anoxia. Predictably, ROS generation also poses a particular problem when plants are re-oxygenated on desubmergence. Antioxidant defences based on reductants such as ascorbate, glutathione and tocopherols or enzymes such as superoxide dismutase and catalases play a vital role in this recovery phase (Jackson and Ram, 2003). Indeed, flooding tolerance of the rice cultivar FR13A has been linked to high-antioxidant defences (Almeida *et al.*, 2003).

It is also relevant to consider the effects of reduced oxygen on soil microbes where the switch to fermentative respiration based on the anaerobic decomposition of organic matter could also contribute to CO_2 production (Crawford, 1992). Such decomposition can also result in the release of redox active minerals, for example, iron and manganese, resulting in chemically reducing conditions and potential accumulation of toxic compounds. Flooding tolerance mechanisms should therefore deal with both the maintenance of primary energetic metabolism as well as the toxic products.

Mechanisms also need to be tailored to the typical submergence period; hours or days, with the latter especially imposing unique demands for a stress tolerance mechanism. Plant species may adopt one of two broad strategies to survive flooding. These have been designated as the low-oxygen 'escape' strategy (LOES) and the low-oxygen quiescence strategy (LOQS; Bailey-Serres and Voesenek, 2008). LOQS is typical of plants that have evolved to survive short-term flooding in metabolic 'quiescence', when growth ceases and metabolism is restricted. The tolerance mechanisms based on 'escape' (LOES) involve physiological and developmental changes through which the plant seeks to return to normoxic conditions. Particularly well-characterized is the hyponastic response, which is characterized by the rapid elongation of underwater stems or leaves growing out of a flood. Another morphological change that can occur is the formation of aerenchyma within roots and leaves through programmed cell death (PCD), effectively forming a 'snorkel' to facilitate gaseous exchange (Drew et al., 2000). Aerenchyma is constitutively formed in deep-water and lowland rice stems and leaf sheaths but in other species formation can be induced by flooding (Steffens et al., 2013). The flooded paddy field system that is typical of lowland rice production in South-east Asia represents an ideal practice to encourage the release of soil nutrients and also reduce weeds, against which rice is a poor competitor. This effective system is totally dependent on rice varieties that form aerenchyma as well as physical barriers that limit O_2 loss from the plant and conversely prevent the entry of soil toxins (Colmer and Voesenek, 2009).

3.2 Ethylene, a Major Regulator of Flooding Tolerance

Ethylene plays diverse and important roles in tolerance to submergence, so that understanding the regulation of its biosynthesis and dependent signalling is a vital part of research into plant responses to flooding (Jackson, 2008) (Fig. 3.1).

Ethylene biosynthesis starts from S-adenosyl-methionine (S-AdoMet), which is converted from the amino acid L-methionine by S-AdoMet synthetase (SAM synthetase; EC 2.5.1.6). The first committed step of ethylene biosynthesis is the conversion of S-AdoMet to 1-aminocyclopropane-1-carboxylic acid (ACC) by ACC synthase (ACS; EC 4.4.1.14) Finally, ACC is oxidized by ACC oxidase (ACO; EC 1.14.17.4) to form ethylene and the by-products CO_2 and cyanide (Wang et al., 2002). Ethylene is perceived by a family of five membrane-bound receptors (ETR1, ETR2, ERS1, ERS2, EIN4; EC 2.7.13.3) that have similarity to bacterial two-component histidine kinases. The ethylene-resistant, loss-of-function mutant *etr1-1*, has a dominant mutation in *ETR1* causing an 80% reduction in the amount of ethylene binding. The ethylene receptors interact with and regulate the constitutive triple response 1 (CTR1) product. This actively suppresses the ethylene responses in the absence of ethylene, an effect which is relieved when ethylene binds to the receptors. CTR1 inactivation results in the stimulation of the endoplasmic reticulum-bound protein ethylene insensitive 2 (EIN2), which in turn activates a family of transcription factors, including the nuclear-localized protein EIN3 and various EIN3-like (EIL) proteins (Schaller

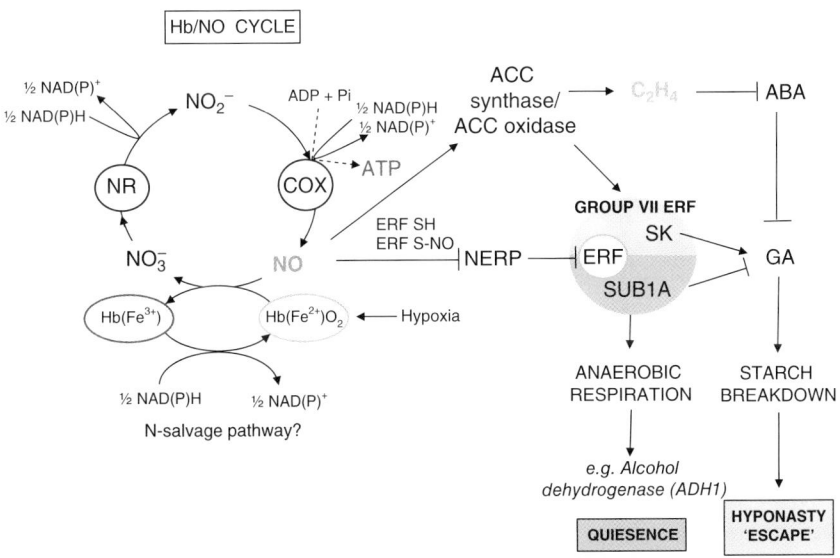

Fig. 3.1. Ethylene, nitric oxide and haemoglobin in submergence tolerance. (See text for details.)

and Kieber, 2002). The EIN3/EIL transcription factors recognize DNA targets, the so-called EIN3-binding site (EBS) or primary ethylene response element (PERE) in the promoters of ethylene response factor (ERFs) genes and stimulate their transcription (Guo and Ecker, 2004).

It will be noted that ethylene production is ACC oxidase- and therefore O_2-dependent, thus it could be questioned how far ethylene biosynthesis is important for flooding tolerance. However, whilst ethylene is continuously being generated by plant cells it is lost to the atmosphere, but the imposition of a diffusion barrier leads to an intercellular condition with increased ethylene content. Interestingly, this is not always accompanied by increases in ethylene biosynthesis, which, apparently, is a reflection of the poor availability of O_2. Therefore, the partial submergence of mature rice results in rapid production of ACC and conversion to ethylene after 4 h (Mekhedov and Kende, 1996). However, in persistent deep submergence, in plants such as *Rumex palustris* and *Rumex acetosa*, ethylene biosynthesis is reduced, although ACC oxidase genes are induced, most likely through a lack of O_2 (Vriezen et al., 1999).

Ethylene plays an important role in certain types of PCD (Navarre and Wolpert, 1999; Steffens and Sauter, 2005; Mase et al., 2012) and so it is unsurprising that it also influences aerenchyma formation. This role for ethylene has been extensively characterized in maize where it is linked to inducible aerenchyma formation (He et al., 1996). Thus, ethylene biosynthesis inhibitors or receptor blockers suppressed aerenchyma formation in hypoxic roots. Aerenchyma formation is enhanced in internodes of deep-water rice by ethylene, which promotes formation of O_2^- (Steffens et al., 2011). This could be via suppression of the ROS scavenger metallothionein MT2b, which occurs in adventitious root emergence in rice in an ethylene-dependent manner (Steffens and Sauter, 2009). More likely is that ethylene activates NADPH oxidase (Wi et al., 2012) and so influences ROS generation via the ROP-ROPGAP4 reciprocal regulatory system (Baxter-Burrell et al., 2002).

Ethylene is also clearly implicated in hyponasty as established by pioneering studies in wetland *Rumex* species, rice and even *Arabidopsis* (Bailey-Serres et al., 2012). Initially studies correlated petiole growth in *Rumex* with ACC and ethylene production (Voesenek et al., 1990a), but this was then related to ethylene entrapment (Voesenek et al., 1993). Hyponasty is not due to cell division but cell expansion leading to unequal growth between the adaxial and abaxial sides of the petiole (Voesenek et al., 1990b). This unequal growth comes about through multiple ethylene-influenced events including the altered expression of expansin (Vriezen et al., 2000) and also a change in cortical microtubules in the abaxial side from longitudinal to transverse orientations whilst the converse occurs in the adaxial side (Polko et al., 2012).

The utility of this ethylene-mediated signalling in plant breeding has been illustrated by the characterization of two quantitative trait loci (QTL) linked to rice flooding tolerance. In rice, this escape mechanism involves submergence-associated internode elongation. This has been linked to the induction of ERF-class transcription factor genes *SNORKEL1* and *2* (*SK1*, *SK2*) and *SUBMERGENCEIA* (*SUB1A*). The ERF transcription factor family have been sub-classified based on phylogeny, chromosome locations and targeted binding motifs (Nakano et al., 2006). *SK1*, *SK2* and *SUB1A* have been classified as group VII ERF as they bind to the MCGGAI(I/L) motif within promoters (Tournier et al., 2003). The SK1/2 and SUB1A sets of ERF contribute to two complementary submergence survival strategies: hypnasty and quiescence, respectively. Further elucidation of ethylene action during submergence has been revealed through the elucidation of the roles of both groups of factors. Increased ethylene will reduce ABA content by inducing the expression of ABA 8'-hydroxylase to form inactive dihydrophaseic acid (Saika et al., 2007). This loss in ABA is an important consequence of ethylene signalling and ABA insensitivity has been linked to greater underwater petiole elongation in *R. palustris* ecotypes (Chen et al., 2010). Suppressing of ABA effects resulted in an increased responsiveness to gibberellic acid (GA). This increased GA sensitivity arises from the suppression of inhibitory DELLA domain-containing proteins SLENDER RICE1

(SLR1) and SLR-LIKE1 (SLRL1; Sun, 2011). GA initiates starch mobilization, which is linked to anaerobic metabolism to maintain elongative growth (Fukao et al., 2006). The ERFs SK1 and SK2 are likely to act by stimulating this GA-mediated stage (Bailey-Serres et al., 2012).

SUB1A is the source of longer-term survival (>2 weeks) with complete submergence and is linked to quiescence (Xu et al., 2006). SUB1A appears to act by maintaining SLRL1 expression and therefore countering the possibly short-term and possibly ultimately retrograde effects of GA signalling when plants are deeply submerged for long periods (Fukao et al., 2011). Additionally, SUB1A also inhibits developmental events linked to flowering as part of the quiescence programme (Pena-Castro et al., 2011). SUB1A reduces carbohydrate catabolism but also increases ADH1 expression so that the activities of enzymes for ethanolic fermentation are elevated during quiescence (Fukao et al., 2006).

SK and SUB1A effects feeding into ABA and GA pathways are not only seen in rice as similar roles have been described in *Rumex* sp. (Benschop et al., 2005; Bailey-Serres and Voesenek, 2008) and *Arabidopsis* (Lee et al., 2011). For example, in *Arabidopsis*, ABA has been shown to antagonize ethylene-induced hyponastic growth (Benschop et al., 2007) whilst submergence increases the expression of the ERFs *HYPOXIA RESPONSIVE ERF1* (*HRE1*) and *HRE2*. *HRE1* is directly associated with increased expression of anaerobic responsive genes such as *ADH1* (Licausi et al., 2010). These mechanisms allow certain accessions of *Arabidopsis* to survive >40 days of submergence, suggesting that flooding tolerance could be an important feature of low-lying plants in temperate zones (Lee et al., 2011).

Recent breakthroughs have established that group VII ERF are part of an O_2 sensing mechanism (Voesenek and Bailey-Serres, 2013). Two groups (Licausi et al., 2010; Gibbs et al., 2011) first observed that *Arabidopsis* mutants in the N-end rule pathway of targeted proteolysis (NERP)-mediated pathway exhibited constitutive low O_2 responsive expression, including *ADH1*. It was found that NERP proteolytically cleaves some group VII ERFs (but not apparently SUB1A) if key cysteines are oxidized. Under anoxic conditions this would not occur, allowing binding to cognate promote-binding sites (Licausi et al., 2010; Gibbs et al., 2011; Voesenek and Bailey-Serres, 2013).

3.3 Nitric Oxide: a Suspected Important Player in Submergence Tolerance?

NO first came to prominence within the context of regulating plant defence during plant pathogen interactions (Delledonne et al., 1998; Durner et al., 1998) but subsequently it was shown to be markedly increased in sunflower leaves under anoxia through the action of nitrate reductase (NR) (Rockel et al., 2002). This NADPH-dependent reduction of NO_2^- by cytosolic NR has now emerged as a main source of NO in plants under aerobic conditions. However, under anoxia higher plant mitochondria, like mammalian mitochondria (Kozlov et al., 1999) and algal mitochondria (Tischner et al., 2004), are sources of NO generation. These mechanisms have been described in detail elsewhere (Gupta et al., 2011a, b).

A mechanism of NO generation during flooding is a peroxisomally located xanthine oxidoreductase (XOR), which reduces nitrite to NO at the expense of NADH under anaerobic conditions (Corpas et al., 2008). However, the importance of this pathway under flooding conditions is not yet known. Instead, the main anoxic mechanism of NO generation is via a mitochondrial-based nitrite reductase activity that uses NO_2^- as a terminal electron acceptor at the site of cytochrome-coxidase (Castello et al., 2006) (Fig. 3.1). The nitrite-reductase activity becomes increasingly important as partial pressures of oxygen are reduced from ambient. It has been demonstrated that mitochondria produce NO when oxygen concentration falls below 1% and NO emission reaches its highest level at anoxia. The IC_{50} (50% inhibitory concentration) is 0.05%, whereas in roots the threshold for nitrite-dependent NO production is 0.5% (Hebelstrup et al., 2012).

The cytochrome-c oxidase/reductase (COX) reduction of NO_2^- to generate NO can be coupled with its oxidation to nitrate by non-symbiotic haemoglobins (nsHb) to form the Hb/NO cycle. Plant Hbs may be subdivided into three classes: I, II and III. Most Hbs found in association with nitrogen-fixing bacteria in root nodules of plants appear to have evolved from class II Hb, which has a relatively low affinity for O_2 (Km ~150 nM) so that this is readily released under low partial pressures of O_2. As such, functions of most of those Hbs called 'symbiotic haemoglobins' are in facilitating oxygen supply to tissues within nitrogen-fixing nodules. However, this requires a high concentration of Hb. Class I haemoglobins are not directly involved in symbiotic association and hence are labelled as non-symbiotic haemoglobins (nsHb). nsHb are found in other tissues at low concentration where the contribution to facilitated oxygen diffusion is negligible (Heckmann et al., 2006). nsHb possess very high affinity for O_2 (2 nM), making them poor oxygen carriers (Smagghe et al., 2009) and NO oxidation is an important role for these proteins. During hypoxic conditions the Hb/NO cycle arises when the oxidation of NO to NO_3^- by oxyhaemoglobin (Hb $(Fe^{2+})O_2$) is coupled to the reduction of NO_3^- and NO_2^- (Dordas et al., 2004). In this Hb/NO cycle excess NAD(P)H is oxidized to maintain electron flow and ATP production under hypoxic conditions (Dordas et al., 2003; Stoimenova et al., 2007). Thus NO generated during submergence could improve the energy status of the plant by adding to the Hb/NO cycle (Igamberdiev and Hill, 2009). Operation of the Hb/O_2 cycle under hypoxic conditions leads to generation of a proton gradient, which subsequently leads to production of 25% to 35% (in comparison to 100% ATP production under aerobic conditions) of ATP under anoxia (Stoimenova et al., 2007). Anaerobic ATP production and NAD(P)H oxidation may act as an alternative to glycolytic fermentation. It was found that anoxic-tolerant rice mitochondria generate more anaerobic ATP than anoxic-intolerant barley (Stoimenova et al., 2007). Haemoglobins may also play a role in nitrogen conservation during hypoxia. In spite of the recycling Hb/NO cycle, major NO emission occurs during hypoxia (Hebelstrup et al., 2012). The rate of NO emission at 0.1% oxygen was 8.3 nmol g^{-1} fresh-weight (FW) h^{-1} equivalent to 0.2 mM (0.2 mmol g^{-1} FW) nitrate lost over the 24 h period. In the absence of snHb this could be expected to be significantly higher. Thus, snHb could be an N-salvage pathway that could aid submergence survival and recovery (Fig. 3.1).

Ethylene is a well-established hormone regulating 'escape' and 'quiescence' strategies of submergence tolerance. Ethylene is derived from 1-aminocyclopropane-1-carboxylic acid (ACC) by ACC synthase (ACS) and ACC oxidase (ACO). Ethylene will relieve abscisic acid (ABA)-mediated suppression of gibberellic acid (GA)-signalling to influence internode elongation. Elongation growth is controlled by GA, which is also linked to starch consumption and elongation growth. Group VII ethylene responsive factors (ERF) represent a key regulatory node governing whether 'escape' or 'quiescence' strategies are followed. The ERFs SK1 and SK2 contribute to GA activation by inhibiting the expression of GA-suppressing DELLA proteins SLENDER RICE1 (SLR1) and SLR-LIKE1 (SLRL1). Conversely, in quiescence, the ERF SUB1A inhibits shoot elongation by maintaining levels of the transcription factors SLR1 and SLRL1. SUB1A will activate fermentative respiration (e.g. inducing the expression of alcohol dehydrogenase, ADH1).

Nitric oxide (NO) is produced during hypoxia through the action of a mitochondrially-located nitrate-NO-reductase (NR). This uses NO_2^- as a terminal electron acceptor for cytochrome-c oxidase/reductase (COX) with consumption of NAD(P) and generation of ATP. NO is oxidized to NO_3^- through the action of nonsymbiotic haemoglobins (nsHb); oxidized haemoglobin (methaemoglobin) is reduced to haemoglobin by NAD(P)H. This nsHb regenerative step forms a key step in the Hb/O_2 cycle, which contributes to hypoxic ATP generation in mitochondria. The scavenging of NO by nsHb could also represent a nitrogen-salvage pathway that prevents excessive loss of N to the environment via NO.

NO can influence ethylene signalling through the activation of ACS and ACO. It is also likely to aid in the stabilization of ERF by protection from the 'N-end rule pathway of targeted proteolysis' (NERP). NERP targets Group VII ERF based on the oxidation status of a key cysteine on the ERF. Under low O_2 this is in a reduced state, which is not targeted by

NERP proteolysis. Reduction of the cysteine by S-nitrosylation (S-S + 2NO → 2SNO) can also prevent proteolysis and thus ERF activation.

It is also relevant to flooding tolerance that hypoxically generated NO plays a role in the induction of alternative oxidase (AOX) (Gupta et al., 2012). NO inhibits aconitase, which converts citrate to isocitrate. This inhibition leads to accumulation of citrate, which can induce AOX. It is well established that AOX can decrease ROS production by preventing over-reduction of the ubiquinone pool, which could otherwise feature during submergence or relief from flooding. Interestingly, it was recently shown that AOX also influences NO levels in leaf tissue (Cvetkovska and Vanlerberghe, 2012).

Relatively few direct assessments of the roles of NO in submergence have been undertaken and more are clearly needed. Removal of NO using a chemical scavenger reduced both the induction of ADH and led to poorer plant survival under low O_2. These effects could be associated solely with the Hb/NO cycle and indeed, significant induction of *nsHb* gene expression upon submergence is a feature of many plant species (van Veen et al., 2013).

Perhaps more importantly, NO generation has been demonstrated to initiate ethylene biosynthesis (Mur et al., 2013) so that it could be considered an upstream trigger for all of the ethylene effects previously described in this chapter. Generation of NO using NO donors and in transgenic plants expressing mammalian nitric oxide synthase (NOS) increased ethylene biosynthesis via elevated expression of ACC synthase and ACC oxidase (Mur et al., 2008; Chun et al., 2012). It is also possible to predict that NO can affect NERP-mediated proteolysis of group VII ERF. NO can affect the redox status of cysteine groups through their oxidation through S-nitrosylation (SH → SNO + H^+) (Gupta, 2011). Suppression of NERP-mediated proteolysis of cysteine in mammals comes about through S-nitrosylation (Hu et al., 2005) and thus, it may be that S-nitrosylation of ERFs in plants will allow their binding to cognate promoters.

If NO does emerge as an important determinant of submergence tolerance, this would suggest that a plant's N-status would influence its effectiveness as a regulator, because it was shown that nitrate nutrition influences NO production (Gupta et al., 2013). Thus, crop nutrition and soil microbial interactions that metabolize and liberate nitrate or nitrite would be a major target in agricultural strategies aiming to combat the effects of flooding.

References

Almeida, A.M., Vriezen, W.H. and van der Straeten, D. (2003) Molecular and physiological mechanisms of flooding avoidance and tolerance in rice. *Russian Journal of Plant Physiology* 50, 743–751.

Bailey-Serres, J. and Voesenek, L.A.C.J. (2008) Flooding stress: acclimations and genetic diversity. *Annual Review of Plant Biology* 59, 313–339.

Bailey-Serres, J., Lee, S.C. and Brinton, E. (2012) Waterproofing crops: effective flooding survival strategies. *Plant Physiology* 160, 1698–1709.

Baxter-Burrell, A., Yang, Z.B., Springer, P.S. and Bailey-Serres, J. (2002) RopGAP4-dependent Rop GTPase rheostat control of *Arabidopsis* oxygen deprivation tolerance. *Science* 296, 2026–2028.

Benschop, J.J., Jackson, M.B., Guhl, K., Vreeburg, R.A.M., Croker, S.J., Peeters, A.J.M. and Voesenek, L.A.C.J. (2005) Contrasting interactions between ethylene and abscisic acid in *Rumex* species differing in submergence tolerance. *Plant Journal* 44, 756–768.

Benschop, J.J., Millenaar, F.F., Smeets, M.E., van Zanten, M., Voesenek, L.A. and Peeters, A.J. (2007) Abscisic acid antagonizes ethylene-induced hyponastic growth in *Arabidopsis*. *Plant Physiology* 143, 1013–1023.

Castello, P.R., David, P.S., McClure, T., Crook, Z. and Poyton, R.O. (2006) Mitochondrial cytochrome oxidase produces nitric oxide under hypoxic conditions: Implications for oxygen sensing and hypoxic signaling in eukaryotes. *Cell Metabolism* 3, 277–287.

Chen, X., Pierik, R., Peeters, A.J., Poorter, H., Visser, E.J., Huber, H., de Kroon, H. and Voesenek, L.A. (2010) Endogenous abscisic acid as a key switch for natural variation in flooding-induced shoot elongation. *Plant Physiology* 154, 969–977.

Chun, H.J., Park, H.C., Koo, S.C., Lee, J.H., Park, C.Y., Choi, M.S., Kang, C.H., Baek, D., Cheong, Y.H., Yun, D.J., Chung, W.S., Cho, M.J. and Kim, M.C. (2012) Constitutive expression of mammalian nitric oxide synthase in tobacco plants triggers disease resistance to pathogens. *Molecules and Cells* 34, 463–471.

Colmer, T.D. and Pedersen, O. (2008) Underwater photosynthesis and respiration in leaves of submerged wetland plants: gas films improve CO_2 and O_2 exchange. *New Phytologist* 177, 918–926.

Colmer, T.D. and Voesenek, L.A.C.J. (2009) Flooding tolerance: suites of plant traits in variable environments. *Functional Plant Biology* 36, 665–681.

Corpas, F.J., Palma, J.M., Sandalio, L.M., Valderrama, R., Barroso, J.B. and Del Rio, L.A. (2008) Peroxisomal xanthine oxidoreductase: characterization of the enzyme from pea (*Pisum sativum* L.) leaves. *Journal of Plant Physiology* 165, 1319–1330.

Crawford, R.M.M. (1992) Oxygen availability as an ecological limit to plant-distribution. *Advances in Ecological Research* 23, 93–185.

Cvetkovska, M. and Vanlerberghe, G.C. (2012) Alternative oxidase modulates leaf mitochondrial concentrations of superoxide and nitric oxide. *New Phytologist* 195, 32–39.

Delledonne, M., Xia, Y., Dixon, R.A. and Lamb, C. (1998) Nitric oxide functions as a signal in plant disease resistance. *Nature* 394(6693), 585–588.

Dordas, C., Hasinoff, B.B., Igamberdiev, A.U., Manac'h, N., Rivoal, J. and Hill, R.D. (2003) Expression of a stress-induced hemoglobin affects NO levels produced by alfalfa root cultures under hypoxic stress. *Plant Journal* 35, 763–770.

Dordas, C., Hasinoff, B.B., Rivoal, J. and Hill, R.D. (2004) Class-1 hemoglobins, nitrate and NO levels in anoxic maize cell-suspension cultures. *Planta* 219, 66–72.

Drew, M.C., He, C.J. and Morgan, P.W. (2000) Programmed cell death and aerenchyma formation in roots. *Trends in Plant Science* 5, 123–127.

Durner, J., Wendehenne, D. and Klessig, D.F. (1998) Defense gene induction in tobacco by nitric oxide, cyclic GMP, and cyclic ADP-ribose. *Proceedings of National Academy of Sciences USA* 95(17), 10328–10333.

Fukao, T., Xu, K.N., Ronald, P.C. and Bailey-Serres, J. (2006) A variable cluster of ethylene response factor-like genes regulates metabolic and developmental acclimation responses to submergence in rice. *Plant and Cell* 18, 2021–2034.

Fukao, T., Yeung, E. and Bailey-Serres, J. (2011) The submergence tolerance regulator SUB1A mediates crosstalk between submergence and drought tolerance in rice. *Plant Cell* 23, 412–427.

Gibbs, D.J., Lee, S.C., Isa, N.M., Gramuglia, S., Fukao, T., Bassel, G.W., Correia, C.S., Corbineau, F., Theodoulou, F.L., Bailey-Serres, J. and Holdsworth, M.J. (2011) Homeostatic response to hypoxia is regulated by the N-end rule pathway in plants. *Nature* 479, 415–418.

Godfray, H.C., Beddington, J.R., Crute, I.R., Haddad, L., Lawrence, D., Muir, J.F., Pretty, J., Robinson, S., Thomas, S.M. and Toulmin, C. (2010) Food security: the challenge of feeding 9 billion people. *Science* 327, 812–818.

Gregory, P.J., Johnson, S.N., Newton, A.C. and Ingram, J.S.I. (2009) Integrating pests and pathogens into the climate change/food security debate. *Journal of Experimental Botany* 60, 2827–2838.

Guo, H.W. and Ecker, J.R. (2004) The ethylene signaling pathway: new insights. *Current Opinion in Plant Biology* 7, 40–49.

Gupta, K.J. (2011) Protein S-nitrosylation in plants: photorespiratory metabolism and no signaling. *Science and Signaling* 4, DOI: 10.1126/scisignal.2001404.

Gupta, K.J., Fernie, A.R., Kaiser, W.M. and van Dongen, J.T. (2011a) On the origins of nitric oxide. *Trends in Plant Science* 16, 160–168.

Gupta, K.J., Igamberdiev, A.U., Manjunatha, G., Segu, S., Moran, J.F. and Neelawarne, B. (2011b) The emerging roles of nitric oxide (NO) in plant mitochondria. *Plant Science* 181, 520–526.

Gupta, K.J., Shah, J.K., Brotman, Y., Jahnke, K., Willmitzer, L., Kaiser, W.M., Bauwe, H. and Igamberdiev, A.U. (2012) Inhibition of aconitase by nitric oxide leads to induction of the alternative oxidase and to a shift of metabolism towards biosynthesis of amino acids. *Journal of Experimental Botany* 63, 1773–1784.

Gupta, K.J., Brotman, Y., Segu, S., Zeier, T., Zeier, J., Persijn, S.T., Cristescu, S.M., Harren, F.J.M., Bauwe, H., Fernie, A.R., Kaiser, W.M. and Mur, L.A.J. (2013) The form of nitrogen nutrition affects resistance against *Pseudomonas syringae* pv. *phaseolicola* in tobacco. *Journal of Experimental Botany* 64, 553–568.

He, C., Finlayson, S.A., Drew, M.C., Jordan, W.R. and Morgan, P.W. (1996) Ethylene biosynthesis during aerenchyma formation in roots of maize subjected to mechanical impedance and hypoxia. *Plant Physiology* 112, 1679–1685.

Hebelstrup, K.H., van Zanten, M., Mandon, J., Voesenek, L.A.C.J., Harren, F.J.M., Cristescu, S.M., Moller, I.M. and Mur, L.A.J. (2012) Haemoglobin modulates NO emission and hyponasty under hypoxia-related stress in *Arabidopsis thaliana*. *Journal of Experimental Botany* 63, 5581–5591.

Heckmann, A.B., Hebelstrup, K.H., Larsen, K., Micaelo, N.M. and Jensen, E.O. (2006) A single hemoglobin gene in *Myrica gale* retains both symbiotic and non-symbiotic specificity. *Plant Molecular Biology* 61, 769–779.

Hu, R.G., Sheng, J., Qi, X., Xu, Z.M., Takahashi, T.T. and Varshavsky, A. (2005) The N-end rule pathway as a nitric oxide sensor controlling the levels of multiple regulators. *Nature* 437, 981–986.

Igamberdiev, A.U. and Hill, R.D. (2009) Plant mitochondrial function during anaerobiosis. *Annals of Botany* 103, 259–268.

Jackson, M.B. (2008) Ethylene-promoted elongation: an adaptation to submergence stress. *Annals of Botany* 101, 229–248.

Jackson, M.B. and Ram, P.C. (2003) Physiological and molecular basis of susceptibility and tolerance of rice plants to complete submergence. *Annals of Botany, London* 91, 227–241.

Kozlov, A.V., Staniek, K. and Nohl, H. (1999) Nitrite reductase activity is a novel function of mammalian mitochondria. *FEBS Letters* 454, 127–130.

Lasanthi-Kudahettige, R., Magneschi, L., Loreti, E., Gonzali, S., Licausi, F., Novi, G., Beretta, O., Vitulli, F., Alpi, A. and Perata, P. (2007) Transcript profiling of the anoxic rice coleoptile. *Plant Physiology* 144, 218–231.

Lee, S.C., Mustroph, A., Sasidharan, R., Vashisht, D., Pedersen, O., Oosumi, T., Voesenek, L.A. and Bailey-Serres, J. (2011) Molecular characterization of the submergence response of the *Arabidopsis thaliana* ecotype Columbia. *New Phytologist* 190, 457–471.

Licausi, F., van Dongen, J.T., Giuntoli, B., Novi, G., Santaniello, A., Geigenberger, P. and Perata, P. (2010) HRE1 and HRE2, two hypoxia-inducible ethylene response factors, affect anaerobic responses in *Arabidopsis thaliana*. *Plant Journal* 62, 302–315.

Mase, K., Mizuno, T., Ishihama, N., Fujii, T., Mori, H., Kodama, M. and Yoshioka, H. (2012) Ethylene signaling pathway and MAPK cascades are required for AAL toxin-induced programmed cell death. *Molecular Plant Microbe Interaction* 25, 1015–1025.

Mayer, J.E., Pfeiffer, W.H. and Beyer, P. (2008) Biofortified crops to alleviate micronutrient malnutrition. *Current Opinion in Plant Biology* 11, 166–170.

Mekhedov, S.L. and Kende, H. (1996) Submergence enhances expression of a gene encoding 1-aminocyclopropane-1-carboxylate oxidase in deepwater rice. *Plant and Cell Physiology* 37, 531–537.

Menegus, F., Cattaruzza, L., Chersi, A., Selva, A. and Fronza, G. (1988) Production and organ distribution of succinate in rice seedlings during anoxia. *Physiologia Plantarum* 74, 444–449.

Menegus, F., Cattaruzza, L., Chersi, A. and Fronza, G. (1989) Differences in the anaerobic lactate-succinate production and in the changes of cell sap pH for plants with high and low resistance to anoxia. *Plant Physiology* 90, 29–32.

Mur, L.A., Laarhoven, L.J.J., Harren, F.J.M., Hall, M.A. and Smith, A.R. (2008) Nitric oxide interacts with salicylate to regulate biphasic ethylene production during the hypersensitive response. *Plant Physiology* 148, 1537–1546.

Mur, L.A., Prats, E., Pierre, S., Hall, M.A. and Hebelstrup, K.H. (2013) Integrating nitric oxide into salicylic acid and jasmonic acid/ethylene plant defense pathways. *Frontiers in Plant Science* 4, 215.

Mustroph, A., Lee, S.C., Oosumi, T., Zanetti, M.E., Yang, H.J., Ma, K., Yaghoubi-Masihi, A., Fukao, T. and Bailey-Serres, J. (2010) Cross-kingdom comparison of transcriptomic adjustments to low-oxygen stress highlights conserved and plant-specific responses. *Plant Physiology* 152, 1484–1500.

Nakano, T., Suzuki, K., Fujimura, T. and Shinshi, H. (2006) Genome-wide analysis of the ERF gene family in *Arabidopsis* and rice. *Plant Physiology* 140, 411–432.

Narsai, R., Howell, K.A., Carroll, A., Ivanova, A., Millar, A.H. and Whelan, J. (2009) Defining core metabolic and transcriptomic responses to oxygen availability in rice embryos and young seedlings. *Plant Physiology* 151, 306–322.

Navarre, D.A. and Wolpert, T.J. (1999) Victorin induction of an apoptotic/senescence-like response in oats. *Plant and Cell* 11, 237–249.

Pedersen, O., Rich, S.M. and Colmer, T.D. (2009) Surviving floods: leaf gas films improve O_2 and CO_2 exchange, root aeration, and growth of completely submerged rice. *Plant Journal* 58, 147–156.

Pedersen, O., Malik, A.I. and Colmer, T.D. (2010) Submergence tolerance in *Hordeum marinum*: dissolved CO_2 determines underwater photosynthesis and growth. *Functional Plant Biology* 37, 524–531.

Pena-Castro, J.M., van Zanten, M., Lee, S.C., Patel, M.R., Voesenek, L.A.J.C., Fukao, T. and Bailey-Serres, J. (2011) Expression of rice SUB1A and SUB1C transcription factors in *Arabidopsis* uncovers flowering inhibition as a submergence tolerance mechanism. *Plant Journal* 67, 434–446.

Penfield, S. (2008) Temperature perception and signal transduction in plants. *New Phytologist* 179, 615–628.

Polko, J.K., van Zanten, M., van Rooij, J.A., Maree, A.F., Voesenek, L.A., Peeters, A.J. and Pierik, R. (2012) Ethylene-induced differential petiole growth in *Arabidopsis thaliana* involves local microtubule reorientation and cell expansion. *New Phytologist* 193, 339–348.

Reggiani, R., Cantu, C.A., Brambilla, I. and Bertani, A. (1988) Accumulation and interconversion of amino-acids in rice roots under anoxia. *Plant and Cell Physiology* 29, 981–987.

Rockel, P., Strube, F., Rockel, A., Wildt, J. and Kaiser, W.M. (2002) Regulation of nitric oxide (NO) production by plant nitrate reductase *in vivo* and *in vitro*. *Journal of Experimental Botany* 53, 103–110.

Saika, H., Okamoto, M., Kushiro, T., Shinoda, S., Jikumaru, Y., Fujimoto, M., Arikawa, T., Takahashi, H., Ando, M., Arimura, S., Miyao, A., Hirochika, H., Kamiya, Y., Tsutsumi, N., Nambara, E. and Nakazono, M. (2007) Ethylene promotes submergence-induced expression of OsABA8oxl, a gene that encodes ABA 8'-hydroxylase in rice. *Plant and Cell Physiology* 48, 287–298.

Schaller, G.E. and Kieber, J.J. (2002) Ethylene. *Arabidopsis Book* 1, e0071.

Setter, T.L. (2012) Analysis of constituents for phenotyping drought tolerance in crop improvement. *Frontiers in Plant Physiology* 3, 180.

Smagghe, B.J., Hoy, J.A., Percifield, R., Kundu, S., Hargrove, M.S., Sarath, G., Hilbert, J.L., Watts, R.A., Dennis, E.S., Peacock, W.J., Dewilde, S., Moens, L., Blouin, G.C., Olson, J.S. and Appleby, C.A. (2009) Review: correlations between oxygen affinity and sequence classifications of plant hemoglobins. *Biopolymers* 91, 1083–1096.

Steffens, B. and Sauter, M. (2005) Epidermal cell death in rice is regulated by ethylene, gibberellin, and abscisic acid. *Plant Physiology* 139, 713–721.

Steffens, B. and Sauter, M. (2009) Epidermal cell death in rice is confined to cells with a distinct molecular identity and is mediated by ethylene and H_2O_2 through an auto amplified signal pathway. *PlantCell* 21, 184–196.

Steffens, B., Geske, T. and Sauter, M. (2011) Aerenchyma formation in the rice stem and its promotion by H_2O_2. *New Phytologist* 190, 369–378.

Steffens, B., Steffen-Heins, A. and Sauter, M. (2013) Reactive oxygen species mediate growth and death in submerged plants. *Frontiers in Plant Science* 4, 179.

Stoimenova, M., Igamberdiev, A.U., Gupta, K.J. and Hill, R.D. (2007) Nitrite-driven anaerobic ATP synthesis in barley and rice root mitochondria. *Planta* 226, 465–474.

Sun, T.P. (2011) The molecular mechanism and evolution of the GA-GID1-DELLA signaling module in plants. *Current Biology* 21, R338–R345.

Tischner, R., Planchet, E. and Kaiser, W.M. (2004) Mitochondrial electron transport as a source for nitric oxide in the unicellular green alga *Chlorella sorokiniana*. *FEBS Letters* 576, 151–155.

Tournier, B., Sanchez-Ballesta, M.T., Jones, B., Pesquet, E., Regad, F., Latche, A., Pech, J.C. and Bouzayen, M. (2003) New members of the tomato ERF family show specific expression pattern and diverse DNA-binding capacity to the GCC box element. *FEBS Letters* 550, 149–154.

van Veen, H., Mustroph, A., Barding, G.A., Vergeer-van Eijk, M., Welschen-Evertman, R.A., Pedersen, O., Visser, E.J., Larive, C.K., Pierik, R., Bailey-Serres, J., Voesenek, L.A. and Sasidharan, R. (2013) Two rumex species from contrasting hydrological niches regulate flooding tolerance through distinct mechanisms. *Plant Cell* doi:10.1105/tpc.113.119016

Voesenek, L.A. and Bailey-Serres, J. (2013) Flooding tolerance: O_2 sensing and survival strategies. *Current Opinion in Plant Biology* 16, 647–653.

Voesenek, L.A., Harren, F.J., Bogemann, G.M., Blom, C.W. and Reuss, J. (1990a) Ethylene production and petiole growth in rumex plants induced by soil waterlogging: the application of a continuous flow system and a laser driven intracavity photoacoustic detection system. *Plant Physiology* 94, 1071–1077.

Voesenek, L.A.C.J., Perik, P.J.M., Blom, C.W.P.M. and Sassen, M.M.A. (1990b) Petiole elongation in rumex species during submergence and ethylene exposure - the relative contributions of cell-division and cell expansion. *Journal of Plant Growth Regulation* 9, 13–17.

Voesenek, L., Banga, M., Thier, R.H., Mudde, C.M., Harren, F., Barendse, G. and Blom, C. (1993) Submergence-induced ethylene synthesis, entrapment, and growth in two plant species with contrasting flooding resistances. *Plant Physiology* 103, 783–791.

Vriezen, W.H., Hulzink, R., Mariani, C. and Voesenek, L.A.C.J. (1999) 1-aminocyclopropane-1-carboxylate oxidase activity limits ethylene biosynthesis in *Rumex palustris* during submergence. *Plant Physiology* 121, 189–195.

Vriezen, W.H., De Graaf, B., Mariani, C. and Voesenek, L.A. (2000) Submergence induces expansin gene expression in flooding-tolerant *Rumex palustris* and not in flooding-intolerant *R. acetosa*. *Planta* 210, 956–963.

Wang, K.L., Li, H. and Ecker, J.R. (2002) Ethylene biosynthesis and signaling networks. *Plant Cell* 14, S131–S151.

Wi, S.J., Ji, N.R. and Park, K.Y. (2012) Synergistic biosynthesis of biphasic ethylene and reactive oxygen species in response to hemibiotrophic *Phytophthora parasitica* in tobacco plants. *Plant Physiology* 159, 251–265.

Xu, K., Xu, X., Fukao, T., Canlas, P., Maghirang-Rodriguez, R., Heuer, S., Ismail, A.M., Bailey-Serres, J., Ronald, P.C. and Mackill, D.J. (2006) Sub1A is an ethylene-response-factor-like gene that confers submergence tolerance to rice. *Nature* 442, 705–708.

Zhang, J.L. and Shi, H.Z. (2013) Physiological and molecular mechanisms of plant salt tolerance. *Photosynthetic Research* 115, 1–22.

4 Monitoring the Activation of Jasmonate Biosynthesis Genes for Selection of Chickpea Hybrids Tolerant to Drought Stress

Palmiro Poltronieri,* Marco Taurino, Stefania Bonsegna,
Stefania De Domenico and Angelo Santino
CNR-ISPA, Institute of Sciences of Food Productions, Lecce, Italy

Abstract

It is apparent that climate change will have great impact on the abiotic as well as biotic stresses to which crops will be exposed. The major effects of climate change will be heat and water deficit together with physical damage due to intense rainfall and perhaps associated wind. Since hormonal homoeostasis controlling plant–pathogen interactions is tightly regulated, the influence of abiotic factors may cause dramatic changes in basal plant defences. Dissection of molecular mechanisms which control plant response to different environmental stresses is extremely important for developing crops with improved tolerance. Complex signalling pathways have evolved in plants to cope with different biotic stresses. Complex interactions among these pathways permit a tight control between development and stress response. Among the different defence mechanisms used by plants, oxylipin metabolism is one of the most important. Oxylipin family consists of fatty acid hydroperoxides, hydroxy-, keto- and oxo-fatty acids, volatiles, aldehydes, divinyl ethers and the plant hormone jasmonic acid. Many of these bioactive compounds participate in various physiological processes, defence mechanisms, adaptation to stresses and communication with other organisms. This review aims to provide new insights on the role of the oxylipins-mediated resistance to multiple stresses in legumes. Our previous results pointed to the involvement of jasmonates in the early signalling of water stress in chickpea and their role in the tolerance mechanism of the drought-tolerant variety. Furthermore, the hormonal response to wounding and salt stress of *Medicago truncatula* roots was also monitored in different tissues (roots, stem and leaves) at different time points from stress onset.

4.1 Introduction

Plants employ several signals to communicate and respond to the various stresses to which they are subjected, including wounding and herbivore attack. Among these, the oxylipin family of signals, comprising a large group of chemicals including fatty acid hydroperoxides, hydroxy-, keto- or oxo-fatty acids, several volatiles and the hormone jasmonic acid (JA), is one of the more important ones. Most of the compounds in this group are volatiles that participate in physiological response, in defence, adaptation to stresses, in communication among plants and microorganisms signalling. Preliminary studies on

---

*E-mail: palmiro.poltronieri@ispa.cnr.it

responses of chickpea varieties which differed in their tolerance to salinity and drought showed that isoforms of HPL, AOS and LOX were involved in tolerance to both stresses. Later studies where abscisic acid (ABA), jasmonates and OPDA contents as well as gene expression comparisons between drought tolerant and responsive varieties were made, further confirmed the earlier results.

There are various levels of regulation of jasmonate signalling and its biosynthesis pathway in roots and nodules of chickpea varieties subjected to salt stress; an additional level of regulation imposed by epigenetics and microRNAs, which in turn involve ABA and nitric oxide (NO) responsive elements in promoters of transcription factor genes.

In this review we present the details of involvement of JA biosynthesis and activation during salinity and water stress in roots of chickpea varieties which show either susceptibility or tolerance responses. An essential trait conferring high tolerance in legume varieties under water stress has been shown to be induction of high levels of JA in the very early stages. Real-time PCR is highly suitable for evaluating the time course expression of specific lipoxygenase (LOX) isoforms in tolerant varieties. RT-PCR could support breeding programmes for the identification of hybrids with improved JA synthesis, able to activate oxylipin specific pathways in an earlier, sustained and prolonged timing during stress perception.

4.2 The Jasmonate Biosynthesis Pathway and Jasmonic Acid Signal Transduction

Most of the compounds included in the oxylipin family are volatile and participate in different defence responses of plants, adaptation to stresses, as well as in communication among plants and microorganisms signalling.

The oxygenation of polyunsaturated fatty acids (PUFAs) gives rise to a variety of oxylipins, such as fatty acid hydroperoxides, hydroxy-, keto- or oxo-fatty acids, aldehydes, divinyl ethers, green leaves volatiles (a series of chemicals belonging to the volatile organic compounds) and to JA. These bioactive compounds participate in defence mechanisms (sensing herbivores, insects and pathogens), environmental stress adaptation and in communication with other organisms (Feussner and Wasternack, 2002; Wasternack, 2007; Hughes et al., 2009; Schaller and Stintzi, 2009; Santino et al., 2013).

The synthesis that leads to JA occurs in a sequential manner; the first steps occur within plastids and the last steps within peroxisomes. Peroxisomes are cell organelles that are ubiquitously present in plants, fungi, yeasts and animals, but their importance is underestimated. However, several novel peroxisome functions have been identified recently which are related to resistance towards various stresses and this has revealed yet unknown mechanisms that allow plants to adapt to adverse environmental conditions. Novel enzyme activities, metabolic pathways and unexpected non-metabolic peroxisome functions have been recently found, such as production of secondary metabolites. For instance, glutathione reductase as well as other proteins have been shown to be specific to peroxisome variants from abiotically stressed plants (Kataya and Reumann, 2010), with a role for glutathione as a major antioxidant (Reumann, 2011).

In the synthesis of JA, divinyl ethers and volatile aldehydes, linolenic acid (18:3) is one of the PUFA substrates used by lipoxygenases. A cytosolic 9-LOX produces 9(S)-hydroperoxy fatty acids, while a plastidial 13-LOX produces 13(S)-hydroperoxy fatty acids. In chloroplasts, in addition to 13-LOX, allene oxide synthase (13-AOS) and allene oxide cyclase (AOC) act sequentially to produce 12-oxophytodienoic acid (OPDA) or dinor-OPDA (Fig. 4.1).

The next step in JA synthesis is the import of OPDA into peroxisomes. OPDA is then reduced by 12-oxophytodienoate reductase 3 (OPR3) to 3-oxo-2(2'-pentenyl)-cyclopentane-1-octanoic acid, which undergoes three cycles of beta-oxidation through an acyl CoA oxidase (ACX), that produces OPC:6, processed by a multifunctional protein (MPF) involved in the synthesis of OPC:4CoA, and by the ketoacyl-CoA thiolase (KAT2) that produces JA-CoA and finally JA.

JA, in the presence of a JA-methyltransferase, can be methylated to form the volatile

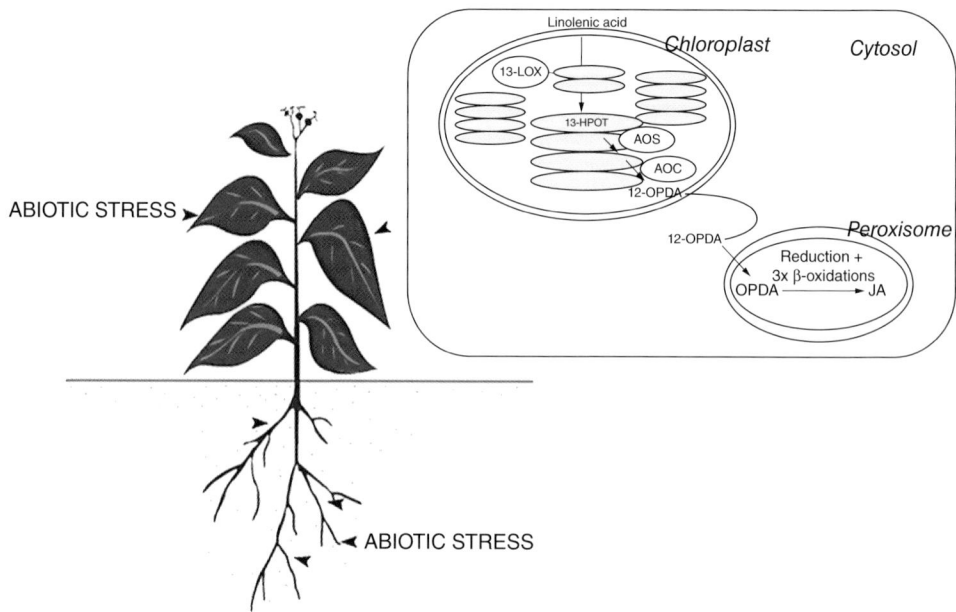

Fig. 4.1. The JA biosynthesis pathway requires the involvement of plastidial and peroxisomal enzymes. OPDA is synthesized inside the chloroplast through the activity of 13-LOX, AOS and AOC, then it moves into peroxisomes, where it undergoes three cycles of beta-oxidation.

compound methyl-jasmonate (Me-JA), freely diffusing across biological membranes and acting at short distances. When JA is converted to 12-hydroxy-JA (12-OH-JA) and 12-hydroxy-JA sulfated forms, its bioactivity is reduced, limiting the inhibition of root growth (Galis et al., 2009).

JA is modified by JAR, JA-amino acid synthetase, to form jasmonoyl derivatives (JA-Ile, JA-Val, JA-Leu) that are stored in organelles and vacuoles. JA-Ile is freely mobile, diffusing through the xylem to roots and to leaves (Koo et al., 2009). JA-Ile is the active hormone derivative responsible for JA activity mediated by JA receptors (Pauwels and Goossens, 2011).

Coronatine, a compound synthesized by *Pseudomonas syringae*, is a JA-Ile mimic that affects the regulation of plant defence responses (Geng et al., 2012). Coronatine insensitive 1 (COI1) has been identified as the receptor for JA-Ile in a study of mutants of the ubiquitin proteasome components (Tiryaki and Staswick, 2002; Lorenzo and Solano, 2005). JA-Ile response is further regulated by nuclear proteins called JASMONATE-ZIM-DOMAIN (JAZ) repressors that bind with a protein partner,

COI1, an F-box protein participating in the SCF (Skp-Cullin-F-box) ubiquitin ligase complex (Pauwels and Goossens, 2009). AtMYC2 is sequestered by JAZ, until JA-Ile binds to COI1. When JA-Ile binds to COI, COI promotes the ubiquitinylation of JAZ proteins, thereby freeing AtMYC2 from repression. Then MYC2, by binding to G-box regions, activates the promoters of JA-regulated genes (Gfeller et al., 2010). The over-expression of the glucosyltransferase UGT76B1 has shown to enhance the JA response, and to delay senescence. UGT76B1, by conjugating glucose to isoleucic acid, directly affects the JA pathway (Schäffner, 2011).

Two oxylipin branches diverge from the main JA synthesis pathway. In the first pathway, hydroperoxides are transformed by divinyl ether synthases (DES) into divinyl ethers. In the second branch, short-lived haemiacetals are produced by hydroperoxide lyases (HPL), that give rise to aldehydes and n-fatty acids (n = 6, 9) (Hughes et al., 2009). These reactive oxylipins are formed during different environmental stresses (Mueller and Berger, 2009) (Fig. 4.2).

In the jasmonate branch, allene oxide cyclase (AOC) is an enzyme that is active in an oligomer form, such as homodimer and heterodimer (Stenzel et al., 2012). A role of different AOC enzymes has been elucidated, showing overlapping and independent functions. Studies on *Arabidopsis thaliana* have revealed a central role of AOC oligomerization in JA synthesis. AOC promoter activities were shown to be correlated with induction of jasmonate-responsive genes in different tissues. In addition, interactions between jasmonates and auxin hormones have been proposed during root-growth regulation.

When the plant senses a pathogen, JA is involved in regulation of gene subsets inducing necrosis, blocking the spreading of microorganisms. In the response against necrotrophic pathogens, there are synergies between the jasmonate and ethylene (ET) signalling pathways. A GCCGCC motif present in promoters is activated either by JA or ET (Memelink, 2009). The induced genes include the *AP2/ERF* family genes (i.e. *AP2*, *ERF* and *DREB*) and other transcription factors regulated by these two hormones (Zarei et al., 2011).

A second type of JA-responsive transcription factors, such as MYC, bind to the G-box sequence in promoters activated only by JA, while repressed by ET (Gfeller et al., 2010). In the *jasmonate-resistant 1* (*jar1*) mutants insensitive to JA, plants are not able to activate a defence response to the necrotrophic fungus *Botrytis cinerea*. *Botrytis* infection causes the synthesis of JA and the expression of *Botrytis* Susceptible 1 (BOS1), a MYB TF involved in ROS production, that mediates either biotic or abiotic stress response (Fujita et al., 2006).

Wounding is a stress involving the JA biosynthesis pathway and JA signalling (Koo et al., 2009). The wounding-induced JA synthesis was preceded by NO production in *Arabidopsis* (Huang et al., 2004). Exogenous NO supply was shown to induce three genes, *lipoxygenase* (*LOX2*), *allene oxide synthase* (*AOS*) and *OPDA Reductase* (*OPR3*) (Huang et al., 2004). However, NO produced also an increase in salicylic acid (SA) that blocks JA production, since in transgenic *NahG* plants (unable to accumulate SA and/or signalling), NO did increase JA production.

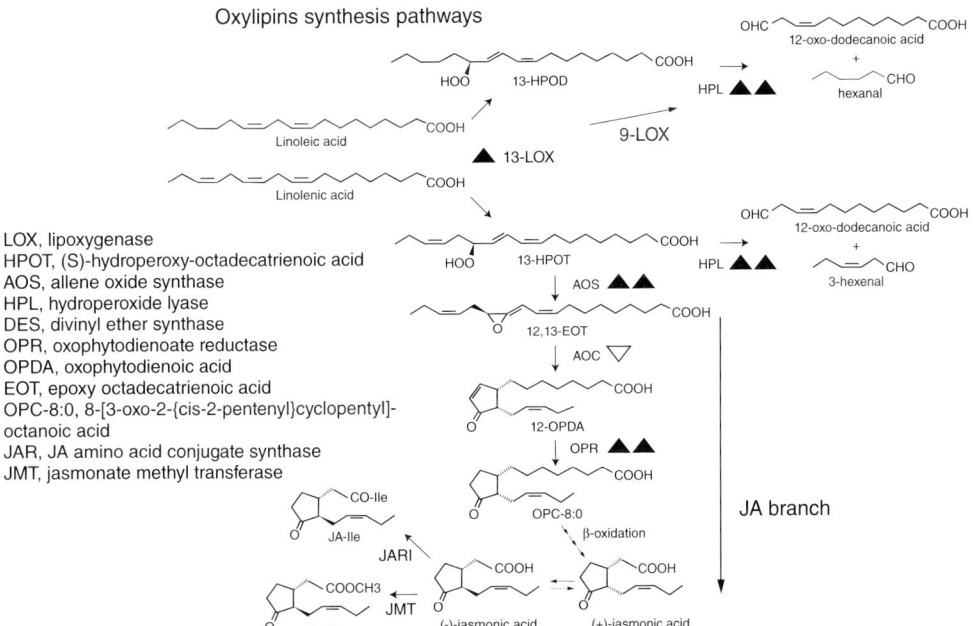

Fig. 4.2. Scheme of the oxylipins synthesis pathways. Several enzymes giving rise to JA, 12-oxododecanoic acid and hexanal have been found up-regulated during abiotic stress response.

4.3 Plant Roots, Hormone Crosstalk and Involvement in Stress Response

Plant growth is based on a well-developed root system, which is essential for water and mineral uptake. Roots are the first organ sensing changes in the soil, being thus able to signal to the plant and to trigger a response to environmental changes. Soil modifications affect root growth, development of lateral roots, resource acquisition and root-to-shoot communication (Seki et al., 2007; Schachtman and Goodger, 2008). Drought and salt stresses elicit root response and production of early signals that are transduced at distance. Thus, plants activate protection mechanisms, such as slowing down growth and resource acquisition, activating osmoprotectants synthesis, water potential preservation and stomatal closure.

There are several signalling compounds (RNAs, lipids, PGPs and peptide factors) involved in root–shoot communication (Seki et al., 2007; Goodger and Schachtman, 2010). Root-produced JA and Me-JA are important in stress response both in plants without symbiotic microorganisms as well as in plant–bacteria symbioses. The involvement of oxylipins in root growth has been recently shown (Vellosillo et al., 2007). The 9-hydroperoxy-derivative of linolenic acid (9-HPOT) produced by 9-LOXs specific to lateral root primordia was found important in lateral root growth in *Arabidopsis*. 9-HPOT was found able to modulate root development through cell wall modification (stimulating callose and pectin deposition) and ROS accumulation. In *Medicago truncatula*, a 9/13-HPL is expressed in *Rhizobium meliloti*-inoculated roots and nodules, mediating the interaction of microorganisms with the plant roots (Mita et al., 2007; Hughes et al., 2009).

MeJA was found at high levels in root tips during soybean germination (Hause and Schaarschmidt, 2009; Oldroyd, 2009). Although the principal JA-derivative functioning in cell-to-cell signalling is JA-Ile, MeJA may function as a mobile molecule that allows rapid storage of JA-compounds in cells surrounding a site of stress sensing. JA, in its methylated form (Me-JA), is involved in the growth of lateral roots (Hsu et al., 2013).

Symbiotic microorganisms improve positively the stress response of plant roots. Endophytic fungi inside plant roots and rhizosphere fungi near plant roots can benefit plants in various ways, including through an improved nutrient supply, protection against pathogens or high temperature and production of phytohormones that may benefit the plant. Plant-growth-promoting (PGP) endophytic bacteria and fungi have the ability to increase root biomass, mitigate salt effects such as heat efflux, modify fatty acid composition, potentiate antioxidant enzymes, maintaining ascorbate in its reduced formed during salt stress (Baltrushat et al., 2008), and improve plant growth synthesizing phytohormones such as 2,3-butanediol, acetoin and indole acetic acid (Taghavi et al., 2010).

Jasmonates induce rhizobium bacteria to express the *nod* gene, and support Nod factor expression through the induction of (iso)flavonoids (Zhang et al., 2007). On its side, Nod factor affects Ca_2^+ spikes in root hairs and inhibits JA through negative feedback (Oldroyd, 2009).

ABA regulates negatively root nodule formation in legumes. Furthermore, NO is involved in nodule formation and function, therefore these two signals may synergize or antagonize depending on specific cases. A *Lotus japonicus enf1* (*enhanced nitrogen fixation 1*) mutant, with increased root nodule number and nitrate synthesis, showed lower ABA sensitivity and also lower nodule NO levels in respect to wild-type roots. Thus, endogenous ABA may control nodulation levels and N_2 fixation by decreasing the nodule synthesis of NO (Hancock et al., 2011).

Redistribution of nutrients and its control in arbuscular mycorrhizal roots is also mediated by jasmonates. In *M. truncatula* and barley, in which a mutualistic symbiosis promotes plant growth, regulation of nutrient exchange between roots and bacteria shows the involvement of JA.

The importance of NO in growth of primary roots (Fernández-Marcos et al., 2012) and development of lateral roots in tomato (Correa-Aragunde et al., 2004) has been established. Involvement of nitrate reductase (NR) in root NO production during osmotic stress was demonstrated in *A. thaliana* (Kolbert *et al.*, 2010). NO signalling has effects on genes and proteins involved in oxylipins synthesis, and supporting information and possible mechanisms will be discussed in the review.

Plants are being continually exposed to NO from the bacteria surrounding the roots. NO synthesis occurs during the oxido-reductive steps ranging from NH_4^+ to NO_3^- that form the nitrogen cycle. Various factors also influence NO production in soil, which include high temperature, oxygen availability, humidity, soil pH and nitrogen status. These factors affect nitrifying and denitrifying bacteria, which can produce NO at differing rates depending upon the conditions. It is well known that during nitrogen metabolism, bacteria assimilate nitrate and reduce it to nitrite (NO_2^-) through a two-electron reduction reaction. Nitrite can be reduced to NO, which has a potential cytotoxic effect, and hence, the accumulation of cellular nitrite can be harmful. Nitrite is removed from the cell by channels and transporters, or reduced to ammonium or N_2 by the activity of assimilatory enzymes. NR and NOS-oxy in bacteria and rhizobia are involved in the production of NO and signalling between bacteria and roots, and have a role in abiotic stress sensing in nodules.

In legumes, leghaemoglobins (lHbs) are found in symbiotic bacteria organelles. Non-symbiotic Hbs are expressed in specific plant tissues, and over-expressed in stressed tissues. These proteins may function as additional O_2 transporters and in buffering of NO, that can be released at later times.

Specific events are triggered locally where stress is perceived. In roots, early timing of JA production may be important in tissue-specific and systemic response to environmental stresses. The activation of genes involved in JA synthesis in roots during drought stress has been demonstrated in chickpea through the identification of LOX and AOC alternatively spliced transcripts specific for the stress-tolerant varieties (Molina *et al.*, 2008, 2011). The expression of JA synthesis genes was found positively related to increased synthesis of JA intermediates, JA and JA-Ile hormones not only in chickpea (De Domenico *et al.*, 2012), but also in tomato and in *Arabidopsis* (Abdala *et al.*, 2003).

4.4 Abiotic Stress Response in Drought and Salt Stresses: Role of Jasmonates

Abiotic stresses affect crop yield and cause yield losses in all major crops up to the extent of 50% or more. Mechanisms of susceptibility and tolerance to these stresses are conferred by complex traits. Different distinct mechanisms are involved in conferring protection to stresses. The traits associated with resistance mechanisms are dependent on signals that often depend on genes shared by different stresses.

Drought is one of the most significant factors that affects crop production. Thus, not only the improvement in water availability but also the need of plant varieties with improved drought stress-tolerance will be the focus of future breeding strategies. It is expected that in the near future environment and climate will be more and more variable. This will require new cultivars with high resilience that are making good use of favourable conditions while withstanding drought, cold or heat peaks.

Molecular tools and genomic studies of responses to abiotic stresses (drought, salinity, cold) in *Arabidopsis* and *Medicago* plants showed how these are characterized by ionic- and osmotic-disequilibrium components, producing specific signalling and stress protection responses (Xiong *et al.*, 2002). These studies showed how early responses are important in plant survival (Shinozaki and Yamaguchi-Shinozaki, 2007).

Drought stress-dependent physiological and biochemical changes in plants include stomata closure, reduction in water evaporation, growth containment and photosynthesis reduction. Many drought-inducible genes have been studied and classified into two major groups: proteins that function directly in abiotic stress-induced response (such as osmoprotectants synthesis); and regulatory proteins involved in signal transduction, and activation of stress-responsive genes. During drought stress, abiotic stress signalling components are up-regulated. The mechanisms underlying plant responses to salinity and drought are highly similar, suggesting that both stresses are sensed by plant cells as water deprivation (Jakab *et al.*, 2005). High salt (NaCl) soil concentrations cause a reduction in water potential, which in turn leads to hyperosmotic and oxidative stress (Borsani et al., 2001). The accumulation of excess NaCl in the apoplast produces an imbalance in nutrients and in solutes (Serrano et al., 1999; Hasegawa et al., 2000).

Plant responses to dehydration, drought and salinity include ionic and osmotic adjustments that trigger signal transduction pathways resulting in the activation of effector signals to adapt the plant and its metabolism. The first signal, as established in *Arabidopsis* and rice, involves stress perception through G-protein-coupled receptors (GPCR), inositol phosphates that regulate the interaction between JAZ-MYC2, and through receptor-like kinases (RLKs).

Plants activate several defence mechanisms that support survival during a number of harmful environmental conditions. In plants, hormones such as ABA, SA, JA and ET are important players involved in their response to environmental stresses and in plant–microbe interactions, positively or antagonistically influencing several families of transcription factors (Fujita et al., 2006). For instance, dehydration-responsive NAC transcription factors, such as RD26 and RD22, are induced by JA, hydrogen peroxide, pathogens, drought, salinity and ABA.

Involvement of ABA not only in several physiological states, such as senescence, seed dormancy and plant development, but also in the signalling of alarm for the occurrence of various stresses is well documented. Stomatal closure in the aerial parts of the plant is regulated by ABA as well as the activity of shoot meristems. ABA accumulates during drought tolerance and determines a reduction in ET synthesis as well as an inhibition of ET-dependent senescence and abscission. ABA can move in the cortex of roots crossing the apoplastic barriers into xylem as ABA glucose ester (ABA-GE), that is stored in microsomes and released by mesophyll cells' glucosidases. A beta-glucosidase gene was found up-regulated in water stress in roots (Schachtman and Goodger, 2008). At the initial stages of water stress, the amount of ABA-GE stored in roots is too low to produce the high ABA increase observed during water stress. Sulfate, mobilized by the action of an early-over-expressed root sulfate transporter, acts as a long distance signal moving through the sap, to induce ABA biosynthesis in leaves. ABA then is transported to roots via phloem where it induces water uptake from soil and expression of stress-resistant genes. Subsequently, ABA is cycled back to leaves via xylem to close the stomata and reduce the transpiration rate. The co-stimulation with ABA, ET, nitric oxide and sulfate produces an additive increase in stomata closure, reinforcing the block of transpiration, for an extended period of drought persistence.

Cytokinin is a plant hormone involved in regulation of growth and development, with an influence also on roots. Other diverse activities of this hormone have also been elucidated, which include crosstalk with other plant hormones as well as environmental stimuli. AP2/ERF transcription factors were identified in particular as responsive to cytokinin, and these are involved in translational control of changes induced by cytokinin. They also stimulate cell proliferation and elongation, and counter-fight senescence signals. Increase in endogenous levels of cytokinin is achieved by over-expressing the *ipt* gene involved in its biosynthesis. It is a stress adaptation which is supported by the delay of drought-induced senescence (Bhargava and Sawant, 2013). Cytokinins negatively regulate root growth and branching. In *Arabidopsis* roots, the degradation of cytokinin was correlated with increased primary root growth as well as branching during drought, thus supporting drought tolerance. Gibberellic acid (GA) promotes growth through the degradation of growth-repressing DELLA proteins in nuclei in *A. thaliana*. The major effect of DELLA is the repression of GA responses. The DELLA family of proteins displays either distinct or overlapping functions. JA-Ile induces DELLA RGA-LIKE3 (RGL3) expression through MYC2 by binding to COI1. Subsequently, RGL3 contributes to enhancing jasmonate (JA)-mediated signals (Wild et al., 2012).

4.5 Chickpea Root Response to Abiotic Stresses

In chickpea (*Cicer arietinum*) there are two principal types of varieties, *desi* types with small-seed and *kabuli* type with large seed. Recently, the chickpea (CDC Frontier, a *kabuli* variety) genome has been sequenced (Varshney *et al.*, 2013). Twenty-nine elite varieties of both *desi* and *kabuli* genotypes were studied and genotyped, by sequencing of 61 *Cicer* accessions from ten countries. The scientists found the presence of admixed genotypes, due to mixed use of *desi* and *kabuli* genotypes in the breeding programmes. The analysis of genome regions containing 122 genes that has potential to be used in selection in modern breeding programmes included a set of 54 genes on chromosome 3 containing the flowering time *CONSTANS* homologue gene. A functional flowering time quantitative trait locus (QTL) was roughly mapped to the same location on Ca3. Selection of varieties and inbred lines is very important in adapting chickpea varieties to different regions.

In previous studies, enhanced expression of the major genes of the jasmonate pathway in root tissues of different plant species under different physiological conditions has been shown, while in some cases transcript increase was confirmed by observed higher levels of JA, JA-Ile and OPDA (De Domenico *et al.*, 2012).

Available data on the involvement of early JA synthesis in chickpea varieties responding positively to abiotic stresses have been obtained studying various varieties: drought-tolerant varieties ILC588 (Molina *et al.*, 2008) and ICC4958 (De Domenico *et al.*, 2012), and drought-sensitive varieties Annigeri (Molina *et al.*, 2008) and ICC1882 (De Domenico *et al.*, 2012), salt stress-tolerant INRAT-93, weakly tolerant ICC6098 variety, and salt partially sensitive varieties Amdoun and ICC4958 (Molina *et al.*, 2011).

Resistance to drought observed in ICC4958 has been shown to be associated with its root system, which is both longer and larger in volume than that of non-tolerant varieties such as Annigeri or ICC1882, while seed mass accumulation, after flowering starts, is faster in ICC4958. This trait permits ICC4958 to accumulate a large seed mass before the soil moisture recedes and drought becomes increasingly severe (ICRISAT, 1992) (Fig. 4.3).

Identification of transcripts through alternative splicing has revealed mechanisms

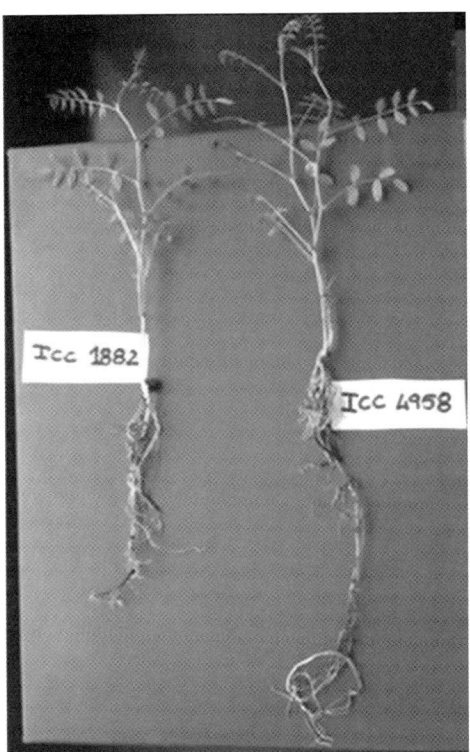

Fig. 4.3. Root system in two chickpea varieties, grown in pots in the same conditions. ICC1882 (drought sensitive; left) and ICC4958 (drought tolerant; right).

of jasmonate biosynthetic pathways in chickpea roots, during drought and salinity (Molina *et al.*, 2008, 2011; De Domenico *et al.*, 2012). Confirmation of the increased enzyme activities in drought-tolerant chickpea varieties has been established through quantification of stress metabolites and hormones (De Domenico *et al.*, 2012). A considerable up-regulation of transcripts of JA biosynthesis gene isoforms was concordant with higher levels of JA, JA-Ile and OPDA measured in roots of tolerant varieties (De Domenico *et al.*, 2012).

SuperSAGE methods have been applied to analyse the drought response in a chickpea variety, ILC588, tolerant to this stress (Molina *et al.*, 2008). The plantlets were left to grow for 28 days, after which they were subjected to dehydration for a period of 6 h. After desiccation there was a loss of turgor in the plants. The roots were excised and immediately frozen using liquid nitrogen. Transcripts were quantified and assigned to specific genes and ontology groups. It was observed that under drought conditions 20 LOX isoforms and splicing variants were identified, corresponding to 11 SNP associated alternative tags (SAAT). Two LOX sequences were highly regulated both during drought (ILC588) and salt stress (INRAT-93), of which STCa-24417 was 25-fold up-regulated (Molina *et al.*, 2011) (Fig. 4.4).

Allene oxide cyclase was found present as five UniTags, varying in expression, from down-regulation to 20-fold up-regulation. This finding that specific isoforms of AOC are also involved in stress response supports the presumption that AOC oligomers/heterodimers produce an increased synthesis of JA (Stenzel *et al.*, 2012). Taqman probes for specific isoforms of several genes in the JA synthesis pathway were designed based on differently spliced isoforms and SAAT sequences, selective enough to discriminate different LOX, AOC and HLP isoforms and spliced variants and these probes have been used to confirm the SuperSAGE studies measured transcripts induced in the roots of drought-tolerant (ILC588), in roots and nodules of salinity-tolerant (INRAT-93) and salt-sensitive varieties (Molina *et al.*, 2011). This study monitored the root and nodule responses to salt stress (at 2 h, 8 h, 24 h and 72 h time intervals), using the salt-tolerant chickpea INRAT-93, the salt-sensitive Amdoun control, the ICC4958 salt-sensitive variety and in the ICC6098 weakly tolerant variety. In this last study, inoculation of seedlings (root length >5 cm) with *Mesorhizobium ciceri* was done and after 3 weeks the plants were transferred to a 5 mM NaCl medium. qRT-PCR assays showed the same results produced by deep SuperSAGE on differential expression of LOX and AOC UniTags. The transcripts over-expressed during salt stress in nodules included several *LOX* and *AOC* isoforms activated in stress-tolerant varieties. The results also showed a higher involvement of ROS scavenging enzymes and signals in the tolerant varieties.

In a French study, using 16K+ microarrays (Mt16KOLI1), the transcripts in salt-treated root apexes were identified in the model legume *M. truncatula* (Gruber *et al.*, 2009; Zahaf *et al.*, 2012) comparing the salt-tolerant TN1.11 variety and the reference Jemalong A17 genotype. The hormonal response of *M. truncatula* roots to salt stress was studied in different tissues (roots, stem and leaves) at different time points from stress onset. Four key genes involved in the oxylipins metabolism, namely *lipoxygenase* (*LOX*), *hydroperoxide lyase* (*HPL*), *allene oxide synthase* (*AOS*) and *allene oxide cyclase* (*AOC*) were up-regulated in the salt-tolerant

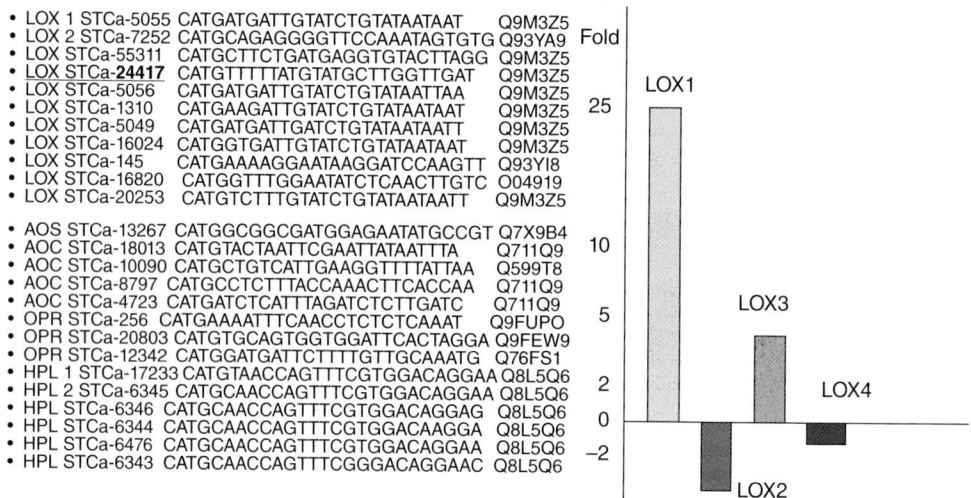

Fig. 4.4. SuperSAGE tags corresponding to different *LOX* splice variants and isoforms.

genotype, under salt stress condition. Comparison of transcription profiles from desiccated young roots using the *Medicago* 16k-microarray (Buitink *et al.*, 2006) with transcripts regulated in drought-tolerant chickpea roots showed differences in the drought response in tolerant varieties in the two species (Molina *et al.*, 2008).

Using the chickpea-specific Taqman probes, we conducted studies on the drought response in the ICC4958 variety, using LOX1, LOX2, AOS, AOC, HLP1, HPL2 and OPR primers. The chickpea drought-tolerant ICC4958 variety and a drought-sensitive variety, ICC1882, were cultivated in pots, then subjected to water stress, maintaining them under the same condition for 72 h (De Domenico *et al.*, 2012).

Early timing and high levels of JA synthesis gene(s) expression in ICC4958 (drought-tolerant) was confirmed by qRT-PCR studies on individual roots for determining the expression of key genes and specific isoforms which were involved in metabolism of oxylipins (De Domenico *et al.*, 2012). AOS and HPL were found rapidly (as soon as 2 h after the onset of stress) and highly (up to 19-fold) induced by drought chickpea variety ICC4958, which was tolerant. This result indicates the involvement of the jasmonate pathway that was more strongly activated and at an earlier timing in drought-tolerant chickpea roots. Results also revealed a sustained activation of lipoxygenase (*lox*1) isoform, which was root-specific, two hydroperoxide lyases (*hpl*1 and *hpl*2), an allene oxide synthase (*aos*) as well as an oxo-phytodienoate reductase (*opr*) gene in the tolerant variety.

Thus, the deepSuperSAGE results (Molina *et al.*, 2011) on the *LOX* 1 and *AOC* transcript up-regulation in ILC588 was demonstrated to be important both in salt and drought stress, indicating its role in stress tolerance mechanisms.

Expression of *LOX* and *AOC* transcripts was observed to vary according to each specific isoform even for 25-fold or higher (Molina *et al.*, 2011). *LOX* 1 transcript was mostly up-regulated in salt-stress in the salt-sensitive Amdoun1 and weakly tolerant variety ICC 6098 but not in the salt-tolerant ones. *AOC* transcripts, on the other hand, were strongly induced early in INRAT-93, which was fairly salt-tolerant.

Confirmation of the increased enzyme activities was obtained through quantification of metabolites and hormones in roots of stressed chickpea varieties. A considerable up-regulation of transcripts of JA biosynthesis gene isoforms was concordant with enhanced levels of JA, JA-Ile and OPDA measured in roots of tolerant varieties (De Domenico *et al.*, 2012). The rapid rise of OPDA and JA-Ile levels concomitant to the induction of *AOS* and *OPR* gene expression in drought-stressed roots in ICC4958 suggests that there may be a coordinate action of JA-Ile and OPDA for the full root response activation to stress in the drought-tolerant ICC4958 variety (Fig. 4.5) (De Domenico *et al.*, 2012). The JA (JA-Ile, JA, OPDA) peaks at 2 h, very early, while ABA starts to accumulate after 24 h.

ABA was shown to increase during drought irrespective of variety. However, the ABA content in drought-stressed roots was 20% higher in the tolerant variety ICC4958. ABA concentration showed a sharp increase within 24 h, after which ABA content remained constant in the tolerant variety, whereas it decreased in the susceptible one. After 72 h from stress onset, ABA levels were about 37% higher in ICC4958 than in ICC1882.

4.6 Nitric Oxide Regulation and Epigenetic Control of Jasmonic Acid Signalling

The intracellular synthesis and containment of JA intermediates occurs in specific and tightly localized reactions, to allow for spatially and timely regulated signalling events.

The experiments (Molina *et al.*, 2011) on salt response in the tolerant chickpea INRAT-93, the salt-sensitive Amdoun control, the ICC4958 salt-sensitive variety and the ICC6098 weakly tolerant variety, were performed with chickpea roots inoculated with chickpea-specific rhizobia.

The over-expression of specific isoforms linked to JA synthesis in nodules and in root apexes (Molina *et al.*, 2011) was higher than in roots, possibly due to nodule-localized activities and involvement of NO in the up-regulation of JA biosynthesis genes. Thus, it is quite probable that a large involvement of bacteria

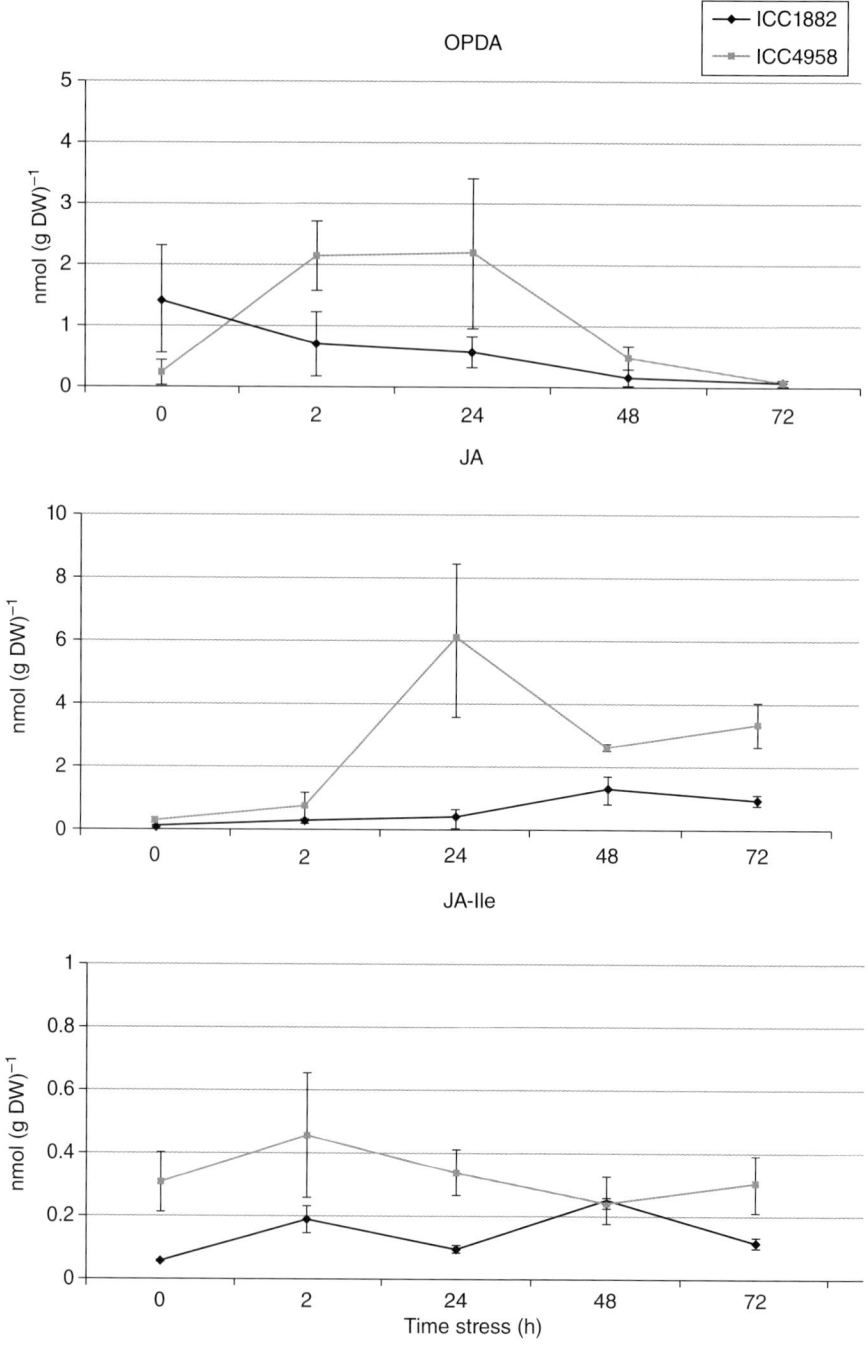

Fig. 4.5. OPDA, JA and JA-Ile concentrations were quantified by HPLC analysis at different timings in chickpea roots of ICC4958 (light line) and ICC1882 (black line) varieties following drought stress.

in stress signalling exists, with NO production, and NO amplification of JA synthesis through specific promoter activation and S-nitrosylation of enzymes and transcription factors. Accordingly, NO–lHb complexes were found associated to radicals production in soybean and *Medicago* nodules (Del Giudice *et al.*, 2011).

NO-responsive promoters were identified bioinformatically, and showed that salicylate- and jasmonate-responsive *cis*-elements were prominent (Palmieri *et al.*, 2008). Allene oxide cyclase (AOC) has been found S-nitrosylated by NO in a cysteine proximal to the catalytic site during the hypersensitive response (HR) (Romero-Puertas *et al.*, 2008; Wang *et al.*, 2009). Nitrosylation may control AOC enzyme activity or AOC oligomerization, described to be important for JA synthesis (Stenzel *et al.*, 2012) with the requirement of specific isoforms to form heterodimers.

In plants, NO-mediated nitrosylation activates transcription factors such as MYB, involved in JA-dependent signalling. SABP3, modulating the SA response and integrating the JA signalling, was nitrosylated by NO during the HR (Wang *et al.*, 2009). It is thus plausible to hypothesize that NO provides a S-nitrosylation control of the R2R3-MYB class of transcription factors (Serpa *et al.*, 2007), inhibiting DNA binding of MYB TFs. Nitrosylation of cysteines in enzymes of the SA/JA synthesis was shown to be involved in JA production and signalling (Stenzel *et al.*, 2012). It was proposed that NO through S-nitrosylation of R2R3-MYB transcription factors controls the JA responses during abiotic stress. Thus, NO- regulated transcription modulates the jasmonate signalling pathway during different abiotic stresses.

Treatment of *Arabidopsis* plants with NO induced key genes of JA biosynthesis such as *AOS* and *LOX* (Huang *et al.*, 2004). However, NO induction of JA-biosynthesis genes did not result in elevated levels of JA in *Arabidopsis* plants subjected to biotic stress (Huang *et al.*, 2004). JA-responsive genes such as *defensin* (PDF1.2) were not induced during biotic stress in this plant, probably indicating that it may be dependent on the type of stress, so that expressed genes may not be paralleled by activation of MYB transcription factors.

Methyl jasmonate (MeJA) stimulated ABA production in rice (Kim *et al.*, 2009). The overexpression of JA carboxyl methyl-transferase (JMT) produced high levels of ABA (Kim *et al.*, 2009). In that study, the drought stress induced the plants to produce MeJA, which in turn stimulated ABA production.

ABA and NO cooperate in many physiological responses including stomatal closure, root formation and seed dormancy. ABA and NO signalling pathways often involve ROS, and have interactions with other hormones and signalling molecules (Hancock *et al.*, 2011).

JA- and ABA-mediated signals during multiple abiotic stresses have yet to be studied. An important crossroad between the signalling of ABA and JA is represented by the *NAC* transcription factors (TF), formed by ATAF, NAM and CUC TFs (Santino *et al.*, 2013). ATAF2 has a role in wounding response, during salinity stress and after JA treatment, while ATAF1 has a role in wounding response, during dehydration and after ABA supplementation. ABA induces plant growth inhibition when the plant is subjected to different abiotic stresses. It is believed that ABA regulates specific microRNAs (possessing an ABA responsive element in their promoters) involved in degradation of transcription factors and hormone signalling elements.

Long-distance signalling is fundamental in plants for the regulation of processes such as leaf development, flowering and pathogen defence. Small RNAs, among them several microRNAs (miRNAs), have been found in various plants. As a prototype of mobile signals, miR399 is a phloem-mobile long distance miRNA (Franco-Zorrilla *et al.*, 2007) responding to phosphate deficiency, moving from leaves to roots via phloem, and targeting PHO2/UBC24, an E2 ubiquitin ligase, thus freeing MYB/PHR1 in the roots.

Several findings have established a fundamental role of miRNAs in response to abiotic stresses in plants and nutrient deprivation (Khraiwesh *et al.*, 2012). Several miRNAs involved in plant growth and development are differentially expressed during stress. These findings imply a control of stress-responsive miRNA on plant growth inhibition and developmental block under stress that is strictly related to hormone signalling.

Several miRNAs target TFs with a role in environmental and hormone responses, such as: miR-159/miR-319 and MYB33, MYB101, TCPs; miR-166 and HD-ZIP TFs; mir-172 and AP2 transcription factors; miR395 and ATP sulfurylase; miR-396 and GRF TFs, miRNA398 and SOD (Sunkar et al., 2006); miR-399 and PHO2/MYB complexes; miR-393 and the auxin-dependent transport inhibitor response 1 (TIR1). TIR1, an auxin receptor, an F-box protein with an inhibitory role, similarly to COI1 in the SCF complex, can be regulated by NO through S-nitrosylation (Terrile et al., 2012).

In particular, ABA signalling promotes the expression of the drought-regulated miR-159, miR-393 and miR-398. ABA signalling acts through the ABA-responsive element (ABRE), present in the promoter of miR-169n, targeting the nuclear factor Y subunit (NF-YA) that is down-regulated by drought in wheat.

The transcription factor TCP4 regulates several genes of the LOX and JA pathway in *Arabidopsis* (Schommer et al., 2008). The transcription factors of the MYB and TCP families of transcription factors are targeted by miR-319. It is proposed that an early activation by TCP4 of JA biosynthesis pathway may be followed by a negative feedback determined by miR-319 binding to TCP4. This coordinated activity may orchestrate timely and localized differential gene expression of *LOX*, *OPR* and *AOS* in roots responding to different stresses.

Epigenetic control of actively transcribed regions of chromatin is orchestrated by protein complexes involving different mediator subunits and several structured RNAs. The *Arabidopsis* mediator subunit MED25, a chromatin modelling complexes partner, epigenetic activator regulating expression of chromatin regions, was shown to control JA and ABA signalling through binding to MYC2 and ABI5 transcription factors (Chen et al., 2012).

Histone acetyltransferases (HATs) have been shown to interact with transcription factors and become involved in activation of stress-responsive genes. ABA down-regulates the expression of AtHD2C while histone deacetylase HDA6 has a role in ABA signalling in salt stress response. Histone modifications, or marks, are responsible for the attraction of specific polycomb complexes (PRC) that repress transcription, or MLL-containing complexes, involved in active transcription.

Small RNAs sustain the memory of JA-mediated response (Galis et al., 2009). Antisense RNAs have also a role to play in activating or maintaining locally the activation of specific genes, through opening promoters and enhancers. Structural RNAs could be involved in the differential, alternate splicing of LOX and AOS isoforms and in the generation of the SAATs up-regulated in ILC588 (Molina et al., 2008) such as STCa-24417, over-expressed 25-fold during tolerance responses to both drought and salt stress (Molina et al., 2011). Such RNAs also mediate processes such as alternative splicing, retention of intron sequences and generation of new codons in TFs.

Thus, TFs are one of the key players during hormone signalling, which, along with NO signals, may positively or negatively regulate stress responses; miRNAs that are controlled by stress-activated ABA in a feedback signalling network also exert a negative response. In this context, an interplay of hormones, signalling pathways and signalling effectors at local and distal tissues has been delineated, with involvement of transcription factors fine-tuned by the levels of specific small RNAs.

Increasing our understanding on physiological, metabolic and molecular aspects of plant response to multiple stresses will be essential for the development and availability of new varieties able to survive in harsh environmental conditions.

4.7 Breeding Strategies

Drought is a limiting factor in growing crops, and drought-stress tolerance is a trait of great importance in breeding. QTLs have been identified and exploited in improvement of yield in maize by marker-assisted selection (MAS) (Landi et al., 2010). MAS is based on the identification of genes that are positively regulated in stress-tolerant varieties and that may confer traits that may be measured early during hybrid selection. Thus, the crossing of tolerant varieties with less-tolerant ones may be followed by induction of water or salt stress and assessment of mRNA expression levels at different timing. The exploitation of real-time PCR with isoform specific Taqman probes could support breeding programmes in the identification of hybrids that express LOX and AOC isoforms at early timing and in a sustained and prolonged activity after stress perception. *LOX1*, *LOX2*, *AOS*, *AOC*, *HLP1*, *HPL2* and *OPR* primers have been shown effective in differentiating the stress response in chickpea, thus these genes could be used as markers to individuate the crosses retaining the characters of the more stress-tolerant parental variety. These studies may lead to new and specific assays and phenotyping techniques to evaluate a species rootstock in order to choose the hybrids better suited to respond to abiotic stresses.

Acknowledgements

The authors thank Elsevier for the permission to use in this book chapter Figs 2 and 5: 'Reproduced from: De Domenico S., Bonsegna S.; Horres R.; Pastor V.; Taurino M.; Poltronieri P.; Imtiaz M.; Kahl G.; Flors V.; Winter P.; Santino A. Transcriptomic analysis of oxylipin biosynthesis genes and chemical profiling reveal an early induction of jasmonates in chickpea roots under drought stress. Plant Physiol. Biochem. 2012, 61, 115-122. Copyright ©2012, published by Elsevier Masson SAS. All rights reserved'.

References

Abdala, G., Miersch, O., Kramell, R., Vigliocco, A., Agostini, E., Forchetti, G. and Alemano, S. (2003) Jasmonate and octadecanoid occurrence in tomato hairy roots. Endogenous level changes in response to NaCl. *Journal of Plant Growth and Regulation* 40, 21–27.

Baltrushat, H., Fodor, J., Harrach, B.D., Niemczyk, E., Barna, B., Gullnet, G., Janeczko, A., Kogel, K.H., Schäfer, P., Schwarczinger, I., Zuccaro, A. and Skoczowski, A. (2008) Salt stress tolerance of barley induced by the root endophyte *Piriformospora indica* is associated with a strong increase in antioxidants. *New Phytologist* 180, 501–510.

Bhargava, S. and Sawant, K. (2013) Drought stress adaptation: metabolic adjustment and regulation of gene expression. *Plant Breeding* 132, 21–32.

Borsani, O., Valpuesta, V. and Botella, M.A. (2001) Evidence for a role of salicylic acid in the oxidative damage generated by NaCl and osmotic stress in *Arabidopsis* seedlings. *Plant Physiology* 126, 1024–1030.

Buitink, J., Leger, J.J., Guisle, I., Vu, B.L., Wuilleme, S., Lamirault, G., Le, B.A., Le, M.N., Becker, A., Kuster, H. and Leprince, O. (2006) Transcriptome profiling uncovers metabolic and regulatory processes occurring during the transition from desiccation-sensitive to desiccation tolerant stages in *Medicago truncatula* seeds. *The Plant Journal* 47, 735–750.

Chen, R., Jiang, H., Li, L., Zhai, Q., Qi, L., Zhou, W., Liu, X., Li, H., Zheng, W., Sun, J. and Li, C. (2012) The *Arabidopsis* mediator subunit MED25 differentially regulates jasmonate and abscisic acid signaling through interacting with the MYC2 and ABI5 transcription factors. *The Plant Cell* 24, 2898–2916.

Correa-Aragunde, N., Graziano, M. and Lamattina, L. (2004) Nitric oxide plays a central role in determining lateral root development in tomato. *Planta* 218, 900–905.

De Domenico, S., Bonsegna, S., Horres, R., Pastor, V., Taurino, M., Poltronieri, P., Imtiaz, M., Kahl, G., Flors, V., Winter, P. and Santino, A. (2012) Transcriptomic analysis of oxylipin biosynthesis genes and chemical profiling reveal an early induction of jasmonates in chickpea roots under drought stress. *Plant Physiology and Biochemistry* 61, 115–122.

Del Giudice, J., Cam, Y., Damiani, I., Fung-Chat, F., Meilhoc, E., Bruand, C., Brouquisse, R., Puppo, A. and Boscari, A. (2011) Nitric oxide is required for an optimal establishment of the *Medicago truncatula*-*Sinorhizobium meliloti* symbiosis. *New Phytologist* 191, 405–417.

Fernández-Marcos, M., Sanz, L. and Lorenzo, O. (2012) Nitric oxide An emerging regulator of cell elongation during primary root growth. *Plant Signalling and Behaviour* 7, 196–200.

Feussner, I. and Wasternack, C. (2002) The lipoxygenase pathway. *Annual Reviews in Plant Biology* 53, 275–297.

Franco-Zorrilla, J.M., Valli, A., Todesco, M., Mateos, I., Puga, M.I., Rubio-Somoza, I., Leyva, A., Weigel, D., García, J.A. and Paz-Ares, J. (2007) Target mimicry provides a new mechanism for regulation of microRNA activity. *Nature Genetics* 39, 1033–1037.

Fujita, M., Fujita, Y., Noutoshi, Y., Takahashi, F., Narusaka, Y., Yamaguchi-Shinozaki, K. and Shinozaki, K. (2006) Crosstalk between biotic and abiotic stress responses: a current view from the points of convergence in the stress signaling networks. *Current Opinions in Plant Biology* 9, 436–442.

Galis, I., Gaquerel, E., Pandey, S.P. and Baldwin, J.T. (2009) Molecular mechanisms underlying plant memory in JA-mediated defence responses. *Plant Cell and Environment* 32, 617–627.

Geng, X., Cheng, J., Gangadharan, A. and Mackey, D. (2012) The coronatine toxin of *Pseudomonas syringae* is a multifunctional suppressor of *Arabidopsis* defense. *The Plant Cell* 24, 4763–4774.

Gfeller, A., Dubugnon, L., Liechti, R. and Farmer, E.E. (2010) Jasmonate biochemical pathway. *Science Signalling* 3(109), cm3.

Goodger, J.Q. and Schachtman, D.P. (2010) Re-examining the role of ABA as the primary long-distance signal produced by water-stressed roots. *Plant Signalling and Behaviour* 5, 1298–1301.

Gruber, V., Blanchet, S., Diet, A., Zahaf, O., Boualem, A., Kakar, K., Alunni, B., Udvardi, M., Frugier, F. and Crespi, M. (2009) Identification of transcription factors involved in root apex responses to salt stress in *Medicago truncatula*. *Molecular Genetics and Genomics* 281, 55–66.

Hancock, J.T., Neill, S.J. and Wilson, I.D. (2011) Nitric oxide and ABA in the control of plant function. *Plant Science* 181, 555–559.

Hasegawa, P.M., Bressan, R.A., Zhu, J.K. and Bohnert, H.J. (2000) Plant cellular and molecular responses to high salinity. *Annual Reviews in Plant Physiology and Plant Molecular Biology* 51, 463–499.

Hause, B. and Schaarschmidt, S. (2009) The role of jasmonates in mutualistic symbioses between plants and soil-born microorganisms. *Phytochemistry* 70, 1589–1599.

Hsu, Y.Y., Chao, Y.Y. and Kao, C.H. (2013) Methyl jasmonate-induced lateral root formation in rice: the role of heme oxygenase and calcium. *Journal of Plant Physiology* 170, 63–69.

Huang, X., Stettmaier, K., Michel, C., Hutzler, P., Mueller, M.J. and Durner, J. (2004) Nitric oxide is induced by wounding and influences jasmonic acid signaling in *Arabidopsis thaliana*. *Planta* 218, 938–946.

Hughes, R.K., De Domenico, S. and Santino, A. (2009) Plant Cytochromes CYP74: biochemical features, endocellular localisation, activation mechanism in plant defence and improvements for industrial applications. *ChemBioChem* 10, 1122–1133.

ICRISAT (1992) ICRISAT Plant Material Description no. 33. ICC 4958. A Drought Resistant Chickpea. Patancheru, Andra Pradesh, India. ISBN 92-9066-232-8.

Jakab, G., Ton, J., Flors, V., Zimmerli, L., Métraux, J.P. and Mauch-Mani, B. (2005) Enhancing *Arabidopsis* salt and drought stress tolerance by chemical priming for its abscisic acid responses. *Plant Physiology* 139, 267–274.

Kataya, A.R.A. and Reumann, S. (2010) *Arabidopsis* glutathione reductase 1 is dually targeted to peroxisomes and the cytosol. *Plant Signaling and Behaviour* 2, 5.

Khraiwesh, B., Zhu, J.K. and Zhu, J. (2012) Role of miRNAs and siRNAs in biotic and abiotic stress responses of plant. *Biochimica et Biophysica Acta* 1819, 137–148.

Kim, E.H., Kim, Y.S., Park, S.H., Koo, Y.J., Choi, Y.D., Chung, Y.Y., Lee, I.J. and Kim, J.K. (2009) Methyl jasmonate reduces grain yield by mediating stress signals to alter spikelet development in rice. *Plant Physiology* 49, 1751–1760.

Kolbert, Z., Ortega, L. and Erdei, L. (2010) Involvement of nitrate reductase (NR) in osmotic stress-induced NO generation of *Arabidopsis thaliana* L. roots. *Journal of Plant Physiology* 167, 77–80.

Koo, A.J., Gao, X., Jones, A.D. and Howe, G.A. (2009) A rapid wound signal activates the systemic synthesis of bioactive jasmonates in *Arabidopsis*. *The Plant Journal* 59, 974–986.

Landi, P., Giuliani, S., Salvi, S., Ferri, M., Tuberosa, R. and Sanguineti, M.C. (2010) Characterization of root-yield-1.06, a major constitutive QTL for root and agronomic traits in maize across water regimes. *Journal of Experimental Botany* 61, 3553–3562.

Lorenzo, O. and Solano, R. (2005) Molecular players regulating the jasmonate signalling network. *Current Opinion in Plant Biology* 8, 532–540.

Memelink, J. (2009) Regulation of gene expression by jasmonate hormones. *Phytochemistry* 70, 1560–1570.

Mita, G., Fasano, P., De Domenico, S., Perrone, G., Epifani, F., Iannacone, R., Casey, R. and Santino, A. (2007) 9-lipoxygenase metabolism is involved in the almond/*Aspergillus carbonarius* interaction. *Journal of Experimental Botany* 58, 1803–1811.

Molina, C., Rotter, B., Horres, R., Udupa, S.M., Besser, B., Bellarmino, L., Baum, M., Matsumura, H., Terauchi, R., Kahl, G. and Winter, P. (2008) SuperSAGE: the drought stress-responsive transcriptome of chickpea roots. *BMC Genomics* 9, 553.

Molina, C., Zaman-Allah, M., Khan, F., Fatnassi, N., Horres, R., Rotter, B., Steinhauer, D., Amenc, L., Drevon, J.J., Winter, P. and Kahl, G. (2011) The salt-responsive transcriptome of chickpea roots and nodules via deep-SuperSAGE. *BMC Plant Biology* 11, 31.

Mueller, M.J. and Berger, S. (2009) Reactive electrophilic oxylipins: pattern recognition and signalling. *Phytochemistry* 70, 1511–1521.

Oldroyd, G.E. (2009) Plant science. Nodules and hormones. *Science* 315, 52–53.

Palmieri, M.C., Sell, S., Huang, X., Scherf, M., Werner, T., Durner, J. and Lindermayr, C. (2008) Nitric oxide-responsive genes and promoters in *Arabidopsis thaliana*: a bioinformatics approach. *Journal of Experimental Botany* 59, 177–186.

Pauwels, L. and Goossens, A. (2011) The JAZ proteins: a crucial interface in the jasmonate signaling cascade. *The Plant Cell* 23, 3089–3100.

Reumann, S. (2011) Toward a definition of the complete proteome of plant peroxisomes: Where experimental proteomics must be complemented by bioinformatics. *Proteomics* 9, 11.

Romero-Puertas, M.C., Campostrini, N., Mattè, A., Righetti, P.G., Perazzolli, M., Zolla, L., Roepstorff, P. and Delledonne, M. (2008) Proteomic analysis of S-nitrosylated proteins in *Arabidopsis thaliana* undergoing hypersensitive response. *Proteomics* 8, 1459–1469.

Santino, A., Taurino, M., De Domenico, S., Bonsegna, S., Poltronieri, P., Pastor, V. and Flors, V. (2013) Jasmonate signalling in plant defense response to multiple (a)biotic stresses. *Plant Cell Reports* 32, 1085–1098.

Schachtman, D.P. and Goodger, J.Q.D. (2008) Chemical root to shoot signalling under drought. *Trends in Plant Science* 13, 281–287.

Schäffner, A.R. (2011) The *Arabidopsis* glucosyltransferase UGT76B1 conjugates isoleucic acid and modulates plant defense and senescence. *The Plant Cell* 23, 4124–4145.

Schaller, A. and Stintzi, A. (2009) Enzymes in jasmonate biosynthesis - structure, function, regulation. *Phytochemistry* 70, 1532–1538.

Schommer, C., Palatnik, J.F., Aggarwal, P., Chételat, A., Cubas, P., Farmer, E.E., Nath, U. and Weigel, D. (2008) Control of jasmonate biosynthesis and senescence by miR319 targets. *PLoS Biol* 6(9), e230.

Seki, M., Umezawa, T., Urano, K. and Shinozaki, K. (2007) Regulatory metabolic networks in drought stress response. *Current Opinion in Plant Biology* 10, 296–302.

Serpa, V., Vernal, J., Lamattina, L., Grotewold, E., Cassia, R. and Terenzi, H. (2007) Inhibition of AtMYB2 DNA-binding by nitric oxide involves cysteine S-nitrosylation. *Biochemical and Biophysical Research Communications* 361, 1048–1053.

Serrano, R., Mulet, J.M., Rios, G., Marquez, J.A., de Larriona, I.F., Leube, M.P., Mendizabal, I., Pascual-Ahuir, A., Proft, M., Ros, R. and Montesinos, C. (1999) A glimpse of the mechanisms of ion homeostasis during salt stress. *Journal of Experimental Botany* 50, 1023–1036.

Shinozaki, K. and Yamaguchi-Shinozaki, K. (2007) Gene networks involved in drought stress response and tolerance. *Journal of Experimental Botany* 58, 221–227.

Stenzel, I., Otto, M., Delker, C., Kirmse, N., Schmidt, D., Miersch, O., Hause, B. and Wasternack, C. (2012) ALLENE OXIDE CYCLASE (AOC) gene family members of *Arabidopsis thaliana*: tissue- and organ-specific promoter activities and *in vivo* heterodimerization. *Journal of Experimental Botany* 63, 6125–6138.

Sunkar, R., Kapoor, A. and Zhu, J.K. (2006) Posttranscriptional induction of two Cu/Zn superoxide dismutase genes in *Arabidopsis* is mediated by downregulation of miR398 and important for oxidative stress tolerance. *The Plant Cell* 18, 2051–2065.

Taghavi, S., can der Leilie, D., Hoffman, A., Zhang, Y.B., Walla, M.D., Vangronsveld, J., Newman, L. and Monchy, S. (2010) Genome sequence of the plant growth promoting endophytic bacterium *Enterobacter* sp. 638. *PLoS Genetics* 6(5), e1000943.

Terrile, M.C., París, R., Calderón-Villalobos, L.I., Iglesias, M.J., Lamattina, L., Estelle, M. and Casalongué, C.A. (2012) Nitric oxide influences auxin signaling through S-nitrosylation of the *Arabidopsis* TRANSPORT INHIBITOR RESPONSE 1 auxin receptor. *The Plant Journal* 70, 492–500.

Tiryaki, I. and Staswick, P.E. (2002) An *Arabidopsis* mutant defective in jasmonate response is allelic to the auxin-signaling mutant axr1. *Plant Physiology* 130, 887–894.

Varshney, R.K., Song, C., Saxena, R.K., Azam, S., Yu, S., Sharpe, A.G., Cannon, S., Baek, J., Rosen, B.D., Tar'an, B., Millan, T., Zhang, X., et al. (2013) Draft genome sequence of chickpea (*Cicer arietinum*) provides a resource for trait improvement. *Nature Biotechnology* 31, 240–248.

Vellosillo, T., Martinez, M., Lopez, M.A., Vicente, J., Cascón, T., Dolan, L., Hamberg, M. and Castresana, C. (2007) Oxylipins produced by the 9-lipoxygenase pathway in *Arabidopsis* regulate lateral root development and defense responses through a specific signaling cascade. *The Plant Cell* 19, 831–846.

Wang, Y.Q., Feechan, A., Yun, B.W., Shafiei, R., Hofmann, A., Taylor, P., Xue, P., Yang, F.Q., Xie, Z.S., Pallas, J.A., Chu, C.C. and Loake, G.J. (2009) S-nitrosylation of AtSABP3 antagonizes the expression of plant immunity. *Journal of Biological Chemistry* 284, 2131–2137.

Wasternack, C. (2007) Jasmonates: an update on biosynthesis, signal transduction and action in plant stress response, growth and development. *Annals of Botany* 100, 681–697.

Wild, M., Davière, J.M., Cheminant, S., Regnault, T., Baumberger, N., Heintz, D., Baltz, R., Genschik, P. and Achard, P. (2012) The *Arabidopsis* DELLA RGA-LIKE3 is a direct target of MYC2 and modulates jasmonate signaling responses. *The Plant Cell* 24, 3307–3319.

Xiong, L., Schumaker, K.S. and Zhu, J.K. (2002) Cell signalling during cold, drought, and salt stress. *The Plant Cell* 14, S165–S183.

Zahaf, O., Blanchet, S., de Zélicourt, A., Alunni, B., Plet, J., Laffont, C., de Lorenzo, L., Imbeaud, S., Ichanté, J.L., Diet, A., Badri, M., Zabalza, A., González, E.M., Delacroix, H., Gruber, V., Frugier, F. and Crespi, M. (2012) Comparative transcriptomic analysis of salt adaptation in roots of contrasting *Medicago truncatula* genotypes. *Molecular Plant* 5, 1068–1081.

Zarei, A., Körbes, A.P., Younessi, P., Montiel, G., Champion, A. and Memelink, J. (2011) Two GCC boxes and AP2/ERF-domain transcription factor ORA59 in jasmonate/ethylene-mediated activation of the PDF1.2 promoter in *Arabidopsis*. *Plant Molecular Biology* 75, 321–331.

Zhang, J., Subramanian, S., Zhang, Y. and Yu, O. (2007) Flavone synthases from *Medicago truncatula* are flavanone-2-hydroxylases and are important for nodulation. *Plant Physiology* 144, 741–751.

5 Genetic Engineering of Crop Plants to Sustain Drought Tolerance

J. Amudha* and G. Balasubramani
Central Institute for Cotton Research, Nagpur, Maharashtra, India

Abstract

The world today is faced with great challenges to produce adequate food, fibre, feed, industrial products and ecosystem services. Under the influence of global climate changes, the situation is getting worse by the destabilization of our ecosystem. With the increasing population, the challenge to develop ecosystem goods and services to meet human needs in the future is very important. Water scarcity, drought conditions and global climate change are major constraints in crop production worldwide. Uncertain rainfall is making conditions worse for farmers. Water stress along with other abiotic stresses is very complex in nature and is a serious challenge that needs to be met urgently in order to sustain and enhance productivity. Agriculture in India and other developing countries is a system which gambles with the monsoon and where irrigation is limited in major parts of the crop cultivation area. Genetic engineering techniques hold great promise for developing crop cultivars with high tolerance to drought. Biotechnological approaches can be utilized; drought and high salinity-tolerant genes can be discovered efficiently and subsequently cloned. Transgenic breeding is a new technology for the development of stress tolerance in crop plants. Drought stress is controlled by multiple polygenes, including signal transduction genes, transcriptional regulation genes and a series of genes for protection, defence and stress tolerance. It is very important to improve the drought tolerance of crops and to evolve plants with various mechanisms for adapting to adverse climatic environments. The introduction of transgenic technology to breeding crops has provided significant benefits to the industry; the first transgenic traits developed and commercialized were designed for insect and herbicide resistance in existing varieties. Eventually, by genetically enhanced technologies, current varieties must be improved or new varieties should be developed that adapt to environmental stresses and have the genetic potential to improve yield factors. This will lead to new levels of sustainable agriculture, with stable yield improvement.

5.1 Introduction

Plants under drought or water-deficit stress show retarded growth during the vegetative stage and also effects on transpiration. As a result of this effect, the water loss and decline in photosynthesis is prevented by an increase in abscisic acid (ABA) concentration and closure of stomata (Chaves and Oliveira, 2004). The CO_2 in the intercellular region declines,

---

*E-mail: jamudhacicr@gmail.com

leading to photo-oxidation of reactive oxygen species (ROS) components, mutation in nucleic acids and protein denaturation caused by de-esterification of membrane lipids. Cellular metabolism is disrupted due to loss of water in the membrane; as a result of this the bilayer structure of the membrane is damaged, membrane proteins are displaced, ion transporter activity is reduced and, finally, denaturation of organelle and cytosolic proteins occurs leading to loss in enzyme activity. The membranes in the plants are protected by the synthesis of osmolytes such as glycine-betaine, mannitol, sucrose, trehalose, proline, fructans, carnitine glutamate, sorbitol and polyols (Chen and Murata, 2002). The gradient uptake of water helps the functional roles of solutes, which also function as chemical chaperones or free-radical scavengers, resulting in stabilizing the membrane proteins. Osmoprotectants, regulatory proteins, kinases or transcription factors are important metabolic proteins produced following abiotic stress. After synthesis, the *trans*-acting factors re-enter the nucleus and bring about the stress responsive promoter activation. These have stress responsive elements (SREs; e.g.: ABRE, ABA responsive element; LTRE, low temperature responsive element; DRE, drought responsive element; HSE, heat-shock element; and ARE, antioxidant responsive element), which are involved in synthesis of the osmolytes, and the TFs (transcription factors) probably bind to these.

5.2 Genetic Engineering Strategies for Stress Tolerance

Drought stress is among the most serious challenges to crop production worldwide. In plants in adverse conditions many stress-related genes are triggered to produce osmolytes, which protect cells against stress-induced damage. The discovery of stress tolerance genes was led by genomic approaches that are used in genetic engineering. In the molecular approach, genes encoding functional proteins such as transporters and chaperones are engineered for drought tolerance. In metabolic engineering, multiple steps are targeted by enzymatic fusions. Signal peptides are attached to make proteins work in their correct organellar location. The TFs, which are regulatory proteins and signalling pathway factors, provide novel routes for engineering drought tolerance through mutations or repression domains. Genes at the mRNA level can be up-regulated and down-regulated by using specific promoters. Sustained agriculture can be achieved in crop plants by transformation with individual genes or combinations of genes/TFs.

5.3 Transcription Factors

Transcription factors play a critical role in regulating cellular and physical changes during abiotic stress in plants. Gene expression regulation by these factors is induced (activators) or repressed (repressors) by RNA polymerase. TFs are grouped based on DNA binding domain into families as: (i) CBF (*cis*-binding factor)/DREB (dehydration responsive element binding) regulon; (ii) NAC; ATAF (*Arabidopsis* transcription activation factor); CUC and ZF-HD (zinc-finger homoeodomain); (iii) AREB/ABF (abscisic acid/abscisic acid binding factor) regulon; and (iv) MYC (myelo-cytomatosis oncogene)/MYB (myelo-blastosis oncogene) regulon. The corresponding *cis*-acting elements regulons are DRE, NACRS, ABRE and MYCRS/MYBRS, respectively. ABA-independent and ABA-dependent signal transduction pathways activate these TFs. The first two regulons are ABA independent and the last two are ABA dependent. These drought-induced transcripts act through specific binding to *cis* acting sequence of down-regulated genes. These genes were classified into several large families as AP2/EREBP, Cys2His2 zinc-finger, NAC, MYB, MYC, bZIP and WRKY. The EREBP transcription factors were classified into subfamilies, the AP2, RAV, ERF and including DREB, which were isolated and characterized (Sakuma *et al.*, 2002). TFs of this element induce important stress-related genes and switch to regulate expression. The DREB subgroups A-1 and A-2, harbouring the DREB1- and DREB2-type genes, respectively, are the largest ones induced in two ABA-independent pathways. The DREB1-type genes (*DREB1A*, *DREB1B* and *DREB1C*) regulate expression of cold-responsive genes, whereas DREB2-type genes (*DREB2A* and *DREB2B*) were mainly involved in osmotic-responsive gene expressions. The functions of genes of the A-5 and A-6 families remain to be determined under stress conditions.

5.3.1 *Cis*-binding factor/dehydration responsive element binding regulon

DREB regulon is a conserved region throughout the whole of the plant kingdom. The conserved CBF/DREB1 transcription factor regulon, a *cis*-acting c-repeat element in the plant kingdom, induces genes producing osmolytes and proteins by RESPONSIVE TO DEHYDRATION 29A (RD29A) promoter. ABA plays a key role in abiotic stress-mediated gene expression induced by DREB protein. DRE contains one sequence A/GCCGAC, a *cis*-acting promoter element dynamic network of genes, which control various biological processes. The AP2/ERF (apetala 2/ethylene responsive factor) TF is the core motif of *cis*-acting element in plants. In *Arabidopsis* c-repeat cold-inducible promoters are referred to as the low-temperature-responsive element (LTRE) (Baker *et al.*, 1994) and cbf 1,2,3 (*cis*-binding factors) were identified (Liu *et al.*, 1998), which lie in tandem repeats. Drought-responsible TFs are given in Table 5.1 and *cis*-acting elements are listed in Table 5.2.

The control of DREB regulon is complicated. The CBF2/DREB1C is a negative regulator of CBF1/DREB1B and CBF3/DREB1A gene expression, but CBF2/DREB1C offers a few target genes. There are two groups of DREBs: the *trans*-active Group I, which are rapidly active on exposure to cold conditions to turn on, and when the proteins of Group I reach a certain level the *trans*-inactive Group II are expressed, and they compete with Group I binding the DRE elements of target genes on

Table 5.1. Drought responsive element/C repeat responsible transcription factors.

Sequence of DRE/CRT	Gene	Specific binding TF	Plant species	References
GGCCGACA/GT	COR15A	DREB1B/CBF1	*Arabidopsis*	Stokinger *et al.*, 1997
TACCGACAT	BN115	BNCBF5	*Brassica napus*	Gao *et al.*, 2002
TGGCCGAC	BN28	BNCBF17	*Brassica napus*	Gao *et al.*, 2002
ACCGAC	RAB17	ZmDREB1 and ZmDREB2	Maize	Kizis and Pagès, 2002
TTGCCGACAT	HVA1	HvCBF1	Barley	Xue, 2002
	RD29A	DREB1A/CBF3 and DREB2A	*Arabidopsis*	Maruyama *et al.*, 2004

Table 5.2. List of *cis*-acting elements.

cis element	Sequence	Gene	Stress condition	Data from
ABRE	PyACGTGGC	EM1A	Water deficit, ABA	Guiltinan *et al.*, 1990
G-box	CACGTG	CHS15	ABA	Loake *et al.*, 1992
DRE	TACCGACAT	RD29A	Water deficit	Yamaguchi-Shinozaki and Shinozaki, 1994
CRT	GGCCGACAT	COR15A	Cold	Baker *et al.*, 1994
LTRE	GGCCGACGT	BN115	Cold	Jiang and Singh, 1996
MYBR, MYCR	TGGTTAG CACATG	RG22	Water deficit, ABA	Abe *et al.*, 1997
HSE	GTGGGCCCTCC	APX1	Water deficit, heat	Storozhenko *et al.*, 1998
SRE	TGACG	GNT35	SA	Garreton *et al.*, 2002
ICEr1, ICEr2	GGACACATGTCAGA, ACTCCG	CBF2/DREB1C	Cold	Zarka *et al.*, 2003
RSRE	CGCGTT	RWR	Water deficit	Walley *et al.*, 2007
NACR	ACACGCATGT	ERD1	Water deficit	Tran *et al.*, 2007

the promoter and decrease their expression, and leads to the DRE-mediated signalling pathway to switch off. ZAT12 TF is parallel to CBFs/DREBs regulon. When ZAT12 is overexpressed in plants freezing tolerance is consistently increased along with a concomitant reduction of cold-induced CBF/DREB genes, confirming that ZAT12 plays a negative role in the regulatory circuit leading to a decline in CBF/DREB expression (Vogel et al., 2005). DREB2 genes are expressed both constitutively as well as under stress conditions even though their target genes (e.g. RD17, RD29A, LEA14 and RD29B) are induced only upon dehydration. DREB2B and DREB2A are not cold induced but are high salinity, heat-shock and dehydration induced and are downregulated. DREB2 and DREB1 are induced acclimation processes by activating transcription in *Arabidopsis*. Many studies, especially those using the transgenic approach for over-expression of stress-induced DREB transcription factors, have shown that the expression of numerous target genes having promoters with DRE elements are activated and the resulting transgenic plants show superior stress tolerance. The expression in transgenic *Arabidopsis* of 35S:AtDREB1A and 35S:OsDREB1A enhanced dehydration tolerance. Enhanced expression resulted in plant growth reduction in the presence of constitutive promoter 35S CaMV and the replacement with desiccation responsive promoter rd 29A. Photosynthesis metabolism in *Arabidopsis* is regulated by STZ factor, which also affects the plant growth and carbohydrate metabolism in abiotic stress. In transgenic rice the over-expression did not retard the growth, with constitutive overexpression of CBF3 and ABF3 increasing the drought tolerance (Oh et al., 2005). In *Arabidopsis* over-expression of AtDREB2A without negative regulatory domain, up-regulates the downstream drought-inducible genes (Sakuma et al., 2006a, b). Two *Brassica* CBF/DREB1 genes (*BNCBF5* and *BNCBF17*) resulted in increased freezing tolerance, photochemical efficiency and photosynthetic capacity (Savitch et al., 2005). Stress tolerance by over-expression of *GhDREB* gene was reported in transgenic wheat. CBF3/DREB1A and STZ/ZAT10 TF is a repressor that functions DLN/EAR-motif (Nakashima et al., 2007). *GhDBP3, GhDREB1L* and *GhDBP2* (Huang and Liu, 2006; Huang et al., 2007, 2008) from cotton were grouped into DREB A-1, A-4 and A-6 subfamilies.

Nuclear factor (NF-Y)

Nuclear factor transcription factors have sequence-specific CCAAT binding specificity. The *Arabidopsis* genome encodes 36 NF-Y subunits (10, 13 and 13 unique genes for A,B and C subunits of NF-Y). AtNF-YB1 regulates genes that do not respond to DREB/CBF and AtNF-YA5 regulates stress response by siRNAs (Gusmaroli et al., 2001; Li et al., 2008).

The *AtNF-YA5* gene produces (nat)-derived siRNA (natsiRNA, Borsani et al., 2005), a natural antisense transcript. Over-expression of the genes increases stress tolerance by stomata closure and post-transcriptional regulation by microRNA 169 (miR169) (Li et al., 2008). The maize orthologue of AtNF-YB1/ZmNF-YB2 has a common regulator and enhances tolerance to drought in inbred transgenic lines under water-limited field conditions compared with control plants. The yield of the transgenic maize plants was improved 50% by low leaf rolling, high photosynthesis and high stomata conductance, leading to cooler leaf temperature under drought stress (Nelson et al., 2007).

5.3.2 NAC, ATAF 1, 2, and CUC2 regulon

NAC regulates ABA-dependent and -independent genes. In various developmental stages this regulon expresses in different tissues (Olsen et al., 2005). The NAC proteins bind specifically to the NAC recognition site CATGTG (NACRS). These proteins could bind to NACRS even as multimers and heterodimerization might facilitate transcriptional activity involved in abiotic and biotic stresses (Tran et al., 2004).

ERD1, a NAC family member, has shown that its expression during dehydration depends on the integrity of both 14-bp rps1 sequence and the putative MYC-like (CATGTG) sequence (Simpson et al., 2003). The NAC *trans*-acting factors and MeJA (methyl jasmonic acid) interact with the above-mentioned putative *cis*-acting motifs found in the ERD1 promoter region (Tran et al., 2007). The RD26 (NAC protein) over-expressing plants are up-regulated by the stress signalling pathway, whereas the ABA-insensitive RD26 repress the genes. Stress-responsive NAC1 (*SNAC1*) gene over-expressed in lowland rice 'Nipponbare' encodes NAM (no apical meristem), ATAF (*Arabidopsis* transcription activation factor), CUC (cup-shaped cotyledon) TF and has a conserved domain (NAC domain) predominantly induced in guard cells. ATAF1 was the first NAC-domain protein identified during drought signalling pathways. The *ataf1* mutant lines showed a sevenfold over-expression in transgenic plants following drought treatment and produces drought-induced genes including *COR47* (*cold regulated 47*; also known as *RD17*), *ERD10* (*early response to dehydration 10*), *KIN1*, *RD22* and *RD29A* (*COR78* or *LTI78* (*low temperature inductive 78*)). ABA-independent pathway transcription factor (ERD1) accumulates transcripts during dehydration and salinity, which belongs to the NAC domain and zinc finger homoeo-domain (ZF-HD). The transcripts produce osmolytes such as sorbitol transporter and exoglucanase, which stabilizes the membrane proteins upon drought conditions (Hu *et al.*, 2006). A group of membrane-bound NAC TFs (designated NTLs) is reported to be closely linked with environmental stresses (Kim *et al.*, 2007). During stress the NTL proteins are released through the membranes by proteolytic cleavage and are transported into the nucleus, where they regulate the drought-induced genes. Salt-inducible NTL member NTL8 regulates NAC proteins through gibberellic acid (GA)-conserved domain interactions. High salinity reduces GA biosynthesis by repressing GA biosynthetic genes, which in turn induces the *NTL8* gene during salt stress preventing seed germination (Kim *et al.*, 2008).

Zinc-finger homoeo-domain regulon

ZF-HDs are classified by type of cysteine and histidine coordinating residues based on folds in the backbone of the domain. The common 'fold groups' of zinc fingers are the Cys2His2-like treble clef and zinc ribbon. ZF proteins have been classified into nine types: C2H2, C8, C6, C3HC4, C2HC, C2HC5, C4, C4HC3 and CCCH (C and H represent cysteine and histidine, respectively; Jenkins *et al.*, 2005; Schumann *et al.*, 2007). ZF protein sequence-specific DNA-binding proteins occur as tandem repeats with two, three, or more fingers comprising the domain of the protein. These tandem arrays bind to DNA major groove and are typically spaced at 3 bp intervals. The α-helix of each domain overlaps with the adjacent helix by specific DNA bases. The hydrophobic region mediates homoeo-DNA binding dimer formation of β-strands by a zinc knuckle. The loop resembles the Cys2His2 classical motif with the helix and β-hairpin. Free proline and ROS-scavenging enzymes are accumulated by pyrroline-5-carboxylate synthetase in plant cells.

5.3.3 Abscisic acid responsive element-binding protein/abscisic acid-binding factor regulon

ABRE-binding protein is a *cis*-acting element A/GCCGAC sequence and regulates gene expression. ABREs are reported in wheat *EM* gene (late embryogenesis) in seed (Guiltinan *et al.*, 1990). A coupling element (CE3) is needed to specify the function of ABRE for the expression of ABA-induced genes. The ABRE's core motif, ACGT, is present in G-boxes of a variety of genes responsive to different environmental and physiological factors, such as anaerobiosis (McKendree and Ferl, 1992), jasmonic acid and salicylic acid (Mason *et al.*, 1993; Qin *et al.*, 1994), auxin and irradiance (Liu *et al.*, 1994). Other *cis*-elements involved in gene expression were suggested to be involved in drought tolerance, based on the mismatch of cell type-specific enrichment (Dinneny *et al.*, 2008). AREB3 of *Arabidopsis* also encodes bZIP-type proteins. RD29B promoter activates ABA stress-inducible AREB1 and AREB2 (Uno *et al.*, 2000). In soybean 131 bZIP genes (ABA-induced) of different groups were identified as group A bZIP and stress-signalling. There are other bZIP-type proteins which belong to subgroups S (GmbZIP44), C (GmbZIP62) and G (GmbZIP78). These proteins up-regulate ERF5, KIN1, COR15A, COR78A and P5CS1 and down-regulate DREB2A and COR47 (Liao *et al.*, 2008). KIN10 and GBF5 (G-box binding factor 5) have a synergistic effect in *Arabidopsis* on DIN6 (dark inducible 6) expression. ABI5 (ABA-insensitive 5) subfamily contains four highly conserved domains along with bZIP binding domain in *Arabidopsis*.

WRKY transcription factors/W box (transcription factors)

The WRKY domain is a 60 amino acid region that is defined by the conserved amino acid sequence WRKYGQK at its N-terminal end, binds specifically to the DNA sequence motif (T)(T)TGAC(C/T), which is known as the W box, and is induced during cold temperature, salt-stress and pathogen defence, seed dormancy and senescence. WRKY TFs play roles during ABA response; some WRKY TFs are ABA-inducible repressors. Seed germination is controlled by WRKY TFs (AfWRKY1/ABF1 and AfWRKY2/ABF2) (Rushton *et al.*, 1995, 2010), which is jointly regulated by the hormones gibberellin (GA) and ABA. In *Arabidopsis*, the target genes for WRKY TFs include DREB1A, DREB2A and ABF2, which have been revealed through promoter-binding studies; these TFs also regulate downstream receptors, the cytoplasmic protein phosphatase, 2C-ABA complex and ABAR–ABA complex and also RD29A and COR47 promoters. WRKY TFs regulate stomata opening during high and cold temperatures, high CO_2 levels, water stress and high ozone concentrations and play an important role in seed germination and to networks that respond to ABA (Jiang and Yu, 2009; Ren *et al.*, 2010) and induce bZIPs, MYBs and ERFs.

5.3.4 Myelocytomatosis oncogene/ myeloblastosis oncogene regulon

Myelocytomatosis (CANNTG) and myeloblastosis (C/TAACNA/G) have *cis* recognition site promoter RD22, which is drought-inducible in *Arabidopsis*. This TF works in the presence of ABA-dependent pathways in the regulation of stress-responsive genes. The plant MYB proteins of the DNA-binding domain consist of two imperfect repeats of about 50 residues (R2, R3). Over-expression of *Arabidopsis* AtMYB60 and AtMYB61 regulates stomata closure to enhance drought tolerance.

Inducer of cbf expression 1a myelocytomatosis type controls drought. The growth of the plant is indirectly regulated by myeloblastosis oncogene (MYB) 15, which regulates DREB1A. Over-expression of SNAC1 up-regulates rice R2R3-MYB (Hu *et al.*, 2006).

5.4 Signalling Factors in Drought Tolerance

Abiotic stress signal transduction systems are stream of TFs, which are involved in the sensing of signalling factor proteins and regulating the cellular components from degradation. In transgenic plants with tobacco MAPK (mitogen-activated protein kinase), NPK1, it is expressed constitutively and activates oxidative signal cascades, which leads to tolerance for abiotic stress. Suppression of signalling factors such as farnesyl transferases (ERA 1), enhances the drought tolerance-regulating closure of stomata by ABA (Cutler *et al.*, 1996). ABA signalling enhances temperature tolerance by antisense down-regulation of the a or b subunits of farnesyl transferase of canola plants (Wang *et al.*, 2005). Engineering signalling factors control the signal output involved in stress resistance and are activated or inactivated during abiotic stress conditions. ABA signalling has specific activation during drought stress in *Arabidopsis* (SRK2A–J/SnRK2.1–10) and rice (SAPK1–10) (Davies *et al.*, 1999). The AAPK (ABA-activated protein kinase) SnRK2 functions in stomata closure in fava bean. The post-translational modification, like phosphorylation by transcription factors AREB 1 and AREB 2, is controlled by protein kinase Snf1. The other signalling factors are 9-*cis*-epoxycarotenoid dioxygenase, SCaBP (SOS3-like calcium-binding protein), SnRK3, CBL1 (calcineurin B-like calcium sensors) and MAPK cascades, which were used for engineering drought-stress tolerance.

5.5 Engineering Functional Proteins for Drought Tolerance

Engineering drought tolerance with functional genes encoding osmolytes in crop plants regulates expression of the drought-related TFs, which in turn produce the functional proteins. In drought-tolerant transgenic crops glycine betaine, polyamines, reducing sugars, mannitol, trehalose and galactinols were produced to maintain the plant water potential. Engineering of drought tolerance with CYP707A3 under stress conditions elevated drought tolerance with reduction of transpiration rate in *Arabidopsis* (Umezawa *et al.*, 2006). Transgenic *Arabidopsis* plants express higher levels of the betaine synthesis genes by co-expression of N-methyltransferase genes, which catalyse the biosynthetic pathway of betaine from glycine of the novel pathway (Waditee *et al.*, 2005). ABA signalling in the cells produces the active metabolites in the cytoplasm. Stress-responsive ROS, LEA proteins (late-embryogenesis-abundant), detoxify the cellular membrane and prevent the plant organelles from degrading by regulating the water potential inside the cell in drought and cold stress. Drought stress by metabolomic engineering leads to sustainable management of the water deficit environment.

5.6 Conclusion

Genetic engineering in crop plants can enhance tolerance to abiotic stress enabling them to overcome climate change, which, in turn, can lead to sustainable agriculture. Biotechnology and molecular biology cutting edge technology tools target genes that regulate the drought stress and enhance the tolerance in plants. Manipulation of plants with regulatory proteins and regulatory RNAs, TFs and chaperones will be a boon to farmers for growing crops in drought-prone areas. ABA-dependent and -independent signalling pathway engineering in over-expression of transcription factors in transgenic crop plants will improve the production of the crop as well as increasing the farmer's income. Transgenic crop plants with high water use efficiency can be cultivated in water-limited areas, thereby withstanding the heat due to global warming and producing more food for the growing population. Hence, genetically modifying plants for drought tolerance is one of the technologies to solve the problem of water stress in agriculture.

References

Abe, H., Yamaguchi-Shinozaki, K., Urao, T., Iwasaki, T., Hosokawa, D. and Shinozaki, K. (1997) Role of *Arabidopsis* MYC and MYB homologs in drought- and abscisic acid-regulated gene expression. *Plant Cell* 9, 1859–1868.

Baker, S.S., Wilhelm, K.S. and Thomashow, M.F. (1994) The 5'-region of *Arabidopsis thaliana* cor15a has cis-acting elements that confer cold-, drought- and ABA-regulated gene expression. *Plant Molecular Biology* 24, 701–713.

Borsani, O., Zhu, J., Verslues, P.E., Sunkar, R. and Zhu, J.K. (2005) Endogenous siRNAs derived from a pair of natural *cis*-antisense transcripts regulate salt tolerance in *Arabidopsis*. *Cell* 123, 1279–1291.

Chaves, M.M. and Oliveira, M.M. (2004) Mechanisms underlying plant resilience to water deficits: prospects for water-saving agriculture. *Journal of Experimental Botany* 55, 2365–2384.

Chen, T.H.H. and Murata, N. (2002) Enhancement of tolerance of abiotic stress by metabolic engineering of betaines and their compatible solutes. *Current Opinion in Plant Biology* 5, 250–257.

Cutler, S., Ghassemian, M., Bonetta, D., Cooney, S. and McCourt, P. (1996) A protein farnesyl transferase involved in abscisic acid signal transduction in *Arabidopsis*. *Science* 273, 1239–1241.

Davies, J.P., Yildiz, F.H. and Grossman, A.R. (1999) Sac3, an Snf1-like serine/threonine kinase that positively and negatively regulates the responses of *Chlamydomonas* to sulfur limitation. *Plant Cell* 11, 1179–1190.

Dinneny, J.R., Long, T.A., Wang, J.Y., Jung, J.W., Mace, D., Pointer, S., Barron, C., Brady, S.M., Schiefelbein, J. and Benfey, P.N. (2008) Cell identity mediates the response of *Arabidopsis* roots to abiotic stress. *Science* 320, 942–945.

Gao, M.J., Allard, G., Byass, L., Flanaganl, A.M. and Singh, J. (2002) Regulation and characterization of four CBF transcription factors from *Brassica napus*. *Plant Molecular Biology* 49, 459–471.

Garreton, V., Carpinelli, J., Jordana, X. and Holuigue, L. (2002) the as-1 promoter element is an oxidative stress-responsive element and salicylic acid activates it via oxidative species. *Plant Physiology* 130, 1516–1526.

Guiltinan, M.J., Marcotte, W.R. and Quatrano, R.S. (1990) A plant leucine zipper protein that recognizes an abscisic acid response element. *Science* 250, 267–271.

Gusmaroli, G., Tonelli, C. and Mantovani, R. (2001) Regulation of the CCAAT-Binding NF-Y subunits in *Arabidopsis thaliana*. *Gene* 264, 173–185.

Hu, H., Dai, M., Yao, J., Xiao, B., Li, X., Zhang, O. and Xiong, L. (2006) Overexpressing a NAM, ATAF, and CUC (NAC) transcription factor enhances drought resistance and salt tolerance in rice. *Proceedings of the National Academy of Sciences USA* 103, 12987–12992.

Huang, B. and Liu, J.Y. (2006) Cloning and functional analysis of the novel gene GhDBP3 encoding a DRE-binding transcription factor from *Gossypium hirsutum*. *Biochimica et Biophysica Acta* 1759, 263–269.

Huang, B., Jin, L.G. and Liu, J.Y. (2007) Molecular cloning and functional characterization of a DREB1/CBF-like gene (GhDREB1L) from cotton. *Science China Life Sciences* 50(1), 7–14.

Huang, B., Jin, L.G. and Liu, J.Y. (2008) Identification and characterization of the novel gene GhDBP2 encoding a DRE-binding protein from cotton (*Gossypium hirsutum*). *Journal of Plant Physiology* 165, 214–223.

Jenkins, T.H., Li, J., Scutt, C.P. and Gilmartin, P.M. (2005) Analysis of members of the *Silene latifolia* Cys2/His2 zinc-finger transcription factor family during dioecious flower development and in a novel stamen-defective mutant ssf1. *Planta* 220, 559–571.

Jiang, C.B. and Singh, J. (1996) Requirement of a CCGAC cis-acting element for cold induction of the BN115 gene from winter *Brassica napus*. *Plant Molecular Biology* 30, 679–684.

Jiang, W. and Yu, D. (2009) *Arabidopsis* WRKY2 transcription factor mediates seed germination and postgermination arrest of development by abscisic acid. *BMC Plant Biology* 9, 96.

Kim, S.G., Kim, S.Y. and Park, C.M. (2007) A membrane-associated NAC transcription factor regulates salt-responsive flowering via FLOWERING LOCUS T in *Arabidopsis*. *Planta* 226, 647–654.

Kim, S.G., Lee, A.K., Yoon, H.K. and Park, C.M. (2008) A membrane-bound NAC transcription factor NTL8 regulates gibberellic acid-mediated salt signaling in *Arabidopsis* seed germination. *Plant Journal* 55, 77–88.

Kizis, D. and Pagès, M. (2002) Maize DRE-binding proteins DBF1 and DBF2 are involved in rab17 regulation through the drought-responsive element in an ABA-dependent pathway. *Plant Journal* 30, 679–689.

Li, W.X., Oono, Y., Zhu, J., He, X.J., Wu, J.M., Iida, K., Lu, X.Y., Cui, X., Jin, H. and Zhu, J.K. (2008) The *Arabidopsis* NFYA5 transcription factor is regulated transcriptionally and posttranscriptionally to promote drought resistance. *Plant Cell* 20, 2238–2251.

Liao, Y., Zou, H., Wei, W., Hao, Y.J., Tian, A.G., Huang, J., Liu, Y.F., Zhang, J.S. and Chen, S.Y. (2008) Soybean GmbZIP44, GmbZIP62 and GmbZIP78 genes function as negative regulator of ABA signaling and confer salt and freezing tolerance in transgenic *Arabidopsis*. *Planta* 228, 225–240.

Liu, Q.M., Kasuga, Y., Sakuma, H., Abe, S., Miura, K., Yamaguchi-Shinozaki, K. and Shinozaki, K. (1998) Two transcription factors, DREB1 and *DREB2*, with an EREBP/AP2 DNA binding domain separate two cellular signal transduction pathways in drought and low-temperature responsive gene expression, respectively, in *Arabidopsis*. *Plant Cell* 10, 1391–1406.

Liu, Z.B., Ulmasov, T., Shi, X., Hagen, G. and Guilfoyle, T. (1994) Soybean GH3 promoter contains multiple auxin-inducible elements. *Plant Cell* 6, 645–657.

Loake, G.J., Faktor, O., Lamb, C.J. and Dixon, R.A. (1992) Combination of H-box [CCTACC (N) 7CT] and G-box (CACGTG) cis-elements is necessary for feed-forward stimulation of a chalcone synthase promoter by the phenylpropanoid-pathway intermediate p-coumaric acid. *Proceedings of the National Academy of Sciences USA* 89, 9230–9234.

Maruyama, K., Sakuma, Y., Kasuga, M., Ito, Y., Seki, M., Goda, H., Shimada, Y., Yoshida, S., Shinozaki, K. and Yamaguchi-Shinozaki, K. (2004) Identification of cold-inducible downstream genes of the *Arabidopsis* DREB1A/CBF3 transcriptional factor using two microarray systems. *Plant Journal* 38, 982–993.

Mason, H.S., DeWald, D.B. and Mullet, J.E. (1993) Identification of a methyl jasmonate-responsive domain in soybean vspB promoter. *Plant Cell* 5, 241–251.

McKendree, W.L. and Ferl, R.J. (1992) Functional elements of the *Arabidopsis* Adh promoter include the G-box. *Plant Molecular Biology* 19, 859–862.

Nakashima, K., Tran, L.S.P., Nguyen, D.V., Fujita, M., Maruyama, K., Todaka, D., Ito, Y., Hayashi, N., Shinozaki, K. and Yamaguchi-Shinozaki, K. (2007) Functional analysis of a NAC-type transcription factor OsNAC6 involved in abiotic and biotic stress-responsive gene expression in rice. *Plant Journal* 51, 617–630.

Nelson, D.E., Repetti, P.P. and Heard, J.E. (2007) Plant nuclear factor Y (NF-Y) B subunits confer drought tolerance and lead to improved corn yields on water-limited acres. *Proceedings of the National Academy of Sciences USA* 104, 16450–16455.

Oh, S.J., Song, S.I., Kim, Y.S., Jang, H.J., Kim, S.Y., Kim, M., Kim, Y.K., Nahm, B.H. and Kim, J.K. (2005) *Arabidopsis* CBF3/DREB1A and ABF3 in transgenic rice increased tolerance to abiotic stress without stunting growth. *Plant Physiology* 138, 341–351.

Olsen, A.N., Ernst, H.A., Lo Leggio, L. and Skriver, K. (2005) DNA binding specificity and molecular functions of NAC transcription factors. *Plant Science* 169, 785–797.

Qin, X.F., Holuigue, L., Horvath, D.M. and Chua, N.H. (1994) Immediate early transcription activation by salicylic acid via the cauliflower mosaic virus as-1 element. *Plant Cell* 6, 863–874.

Ren, X., Chen, Z., Liu, Y., Zhang, H., Zhang, M., Liu, Q., Hong, X., Zhu, J.K. and Gong, Z. (2010) ABO3, a WRKY transcription factor, mediates plant responses to abscisic acid and drought tolerance in *Arabidopsis*. *Plant Journal* 63, 417–429.

Rushton, P.J., Macdonald, H., Huttly, A.K., Lazarus, C.M. and Hooley, R. (1995) Members of a new family of DNA-binding proteins bind to a conserved cis-element in the promoters of alpha-Amy2 genes. *Plant Molecular Biology* 29, 691–702.

Rushton, P.J., Somssich, I.E., Ringler, P. and Shen, Q.J. (2010) WRKY transcription factors. *Trends in Plant Science* 15, 247–258.

Sakuma, Y., Liu, Q., Dubouzet, J.G., Abe, H., Shinozaki, K. and Yamaguchi-Shinozaki, K. (2002) DNA-binding specificity of the AP2/ERF domain of *Arabidopsis* DREBs, transcription factors involved in dehydration- and cold-inducible gene expression. *Biochemical and Biophysical Research Communications* 290, 998–1009.

Sakuma, Y., Maruyama, K., Qin, F., Osakabe, Y., Shinozaki, K. and Yamaguchi-Shinozaki, K. (2006a) Dual function of an *Arabidopsis* transcription factor DREB2A in water-stress responsive and heat-stress-responsive gene expression. *Proceedings of the National Academy of Sciences USA* 103, 18822–18827.

Sakuma, Y., Maruyama, K., Osakabe, Y., Qin, F., Seki, M., Shinozaki, K. and Yamaguchi-Shinozaki, K. (2006b) Functional analysis of an *Arabidopsis* transcription factor, DREB2A, involved in drought-responsive gene expression. *Plant Cell* 18, 1292–1309.

Savitch, L.V., Allard, G., Seki, M., Robert, L.S., Tinker, N.A., Huner, N.P.A., Shinozaki, K. and Singh, J. (2005) The effect of overexpression of two *Brassica* CBF/DREB1-like transcription factors on photosynthetic capacity and freezing tolerance in *Brassica napus*. *Plant Cell Physiology* 46, 1525–1539.

Schumann, U., Prestele, J., O'Geen, H., Brueggeman, R., Wanner, G. and Gietl, C. (2007) Requirement of the C3HC4 zinc RING finger of the *Arabidopsis* PEX10 for photorespiration and leaf peroxisome contact with chloroplasts. *Proceedings of the National Academy of Sciences USA* 104, 1069–1074.

Simpson, S.D., Nakashima, K., Narusaka, Y., Seki, M., Shinozaki, K. and Yamaguchi-Shinozaki, K. (2003) Two different novel cis-acting elements of erd1, a clpA homologous *Arabidopsis* gene, function in induction by dehydration stress and dark-induced senescence. *Plant Journal* 33, 259–270.

Somerville, C. and Briscoe, J. (2001) Genetic engineering and water. *Science* 292, 2217.

Stockinger, E.J., Gilmour, S.J. and Thomashow, M.F. (1997) *Arabidopsis thaliana* CBF1 encodes an AP2 domain-containing transcriptional activator that binds to the C-repeat/DRE, a cis-acting DNA regulatory element that stimulates transcription in response to low temperature and water deficit. *Proceedings of the National Academy of Sciences USA* 94, 1035–1040.

Storozhenko, S., Pauw, P.D., Montagu, M.V., Inzé, D. and Kushnir, S. (1998) The heat-shock element is a functional component of the *Arabidopsis* APX1 gene promoter. *Plant Physiology* 118, 1005–1014.

Tran, L.S., Nakashima, K., Sakuma, Y., Simpson, S.D., Fujita, Y., Maruyama, K., Fujita, M., Seki, M., Shinozaki, K. and Yamaguchi-Shinozaki, K. (2004) Isolation and functional analysis of *Arabidopsis* stress-inducible NAC transcription factors that bind to a drought-responsive cis-element in the *early responsive to dehydration stress1* promoter. *Plant Cell* 16, 2481–2498.

Tran, L.S., Nakashima, K., Sakuma, Y., Osakabe, Y., Qin, F., Simpson, S.D., Maruyama, K., Fujita, Y., Shinozaki, K. and Yamaguchi-Shinozaki, K. (2007) Co-expression of the stress-inducible zinc finger homeodomain ZFHD1 and NAC transcription factors enhances expression of the ERD1 gene in *Arabidopsis*. *Plant Journal* 49, 46–63.

Umezawa, T., Okamoto, M., Kushiro, T., Nambara, E., Oono, Y., Seki, M., Kobayashi, M., Koshiba, T., Kamiya, Y. and Shinozaki, K. (2006) CYP707A3, a major ABA 8'-hydroxylase involved in dehydration and rehydration response in *Arabidopsis thaliana*. *The Plant Journal* 46, 171–182.

Uno, Y., Furihata, T., Abe, H., Yoshida, R., Shinozaki, K. and Yamaguchi-Shinozaki, K. (2000) *Arabidopsis* basic leucine zipper transcription factors involved in an abscisic acid-dependent signal transduction pathway under drought and high-salinity conditions. *Proceedings of the National Academy of Sciences USA* 97, 11632–11637.

Vogel, J.T., Zarka, D.G., Van Buskirk, H.A., Fowler, S.G. and Thomashow, M.F. (2005) Roles of the CBF2 and ZAT12 transcription factors in configuring the low temperature transcriptome of *Arabidopsis*. *The Plant Journal* 412, 195–211.

Waditee, R., Bhuiyan, M.N., Rai, V., Aoki, K., Tanaka, Y., Hibino, T., Suzuki, S., Takano, J., Jagendorf, A.T. and Takabe, T. (2005) Genes for direct methylation of glycine provide high levels of glycinebetaine and abiotic-stress tolerance in *Synechococcus* and *Arabidopsis*. *Proceedings of the National Academy of Sciences USA* 102, 1318–1323.

Walley, J.W., Coughlan, S., Hudson, M.E., Covington, M.F. and Kaspi, R. (2007) Mechanical stress induces biotic and abiotic stress responses via a novel *cis*-element. *PLoS Genetics* 3, 172.

Wang, Y., Ying, J., Kuzma, M., Chalifoux, M., Sample, A., McArthur, C., Uchacz, T., Sarvas, C., Wan, J. and Dennis, D.T. (2005) Molecular tailoring of farnesylation for plant drought tolerance and yield protection. *Plant Journal* 43, 413–424.

Xue, G.P. (2002) Characterisation of the DNA-binding profile of barley HvCBF1 using an enzymatic method for rapid, quantitative and high-throughput analysis of the DNA-binding activity. *Nucleic Acids Research* 30, 77.

Yamaguchi-Shinozaki, K. and Shinozaki, K. (1994) A novel cis-acting element in an *Arabidopsis* gene is involved in responsiveness to drought, low-temperature, or high-salt stress. *Plant Cell* 6, 251–264.

Zarka, D.G., Jonathan, T., Vogel, J.T., Cook, D. and Thomashow, M.F. (2003) Cold induction of *Arabidopsis* CBF genes involves multiple ICE (Inducer of CBF Expression) promoter elements and a cold-regulatory circuit that is desensitized by low temperature. *Plant Physiology* 133, 910–918.

6 Physiology and Biochemistry of Salt Stress Tolerance in Plants

André Dias de Azevedo Neto[1]* and Elizamar Ciríaco da Silva[2]

[1]*Laboratory of Biochemistry, Centre of Exact and Technological Sciences, Federal University of Recôncavo of Bahia, Cruz das Almas, Brazil;* [2]*Laboratory of Applied Botany, Department of Biology, Federal University of Sergipe, Aracaju, Brazil*

Abstract

Salinity is one the major environmental stresses affecting crop production worldwide. The salt effects on plants include osmotic stress, ion toxicity, nutrient imbalance and deficiencies, resulting in membrane damage, decreased cell expansion and division, changes in metabolic processes, oxidative stress and genotoxicity. Thus plant salt tolerance is a highly complex phenomenon that involves alterations in physiological and biochemical processes, which may result in morphological and developmental changes. In this scenario, the regulation of uptake, transport and compartmentation of Na^+ and Cl^-, biosynthesis of compatible solute and specific proteins, reduction of reactive oxygen species formation and increase of antioxidant defence system have been related as important mechanisms for salt tolerance. In this chapter, we give an overview of the physiological, biochemical and molecular mechanisms underlying salt tolerance, combining knowledge from classic physiology with recent findings. Special emphasis will be given on salt signal perception and transduction and mechanisms related to maintenance of osmotic, ionic, biochemical and redox homoeostasis in salt-stressed plants. A fundamental biological knowledge in conjunction with understanding about the effects of salt stress on plants is essential to supply additional information for a thorough analysis of the plant salt-tolerance mechanisms and reduce the deleterious effects of salinity on plants, improving crop productivity important to agricultural sustainability.

6.1 Introduction

The expansion of agriculture has led to the increased use of marginal areas for crop production, such as saline soils. In addition, the inadequate use of irrigation and drainage practices has induced soil salinization. Most researchers believe that the solution for the salinity problems in agricultural production depends on understanding the physiology and biochemistry of plants grown under these conditions.

In this scenario, it is quite natural that a knowledge of salinity tolerance mechanisms is required for the development of cultivars that produce economically under salt conditions. The regulation of uptake, transport and compartmentalization of Na^+ and Cl^-, biosynthesis of compatible solute and specific proteins, reduction of reactive oxygen species (ROS) formation and increase of antioxidant defence system have been related as important mechanisms for salt tolerance. In addition, such knowledge may contribute to the

*E-mail: andre@ufrb.edu.br

development of crop management techniques that increase plant salt tolerance. Despite the importance of these studies, the research has only started to show promising results in the last five decades.

In this chapter, we give an overview of the physiological, biochemical and molecular mechanisms underlying salt tolerance, combining knowledge from classic physiology with recent findings. Special emphasis will be given on salt signal perception and transduction and mechanisms related to maintenance of osmotic, ionic, biochemical and redox homoeostasis in salt-stressed plants.

6.2 Salinity Effects on Growth and Development

The effects of salt-induced metabolism alterations on growth and development of plants are the result of the plant and stress interactions (Fig. 6.1). In this figure, it can be observed that the stress characteristics affecting the plants are the ion composition and concentration of cultivation media (soil or nutrient solution), the time of exposure and the number of exposures to stress, if stress was imposed gradually or abruptly, and the combination of stresses. The plant characteristics related to stress response are the genotype, the development stage, and organ or tissue submitted to stress. The relationship between stress and plant characteristics can result in tolerance or susceptibility, that is, the life or death of the plant (Bray et al., 2000).

Salinity has two components that are responsible for stress: the osmotic and the ionic component (Fig. 6.2). In the short term salinity alters water and nutrient absorption and the membrane integrity (Läuchli and Grattan, 2007; Munns and Tester, 2008). These changes will affect the osmotic and ionic homoeostasis and trigger a sequence of reactions that lead to changes in the metabolism, gas exchanges, hormonal balance and reactive oxygen species (ROS) production, reducing cell expansion and division. As a result, the vegetative and reproductive growth is reduced and the tissue senescence is increased. Thus, a sequence of events that occurs from stress exposure to observation of effects on plants will determine the response to salinity and salt tolerance is related to the rate at which salt reaches toxic levels in leaves. Timescale is minutes, hours, days, weeks or months, depending on the stress level, the species studied and the parameter evaluated.

Recognition of the importance of time was the basis of the two-phase model describing

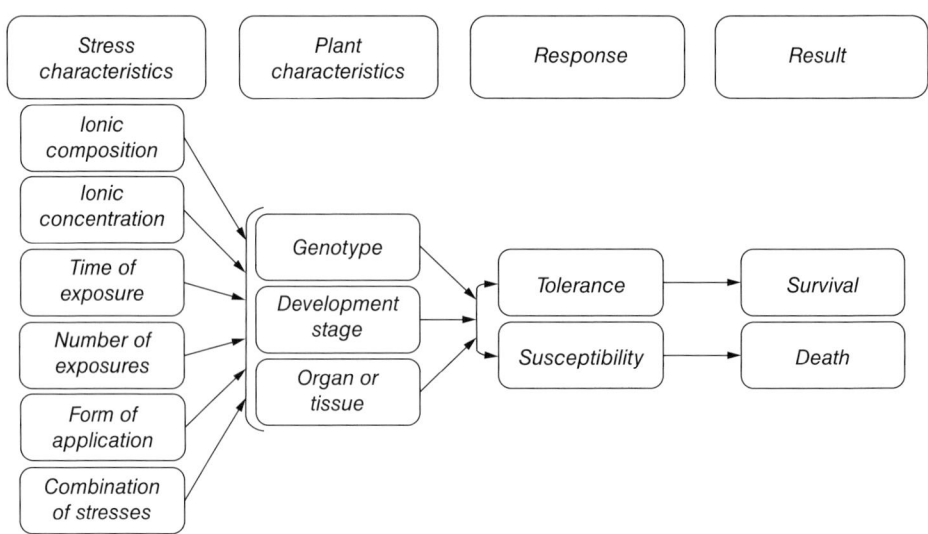

Fig. 6.1. Factors that determine how plants respond to salt stress (adapted from Bray et al., 2000).

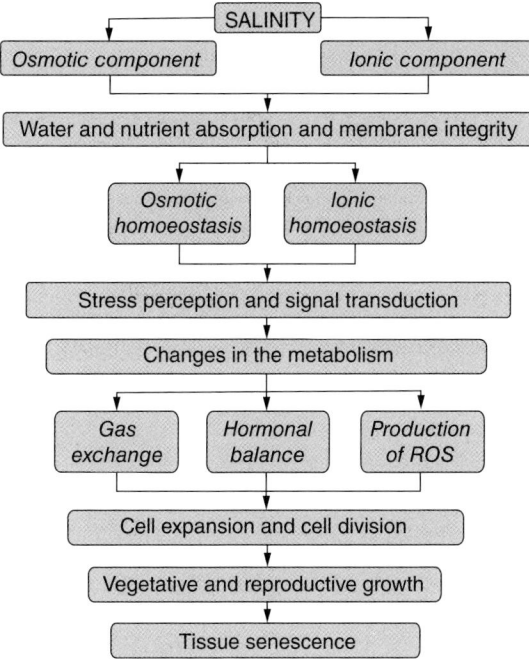

Fig. 6.2. Physiological and biochemical changes occurring in plants when submitted to salt stress (adapted from Azevedo Neto et al., 2008).

the osmotic and ionic effects of salt stress on plant growth (Munns et al., 1995; Munns, 2002). This is very important when screening plants for salt tolerance. The first phase of salinity-induced growth reduction is quickly apparent, and results from the salt in the root medium. It is essentially a water-stress or osmotic phase, for which there is surprisingly little genotypic variation. Growth reduction is presumably regulated by hormonal signals coming from the roots. Then there is a second phase of growth reduction, which takes time to develop, and results from internal injury. It is due to accumulation of salts to excessive levels in leaves, exceeding the ability of the cells to compartmentalize Na^+ and Cl^- in the vacuole. This will inhibit the growth of younger leaves by a reduced carbohydrate supply to the growing cells. It is also possible that the metabolic disorders are the result of acclimatory changes required for plants to withstand the salt stress.

As cells are exposed to salinity, carbon flux can be altered to support the low-molecular-weight organic solute biosynthesis and the production of energy necessary for this biosynthesis and other essential metabolic processes required for salt-stress acclimation. Such a situation may result in a substantial partitioning of carbon away from growth processes (Binzel et al., 1985). This is what clearly separates species and genotypes that differ in the ability to tolerate saline soil, and is considered the second phase of the growth response to salt stress.

6.3 Stress Perception

In plant cells, the plasma membrane is the site where the processing of the signals involved in the response to salt stress occurs. This processing is performed by receptors that either recognize the osmotic or the ionic component. Direct or indirect mechanisms of perception can be used by a cell to sense the osmotic stress. A direct mechanism for determining the water activity would be through a direct osmosensor, which would act as a chemosensor, that is, as a classical ligand-specific receptor (Wood *et al.*, 2001). However, osmotic stress changes several cellular properties. Thus, an indirect osmosensor could detect some cellular properties affected by osmotic changes in the external media. Potentially significant properties include cell volume, turgor pressure and membrane strain as well as the activities of organic and inorganic solutes or the macromolecular crowding in the periplasm or cytoplasm (Wood *et al.*, 2001). In this context, histidine kinases and their response regulators are promising candidates as receptors for salt and osmotic stress. Several biochemical activities are related to histidine kinases: ATP binding, autophosphorylation, phosphor-donor for their response regulators and, in many cases, catalysis of the hydrolytic dephosphorylation of their response regulator (Amin *et al.*, 2013). Histidine kinases receptors have been identified as osmosensors in prokaryotes and in yeast as well as in other organisms (Maeda *et al.*, 2006; Heermann *et al.*, 2009; Heermann and Jung, 2010).

In *Arabidopsis thaliana*, a plasma membrane protein, AtHK1 (*Arabidopsis thaliana* histidine kinase 1), was proposed as the osmosensor as early as the late 1930s (Urao *et al.*, 1999; Tran *et al.*, 2007). During seed maturation, this protein is also involved in the drying process regulation. This process seems to be related to drought tolerance (Wohlbach *et al.*, 2008). Similar osmosensors have also been identified in the woody plant *Populus deltoides* and other plant species (Chefdor *et al.*, 2006). The osmotic stress activates the synthesis of abscisic acid (ABA), which can up-regulate the transcription of *AtNHX1*, the gene responsible for coding the vacuolar Na^+/H^+ exchanger (Shi and Zhu, 2002; Yokoi *et al.*, 2002b). Thus, it has been hypothesized that in the perception machinery, histidine kinases are probably very early elements. Besides their role in hormone signalling, they play key roles in regulating the responses to salt and osmotic stresses. Although little is known about how Na^+ is sensed in any cellular system, in recent years there has been significant progress with regard to plant response under salt stress. In mutants of *A. thaliana* with salt hypersensitivity, the discovery of *Salt Overly Sensitive* (*SOS*) led to a better understanding of signal perception and transduction. The *SOS* mutants *1*, *2* and *3* are involved in a salt-induced signalling pathway, and a number of intermediates of this signalling pathway have been identified (Qiu *et al.*, 2004). The SOS1 protein is a plasma membrane Na^+/H^+ antiporter that has 10–12 transmembrane domains and a long C-terminal tail, which extends into the cytoplasm of the cell and interacts with RCD1, which has been shown to regulate oxidative stress responses in *Arabidopsis*. This indicates that there is some level of cross-talk between the pathways of tolerance to salinity and oxidative stress (Katiyar-Agarwal *et al.*, 2006). SOS2 is a serine/threonine protein kinase, a member of the sucrose non-fermenting 1-related protein family (Plett *et al.*, 2010; Quintero *et al.*, 2011). SOS3 is a myristoylated calcium-binding protein that responds to increases in the cytosolic concentrations of Ca^{2+} (Liu and Zhu, 1997; Ishitani *et al.*, 2000; Gong *et al.*, 2004).

Before entering the cell, extracellular Na^+ can be sensed by a transmembrane receptor. Alternatively, membrane proteins or a number of Na^+-sensitive cytoplasmic enzymes may sense intracellular Na^+ after entering the cell. However, both mechanisms can also be used by the cell to sense the Na^+ (Zhu, 2003; Chinnusamy *et al.*, 2005). The Na^+/H^+ antiporter *SOS1* is a possible Na^+ sensor in the plasma membrane. In *Arabidopsis* cells this antiporter is also essential for Na^+ efflux to apoplast (Shi *et al.*, 2000, 2002; Qiu *et al.*, 2002). Recently, a protein containing arabinogalactan residue (SOS5) was located on the outer surface of plasma membranes. This has been identified as a candidate for detecting Na^+_{ext} (Mahajan *et al.*, 2008; Türkan and Demiral, 2009). Some studies have reported that in addition to its importance for normal cell expansion, SOS5 encodes a protein of cell surface adhesion (Shi *et al.*, 2003; Plett *et al.*, 2010).

6.4 Signal Transduction

The perception of both osmotic and ionic stress components induces an increase in Ca^{2+}_{cyt}, a secondary messenger. Salt-induced osmotic stress triggers a short period of time (1 min) increase in Ca^{2+}_{cyt}, which results from the influx across the plasma membrane. The osmotic signal also up-regulates the ABA biosynthesis (Jia et al., 2002; Xiong and Zhu, 2003) and causes accumulation of ROS (Smirnoff, 1993; Hernández et al., 2001). ABA and ROS also regulate ionic and osmotic homoeostasis as well as stress-damage control and repair processes (Chinnusamy et al., 2005).

Na$^+$ sensors also control the Ca$^{2+}_{cyt}$, which exerts at least two roles in salt tolerance: a critical signalling function (the SOS signalling pathway) for the regulation of Na$^+$ homoeostasis leading to plant acclimation, and an inhibitory effect on the Na$^+$ entry system. The increase in Ca$^{2+}_{cyt}$ activates SOS3, the calcium sensor protein. Then, SOS3 binds to and activates SOS2, the serine/threonine protein kinase. The attachment of SOS3/SOS2 protein kinase complex phosphorylates and activates SOS1 and NHX1, a tonoplast Na$^+$/H$^+$ antiporter. This results in Na$^+$ efflux and vacuolar compartmentalization, contributing to Na$^+$ ion homoeostasis (Halfter et al., 2000; Ishitani et al., 2000; Liu et al., 2000; Quintero et al., 2002).

Several results suggest that the SOS pathway includes not only regulation of SOS1 and NHX1 activities, but also the regulation of other transporters. SOS3/SOS2 protein kinase complex probably down-regulates the activity of HKT1, a low-affinity Na$^+$ transporter (Mahajan and Tuteja, 2005), which mediates Na$^+$ entry into the root cells of *Arabidopsis* (Uozumi et al., 2000; Zhu, 2002). Activated SOS2 regulates the activity of vacuolar H$^+$/Ca^{2+} antiporter CAX1, independently of SOS3, resulting in Ca^{2+} homoeostasis maintenance. Similarly, SOS2 also interacts with ABI2 that negatively regulates ion homoeostasis after a period of stress either by dephosphorylating SOS2 or the proteins that are phosphorylated by SOS2 (Zhu, 2002, 2003).

Other Ca^{2+}-dependent signalling proteins seem to be involved in the abiotic stress acclimation. In *A. thaliana* a parallel pathway to mitogen-activated protein kinase (MAPK) signalling is provided by Ca^{2+}-dependent protein kinase 3 (CPK3) (Mehlmer et al., 2010; Wurzinger et al., 2011). The Ca^{2+} signalling is recognized as important as the case of SOS1, although the components that trigger the stress-induced Ca^{2+} signalling still remain unknown.

Salt-dependent protein–nucleic acid interactions have recently been suggested as a new osmoperception mechanism (Novak et al., 2011). In *Synechocystis* sp. PCC6803, a moderately halotolerant cyanobacterium, glucosylglycerol is synthesized as compatible solute. The internal salt concentration could serve as a trigger for activation and regulation of the key enzyme of the glucosylglycerol pathway (GgpS). This enzyme is non-competitively inhibited via a salt-dependent electrostatic interaction at low salt concentrations. An increase in salt concentration releases GgpS and consequently, the accumulation of glucosylglycerol and the acclimatization to salt stress. It is still unclear if this mechanism is also conserved in plants, but this question should be answered by future research.

6.5 Ion Homoeostasis

6.5.1 Sodium uptake

The increase in concentration of salts in the soil solution increases the flow of ions toward the epidermal cells of roots, resulting in a high ionic concentration in the apoplast. Sodium entry into the root cells is a passive process mediated by ion channels or uniporter-type transporters. The main systems for transporting Na$^+$ into the plant cells are high-affinity potassium transporters (HKT), low-affinity cation transporters (LCT), voltage insensitive cation (VIC) channels and non-selective cation (NSCC) channels (Apse and Blumwald, 2007). Although the role of each transport system may vary within species and/or growth conditions, there is evidence suggesting a coordinated action of all transport systems mediating Na$^+$ uptake into the roots.

HKT-type transporters have been characterized in several plant and bacterial species and seem to operate as Na$^+$-selective uniporters and as Na$^+$/K$^+$ symporters (Garciadeblas et al., 2003; Horie and Schroeder, 2004). In rice, the expression of some members of HKT family supports its potential role in root influx of Na$^+$. OsHKT1 is a Na$^+$ transporter and OsHKT2 a Na$^+$/K$^+$ co-transporter. In leaves and roots of rice cultivars, these transcripts were detected. Under salt stress, OsHKT1 transcript was significantly down-regulated in Pokkali (salt-tolerant) but not in

BRRI Dhan29 (salt-sensitive) rice cultivars, suggesting a reduced influx of Na$^+$ in the salt-tolerant cultivar. The OsHKT2 expression was also induced by NaCl stress in both cultivars. However, in salt-tolerant, OsHKT2 transcript was induced immediately after NaCl stress, and in salt-sensitive the induction of OsHKT2 was quite low when compared to cv. Pokkali (Kader et al., 2006). As a result, a salt-tolerant rice cultivar contains less Na$^+$ in both roots and shoots than a salt-sensitive rice cultivar (Golldack et al., 2003; Kader and Lindberg, 2005). In wheat at least two transport modes have been shown by TaHKT1, a Na$^+$/K$^+$ symporter and Na$^+$ influx at high Na$^+$ concentrations (Rubio et al., 1995). TaHKT1 down-regulation reduced Na$^+$ accumulation in the roots, resulting in increased salt tolerance (Laurie et al., 2002).

Although the HKT-type transporters represent an important pathway for the Na$^+$ influx into plant cells, ion channels are considered to be the main pathway mediating the transport of this ion. Thus, the evidence supports the hypothesis that non-selective cation (NSCC) channels are the main transport system for Na$^+$ entry into the root cells (Tester and Davenport, 2003). Although there are many candidate genes that could encode NSCC channels, their identity remains uncertain. There are two families of NSCC that could be NSCC channels, the CNGCs (cyclic nucleotide-gated channels) (Leng et al., 2002) and GLRs (glutamate-activated channels) (Tester and Davenport, 2003; Demidchik et al., 2004).

6.5.2 Sodium extrusion and compartmentation

Ion homoeostasis is an essential determinant for plant salt tolerance. Beyond water deficit, salt-induced changes in K$^+$/Na$^+$ ratios impose an ion-specific stress. These changes in K$^+$/Na$^+$ ratios is due to the Na$^+$ influx through the K$^+$ transport systems. A physiological K$^+$ concentration between 100 and 150 mM is necessary for in vitro protein synthesis, but is inhibited by Na$^+$ concentrations above 100 mM (Wyn Jones and Pollard, 1983). Therefore, Na$^+$ extrusion and/or Na$^+$ compartmentation are two fundamental processes for detoxification of cytosolic Na$^+$, cell osmotic adjustment and maintenance of a high K$^+$/Na$^+$ ratio in the cytosol, which are required for salt-stress tolerance.

However, for Na$^+$ extrusion and compartmentation, plant cells use the secondary active transport. Na$^+$/H$^+$ exchangers (antiporters) mediate the Na$^+$ compartmentation within the vacuole and the Na$^+$ extrusion from the cell. In plants, the H$^+$-ATPase in plasma membrane and two H$^+$-ATPases (V-ATPase and the V-PPiase) in tonoplast are the primary mechanism for Na$^+$ extrusion (Sussman, 1994). The proton translocation out by these electrogenic pumps will therefore generate an electrochemical H$^+$ gradient. This proton motive force enables the operation of Na$^+$/H$^+$ antiport systems involving the downhill movement of H$^+$ along its electrochemical gradient to the Na$^+$ extrusion against its electrochemical gradient.

Under physiological conditions, this electroneutral exchange of Na$^+$ for H$^+$ is the only mode of transport that has been determined for Na$^+$ efflux in plants. Two antiporter systems carry out the removal of Na$^+$ from cytosol, a Na$^+$/H$^+$ antiporter located in the plasma membrane (SOS1) and another located in the tonoplast (NHX1) (Blumwald et al., 2000).

In the root system, SOS1 can be found in plasma membrane of epidermal cells excluding the cytosolic Na$^+$ and in plasma membrane of cells surrounding the stele or on parenchyma cells adjacent to xylem. In the latter case, under high transpiration conditions the SOS1-mediated Na$^+$ efflux may lead to a higher xylem Na$^+$ concentration, which could lead to the dangerous Na$^+$ accumulation in the shoot. Thus, this transport system is a Na$^+$ exclusion mechanism that may be related directly with another mechanism regulating the transport and distribution of Na$^+$ within the plant. Beyond the Na$^+$/H$^+$ antiporter activity in the plasma membrane, SOS1 is a possible sensor of extracellular Na$^+$.

Similarly to SOS1, NHX1 is able to realize Na$^+$/H$^+$ active secondary transport using the electrochemical H$^+$ gradient generated across the tonoplast by the two H$^+$ pumps, the V-ATPase and V-PPiase. Vacuolar sequestration of Na$^+$ is an important mechanism for salinity tolerance by reducing the Na$^+$ concentrations in the cytosol and its toxic effects on metabolism during plant growth under salt-stress conditions (Blumwald et al., 2000; Anil et al., 2007). In transgenic *Arabidopsis*, the over-expression of AtNHX1 increased salt (200 mM) tolerance compared with wild-type plants (Apse et al., 1999). In salt-tolerant cultivars of cotton (Wu et al., 2004) and rice (Anil et al., 2007), a correlation between the gene expression encoding NHX antiporters and the level of salt tolerance was shown. In roots and shoots of salt-tolerant wheat genotypes, the higher expression of vacuolar Na^+/H^+ antiporters also was suggested to be related to salt tolerance (Saquib et al., 2005).

6.6 Osmotic Adjustment

Salt stress affects plants for several reasons, due to the osmotic stress imposed and ion-specific effects. The salt-tolerance mechanisms are characterized to minimize osmotic stress or ion disequilibrium or to minimize the secondary effects caused by this stress. One of these mechanisms recognized to confer salt plant tolerance is osmoregulation, which is recognized as being a cell response when the water content is reduced (Yokoi et al., 2002a) in order to avoid salt-induced dehydration.

Osmoregulation or osmotic adjustment is a net increase in cell solute content, leading to decreases in cell water potential without any decrease in cell volume or turgor (Taiz and Zeiger, 2006). Some controversies regarding the use of this term have been reported, because while osmoregulation is usually discussed as the capacity of cells to decrease their osmotic potential in response to external water stress, osmotic homoeostasis is undoubtedly a more general aspect including changes in cell as well as whole plant physiology and biochemistry (Wyn Jones and Gorham, 1983). Considering that the term osmoregulation is widely reported in the literature and is part of common sense, many particulars relating to the control of solute concentrations at the cellular and whole plant levels must be considered within the general term 'osmoregulation'.

This phenomenon is regulated and maintained by the accumulation of organic compounds of low molecular mass called compatible solutes and inorganic ions (Strange, 2004), making it possible to maintain water absorption and cell turgor pressure, which might contribute to the maintenance of physiological processes, such as stomatal opening, photosynthesis, and cell division and expansion (Serraj and Sinclair, 2002). Osmotic adjustment is thus a significant mechanism of plant acclimation to salinity or drought conditions (Taiz and Zeiger, 2006) and occurs in both halophytes and glycophytes. Under salinity and drought conditions, the K^+ and the organic compounds (compatible solutes) are of considerable importance for the osmotic adjustment in cells with little vacuolation, while K^+, Na^+ and Cl^- ions are the main solutes in highly vacuolated cells (Wyn Jones and Gorham, 1983).

In vacuolated cells subjected to salinity, subcellular compartmentalization is a key factor for solute accumulation. High concentrations of ions into the cytosol can severely inhibit enzymes of plant cells. Thus, an important mechanism of protection is the ion compartmentalization into the vacuoles, which contribute to osmotic adjustment without affecting the enzyme activities either in the cytosol or subcellular organelles. In these cells, the synthesis and accumulation of organic compounds is responsible for maintaining the water potential equilibrium between the vacuole and the cytoplasm (Taiz and Zeiger, 2006). Furthermore, it is well known that the accumulation of organic solutes in the cytoplasm can protect cell membranes, proteins and metabolic machinery that can preserve the subcellular structure of damage resulting from cell dehydration (Serraj and Sinclair, 2002). As the volume of the cytoplasm in the mesophyll cells and mature root cortex cells is, on average, 10% of the total cell volume, the amount of carbon required for organic solute synthesis for osmoregulation is relatively small.

Although organic compounds and inorganic ions play an significant role in the higher plants to maintain the growth processes under salinity conditions, their relative contribution varies among species, cultivars of the same species, among tissues and organs of the same plant and even between different structures of the same cell (Ashraf and Harris, 2004). Thus, the ability to accumulate and compartmentalize inorganic solutes, and the capacity to synthesize and accumulate organic solutes undoubtedly represents an additional factor favouring the growth and development of the plants in saline environments.

The inorganic ions are important for participating in the preservation of plant water potential. The ion storage in the vacuole allows the plant to maintain cell turgor without spending energy on the organic solute synthesis (Martinoia et al., 1986). The energy cost to accumulate 1 mol of NaCl is about seven ATP, while to synthesize organic solutes is much higher (approximately 34 for mannitol, 41 for proline, 50 for glycine betaine and 52 for sucrose) (Raven, 1985; Munns and Tester, 2008). However, this ion accumulation can induce cell toxicity, mineral nutrient deficiencies, or both (Greenway and Munns, 1980; Termaat and Munns, 1986).

The salinity induces an increase in the sodium and chloride ion concentrations both in halophytes (Flowers et al., 1977) and in glycophytes (Greenway and Munns, 1980). In general, excess of sodium in the root zone causes a disequilibrium in several processes such as absorption, transport assimilation of other ions such as K^+, Ca^{2+} and Mg^{2+} (Marschner, 2012). Maize plants submitted to salt stress show decreases in K^+, Ca^{2+} and Mg^{2+} content in shoots as well as in roots (Azevedo Neto and Tabosa, 2000). The increase in the Na^+ concentration with concomitant reduction in the concentrations of K^+, Ca^{2+} and Mg^{2+} increases excessively the relations of Na^+/K^+, Na^+/Ca^{2+} and Na^+/Mg^{2+}, and may cause disturbances in ion homoeostasis.

The principal characteristic of halophytes to osmotic adaptation is the absorbance of ions counterbalanced by organic solutes in the cytoplasm (Breckle, 2002). In terrestrial halophytes, high salt tolerance is primarily achieved by the inclusion of salts and their use for maintaining the leaf turgidity or replacement of Na^+ by K^+ in various metabolic functions (Marschner, 2012). In glycophytes, the main determinant of salt tolerance is the salt exclusion from the shoot (Binzel et al., 1988; Robinson et al., 1997; Munns, 2002), which may occur by the ability to limit the absorption and/or transport the ions (mainly Na^+ and Cl^-) from the root to the shoot, as noted by Chartzoulakis et al. (2002) in six olive cultivars.

A metabolic change common to most plants is the accumulation of organic compounds of low molecular mass. These solutions can include: (i) organic acids (malate, oxalate, etc.); (ii) polyhydroxy compounds such as soluble carbohydrates (glucose, fructose, sucrose, trehalose and raffinose), straight-chain polyols (glycerol, mannitol and sorbitol) and cyclic polyols (inositol, ononitol and pinitol); and (iii) zwitterionic alkylamines such as protein amino acids (arginine, glycine, serine, etc.), non-protein amino acids (citrulline, ornithine, etc.), amino acids (proline and hydroxyproline), amides (asparagine or glutamine), betaines (glycine betaine, alanine betaine or proline betaine) and polyamines (putrescine, spermidine and spermine). Under nitrogen deficiency, plants can often accumulate dimethylsulfonium propionate, a tertiary sulfonium compound equivalent to betaines (Rabe, 1990; Ashraf and Harris, 2004; Zhu, 2007; Flowers and Colmer, 2008).

Since organic solutes, even at high concentrations, do not interfere with normal cellular metabolism (Hasegawa et al., 2000; Sairam and Tyagi, 2004), these compounds are called compatible solutes, osmolytes or compatible metabolites. Although the molecules of these compounds are not heavily charged, they are polar, highly soluble and have a large hydration shell. Thus, they are easily solubilized and can directly interact with the macro-molecules (Sairam and Tyagi, 2004) and they do not interfere with protein function and structure, thus alleviating the inhibitory effects of high ion concentrations on enzyme activity (Bohnert and Shen, 1999). In fact, compatible solutes have the ability to preserve the enzyme activity in saline solutions and these osmolytes have very less effect on pH or charge balance of the cytosol or lumenal compartments of organelles (Parida and Das, 2005). Although there is a high energetic cost to synthetize these organic compounds at the expense of plant growth, it is offset by the ability of these compounds to make the plants more tolerant to high salinity and thus allow plant survival (Munns and Tester, 2008).

At high concentrations, the compatible solutes certainly work to osmotic adjustment. The highest concentrations of these solutes reside mainly in the cytosol, promoting water balance between the apoplast, cytoplasm and vacuole (Bray *et al.*, 2000; Zhu, 2001, 2002). Under stress conditions, these solutes can be accumulated in such large quantities as from 5 to 10% of the dry weight of the tissue (Naidu *et al.*, 1992). Besides their strictly osmotic role, some organic solutes can assist in stabilizing proteins, protein complexes and membranes, maintenance of osmotic and ionic homoeostasis, and as a store of carbon and nitrogen (Bohnert and Shen, 1999; Bray *et al.*, 2000). In addition, they can contribute to cytosolic pH control and detoxification of NH_4^+ excess (Gilbert *et al.*, 1998). Current models suggest that small amounts of compatible solutes may also protect plants by removing free oxygen radicals generated by secondary oxidative stress (Smirnoff and Cumbes, 1989; Zhu, 2001, 2002).

Among the organic solutes that can be accumulated in plants grown under stress, proline has undoubtedly received more attention. It is stated that the accumulation of proline has action as a protein-compatible hydrotrope and radical scavenger improving adaptation to stresses such as drought and salinity (Türkan and Demiral, 2009). In addition, conditions of enhanced proline synthesis alleviate cytoplasmic acidosis and sustain $NADP^+/NADPH$ ratios at required levels for metabolism in plants subjected to drought and salt stresses (Hare and Cress, 1997; Türkan and Demiral, 2009).

The literature suggests that high proline concentration in the leaves has a great contribution to cell osmotic adjustment of halophytes as a whole. However, in glycophytes proline accumulation is usually too small to significantly reduce the osmotic potential of the whole cell, but when partitioned exclusively into the cytoplasm it could act as an osmolyte and promote the balance in water potential between the cytosol and the vacuole (Munns and Tester, 2008). In dwarf-cashew seedlings, proline content demonstrated a remarkabe increase in leaves and roots at higher saline conditions rather than soluble carbohydrates and soluble amino acids. However, quantitatively proline accumulation was much lower than these other organic solutes and apparently not sufficient to provide osmotic adjustment (Abreu *et al.*, 2008).

Proline accumulation in plant tissue was first observed by Kemble and MacPherson (1954), resulting in the increase of studies with proline in plants under stress. In addition, studies have also revealed that several other organic solutes accumulate in cells of stressed plants. However, while many studies indicate a positive correlation between proline accumulation and acclimation to drought and salinity, this is not corroborated by other studies (details in reviews of Delauney and Verma, 1993; Hare and Cress, 1997; Hare *et al.*, 1998, 1999; Ashraf and Harris, 2004). Thus, to date the question still remains as to whether the accumulation of proline in plant tissues provides an adaptive advantage to stressed plants or it is merely an incidental consequence of other metabolic changes induced by stress. In this scenario, as suggested by Delauney and Verma (1993), the absence of a positive correlation between proline accumulation and osmoregulation in some species does not negate an adaptive role for the proline. In these species, this may reflect the predominance of other adaptive mechanisms, such as morphological changes (e.g. development of deeper root systems), developmental (e.g. reduction of flowering time), physiological (e.g. ion sequestration in the vacuole) or biochemical (e.g. preferential synthesis and accumulation of other organic solutes).

Glycine-betaine (GB) is a quaternary ammonium compound (QAC) that functions as an effective organic solute for osmoregulation in plants facing salt stress (Ashraf and Harris, 2004). It is an amphoteric compound and is highly water soluble. The molecular features of glycine-betaine allow the macromolecular interaction with both the hydrophobic and the hydrophilic domains (Ohnishi and Murata, 2006). In addition to its role as osmolyte, it is believed to be an osmoprotectant (Ashraf and Harris, 2004). It also plays an important role to preserve thylakoid membrane and maintain the integrity of plasma membrane (Rhodes and Hanson, 1993; Ashraf and Harris, 2004), thereby maintaining photosynthetic efficiency (Robinson and Jones, 1986) by stabilizing the association of the extrinsic PSII complex protein (Murata et al., 1992; Ashraf and Harris, 2004). Studies carried out by Ohnishi and Murata (2006) on the cyanobacterium *Synechococcus* sp. suggest that GB protects PSII against photoinhibition, counteracting the inhibitory effects of salinity, demonstrating a fast PSII photodamage repair.

Some evidence also shows that salt-tolerant species accumulate high levels of GB while moderately and sensitive species generally accumulate intermediate and low levels or no GB (Sairam and Tyagi, 2004).

Among various organic solutes, soluble carbohydrates form the major contribution to osmotic adjustment in glycophytes subjected to saline environments, contributing up to 50% of the total osmotic potential (Ashraf and Harris, 2004). Thus, the accumulation of sugars has been related to confer salt tolerance in plants. Salt-tolerant cultivars of sunflower (Ashraf and Tufail, 1995), tomato (Amini and Ehsanpour, 2005) and barley (Khosravinejad et al., 2009) increased the soluble sugars accumulation more than salt-sensitive cultivars. Besides its role as osmolyte, soluble sugars protect cells during stress maintaining hydrophilic interactions in membrane and proteins, thus preventing protein denaturation (Khosravinejad et al., 2009).

In summary, the evaluation of the functional significance of the proline accumulation process and other compatible solutes should be done holistically always in the context of other available mechanisms, which favours the acclimatization and adaptation processes to stresses (Hare and Cress, 1997).

6.7 Antioxidative System

It is well established that when plants are subjected to high light irradiance, wounding, elevated or low temperature, ozone, drought, flooding, herbicides, mineral nutrient deficiency or mineral ion toxicity, the production of ROS is increased (Bray et al., 2000). Therefore, it is not startling that oxidative stress is also an important component of the salinity damage to plants. As in ROS generation during a number of abiotic stresses, salinity also damages photosynthetic apparatus, leading to higher leakage of electrons to O_2 in the absence of other acceptors (Alscher et al., 1997; Mittler, 2002).

During salt stress, stomatal closure reduces the internal CO_2 concentration, leading to a reduced NADPH consumption by the Calvin cycle. In this scenario, the photosynthetic electron transfer leads to an over-reduction of ferrodoxine, and electrons may be transferred from photosystem I to dioxygen increasing the production of superoxide radicals ($^{\bullet}O_2^{-}$) by the Mehler reaction. In addition, the increased photorespiration and other physiological reactions lead to an increase in H_2O_2 production, which initiates chain reactions that produce the highly reactive hydroxyl radical ($^{\bullet}OH$) (Azevedo Neto et al., 2008). When produced in excess, these cytotoxic ROS can alter the cell metabolism by oxidative damage of biomolecules such as lipids, proteins and nucleic acids (Azevedo Neto et al., 2008).

The concerted action of a complex antioxidant system is required for efficient destruction of $^{\bullet}O_2^{-}$ and H_2O_2. This system consists of both enzymatic and non-enzymatic antioxidant defences (Fig. 6.3). Non-enzymatic defences capable of removing ROS are composed of both hydrophilic and lipophilic compounds. The former include ascorbate (ASC) and reduced glutathione (GSH) and the latter include tocopherols and carotenoids. Enzymatic defences include superoxide dismutases (SOD), catalases (CAT) and peroxidases (POX). In addition to these enzymes, glutathione reductase (GR), monodehydroascorbate reductase (MDHAR), and dehydroascorbate reductase (DHAR) are needed for the regeneration of antioxidants in their active forms (Azevedo Neto et al., 2008).

Several studies have now shown that plants with high levels of antioxidants, either constitutive or induced, show greater tolerance to ROS-induced oxidative damage (Parida and Das, 2005). Several researchers, working with different crop species, reported that SOD, CAT, APX, GPX, MDHAR, DHAR and GR activities change with salt stress, suggesting that resistance to oxidative stress may be involved in salinity tolerance (Gosset et al., 1994; Gueta-Dahan et al., 1997; Hernández et al., 2000; Mittova et al., 2002a; Azevedo Neto et al., 2006).

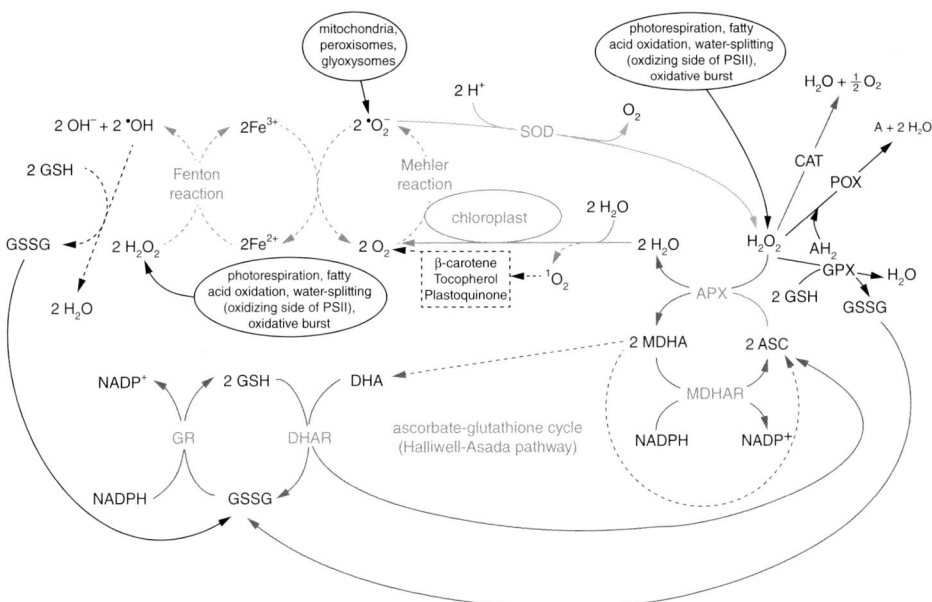

Fig. 6.3. Generation of reactive oxygen species in cell organelles such as mitochondria, peroxisomes, glyoxysomes and chloroplast. These ROS are detoxified by enzymes such as CAT, POX, GPX and APX. ASC and GSH participate in the cyclic transfer of reducing equivalents in the removal of H_2O_2 in the ascorbate–glutathione cycle, using NADPH as the reducing power. Non-enzymatic pathways are indicated by dotted lines. APX, ascorbate peroxidase; ASC, ascorbate; AH2, oxidizable substrate; DHA, dehydroascorbate; DHAR, dehydroascorbate reductase; GPX, glutathione peroxidase; POX, non-specific peroxidase; GR, glutathione reductase; GSH, reduced glutathione; GSSG, oxidized glutathione; hydrogen peroxide, H_2O_2; hydroxyl radical, $^\bullet OH$; MDHA, monodehydroascorbate; MDHAR, monodehydroascorbate reductase; SOD, superoxide dismutase; superoxide radical, $^\bullet O_2^-$ (adapted from Azevedo Neto et al., 2008).

Most studies dealing with salt stress-induced antioxidant enzyme system were made in leaves and enzyme activities determined at a certain time during the development of the stressed and control plants. The time course of the antioxidant enzyme system, studied in both leaves and roots from control and salt-stressed maize plants differing in salt tolerance, has shown that the activities of some antioxidant enzymes fluctuate along plant development; that is, on one day, the activity of an enzyme may be higher in salt-stressed tissues than in the control and, on another day, it may be reversed (Azevedo Neto et al., 2006). Therefore, in studies related to stress-induced antioxidant enzyme system, the time course of enzyme activity and not the measurement made at only one time along plant development should be considered.

In leaves of salt-stressed maize plants, the activities of SOD, APX, GPX and GR increased with time (Azevedo Neto et al., 2006). However, in the salt-tolerant genotype the increases in these enzyme activities were more pronounced than in the salt-sensitive one. Leaf CAT was also studied, but salt stress had no significant effect on its activity when the enzyme was extracted from the salt-tolerant genotype, but it was significantly reduced in the salt-sensitive genotype. These results strongly suggest a relation between leaf antioxidant enzyme system and salinity tolerance. In roots of the salt-tolerant genotype, salt stress reduced the activities of CAT and SOD, but the activities of APX, GPX and GR were little affected by salt-stress. In roots of the salt-sensitive genotype, the activities of all antioxidant enzymes studied were inhibited by salinity.

H_2O_2, the product of SOD activity, is itself a ROS, and needs to be eliminated. In plants, intracellular levels of H_2O_2 are regulated mainly by CAT, APX and GPX. In leaves of the salt-sensitive maize genotype, salt stress reduced only CAT activity, but the activities of CAT, APX and GPX were substantially reduced in roots, indicating that the H_2O_2 scavenging system is less effective in the salt-sensitive genotypes. Salt stress increased GR activity in leaves of both tolerant and sensitive genotypes, but in roots, GR activity was reduced only in the salt-sensitive genotype, suggesting a decreased GSH turnover rate. Considering that salinity also reduced APX activity in roots of this genotype, it was suggested that the ascorbate–glutathione cycle in roots of the salt-sensitive maize genotype is less active and that in roots of this plant species this pathway may play a key role in salt tolerance. Rios-Gonzalez et al. (2002) also investigated the salinity effects on SOD, CAT, GPX and GR activities in maize seedlings, and reported that the antioxidative enzymes exhibit higher activities in salt-treated plants.

Comparing antioxidant defence mechanisms in salt-tolerant and salt-sensitive rice varieties, Dionisio-Sese and Tobita (1998) reported that in the salt-sensitive rice varieties salt stress reduced SOD and increased GPX activity. In contrast, in the salt-tolerant rice variety salt stress induced a slight increase in SOD and a slight decrease in GPX activity. In leaves of rice plants, salt stress enhanced the activities of SOD, APX and GPX and decreased CAT activity (Lee et al., 2001). On the other hand, salt stress had little effect on GR activity.

In salt-stressed tomatoes the activities of antioxidant enzymes were also studied by several researchers. Shalata and Tal (1998) evaluated the activity of the antioxidant system in *Lycopersicon esculentum* (a salt-sensitive cultivated tomato) and in *Lycopersicon pennellii* (a wild salt-tolerant). They reported that, at 100 mM NaCl, SOD, CAT and APX activities were higher in salt-tolerant but GR activity lower in salt-tolerant than in salt-sensitive species. The authors also reported that the constitutive level of lipid peroxidation and the CAT and GR activities in the salt-tolerant species were lower, whereas SOD and APX activities were inherently higher than those in the salt-sensitive species. Working with the same tomato species, Mittova et al. (2000) concluded that the maintenance of high SOD to APX activity was responsible for high salt tolerance of the wild species. In other studies with the same tomato species, the activity of the antioxidative systems in root plastids, chloroplasts, mitochondria and peroxisomes were also investigated (Mittova et al., 2002a). They reported that the better protection of the root plastids to oxidative stress in the wild salt-tolerant tomato was correlated with the increase in SOD, APX and GPX activities. They also found increased activities of SOD, APX, MDHAR and in several isoforms of GPX in chloroplasts from leaves of salt-tolerant species (Mittova et al., 2002b). These data indicate that the salt-induced oxidative stress is effectively alleviated in chloroplasts of salt-tolerant plants by the selective up-regulation of a set of antioxidative enzymes. It was reported later that SOD activity decreased whereas APX and GPX activities remained at control level in mitochondria of salt-sensitive tomato. In the salt-tolerant species salinity increased mitochondrial SOD, APX, MDH and MDHAR activities, and peroxisomal SOD, APX, MDHAR and CAT activities. In peroxisomes of salt-sensitive plants, the activities of all these enzymes remained at control level (Mittova et al., 2003).

In salt-tolerant cotton (*Gossypium hirsutum* L.) cultivars, salt-stress led to a considerable increase in the CAT, GPX and GR activities, whereas the activities of these enzymes remained unchanged or decreased in the salt-sensitive cultivars. The salt-tolerant cultivars also had a higher reduced/oxidized ascorbate ratio and a higher reduced/oxidized glutathione ratio than the salt-sensitive plants when subjected to saline stress (Gossett et al., 1994). It was suggested by the same authors that in cotton, the salt tolerance was associated with the high levels of antioxidants and an active ascorbate–glutathione cycle. Salt stress has also been reported to increase SOD, GPX and GR activities in a salt-tolerant cotton cultivar, but salt treatment did not affect SOD activity in the salt-sensitive plants (Meloni et al., 2003). The SOD activity from cotton cell cultures grown in a saline medium was higher in salt-tolerant cells than in those moderately salt-tolerant or salt-sensitive (Garratt *et al.*, 2002).

Salt stress also increased the activities of SOD and CAT in leaves and roots of salt-tolerant and salt-sensitive sorghum genotypes, but the increases were higher in the salt-tolerant genotype (Costa *et al.*, 2005). They also reported that responses of salt-sensitive and salt-tolerant genotypes differed with respect to APX and GPX activities in leaves and roots. In addition, the salt-sensitive genotype showed a higher SOD/H_2O_2-scavenging enzymes (CAT, APX and GPX) activity ratio. Salt stress did not affect the activity of GR in both genotypes. The authors hypothesized that one of the factors determining the increased tolerance to salinity in the salt-tolerant genotype was the more efficient antioxidant system, and concluded that the SOD/H_2O_2-scavenging enzyme activity ratio could be used as a biochemical marker for salt tolerance in sorghum.

Generally, in shoots the activities of APX, MDHAR, DHAR and GR antioxidative enzymes tend to increase, whereas in roots these activities tend to decrease (Meneguzzo and Navari-Izzo, 1999; Hernández *et al.*, 2000). In addition, different types of SOD may be affected differently. It was reported that salinity increased Mn-SOD activity in wheat, but did not affect Cu/Zn-SOD, whereas in pea salinity enhanced activities of both Cu/Zn-SOD and Mn-SOD and this effect was coupled to increases in the APX and MDHAR activities. Salt stress also increased the activities of GR and DHAR, and a decrease in both reduced/oxidized ascorbate and reduced/oxidized glutathione ratios (Hernández *et al.*, 1999). Activities of antioxidative enzymes such as CAT, GPX, SOD and GR were affected by salinity in both salt-sensitive and salt-tolerant cultivars of mulberry, with higher activity being registered in the salt-tolerant ones (Sudhakar *et al.*, 2001).

It is well established that photosynthesis and respiration can be limited by osmotic and ionic effects resulting from salinity leading to increased ROS production. This limitation is responsible for the secondary oxidative stress, which can damage cell structure and metabolism. It is also known that plant responses to salt stress are multigenic, involving both osmotic and ionic homoeostasis, as well as cell detoxification. The efficiency of the latter process is dependent upon the plant's antioxidative defence mechanisms. Although plants have adapted during evolution to prevailing environmental conditions, the level of 'natural' tolerance to oxidants varies widely among species. Differences in tolerance can be related to modifications in the constitutive levels of antioxidative enzymes or non-enzymatic antioxidants. These can be age-dependent or species-dependent but always due to modifications in gene expression, which may explain, at least partially, the differences in the salt tolerance among species or genotypes.

In view of the considerable variation in the protective mechanisms against ROS among plant species and cultivars from the same species, further work is needed to establish the general validity of this mechanism for salt tolerance. Studies using modern genetic engineering techniques have shown improvement in salinity tolerance in several crops through over-expression of specific enzymes for scavenging ROS (Tanaka *et al.*, 1999; Badawi *et al.*, 2004). The combinations of such techniques with those used in classical plant breeding may provide the basis for a breeding programme using antioxidant compounds and/or antioxidative enzymes as a means to increase plant salt tolerance.

6.8 Cross-Tolerance

Cellular homoeostasis is achieved by the coordinated action of many biochemical pathways. However, under stress conditions different pathways can be affected and this could lead to disrupted coupling of biochemical pathways with cellular homoeostasis (Rizhsky *et al.*, 2002). This disruption in the pathways results in an increased electron flow to the reduction of oxygen and ROS generation. ROS formation could lead to changes in intracellular redox homoeostasis (Bowler and Fluhr, 2000), and it is now widely accepted that plant metabolism, morphology and development are governed by redox signals (Foyer and Noctor, 2003). Considering that many stress conditions may cause cellular redox imbalance, some researchers have proposed that cross-tolerance is mediated by defence responses to oxidative stress (Bowler et al., 1992).

Thus, cross-tolerance may be extremely important for crop production because the plants may be bred to tolerate more than one stress. Additionally, cross-tolerance allows us to compare and contrast individual responses and to examine the roles of common signal-transducing molecules. The first evidence for cross-tolerance came from *Chlorella*, a unicellular green alga. It was found that low temperature-induced damage was reduced by the previous growth on sub-lethal concentrations of sulfite (Rabinowich and Fridovich, 1985) and paraquat (a herbicide that generates oxidative stress) (Clare et al., 1984). Such phenomena have now been reported for several plant species, for example, water stress induces chilling resistance in rice (Takahashi et al., 1994), salt stress stimulates cold hardiness in potato and spinach seedlings (Ryu et al., 1995), mechanical stress increases chilling tolerance in leaves of tomato (Keller and Steffen, 1995) and ozone pretreatment can induce pathogen resistance in tobacco and *Arabidopsis* (Yalpani et al., 1994; Sharma et al., 1996).

Cross-tolerance has emphasized the unity of different stresses and underlined their common feature of enhanced ROS production. A few years ago, hydrogen peroxide was viewed mainly as a toxic cellular metabolite. However, because it is relatively stable and diffusible through membranes, it is a perfect candidate to act as a signalling molecule during stress responses, and it is now clear that in cells of both plant and animal, it may also function as a signalling molecule (Neill et al., 2002). In this scenario, rather than a system designed to extinguish ROS, the antioxidative system may have developed primarily to allow for adjustment of the cellular redox state and to enable redox signalling. A wide variety of abiotic and biotic stresses increases the cell H_2O_2 production, thus it has been suggested that this molecule plays a dual role in plants: at low concentrations, H_2O_2 triggers tolerance against various abiotic stresses acting as a messenger molecule involved in signalling for acclimation; and at high concentrations H_2O_2 orchestrates programmed cell death (Dat et al., 2000; Van Breusegem et al., 2001). Thus, it seems likely that cross-tolerance and plant acclimation to biotic and abiotic stresses may be favoured by the accumulation of H_2O_2 in specific tissues and in appropriate amounts (Bowler and Fluhr, 2000).

Evidences for a signalling role of H_2O_2 have been reported in several studies (Foyer et al., 1997). The experimental generation of H_2O_2 or the exogenous application in plant tissues has been shown to act as a signal for induction of gene expression of CAT (Prasad et al., 1994; Polidoros and Scandalios, 1999), APX (Van Breusegem et al., 2001), GPX and GR (Janda et al., 1999). The synthesis of heat-shock proteins (Hsps) and activation of the mitogen-activated protein kinase cascade also were induced by changes in H_2O_2 homoeostasis (Kovtun et al., 2000; Van Breusegem et al., 2001). Additionally, Desikan et al. (2001) identified 175 H_2O_2-regulated non-redundant expressed sequence tags.

In maize seedlings, it has been shown that chilling stress increased the endogenous H_2O_2 production, and chilling stress tolerance was increased by exogenous application of H_2O_2 (Prasad et al., 1994). The authors suggested that the increase in the antioxidant system activity prevented the accumulation of ROS during chilling stress and increased the plant tolerance. Nodal potato explants sub-cultured from H_2O_2-treated microplants (acclimated) were resistant to a 15 h heat shock at 42°C, a normally lethal treatment (Lopez-Delgado et al., 1998). In leaves of *Arabidopsis* the injection of H_2O_2 protected against the effect of photo-bleaching induced by subsequent exposure to light excess (Karpinski et al., 1999). In rice seedlings the H_2O_2 pretreatment induced acclimation to both salt and heat stresses (Uchida et al., 2002). Similarly, exogenous application of H_2O_2 in maize seedlings induced, simultaneously, multi-tolerance to heat, chilling, drought and salt stresses (Gong et al., 2001).

There is important evidence that H_2O_2 pretreatment can induce salt acclimation in a salt-sensitive maize genotype (Azevedo Neto et al., 2005). It was shown that this pretreatment induced SOD activity in leaves of acclimated-stressed plants, suggesting that H_2O_2

pretreated plants had a better $\cdot O_2^-$ scavenging ability. The activities of CAT and GPX also increased in H_2O_2 pretreated plants, indicating a higher ability to remove the H_2O_2 resulting from the increase in SOD activity. The authors also showed that, in leaves, salinity appeared to induce APX and GR activities, while H_2O_2-induced acclimation appeared to induce CAT and GPX activities.

It was concluded that the ability of plants to cope with the secondary oxidative stress resulting from salt exposure may be increased by H_2O_2 pretreatment, which, directly or indirectly, induces the activity of several antioxidative enzymes. In addition, they also concluded that the balance between the activity of H_2O_2 production by SOD and the activity of different H_2O_2-scavenging enzymes plays a key role in maintaining the steady-state level of H_2O_2 and $\cdot O_2^-$ in plant cells.

Taken together, the results provide additional evidence that H_2O_2 is involved as a signalling molecule in plant acclimation to salt stress, and that plants could make use of common pathways and components in the stress–response relationship. The similarities among intermediate signalling molecules used by plants as a result of diverse stresses actually suggest that there are intracellular signalling networks instead of linear pathways (Genoud and Métraux, 1999). Whether only a few intermediate signalling molecules can interact in a coordinated way, these networks can allow specific responses of plant cells to several signals, such as salinity, H_2O_2, cold, heat, drought, waterlogging, wounding, pathogens and so on.

6.9 Conclusions

Salinity is an important environmental problem that reduces crop production around the world and it is expected that in the coming decades it will become a larger problem. In this scenario, the understanding of cell responses to salt stress in coordination with whole-plant responses is essential to our knowledge of how plants can tolerate this abiotic stress. In recent decades, our knowledge of salt-tolerance mechanisms has expanded considerably; however, individual plant genotypes exhibit distinct salt tolerance. Therefore, comparisons of salt-tolerance mechanisms in different species and cultivars are fundamental to understand the regulatory points related to mechanisms providing salinity tolerance. Transcriptome, proteome and metabolome analyses, sequencing of entire genomes in plants, bioinformatics analyses and functional studies, are powerful molecular tools that will enable the clarification of the mechanisms of salt tolerance and to develop salt-tolerant cultivars able to cope with the increasing soil salinity.

References

Abreu, C.E.B., Prisco, J.T., Nogueira, A.R.C., Bezerra, M.A., Lacerda, C.F. and Gomes-Filho, E. (2008) Physiological and biochemical changes occurring in dwarf-cashew seedlings subjected to salt stress. *Brazilian Journal of Plant Physiology* 20, 105–118.

Alscher, R.G., Donahue, J.L. and Cramer, C.L. (1997) Reactive oxygen species and antioxidants: relationship in green cells. *Physiologia Plantarum* 100, 224–233.

Amin, M., Porter, S.L. and Soyer, O.S. (2013) Split histidine kinases enable ultrasensitivity and bistability in two-component signaling networks. *PLoS Computational Biology* 9, 1–12.

Amini, F. and Ehsanpour, A.A. (2005) Soluble proteins, proline, carbohydrates and Na^+/K^+ changes in two tomato (*Lycopersicon esculentum* Mill.) cultivars under in vitro salt stress. *American Journal of Biochemistry and Biotechnology* 1, 204–208.

Anil, V.S., Krishnamurthy, H. and Mathew, M. (2007) Limiting cytosolic Na^+ confers salt tolerance to rice cells in culture: a two photon microscopy study of SBFI-loaded cells. *Physiologia Plantarum* 129, 607–621.

Apse, M.P. and Blumwald, E. (2007) Na^+ transport in plants. *FEBS Letters* 581, 2247–2254.

Apse, M.P., Aharon, G.S., Snedden, W.A. and Blumwald, E. (1999) Overexpression of a vacuolar Na^+/H^+ antiport confers salt tolerance in Arabidopsis. *Science* 285, 1256–1258.

Ashraf, M. and Harris, P.J.C. (2004) Potential biochemical indicators of salinity tolerance in plants. *Plant Science* 166, 3–16.

Ashraf, M. and Tufail, M. (1995) Variation in salinity tolerance in sunflower (*Helianthus annuus* L.). *Journal of Agronomy and Soil Science* 174, 351–362.

Azevedo Neto, A.D. and Tabosa, J.N. (2000) Estresse salino em plântulas de milho: Parte I - Análise do crescimento. *Revista Brasileira de Engenharia Agrícola e Ambiental* 1, 159–164.

Azevedo Neto, A.D., Prisco, J.T., Enéas-Filho, J., Medeiros, J.R. and Gomes-Filho, E. (2005) Hydrogen peroxide pre-treatment induces salt-stress acclimation in maize plants. *Journal of Plant Physiology* 162, 1114–1122.

Azevedo Neto, A.D., Prisco, J.T., Enéas-Filho, J., Abreu, C.E.B. and Gomes-Filho, E. (2006) Effect of salt stress on antioxidative enzymes and lipid peroxidation in leaves and roots of salt-tolerant and salt-sensitive maize genotypes. *Environmental and Experimental Botany* 56, 87–94.

Azevedo Neto, A.D., Gomes-Filho, E. and Prisco, J.T. (2008) Salinity and oxidative stress. In: Khan, N.A. and Singh, S. (eds) *Abiotic Stress and Plant Responses*. I K International, New Delhi, India, pp. 57–82.

Badawi, G.H., Yamauchi, Y., Shimada, E., Sasaki, R., Naoyoshi, K., Tanaka, K. and Tanaka, K. (2004) Enhanced tolerance to salt stress and water deficit by overexpressing superoxide dismutase in tobacco (*Nicotiana tabacum*) chloroplasts. *Plant Science* 166, 919–928.

Binzel, M.L., Hasegawa, P.M., Handa, A.K. and Bressan, R.A. (1985) Adaptation of tobacco cells to NaCl. *Plant Physiology* 79, 118–125.

Binzel, M.L., Hess, F.D., Bressan, R.A. and Hasegawa, P.M. (1988) Intercellular compartimentation of ions in salt adapted tobacco cells. *Plant Physiology* 86, 607–614.

Blumwald, E., Aharon, G.S. and Apse, M.P. (2000) Na^+ transport in plant cells. *Biochimica et Biophysica Acta* 1465, 140–151.

Bohnert, H.J. and Shen, B. (1999) Transformation and compatible solutes. *Scientia Horticulturae* 78, 237–260.

Bowler, C. and Fluhr, R. (2000) The role of calcium and activated oxygens as signals for controlling cross-tolerance. *Trends in Plant Science* 5, 241–246.

Bowler, C., Van Montagu, M. and Inzé, D. (1992) Superoxide dismutase and stress tolerance. *Annual Review of Plant Physiology and Plant Molecular Biology* 43, 83–116.

Bray, E.A., Bailey-Serres, J. and Weretilnyk, E. (2000) Responses to abiotic stress. In: Buchanan, B.B., Gruissem, W. and Jones, R.L. (eds) *Biochemistry and Molecular Biology of Plants*. American Society of Plant Physiologists, Rockville, Maryland, pp. 1158–1203.

Breckle, S.W. (2002) Salinity, halophytes and salt affected natural ecosystem. In: Läunchli, A. and Lüttge, U. (eds) *Salinity: Environment-Plants-Molecules*. Kluwer Academic Publishers, New York, pp. 53–80.

Chartzoulakis, K., Loupassaki, M., Bertaki, M. and Androulakis, I. (2002) Effect of NaCl salinity on growth, ion content and CO_2 assimilation of six olive cultivars. *Scientia Horticulturae* 96, 235–247.

Chefdor, F., Bénédetti, H., Depierreux, C., Delmotte, F., Morabito, D. and Carpin, S. (2006) Osmotic stress sensing in *Populus*: components identification of a phosphorrelay system. *FEBS Letters* 580, 77–81.

Chinnusamy, V., Jagendorf, A. and Zhu, J.-K. (2005) Understanding and improving salt tolerance in plants. *Crop Science* 45, 437–448.

Clare, D.A., Rabinowitch, H.D. and Fridovich, I. (1984) Superoxide dismutase and chilling injury in *Chlorella ellipsoidea*. *Archives of Biochemistry and Biophysics* 231, 158–163.

Costa, P.H.A., Azevedo Neto, A.D., Bezerra, M.A., Prisco, J.T. and Gomes-Filho, E. (2005) Antioxidant-enzymatic system of two sorghum genotypes differing in salt tolerance. *Brazilian Journal of Plant Physiology*, 17, 353–361.

Dat, J., Vandenabeele, S., Vranová, E., Van Montagu, M., Inzé, D. and Van Breusegen, F. (2000) Dual action of the active oxygen species during plant stress responses. *Cellular and Molecular Life Sciences* 57, 779–795.

Delauney, A.J. and Verma, D.P.S. (1993) Proline biosynthesis and osmoregulation in plants. *The Plant Journal* 4, 215–223.

Demidchik, B., Essah, P.A. and Tester, M. (2004) Glutamate activates cation currents in the plasma membrane of *Arabidopsis* root cells. *Planta* 219, 167–175.

Desikan, R., Mackerness, S. A.-H., Hancock, J.T. and Neill, S.J. (2001) Regulation of the *Arabidopsis* transcriptome by oxidative stress. *Plant Physiology* 127, 159–172.

Dionisio-Sese, M.L. and Tobita, S. (1998) Antioxidant responses of rice seedlings to salinity stress. *Plant Science* 135, 1–9.

Flowers, T.J. and Colmer, T.D. (2008) Salinity tolerance of halophytes. *New Phytologist* 179, 945–963.

Flowers, T.J., Troke, P.F. and Yeo, A.R. (1977) The mechanism of salt tolerance in halophytes. *Annual Review of Plant Physiology* 28, 89–121.

Foyer, C.H. and Noctor, G. (2003) Redox sensing and signalling associated with reactive oxygen in chloroplasts, peroxisomes and mitochondria. *Physiologia Plantarum* 119, 355–364.

Foyer, C.H., Lopez-Delgado, H., Dat, J.E. and Scott, I.M. (1997) Hydrogen peroxide- and glutathione-associated mechanisms of acclimatory stress tolerance and signalling. *Physiologia Plantarum* 100, 241–254.

Garciadeblas, B., Senn, M.E., Banuelos, M.A. and Rodrıguez-Navarro, A. (2003) Sodium transport and HKT transporters: the rice model. *The Plant Journal* 34, 788–801.

Garratt, L.C., Janagoudar, B.S., Lowe, K.C., Anthony, P., Power, J.B. and Davey, M.R. (2002) Salinity tolerance and antioxidant status in cotton cultures. *Free Radical Biology & Medicine* 33, 502–511.

Genoud, T. and Métraux, J.-P. (1999) Crosstalk in plant cell signaling: structure and function of the genetic network. *Trends in Plant Science* 4, 503–507.

Gilbert, G.A., Gadush, M.V., Wilson, C. and Madore, M.A. (1998) Amino acid accumulation in sink and source tissues of *Coleus blumei* Benth during salinity stress. *Journal of Experimental Botany* 49, 107–114.

Golldack, D., Quigley, F., Michalowski, C.B., Kamasani, U.R. and Bohnert, H.J. (2003) Salinity stress-tolerant and -sensitive rice (*Oryza sativa* L.) regulate AKT-type potassium channel transcripts differently. *Plant Molecular Biology* 51, 71–81.

Gong, D., Guo, Y., Schumaker, K.S. and Zhu, J.-K. (2004) The SOS3 family of calcium sensors and SOS2 family of protein kinases in *Arabidopsis*. *Plant Physiology* 134, 919–926.

Gong, M., Chen, B., Li, Z.-G. and Guo, L.-H. (2001) Heat-shock-induced cross adaptation to heat, chilling, drought and salt in maize seedlings and involvement of H_2O_2. *Journal of Plant Physiology* 158, 1125–1130.

Gosset, D.R., Millhollon, E.P. and Lucas, M.C. (1994) Antioxidant response to NaCl stress in salt-tolerant and salt-sensitive cultivars of cotton. *Crop Science* 34, 706–714.

Greenway, H. and Munns, R. (1980) Mechanism of salt tolerance in nonhalophytes. *Annual Review of Plant Physiology* 31, 149–190.

Gueta-Dahan, Y., Yaniv, Z., Zilinskas, B.A. and Ben-Hayyim, G. (1997) Salt and oxidative stress: similar and specific responses and their relation to salt tolerance in Citrus. *Planta* 203, 460–469.

Halfter, U., Ishitani, M. and Zhu, J.-K. (2000) The Arabidopsis SOS2 protein kinase physically interacts with and is activated by the calcium-binding protein SOS3. *Proceedings of National Academy of Sciences USA* 97, 3735–3740.

Hare, P.D. and Cress, W.A. (1997) Metabolic implications of stress-induced proline accumulation in plants. *Plant Growth Regulation* 21, 79–102.

Hare, P.D., Cress, W.A. and Van Staden, J. (1998) Dissecting the roles of osmolyte accumulation during stress. *Plant, Cell and Environment* 21, 535–553.

Hare, P.D., Cress, W.A. and Van Staden, J. (1999) Proline synthesis and degradation: a model system for elucidating stress-related signal transduction. *Journal of Experimental Botany* 50, 413–434.

Hasegawa, P.M., Bressan, R.A., Zhu, J.K. and Bohnert, H.J. (2000) Plant cellular and molecular responses to high salinity. *Annual Review of Plant Physiology and Plant Molecular Biology* 51, 463–499.

Heermann, R. and Jung, K. (2010) The complexity of the 'simple' two-component system KdpD/KdpE in *Escherichia coli*. *FEMS Microbiology Letters* 304, 97–106.

Heermann, R., Lippert, M.-L. and Jung, K. (2009) Domain swapping reveals that the N-terminal domain of the sensor kinase KdpD in *Escherichia coli* is important for signaling. *BMC Microbiology* 9, 133–145.

Hernández, J.A., Campillo, A., Jimenez, A., Alacon, J.J. and Sevilla, F. (1999) Response of antioxidant systems and leaf water relations to NaCl stress in pea plants. *New Phytologist* 141, 241–251.

Hernández, J.A., Jiménez, A., Mullineaux, P. and Sevilla, F. (2000) Tolerance of pea (*Pisum sativum* L.) to long-term salt stress is associated with induction of antioxidant defences. *Plant, Cell & Environment* 23, 853–862.

Hernández, J.A., Ferrer, M.A., Jimenez, A., Barcelo, A.R. and Sevilla, F. (2001) Antioxidant systems and $O_2^{·-}/H_2O_2$ production in the apoplast of pea leaves. Its relation with salt-induced necrotic lesions in minor veins. *Plant Physiology* 127, 817–831.

Horie, T. and Schroeder, J.I. (2004) Sodium transporters in plants. Diverse genes and physiological functions. *Plant Physiology* 136, 2457–2462.

Ishitani, M., Liu, J., Halfter, U., Kim, C.-S., Shi, W. and Zhu, J.-K. (2000) SOS3 function in plant salt tolerance requires N-myristoylation and calcium binding. *The Plant Cell* 12, 1667–1677.

Janda, T., Szalai, G., Tari, I. and Páldi, E. (1999) Hydroponic treatment with salicylic acid decreases the effects of chilling injury in maize (*Zea mays* L.) plants. *Planta* 208, 175–180.

Jia, W., Wang, Y., Zhang, S. and Zhang, J. (2002) Salt-stress-induced ABA accumulation is more sensitively triggered in roots than in shoots. *Journal of Experimental Botany* 53, 2201–2206.

Kader, M.A. and Lindberg, S. (2005) Uptake of sodium in protoplasts of salt-sensitive and salt-tolerant cultivars of rice, *Oryza sativa* L. determined by the fluorescent dye SBFI. *Journal of Experimental Botany* 56, 3149–3158.

Kader, M.A., Seidel, T., Golldack, D. and Lindberg, S. (2006) Expressions of OsHKT1, OsHKT2 and OsVHA are differently regulated under NaCl stress in salt-sensitive and salt-tolerant rice (*Oryza sativa* L.) cultivars. *Journal of Experimental Botany* 57, 4257–4268.

Karpinski, S., Reynolds, H., Karpinska, B., Wingsle, G., Creissen, G. and Mullineaux, P. (1999) Systemic signaling and acclimation in response to excess excitation energy in *Arabidopsis*. *Science* 284, 654–657.

Katiyar-Agarwal, S., Zhu, J., Kim, K., Agarwal, M., Fu, X., Huang, A. and Zhu, J.-K. (2006) The plasma membrane Na^+/H^+ antiporter SOS1 interacts with RCD1 and functions in oxidative stress tolerance in *Arabidopsis*. *Proceedings of National Academy of Sciences USA* 103, 18816–18821.

Keller, E. and Steffen, K.L. (1995) Increased chilling tolerance and altered carbon metabolism in tomato leaves following application of mechanical stress. *Physiologia Plantarum* 93, 519–525.

Kemble, A.R. and MacPherson, H.T. (1954) Liberation of amino acids in perennial rye grass during wilting. *Biochemical Journal* 58, 46–59.

Khosravinejad, F., Heydari, R. and Farboodnia, T. (2009) Effect of salinity on organic solutes contents in barley. *Pakistan Journal of Biological Sciences* 12, 2, 158–162.

Kovtun, Y., Chiu, W.-L., Tena, G. and Shenn, J. (2000) Functional analysis of oxidative stress-activated mitogen-activated protein kinase cascade in plants. *Proceedings of National Academy of Sciences USA* 97, 2940–2945.

Läuchli, A. and Grattan, S.R. (2007) Plant growth and development under salinity stress. In: Jenks, M.A., Hasegawa, P.M. and Jain, S.M. (eds) *Advances in Molecular Breeding Toward Drought and Salt Tolerant Crops*. Springer, Dordrecht, the Netherlands, pp. 1–32.

Laurie, S., Feeney, K.A., Maathuis, F.J.M., Heard, P.J., Brown, S.J. and Leigh, R.A. (2002) A role for HKT1 in sodium uptake by wheat roots. *Plant Journal* 32, 139–149.

Lee, D.H., Kim, Y.S. and Lee, C.B. (2001) The inductive responses of the antioxidant enzymes by salt stress in the rice (*Oryza sativa* L.). *Journal of Plant Physiology* 158, 737–745.

Leng, Q., Mercier, R.W., Hua, B.-G., Fromm, H. and Berkowitz, G.A. (2002) Electrophysiological analysis of cloned cyclic nucleotide-gated ion channels. *Plant Physiology* 128, 400–410.

Liu, J. and Zhu, J.-K. (1997) An Arabidopsis mutant that requires increased calcium for potassium nutrition and salt tolerance. *Proceedings of National Academy of Sciences USA* 94, 14960–14964.

Liu, J., Ishitani, M., Halfter, U., Kim, C.-S. and Zhu, J.-K. (2000) The *Arabidopsis thaliana* SOS2 gene encodes a protein kinase that is required for salt tolerance. *Proceedings of National Academy of Sciences USA* 97, 3730–3734.

Lopez-Delgado, H., Dat, J.F., Foyer, C.H. and Scott, I.M. (1998) Induction of thermotolerance in potato microplants by acetylsalicylic acid and H_2O_2. *Journal of Experimental Botany* 49, 713–720.

Maeda, S., Sugita, C., Sugita, M. and Omata, T. (2006) A new class of signal transducer in His-Asp phosphorelay systems. *Journal of Biological Chemistry* 281, 37868–37876.

Mahajan, S. and Tuteja, N. (2005) Cold, salinity and drought stresses: an overview. *Archives of Biochemistry and Biophysics* 444, 139–158.

Mahajan S., Pandey, G.K. and Tuteja, N. (2008) Calcium and salt stress signaling in plants: shedding light on SOS pathway. *Archives of Biochemistry and Biophysics* 471, 146–158.

Marschner, P. (2012) *Marschner's Mineral Nutrition of High Plants*. Academic Press, London.

Martinoia, E., Schramm, M.J., Kaiser, G., Kaiser, W.M. and Heber, U. (1986) Transport of anions in isolated barley vacuoles: I. Permeability to anions and evidence for a Cl⁻-uptake system. *Plant Physiology* 80, 895–901.

Mehlmer, N., Wurzinger, B., Stael, S., Hofmann-Rodrigues, D., Csaszar, E., Pfister, B., Bayer, R. and Teige, M. (2010) The Ca^{2+}-dependent protein kinase CPK3 is required for MAPK-independent salt-stress acclimation in *Arabidopsis*. *The Plant Journal* 63, 484–498.

Meloni, D.A., Oliva, M.A., Martinez, C.A. and Cambraia, J. (2003) Photosynthesis and activity of superoxide dismutase, peroxidase and glutathione reductase in cotton under salt stress. *Environmental and Experimental Botany* 49, 69–76.

Meneguzzo, S. and Navari-Izzo, I. (1999) Antioxidative responses of shoots and roots of wheat to increasing NaCl concentrations. *Journal of Plant Physiology* 155, 274–280.

Mittler, R. (2002) Oxidative stress, antioxidants, and stress tolerance. *Trends in Plant Science* 9, 405–410.

Mittova, V., Volokita, M., Guy, M. and Tal, M. (2000) Activities of SOD and the ascorbate-glutathione cycle enzymes in subcellular compartments in leaves and roots of the cultivated tomato and its wild salt-tolerant relative *Lycopersicon pennellii*. *Physiologia Plantarum* 110, 42–51.

Mittova, V., Guy, M., Tal, M. and Volokita, M. (2002a) Response of the cultivated tomato and its wild salt-tolerant relative *Lycopersicon pennellii* to salt-dependent oxidative stress: increased activities of antioxidant enzymes in root plastids. *Free Radical Research* 36, 195–202.

Mittova, V., Tal, M., Volokita, M. and Guy, M. (2002b) Salt stress induces up-regulation of an efficient chloroplast antioxidant system in the salt-tolerant wild tomato species *Lycopersicon pennellii* but not in the cultivated species. *Physiologia Plantarum* 115, 393–400.

Mittova, V., Tal, M., Volokita, M. and Guy, M. (2003) Up-regulation of the leaf mitochondrial and peroxisomal antioxidative systems in response to salt-induced oxidative stress in the wild salt-tolerant tomato species *Lycopersicon pennellii*. *Plant, Cell & Environment* 26, 845–856.

Munns, R. (2002) Comparative physiology of salt and water stress. *Plant, Cell & Environment* 28, 239–250.

Munns, R. and Tester, M. (2008) Mechanisms of salinity tolerance. *Annual Review of Plant Biology* 59, 651–681.

Munns, R., Schachtman, D. and Condon, A. (1995) The significance of a two-phase growth response to salinity in wheat and barley. *Functional Plant Biology* 22, 561–569.

Murata, N., Mohanty, P.S., Hayashi, H. and Papageorgiou, G.C. (1992) Glycinebetaine stabilizes the association of extrinsic proteins with the photosynthetic oxygen-evolving complex. *FEBS Letters* 296, 187–189.

Naidu, B.P., Paleg, L.G. and Jones, G.P. (1992) Nitrogenous compatible solutes in drought-stressed *Medicago* spp. *Phytochemistry* 31, 1195–1197.

Neill, S.J., Desikan, R., Clarke, A., Hurst, R.D. and Hancock, J.T. (2002) Hydrogen peroxide and nitric oxide as signalling molecules in plants. *Journal of Experimental Botany* 53, 1237–1247.

Novak, J.F., Stirnberg, M., Roenneke, B. and Marin, K. (2011) A novel mechanism of osmosensing, a salt-dependent protein-nucleic acid interaction in the cyanobacterium *Synechocystis* species PCC 6803. *Journal of Biological Chemistry* 286, 3235–3241.

Ohnishi, N. and Murata, N. (2006) Glycinebetaine counteracts the inhibitory effects of salt stress on the degradation and synthesis of D1 protein during photoinhibition in *Synechococcus* sp. PCC 7942. *Plant Physiology* 141, 758–765.

Parida, A.K. and Das, A.B. (2005) Salt tolerance and salinity effects on plants: a review. *Ecotoxicology and Environmental Safety* 60, 324–349.

Plett, D., Berger, B. and Tester, M. (2010) Genetic determinants of salinity tolerance in crop plants. In: Jenks, M.A. and Wood, A.J. (eds) *Genes for Plant Abiotic Stress*. Wiley-Blackwell, Oxford, UK, pp. 83–111.

Polidoros, A. and Scandalios, J. (1999) Role of hydrogen peroxide and different classes of antioxidants in the regulation of catalase and glutathione-S-transferase gene expression in maize (*Zea mays* L.). *Physiologia Plantarum* 106, 112–120.

Prasad, T.K., Anderson, M.D., Martin, B.A. and Stewart, C.R. (1994) Evidence for chilling-induced oxidative stress in maize seedlings and a regulatory role for hydrogen peroxide. *The Plant Cell* 6, 65–74.

Qiu, Q.-S., Guo, Y., Dietrich, M.A., Schumaker, K.S. and Zhu, J.-K. (2002) Regulation of SOS1, a plasma membrane Na^+/H^+ exchanger in *Arabidopsis thaliana*, by SOS2 and SOS3. *Proceedings of National Academy of Sciences USA* 99, 8436–8441.

Qiu, Q.-S., Guo, Y., Quintero, F.J., Pardo, J.M., Schumaker, K.S. and Zhu, J.-K. (2004) Regulation of vacuolar Na^+/H^+ exchange in *Arabidopsis thaliana* by the salt-overly sensitive (SOS) pathway. *Journal of Biological Chemistry* 279, 207–215.

Quintero, F.J., Ohta, M., Shi, H., Zhu, J.-K. and Pardo, J.M. (2002) Reconstitution in yeast of the Arabidopsis SOS signaling pathway for Na^+ homeostasis. *Proceedings of National Academy of Sciences USA* 99, 9061–9066.

Quintero, F.J., Martinez-Atienza, J., Villalta, I., Jiang, X., Kim, W.Y., Ali, Z., Fujii, H., Mendoza, I., Yun, D.J., Zhu, J.-K. and Pardo, J.M. (2011) Activation of the plasma membrane Na/H antiporter Salt-Overly-Sensitive 1 (SOS1) by phosphorylation of an auto-inhibitory C-terminal domain. *Proceedings of National Academy of Sciences USA* 108, 2611–2616.

Rabe, E. (1990) Stress physiology: the functional significance of the accumulation of nitrogen-containing compounds. *Journal of Horticultural Science* 65, 231–243.

Rabinowich, H.D. and Fridovich, I. (1985) Growth of *Chlorella sorokiniana* in the presence of sulfite elevates cell content of superoxide dismutase and imparts resistance towards paraquat. *Planta* 164, 524–528.

Raven, J.A. (1985) Regulation of pH and generation of osmolarity in vascular plants: a cost-benefit analysis in relation to efficiency of use of energy, nitrogen and water. *New Phytologist* 101, 25–77.

Rhodes, D. and Hanson, A.D. (1993) Quaternary ammonium and tertiary sulfonium compounds in higher plants. *Annual Review of Plant Physiology and Plant Molecular Biology* 44, 357–384.

Rios-Gonzalez, K., Erdei, L. and Lips, S.H. (2002) The activity of antioxidant enzymes in maize and sunflower seedlings as affected by salinity and different nitrogen sources. *Plant Science* 162, 923–930.

Rizhsky, L., Liang, H. and Mittler, R. (2002) The combined effect of drought stress and heat shock on gene expression in tobacco. *Plant Physiology* 130, 1143–1151.

Robinson, M.F., Véry, A.-A., Sanders, D. and Mansfield, T.A. (1997) How can stomata contribute to salt tolerance? *Annals of Botany* 80, 387–393.

Robinson, S.P. and Jones, J.P. (1986) Accumulation of glycinebetaine in chloroplasts provides osmotic adjustment during salt stress. *Australian Journal of Plant Physiology* 13, 659–668.

Rubio, F., Gassmann, W. and Schroeder, J.I. (1995) Sodium-driven potassium uptake by the plant potassium transporter HKT1 and mutations conferring salt tolerance. *Science* 270, 1660–1663.

Ryu, S.B., Costa, A., Xin, Z. and Li, P.H. (1995) Induction of cold hardiness by salt stress involves synthesis of cold and abscisic acid-responsive proteins in potato (*Solanum commersonii* Dun). *Plant & Cell Physiology* 36, 1245–1251.

Sairam, R.K. and Tyagi, A. (2004) Physiology and molecular biology of salinity stress tolerance in plants. *Current Science* 86, 407–421.

Saqib, M., Zorb, C., Rengel, Z. and Schubert, S. (2005) The expression of the endogenous vacuolar Na^+/H^+ antiporters in roots and shoots correlates positively with the salt tolerance of wheat (*Triticum aestivum* L.). *Plant Science* 169, 959–965.

Serraj, R. and Sinclair, T.R. (2002) Osmolyte accumulation: can it really help increase crop yield under drought conditions? *Plant, Cell and Environment* 25, 333–341.

Shalata, A. and Tal, M. (1998) The effect of salt stress on lipid peroxidation and antioxidants in the leaf of the cultivated tomato and its wild salt-tolerant relative *Lycopersicon pennellii*. *Physiologia Plantarum* 104, 169–174.

Sharma, Y.K., Leon, J., Raskin, I. and Davis, K.R. (1996) Ozone-induced responses in *Arabidopsis thaliana*: the role of salicylic acid in the accumulation of defense-related transcripts and induced resistance. *Proceedings of National Academy of Sciences USA* 93, 5099–5104.

Shi, H. and Zhu, J.-K. (2002) Regulation of expression of the vacuolar Na^+/H^+ antiporter gene AtNHX1 by salt stress and ABA. *Plant Molecular Biology* 50, 543–550.

Shi, H., Ishitani, M., Kim, C. and Zhu, J.-K. (2000) The *Arabidopsis thaliana* salt tolerance gene SOS1 encodes a putative Na^+/H^+ antiporter. *Proceedings of National Academy of Sciences USA* 97, 6896–6901.

Shi, H., Quintero, F.J., Pardo, J.M. and Zhu, J.-K. (2002) The putative plasma membrane $Na+/H+$ antiporter SOS1 controls long-distance Na^+ transport in plants. *The Plant Cell* 14, 465–477.

Shi, H., Kim, Y., Guo, Y., Stevenson, B. and Zhu, J.-K. (2003) The *Arabidopsis* SOS5 locus encodes a putative cell surface adhesion protein and is required for normal cell expansion. *The Plant Cell* 15, 19–32.

Smirnoff, N. (1993) The role of active oxygen in the response of plants to water deficit and desiccation. *New Phytologist* 125, 27–58.

Smirnoff, N. and Cumbes, Q.J. (1989) Hydroxyl radical scavenging activity of compatible solutes. *Phytochemistry* 28, 1057–1060.

Strange, K. (2004) Cellular volume homeostasis. *Advances in Physiology Education* 28, 155–159.

Sudhakar, C., Lakshmi, A. and Giridarakumar, S. (2001) Changes in the antioxidant enzyme efficacy in two high yielding genotypes of mulberry (*Morus alba* L.) under NaCl salinity. *Plant Science* 161, 613–619.

Sussman, M.R. (1994) Molecular analysis of proteins in the plant plasma membrane. *Annual Review of Plant Physiology and Plant Molecular Biology* 45, 211–234.

Taiz, L. and Zeiger, E. (2006) *Plant Physiology*, 4th edn. Sinauer Associates, Sunderland, Massachusetts.

Takahashi, R., Joshee, N. and Kitagawa, Y. (1994) Induction of chilling resistance by water stress, and expression of water stress-regulated genes in rice. *Plant Molecular Biology* 26, 339–352.

Tanaka, Y., Hibino, T., Hayashi, Y., Tanaka, A., Kishitani, S., Takabe, T., Yokota, S. and Takabe, T. (1999) Salt tolerance of transgenic rice overexpressing yeast mitochondria Mn-SOD in chloroplasts. *Plant Science* 148, 131–138.

Termaat, A. and Munns, R. (1986) Use of concentrated macronutrient solutions to separate osmotic from NaCl-specific effects on plant growth. *Australian Journal of Plant Physiology* 13, 509–522.

Tester, M. and Davenport, R. (2003) Na^+ tolerance and Na^+ transport in higher plants. *Annals of Botany* 91, 503–527.

Tran, L.S., Urao, T., Qin, F., Maruyama, K., Kakimoto, T., Shinozaki, K. and Yamaguchi-Shinozaki, K. (2007) Functional analysis of AHK1/ATHK1 and cytokinin receptor histidine kinases in response to abscisic acid, drought, and salt stress in *Arabidopsis*. *Proceedings of National Academy of Sciences USA* 104, 20623–20628.

Türkan, I. and Demiral, T. (2009) Recent developments in understanding salinity tolerance. *Environmental and Experimental Botany* 67, 2–9.

Uchida, A., Jagendorf, A.T., Hibino, T., Takabe, T. and Takabe, T. (2002) Effects of hydrogen peroxide and nitric oxide on both salt and heat stress tolerance in rice. *Plant Science* 163, 515–523.

Uozumi, N., Kim, E.J., Rubio, F., Yamaguchi, T., Muto, S., Tsuboi, A., Bakker, E.P., Nakamura, T. and Schroeder, J.I. (2000) The *Arabidopsis HKT1* gene homolog mediates inward Na^+ currents in *Xenopus laevis* oocytes and Na^+ uptake in *Saccharomyces cerevisiae*. *Plant Physiology* 122, 1249–1260.

Urao, T., Yakubov, B., Satoh, R., Yamaguchi-Shinozaki, K., Seki, M., Hirayama, T. and Shinozaki, K. (1999) A transmembrane hybrid-type histidine kinase in *Arabidopsis* functions as an osmosensor. *The Plant Cell* 11, 1743–1754.

Van Breusegem, F., Vranová, E., Dat, J.F. and Inzé, D. (2001) The role of active oxygen species in plant signal transduction. *Plant Science* 161, 405–414.

Wohlbach, D.J., Quirino, B.F. and Sussman, M.R. (2008) Analysis of the *Arabidopsis* histidine kinase ATHK1 reveals a connection between vegetative osmotic stress sensing and seed maturation. *The Plant Cell* 20, 1101–1117.

Wood, J.M., Bremer, E., Csonka, L.N., Kraemer, R., Poolman, B., van der Heide, T. and Smith, L.T. (2001) Osmosensing and osmoregulatory compatible solute accumulation by bacteria. *Comparative Biochemistry and Physiology. Part A: Molecular & Integrative Physiology* 130, 437–460.

Wu, Y.Y., Chen, G.D., Meng, Q.W. and Zheng, C.C. (2004) The cotton GhNHX1 gene encoding a novel putative tonoplast Na^+/H^+ antiporter plays an important role in salt stress. *Plant & Cell Physiology* 45, 600–607.

Wurzinger, B., Mair, A., Pfister, B. and Teige, M. (2011) Cross-talk of calcium-dependent protein kinase and MAP kinase signalling. *Plant Signaling & Behavior* 6, 8–12.

Wyn Jones, R.G. and Gorham, J. (1983) Osmoregulation. In: Lange, D.L., Nobel, P.S., Osmond, C.B. and Ziegler, H. (eds) *Encyclopedia of Plant Physiology: physiological plant ecology. III. Response to chemical and biological environment*. Springer-Verlag, Berlin, pp. 35–58.

Wyn Jones, R.G. and Pollard, A. (1983) Proteins, enzymes and inorganic ions. In: Lauchli, A. and Person, A. (eds) *Encyclopedia of Plant Physiology*, New Series, 15B. Springer-Verlag, Berlin, pp. 528–562.

Xiong, L. and Zhu, J.-K. (2003) Regulation of abscisic acid biosynthesis. *Plant Physiology* 133, 29–36.

Yalpani, N., Enyedi, A.J., León, J. and Raskin, I. (1994) Ultraviolet light and ozone stimulate accumulation of salicylic acid, pathogenesis-related proteins and virus resistance in tobacco. *Planta* 193, 372–376.

Yokoi, S., Bressan, R.A. and Hasegawa, P.M. (2002a) Salt stress tolerance of plants. *JIRCAS Working Report*, 25–33.

Yokoi, S., Quintero, F.J., Cubero, B., Ruiz, M.T., Bressan, R.A., Hasegawa, P.M. and Pardo, J.M. (2002b) Differential expression and function of *Arabidopsis thaliana* NHX Na^+/H^+ antiporters in the salt stress response. *The Plant Journal* 30, 529–539.

Zhu, J.-K. (2001) Plant salt tolerance. *Trends in Plant Science* 6, 66–71.

Zhu, J.-K. (2002) Salt and drought stress signal transduction in plants. *Annual Review of Plant Biology* 53, 247–273.

Zhu, J.-K. (2003) Regulation of ion homeostasis under salt stress. *Current Opinion in Plant Biology* 6, 441–445.

Zhu, J.-K. (2007) Plant salt stress. In: *Encyclopedia of Life Sciences*. John Wiley & Sons, New York, pp. 1–3.

7 Sugarcane (*Saccharum* sp.) Salt Tolerance at Various Developmental Levels

Ch.B. Gandonou,[1,2]* and N. Skali-Senhaji[1]

[1]*Laboratoire de Biologie et Santé, Faculté des Sciences de Tétouan, Université Abdelmalek Essaâdi, Tétouan, Morocco;* [2]*Laboratoire de Physiologie Végétale et d'Etude des Stress Environnementaux, Faculté des Sciences et Techniques, Université d'Abomey-Calavi, Cotonou, République du Bénin*

Abstract

Salt-stress affects plant growth and development at different stages. In this work, we evaluated the level of salinity tolerance of five sugarcane (*Saccharum* sp.) varieties: CP66-346, CP65-357, CP70-321, CP59-73 and NCo310 by using different NaCl concentrations (0, 17, 34, 68 and 102 mM). This evaluation was based on the *in vitro* bud emergency, young plants' survival and growth in hydroponic system, and finally on the aspect and the growth of calli issued from foliar explants. NaCl stress effects result in a reduction of the final bud emergency percentage. At bud emergence stage, varieties CP66-346 and CP59-73 appeared to be the most salt tolerant while NCo310 behaved as the most salt sensitive. Young plants' survival and growth are also reduced by salinity and at this stage variety CP66-346 seems to be the most tolerant while CP65-357 and CP70-321 are the most sensitive. Salinity causes calli necrosis and reduces their growth; varieties NCo310 and CP70-321 appeared to be the most salt tolerant while CP65-357 seems to be the most sensitive. These results indicate that the salt tolerance of a variety depends on the stage of development and the level considered. Consequently, salt tolerance of a given cultivar at whole plant level does not guarantee salt tolerance of tissue or cell cultures issued from this cultivar. Bud emergency stage seems to be the most tolerant stage. Variety CP66-346 appeared to be a salt-tolerant variety at both bud emergence and young plant stages.

7.1 Introduction

Salinity is a major abiotic stress increasingly affecting plant health and survival worldwide (Sakhanokho and Kelly, 2009). The cultivable areas affected by this stress were estimated to be 900 million ha (Flowers, 2004), half of which is irrigated (Zhu, 2001); this area increases continuously due to bad agricultural practices. Sugarcane culture which is generally carried out under strong irrigation is confronted with this problem. This plant is classified as a moderately salt-sensitive species with a threshold electrical conductivity of paste saturated extract of about 1.7 dS m^{-1} beyond which production decreases (Ayers and Westcott, 1989). This complex abiotic stress which presents osmotic and ionic components, poses a threat to agriculture (Munns *et al.*, 2006). It causes many metabolic disturbances in higher plants without being always possible to distinguish effects due to the osmotic component from those related to specific ion toxicity. However, it is largely

*E-mail: ganchrist@hotmail.com

known that a substantial variation exists in salt sensitivity of various species and of various cultivars and ecotypes of the same species. In addition, several authors reported that NaCl affects seed germination (Cramer, 1994; Ghoulam and Farès, 2001; Debez *et al.*, 2004), plant growth and survival (Lutts *et al.*, 1995; Wang *et al.*, 1997; Almansouri *et al.*, 2001; Aghaei *et al.*, 2008; Shafi *et al.*, 2011) and cell growth and necrosis (Arzani and Mirodjagh, 1999; Basu *et al.*, 2002; Alvarez *et al.*, 2003; Htwe *et al.*, 2011). It is generally admitted that the salt-tolerance of a given genotype depends on the developmental stage and the selected organization level (Lutts *et al.*, 1995); therefore, a genotype tolerant at the germination stage can appear rather sensitive at the young plant stage and/or at cellular level. Salt effects on sugarcane plants are generally studied either at the stage of germination (i.e. bud emergence) (Kumar and Naidu, 1993; Chowdhury *et al.*, 2001; Akhtar *et al.*, 2003; Gandonou *et al.*, 2008, 2011), whole plant level (Chowdhury *et al.*, 2001; Wahid, 2004; Sebastian *et al.*, 2009; Gandonou *et al.*, 2012), or at cellular level (Gonzalez *et al.*, 1995; Gandonou *et al.*, 2005a; Errabii *et al.*, 2006). Little work has dealt with the study of sugarcane NaCl tolerance combined with germination stage, whole plant level and cellular level. The aim of this study is to compare the average level of salt tolerance of five sugarcane cultivars at germination stage, young plant level and at cellular level in order to check if the average level of tolerance of a cultivar is the same at all three stages.

7.2 Material and Methods

7.2.1 Plant material

The experimental plant materials used are sugarcane cvs NCo310, CP70-321, CP65-357, CP59-73 and CP66-346 used by Gandonou *et al.* (2012).

7.2.2 Salt concentrations

NaCl was used as salt. For germination stage and *in vitro* callus culture stage studies, five NaCl concentrations were used: 0, 17, 34, 68 and 102 mM (0, 1, 2, 4 and 6 g l^{-1}, respectively) (Gandonou *et al.*, 2005a, 2011) while for young plant-level study, only the first four NaCl concentrations were used because 102 mM NaCl appeared to be high enough to prevent young plant survival (Gandonou *et al.*, 2012).

7.2.3 Germination stage study

The study was done *in vitro*. Young single bud setts (approximately 4 cm) were taken from the top of each plant and wiped with cotton saturated with ethanol 70%. Setts' disinfection conditions, medium composition and culture conditions are those described by Gandonou *et al.* (2008, 2011). Final germination percentage was determined for each variety and each NaCl concentration after 8 days.

7.2.4 Young plants-level study

The study was done in hydroponic medium. Stalk disinfection and germination, media composition, plant growth conditions, plant survival and growth determination methods are those described by Gandonou *et al.* (2012). Relative height growth of plants (RHG) was calculated as described by the former authors.

7.2.5 Cellular-level study

Calli were induced from young leaf cylinders. Medium composition, callus cultures and *in vitro* salt treatment conditions, calli necrosis and relative fresh weight growth (RFWG) determination methods are those described by Gandonou *et al.* (2005a, b).

7.2.6 Statistical analysis

All the experiments were repeated twice independently. The number of germinated buds, the number of dead plants and the number of necrotic calli were analysed as binomial distribution variates. For plants and calli growth,

1-way or 2-way analysis of variance (ANOVA) were used to study the main effects of cultivars and/or stress intensity. All analyses were carried out using SAS program (SAS Institute, 1992).

7.3 Results

7.3.1 Effect of NaCl stress at germination stage

Practically no effect of salt stress was observed on final germination for the varieties CP59-73 and CP66-346 (Fig. 7.1). The final bud germination rate of variety CP66-346 was reduced by about 4% in presence of 17 and 68 mM of NaCl; a slight increase (not significant) was observed at 34 mM. For variety CP59-73, bud germination showed a slight reduction (3%, not significant) only at 34 mM of NaCl. The highest bud germination rate reduction was observed for cv. NCo310 with a reduction of 10% at 17, 34 and 68 mM NaCl. For cv. CP70-321, bud germination rate reduction under salt stress was about 6% at the three NaCl concentrations above while for cv. CP65-357, bud germination rate showed a reduction of 3%, 9% and 16% at 17, 34 and 68 mM of NaCl, respectively. These reductions were not significant ($p < 0.05$).

Table 7.1 presents the percentages of final bud germination when means values were calculated from data collected for the three doses of NaCl (17, 34 and 68 mM) and expressed as percentage of that of the control. These data showed that salt stress reduced bud germination by about 10% compared to control for cv. NCo310 while this reduction was about 6% for cvs CP70-321 and CP65-357; no reduction was observed for CP59-73 and CP66-346.

Thus, cvs CP59-73 and CP66-346 appeared to be the most salt tolerant at germination stage while NCo310 behaved as the most salt-sensitive cultivar.

7.3.2 Effect of NaCl stress at whole plant level

Plant survival

The reduction of survival (plant death) due to the average effect of salt stress was lower for cv. CP66-346 (5%) and higher for cvs CP70-321 and CP65-357 (21.67% and 20%, respectively) (Table 7.2); this reduction was intermediary for cvs CP59-73 and NCo310 (13.33% and 16.67%, respectively). Thus, cv. CP66-346 presented the highest survival while CP70-321 and CP65-357 showed the lowest survival

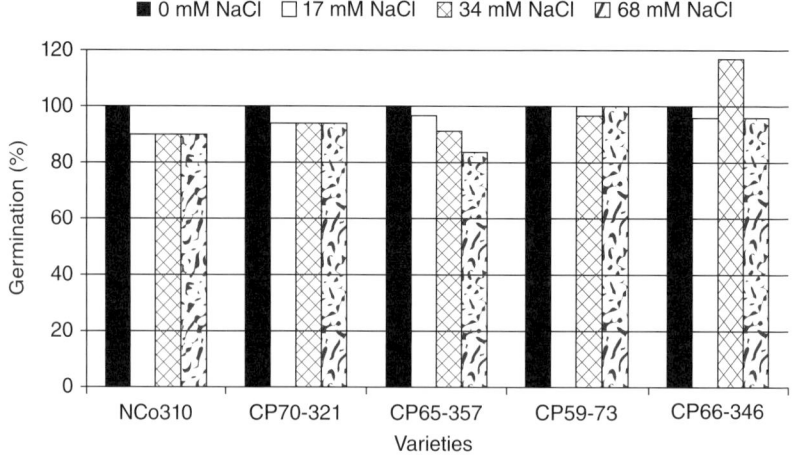

Fig. 7.1. Effect of NaCl salinity on five sugarcane variety buds *in vitro* germination percentage (cvs NCo310, CP70-321, CP65-357, CP59-73 and CP66-346) after 8 days of culture: germination percentages in presence of NaCl were expressed as percentage of that of the control (Gandonou et al., 2011).

Table 7.1. Mortality percentages of plants of five sugarcane cultivars as affected by different NaCl concentrations (Gandonou et al., 2012) (n = 20 or n = 30).[a]

NaCl concentration (mM)	Cultivars				
	NCo310	CP70-321	CP65-357	CP59-73	CP66-346
0	0 a	0 a	0 a	0 a	0 a
17	10 ab	10 a	10 ab	0 a	0 a
34	10 ab	10 ab	20 bc	10 ab	5 ab
68	30 bc	45 c	30 bc	30 bc	10 ab

[a] Values followed with same letter are not significantly different at $p < 0.05$.

Table 7.2. Germination percentages, mortality percentages, relative height growth of plants, callus necrosis percentage and callus relative fresh weight growth of five sugarcane cultivars (CP65-357, NCo310, CP70-321, CP59-73 and CP66-346) as affected by NaCl stress (Gandonou et al., 2005a, 2011, 2012).

		Cultivars				
		NCo310	CP70-321	CP65-357	CP59-73	CP66-346
Bud germination rate (%)	0 NaCl	100	100	100	100	100
	+ NaCl	90.78	94.12	94.29	98.96	103.08
Plant mortality (%)	0 NaCl	0 a	0 a	0 a	0 a	0 a
	+ NaCl	16.67 b	21.67 b	20 b	13.33 b	5 a
Plant growth (%)	0 NaCl	100 a	100 a	100 a	100 a	100 a
	+ NaCl	75.65 b	50.99 b	63.31 b	65.44 b	80.47 a
Callus necrosis (%)	0 NaCl	0 a	0 a	2.86 a	0 a	ND
	+ NaCl	8.33 ab	2.86 a	26.31 b	0 a	ND
Callus growth (%)	0 NaCl	100 a	100 a	100 a	100 a	ND
	+ NaCl	75.69 b	73.33 b	49.09 b	66.83 b	ND

0 NaCl, control; + NaCl, presence of NaCl: data in presence of NaCl were expressed as the average of the three (or four) values obtained in the presence of the three (or four) NaCl concentrations (17; 34, 68 and 102[a] mM) expressed in percentage of that of control.
[a] This concentration was used only for cellular level study.

in the presence of NaCl compared to NCo310 and CP59-73.

On the basis of plant survival criterion, cv. CP66-346 appeared to be the most salt-tolerant at the whole-plant level, while cvs CP70-321 and CP65-357 were the most salt-sensitive.

Plant growth

NaCl stress reduced significantly plant RHG for all cultivars (Fig. 7.2). For cv. NCo310, RHG reduction was significant ($p < 0.05$) for 17 mM NaCl (reduction was not significant at 34 mM NaCl) and 68 mM NaCl. These reductions correspond to 25%, 16% and 33% of the control, respectively. For cv. CP70-321, RHG reduction corresponds to 27%, 50% and 70% of control at 17, 34 and 68 mM NaCl, respectively. The reduction observed was significant ($p < 0.001$) at all NaCl concentrations used (Fig. 7.2). In cv. CP65-357, RHG reduction under salt stress was significant ($p < 0.01$) starting from 17 mM NaCl (Fig. 7.2) and corresponds to a reduction of growth of 29%, 40% and 41% compared to the control at 17, 34 and 68 mM NaCl, respectively. For CP66-346, RHG reduction was significant ($p < 0.001$) at 34 and 68 mM NaCl (Fig. 7.2) and corresponds to a growth reduction of 4%, 10% and 45% compared to the control at 17, 34 and 68 mM NaCl, respectively. For cv. CP59-73, plant RHG reduction was about 6%, 48% and 50% in the presence of 17, 34 and 68 mM of NaCl, respectively; this reduction was significant ($p < 0.001$) at 34 mM and 68 mM NaCl (Fig. 7.2). There is, thus, a significant difference in the behaviour of the studied cultivars. It is important to note that contrary to cvs NCo310 and CP65-357 for

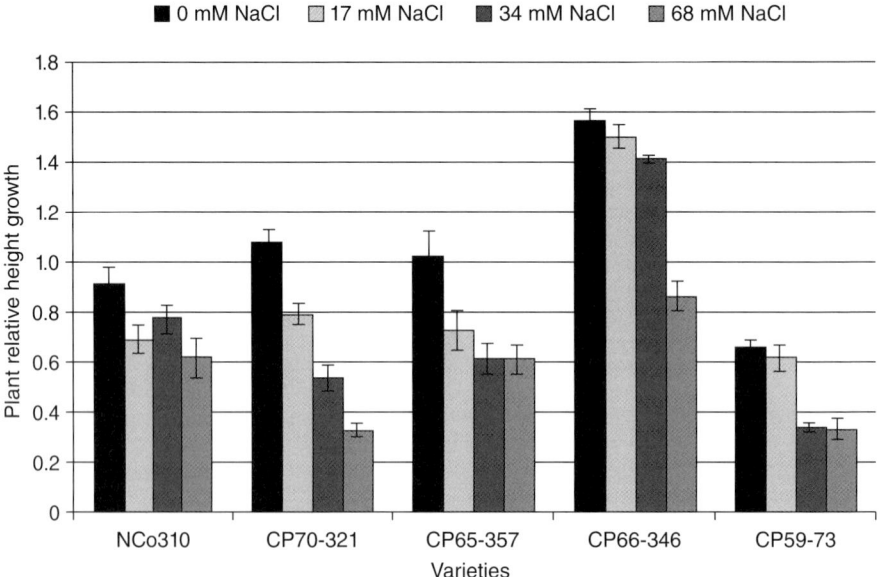

Fig. 7.2. Plant relative height growth of five sugarcane cultivars as affected by different NaCl concentrations (n = 4; vertical bars are standard errors): values within cultivar with the same letter are not significantly different at p < 0.05 (Gandonou et al., 2012).

which plant growth was significantly affected starting from 17 mM of NaCl (but not at 34 mM for NCo310), the growth of CP59-73 and CP66-346 was significantly affected only starting from 34 mM of NaCl.

The reduction of plant growth due to the average effect of salt stress was lower for cvs CP66-346 and NCo310 (19.53% and 24.35%, respectively) and higher for cv. CP70-321 (approximately 49%); this reduction was intermediary for cvs CP59-73 and CP65-357 (34.56% and 36.69%, respectively) (Table 7.2). Thus cvs CP66-346 and NCo310 presented a higher growth rate in the presence of NaCl compared to cvs CP59-73, CP65-357 and CP70-321.

On the basis of plant growth criterion, cv. CP66-346 appeared to be the most salt tolerant at the young plant stage (whole-plant level) while CP70-321 and CP65-357 behaved as the most salt-sensitive cultivars.

7.3.3 Effect of NaCl stress at cellular level

In the culture conditions used, cv. CP66-346 did not produce callus. Thus, the salt stress effect at cellular level was studied only for the four cultivars that produced calli in the culture conditions.

The addition of NaCl to culture medium caused an increase in calli necrosis for all cultivars (Table 7.3) and significant difference in calli necrosis was observed among genotypes. No callus of CP59-73 showed necrosis under NaCl concentrations used. For CP70-321, no callus showed necrosis below 68 mM NaCl, while for NCo310, the first callus necrosis was observed at 34 mM NaCl. For NCo310 calli, the effect of NaCl was significant (p < 0.05) only at the highest dose of NaCl used (102 mM) while this effect was significant (p < 0.05) at 68 mM for CP65-357 calli. These results revealed significant differences among cultivars for callus necrosis percentages.

The increase in callus necrosis due to the average effect of salt stress was lower for cvs CP59-73 and CP70-321 (0% and 2.86%, respectively) and higher for cv. CP65-357 (approximately 26%); this reduction was intermediary for cv. NCo310 (8.33%) (Table 7.2).

On the basis of callus necrosis criterion, cv. CP59-73 appeared to be the most salt-tolerant at cellular level followed by CP70-321; CP65-357 was the most salt-sensitive.

Callus RFWG decreased as the concentration of NaCl increased in the culture medium (Fig. 7.3). The highest reduction was observed in cv. CP65-357 with a reduction of callus growth about 27%, 45%, 59% and 63% at 17, 34, 68 and 102 mM NaCl, respectively. The reduction was significant ($p<0.001$) from 17 mM NaCl. Cultivars NCo310 and CP70-321 showed the lowest reduction with a reduction of 4%, 21%, 28% and 44%, and 10%, 21%, 35% and 41% respectively, at the same NaCl concentrations. The reduction observed was significant ($p<0.01$ in the case of NCo310, and $p<0.001$ in the case of CP70-321) only from 34 mM NaCl. Cultivar CP59-73 showed an intermediary behaviour with callus growth reduction about 12%, 28%, 36% and 56% in the presence of 17, 34, 68 and 102 mM NaCl, respectively. This reduction was significant ($p<0.001$) only from 34 mM NaCl.

The reduction of callus growth due to the average effect of salt stress was lower for cvs NCo310 and CP70-321 (24.31% and 26.67%, respectively) and higher for cv. CP65-357 (approximately 51%); this reduction was intermediary for cv. CP59-73 (33.17%) (Table 7.3). Thus, cvs NCo310 and CP70-321 presented the highest callus growth in the presence of NaCl while cv. CP65-357 presented the lowest growth; CP59-73 was intermediate.

Table 7.3. Necrosis percentages of callus obtained from four sugarcane varieties (CP65-357, NCo310, CP70-321 and CP59-73) as affected by different NaCl concentrations (partially from Gandonou et al., 2005a).[a]

NaCl concentrations (mM)	Cultivars			
	CP65-357	NCo310	CP70-321	CP59-73
0	2.86 ab	0 a	0 a	0 a
17	8.57 ab	0 a	0 a	0 a
34	13.33 bc	10.00 ab	0 a	0 a
68	40 cd	6.67 ab	2.86 ab	0 a
102	43.33 d	16.67 bc	8.57 ab	0 a

[a] Values within columns followed by the same letter do not differ significantly at $p=0.05$.

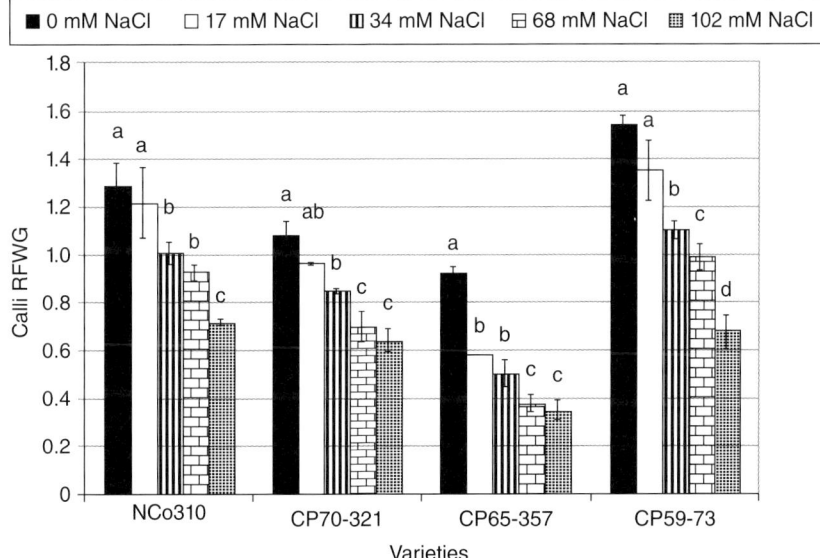

Fig. 7.3. Relative fresh weight growth (RFWG) of callus obtained from four sugarcane cultivars (CP65-357, NCo310, CP70-321 and CP59-73) as affected by different NaCl concentrations: vertical bars are standard errors (n=4): values within cultivar with the same letter are not significantly different at $p<0.05$ (partially from Gandonou et al., 2005a).

On the basis of the criterion of callus growth, cvs NCo310 and CP70-321 appeared to be the most salt-tolerant cultivars at cellular level while cv. CP65-357 appeared as the most salt-sensitive.

7.4 Discussion

NaCl stress reduced bud emergence rate, young plants' survival and growth and callus growth. It also enhanced callus necrosis. Similar results were reported in several plants including rice (Lutts *et al.*, 1995, 1996; Basu *et al.*, 2002), durum wheat (Karadimova and Dyambova, 1993; Arzani and Mirodjagh, 1999), sunflower (Alvarez *et al.*, 2003) and sugarcane (Kumar and Naidu, 1993; Gonzalez *et al.*, 1995; Chowdhury *et al.*, 2001; Akhtar *et al.*, 2003; Hussain *et al.*, 2004; Gandonou *et al.*, 2005a, 2008). However, our results show a difference among cultivars' behaviour according to the stage or level of the study and the criterion used. For germination stage, cvs CP59-73 and CP66-346 appeared to be more salt tolerant whereas cv. NCo310 behaved as the most salt sensitive. Cultivars CP65-357 and CP70-321 were intermediaries or moderately sensitive. At young-plant stage, cv. CP66-346 was the most tolerant whereas cvs CP70-321 and CP65-357 behaved as the most sensitive. Cultivars CP59-73 and NCo310 appeared as moderately sensitive when based on plant survival rate criterion. Using plant growth criterion, cvs CP66-346 and NCo310 behaved as the most salt-tolerant whereas cv. CP70-321 was the most sensitive. Cultivars CP59-73 and CP65-357 were intermediaries. At cellular level, cvs CP59-73 and CP70-321 were the most tolerant whereas CP65-357 behaved as the most sensitive. Cultivar NCo310 appeared as moderately tolerant when based on callus necrosis criterion. Using callus growth criterion, cvs NCo310 and CP70-321 appeared as the most salt-tolerant whereas cv. CP65-357 was the most sensitive and cv. CP59-73 was intermediate.

Considering young plant stage, cv. CP65-357, which behaved as salt-sensitive based on plant survival criterion, appeared as moderately tolerant when plant growth was used as criterion. The same tendency was observed for cv. NCo310, which appeared as moderately sensitive when plant survival rate criterion was used, and as tolerant based on plant growth criterion. In the case of cellular level study, similar observations can be made for cvs CP59-73 and NCo310. These observations indicated that the relative salinity tolerance of the investigated sugarcane cultivars changed as a function of the criterion considered. These results are in agreement with those reported in rice by Lutts *et al.* (1995). For example, these authors have observed that at young seedling stage, cv. IR 31785 was the most sensitive variety according to root dry weight while it was viewed as rather resistant when root elongation was used as a criterion.

Overall, at the germination stage, among the investigated cultivars, cvs CP59-73 and CP66-346 were the most salt tolerant whereas cv. NCo310 appeared as the most salt sensitive; while at young plant stage, cv. CP66-346 was the most salt tolerant whereas cv. CP70-321 was the most salt sensitive. At the cellular level, cvs CP70-321 and NCo310 were the most salt-tolerant whereas cv. CP65-357 was the most salt-sensitive. These data indicated variability in the relative order of tolerance of sugarcane cultivars according to the stage of development as reported previously (Heenan *et al.*, 1988; Aslam *et al.*, 1993; Lutts *et al.*, 1995; Foolad, 2004) and that salt tolerance at a specific stage does not guarantee tolerance at another stage. Cultivar NCo310, for example, was sensitive during germination whereas it was rather tolerant at young plant stage and cellular level. A similar tendency was observed for cv. CP70-321, which was salt-sensitive at young plant stage but appeared as tolerant at cellular level.

Considering the four cultivars studied at both young plant stage and cellular level (i.e. cvs CP65-357, CP70-321, NCo310 and CP59-73), three tendencies were observed:

- Cultivar CP65-357 behaved as salt sensitive either at young plant stage or at cellular level;
- Cultivars CP59-73 and NCo310 behaved overall as moderately tolerant either at young plant stage or at cellular level; and
- Cultivar CP70-321, which was the most sensitive at young plant stage, appeared as the most tolerant at cellular level.

These findings revealed two types of correlation between salt tolerance of whole plant and salt tolerance of tissue or cell cultures derived from that plant. Positive correlation characterized by the same behaviour between whole plant and tissue or cell cultures as observed in this study for cvs CP65-357, CP59-73 and NCo310, and negative correlation characterized by a different behaviour between whole plant and tissue or cell cultures as observed in this study for cv. CP70-321. These findings corroborate data reported by Mills and Tal (2004) in tomato, which revealed these two types of correlation between whole-plant salt tolerance and salt tolerance of tissue or cell cultures issued from this plant. The fact that cvs NCo310 and CP59-73 are relatively tolerant, either at whole-plant stage or at cellular level, can be explained by the existence of a cellular basis, at least partial, of the tolerance of these cultivars.

Considering each specific stage, young-plant stage was more sensitive to salt than germination since the average effect of salt stress was significant and more accentuated at young-plant stage corroborating previous observations in rice (Akbar and Neue, 1987; Lutts et al., 1995). Although the high NaCl concentration (102 mM) was used only in the case of cellular study, the average effect of NaCl was similar to that obtained at young-plant stage. This observation indicates that cellular level was more salt tolerant than at the young-plant stage.

This study underlined the variability of relative salt-stress tolerance for some sugarcane cultivars during development and at cellular level. It indicated that salinity tolerance at different development stages does not behave as an interdependent characteristic. Germination stage appeared as the most salt-tolerant stage in sugarcane. For the first time, we have demonstrated that, in sugarcane, salt tolerance of a given cultivar at whole-plant level does not guarantee salt tolerance of tissue or cell cultures issued from this cultivar.

Acknowledgements

This research was financially supported by 'Programme d'Appui à la Recherche Scientifique (PARS AGRO 180)' from the 'Ministère de l'Enseignement Supérieur, de la Formation des Cadres et de la Recherche Scientifique' of Morocco. The authors thank Mr Mohamed El Ghrassli (CTCS, Morocco) for plant material provision.

References

Aghaei, K., Ehsanpour, A.A., Balali, G. and Mostajeran, A. (2008) In vitro screening of potato (Solanum tuberosum L.) cultivars for salt tolerance using physiological parameters and RAPD analysis. American-Eurasian Journal of Agricultural and Environmental Sciences 3(2), 159–164.

Akbar, M.S. and Neue, H.U. (1987) Effect of Na/Ca and Na/K ratios in saline culture solution on the growth and mineral nutrition of rice (Oryza sativa L.). Plant and Soil 104, 57–62.

Akhtar, S., Wahid, A. and Rasul, E. (2003) Emergence, growth and nutrient composition of sugarcane sprouts under NaCl salinity. Biologia Plantarum 46(1), 113–116.

Almansouri, M., Kinet, J.M. and Lutts, S. (2001) Effect of salt and osmotic stresses on germination in durum wheat (Triticum durum Desf.). Plant and Soil 231, 243–254.

Alvarez, I., Tomaro, L.M. and Benavides, P.M. (2003) Changes in polyamines, proline and ethylene in sunflower calluses treated with NaCl. Plant Cell, Tissue and Organ Culture 74(1), 51–59.

Arzani, A. and Mirodjagh, S.S. (1999) Response of durum wheat cultivars to immature embryo culture, callus induction and in vitro salt stress. Plant Cell, Tissue and Organ Culture 58, 67–72.

Aslam, M., Qureshi, R.H. and Ahmed, N. (1993) A rapid screening technique for salt-tolerance in rice (Oryza sativa L.). Plant and Soil 150, 99–107.

Ayers, R.S. and Westcott, D.W. (1989) Water quality for agriculture. FAO Irrigation and Drainage Paper, No. 29 (rev. 1).

Basu, S., Gangopadhyay, G. and Mukherjee, B.B. (2002) Salt tolerance in rice in vitro: implication of accumulation of Na^+, K^+ and proline. Plant Cell, Tissue and Organ Culture 69, 55–64.

Chowdhury, M.K.A., Miah, M.A.S., Ali, S., Hossain, M.A. and Alam, Z. (2001) Influence of sodium chloride salinity on germination and growth of sugarcane (*Saccharum officinarum* L.). *Sugar Cane International* 7, 15–16.

Cramer, G.R. (1994) Response of maize (*Zea mays* L.) to salinity. In: Pessarakli, M. (ed.) *Handbook of Plant and Crop Stress*. Marcel Dekker, New York, pp. 449–459.

Cramer, G.R. (2003) Differential effects of salinity on leaf elongation kinetics of three grass species. *Plant and Soil* 253, 233–244.

Debez, A., Ben Hamed, K., Grignon, C. and Abdelly, C. (2004) Salinity effects on germination, growth, and seed production of the halophyte *Cakile maritima*. *Plant and Soil* 262, 179–189.

Errabii, T., Gandonou, Ch.B., Essalmani, H., Abrini, J., Idaomar, M. and Skali-Senhaji, N. (2006) Growth, proline and ion accumulation in sugarcane callus cultures under drought-induced osmotic stress and its subsequent relief. *African Journal of Biotechnology* 5(16), 1488–1493.

Flowers, T.J. (2004) Improving crop salt tolerance. *Journal of Experimental Botany* 55(396), 307–319.

Foolad, M.R. (2004) Recent advances in genetics of salt tolerance in tomato. *Plant Cell, Tissue and Organ Culture* 76, 101–119.

Gandonou, Ch., Abrini, J., Idaomar, M. and Skali-Senhaji, N. (2005a) Response of sugarcane (*Saccharum* sp.) varieties to embryogenic callus induction and *in vitro* salt stress. *African Journal of Biotechnology* 4(4), 350–354.

Gandonou, Ch., Abrini, J., Idaomar, M. and Skali-Senhaji, N. (2005b) Effects of NaCl on growth, ions and proline accumulation in sugarcane (*Saccharum* sp.) callus culture. *Belgian Journal of Botany* 138(2), 173–180.

Gandonou, Ch.B., Agbangla, C., Ahanhanzo, C., Errabii, T., Idaomar, M., Abrini, J. and Skali-Senhaji, N. (2008) *In vitro* culture techniques as a tool of sugarcane bud germination study under salt stress. *African Journal of Biotechnology* 7(20), 3680–3682.

Gandonou, Ch.B., Ahanhanzo, C., Agbangla, C., Errabii, T., Idaomar, M., Abrini, J. and Skali-Senhaji, N. (2011) Effect of NaCl on *in vitro* sugarcane (*Saccharum* sp.) bud emergence. *African Journal of Biotechnology* 10(4), 539–544.

Gandonou, Ch.B., Gnancadja, L.S., Abrini, J. and Skali-Senhaji, N. (2012) Salinity tolerance of some sugarcane (*Saccharum* sp.) cultivars in hydroponic medium. *International Sugar Journal* 114(1359), 190–196.

Ghoulam, C. and Farès, K. (2001) Effect of salinity on seed germination and early seedling growth of sugar beet (*Beta vulgaris* L.). *Seed Science and Technology* 29, 357–364.

Gonzalez, V., Castroni, S. and Fuchs, M. (1995) Evaluacion de la reaccion de genotipos de caňa de azucar a diferentes concentraciones de NaCl. *Agronomia Tropical* 46(2), 219–232.

Heenan, D.P., Lewin, L.G. and McCaffery, D.W. (1988) Salinity tolerance in rice varieties at different growth stages. *Australian Journal of Experimental Agriculture* 28, 343–351.

Htwe, N.N., Maziah, M., Ling, H.C., Zaman, F.Q. and Zain, A.M. (2011) Responses of some selected Malaysian rice genotypes to callus induction under *in vitro* salt stress. *African Journal of Biotechnology* 10(3), 350–362.

Hussain, A., Khan, Z.I., Ashraf, M., Rashid, H.M. and Akhtar, M.S. (2004) Effect of salt stress on some growth attributes of sugarcane cultivars CP-77-400 and COJ-84. *International Journal of Agriculture and Biology* 6(1), 188–191.

Karadimova, M. and Djambova, G. (1993) Increased NaCl-tolerance in wheat (*Triticum aestivum* L. and *T. durum* Desf.) through *in vitro* selection. *In Vitro Cellular and Developmental Biology* 29P, 180–182.

Kumar, S. and Naidu, M.K. (1993) Germination of sugarcane setts under saline conditions. *Sugar Cane* 4, 2–5.

Lutts, S., Kinet, J.M. and Bouharmont, J. (1995) Changes in plant response to NaCl during development of rice (*Oryza sativa* L.) varieties differing in salinity resistance. *Journal of Experimental Botany* 46, 1843–1852.

Lutts, S., Kinet, J.M. and Bouharmont, J. (1996) Effects of various salts and of mannitol on ion and proline accumulation in relation to osmotic adjustment in rice (*Oryza sativa* L.) callus cultures. *Journal of Plant Physiology* 149, 186–195.

Mills, D. and Tal, M. (2004) The effect of ventilation on *in vitro* response of seedlings of the cultivated tomato and its wild salt-tolerant relative *Lycopersicon pennellii* to salt stress. *Plant Cell, Tissue and Organ Culture* 78, 209–216.

Munns, R., James, R.A. and Lauchli, A. (2006) Approaches to increasing the salt tolerance of wheat and other cereals. *Journal of Experimental Botany* 57, 1025–1043.

Sakhanokho, H.F. and Kelley, R.Y. (2009) Influence of salicylic acid on *in vitro* propagation and salt tolerance in *Hibiscus acetosella* and *Hibiscus moscheutos* (cv 'Luna Red'). *African Journal of Biotechnology* 8(8), 1474–1481.

SAS Institute (1992) *SAS/STAT User's guide*, Vol. 1; Release 6.03. SAS Institute Inc., Cary, North Carolina.

Sebastian, S.P., Udayasoorian, C., Jayabalakrishnan, R.M. and Parameswari, E. (2009) Performance of sugarcane varieties under organic amendments with poor quality irrigation water. *Australian Journal of Basic and Applied Sciences* 3(3), 1674–1684.

Shafi, M., Bakht, J., Khan, M.J., Khan, M.A. and Raziuddin (2011) Role of abscisic acid and proline in salinity tolerance of wheat genotypes. *Pakistan Journal of Botany* 43(2), 1111–1118.

Wahid, A. (2004) Analysis of toxic and osmotic effects of sodium chloride on leaf growth and economic yield of sugarcane. *Botanical Bulletin of Academia Sinica* 45, 133–141.

Wang, L.W., Showalter, A.M. and Ungar, I.A. (1997) Effect of salinity on growth, ion content, and cell wall chemistry in *Atriplex prostrata* (Chenopodiaceae). *American Journal of Botany* 84(9), 1247–1255.

Zhu, J.K. (2001) Plant salt tolerance. *Trends in Plant Science* 6(2), 66–71.

8 The Impact of Ozone Pollution on Plant Defence Metabolism: Detrimental Effects on Yield and Quality of Agricultural Crops

Fernanda Freitas Caregnato,[1]* Rafael Calixto Bortolin,[1] Armando Molina Divan Junior[2] and José Cláudio Fonseca Moreira[1]

[1]*Department of Biochemistry, Center for Oxidative Stress Research, Federal University of Rio Grande do Sul (UFRGS), Porto Alegre; [2]Laboratory of Plant Bioindication, Center of Ecology, UFRGS, Porto Alegre, Brazil*

Abstract

Over the past decades, research on the negative effects of air pollutants on agricultural crops and agro-ecosystems point out for emission reduction strategies, with practical recommendations to increase the sustainability of agricultural and land management in an environment that is constantly changing. Agricultural production will need to keep pace with the growing food demand, which depends on many factors, including the future levels of air pollution, such as tropospheric ozone. The risk of negative effects of ozone on crop productivity created the need to improve our understanding on the mechanisms underlying ozone toxicity, and biotechnological advances are now starting to provide us with the necessary knowledge to safely develop and/or select crops varieties better adapted to ozone stress. Ozone phytotoxicity arises mainly because of its high oxidation potential to generate reactive oxygen species (ROS) in exposed plant tissue. After entering leaf stomata, ozone rapidly degrades into various ROS species, and plants reduce the oxidative damage by activation of antioxidant enzymes and accumulation of molecules that effectively scavenge ROS. If ROS production exceeds the plant's capacity to detoxify it, deleterious effects at the cellular level may occur. The balance between the production and the scavenging of activated oxygen is thus crucial to plant growth maintenance and overall environmental stress tolerance. However, alterations in plant metabolism may lead to reduced crop yield and quality, directly or indirectly by exposing susceptible plants to stress factors. Secondary metabolites are constitutively synthesized and are of interest for human health and nutrition, especially because some of them are major sources of biologically active substances. However, they are also well known as plant defence molecules and their concentrations can be influenced by abiotic stresses such as ozone. Increased accumulation of plant secondary metabolites in leaves of forest trees in response to ozone exposure has been reported in several studies, while the changes on crop plants composition and nutritional quality need to be further studied and discussed to guide our efforts to select ozone-tolerant crops in an attempt to provide a secure food supply for a developing world.

8.1 Introduction

Tropospheric ozone is a major secondary air pollutant formed by the chemical reaction between nitrogen oxides (NOx) and volatile organic compounds (VOCs) in the presence of sunlight. Ground-level ozone concentrations have significantly increased since pre-industrial

---

*E-mail: ffcaregnato@gmail.com

times, and in the northern hemisphere the mean ozone concentrations have increased from 10–15 ppb to current levels above 40 ppb (Vingarzan, 2004; Ashmore, 2005). According to the modelling studies presented on the Intergovernmental Panel on Climate Change (IPCC, 2007), projections based upon scenarios with high emissions of primary pollutant species deriving from anthropogenic activity (NOx, CH_4, CO and VOCs) indicate that concentrations of tropospheric ozone might increase throughout the 21st century, and simulations for the period of 2015 through 2050 indicate an increase in ozone levels of 20 to 25%, whereas through 2100 the ozone levels below 250 mb (an altitude around 10 km) may grow by 40 to 60%. Therefore, ozone concentrations will probably exceed the internationally accepted environmental criteria (ranging around 40–50 ppb), which represents a significant risk for human health, natural vegetation and crop production (WHO, 2005).

On a global scale, pollution by ozone was considered largest in central Europe and eastern USA, but recent trends in ozone concentration obtained through global photochemical modelling studies performed for the Hemispheric Transport of Air Pollution 2010 assessment indicated reductions in peak surface ozone levels in North America and Europe (Dentener et al., 2010). These changes are likely to have been due to effective emission controls on primary air pollutants over the past two decades in response to the Clean Air Act in the USA and the Long-Range Transboundary Air Pollution Convention and European Union targets in Europe (Collins et al., 2000; Vingarzan, 2004; Ashmore, 2005). However, in several developing countries we observe a different scenario, and emissions of ozone precursors are going upward as a consequence of rapid urbanization and industrialization across these regions (UNEP, 1999). The concentrations of air pollutants in some cities located in South Asia, India and Latin America often exceed the thresholds of toxicity to human and ecosystem health (Emberson et al., 2001; Agrawal et al., 2003; Ashmore, 2005).

During the past decades the impacts of ozone have assumed great concern, and tropospheric ozone is now recognized as the most harmful air pollutant to crop plants and ecosystems. Despite control measures intended to reduce ozone pollution, current ground-level ozone concentrations in several countries worldwide leads to growth and yield impairment of many agricultural and horticultural plants, affecting crop productivity in regions where the agricultural production is the dominant economic activity (Booker et al., 2009; Rai and Agrawal, 2012). Data collected from large-scale experimental studies conducted in filtration and fumigation chamber experimental studies performed by the North American Crop Loss Assessment Network (NCLAN) and the European Open Top Chamber Programme (EOTC) have estimated that the yields of about one-third of US crops were reduced by 10% due to ambient ozone in the 1980s (EPA, 1996), whereas the European Union (EU) may have lost more than 5% of their wheat yield due to ozone exposure concentrations during the 1990s (Krupa et al., 1998). Recently, Avnery et al. (2011), estimating the global yield reductions of three key staple crops due to surface ozone exposure using hourly ozone concentrations simulated by the Model for Ozone and Related Chemical Tracers version 2.4 (MOZART-2), found that detrimental impacts of ozone were already responsible for reductions of global yields for maize (ranging from 2.2 to 5.5%), wheat (3.9–15%) and soybean (8.5–14%) in 2000.

The increasing emissions of reactive VOCs and NOx in urban areas have significantly increased ozone concentrations in rural areas, and nowadays ozone levels are found to be higher in agricultural land than in cities (Ainsworth et al., 2012). This is the case for many regions located in the major crop-growing areas of Asia, India, Africa and Latin America. According to Emberson et al. (2009), ambient ozone concentrations in South Asia range between 35 and 75 ppb (4–8 h growing season mean), and the modelling-based studies performed by the authors suggest that yield losses of 5–20% of important crops are predicted to become common in Asian areas experiencing elevated ozone concentrations. Using HANK model for ozone concentration, Mittal et al. (2007) reported ozone levels varying from 25 to 100 ppb in the Indian subcontinent (Afghanistan, parts of South-east Asian countries, and parts of China and Sri Lanka). The magnitude of potential risk of ozone to plant productivity and food safety in India was revised by Oksanen et al. (2013), and as showed by Sarkar and Agrawal (2012) current ozone concentrations severely affect growth, reproductive, physiological, molecular and yield parameters on two Indian rice cultivars. In Latin America, where crop and livestock production continues to expand, data concerning the effects of ozone on yield losses are still scarce. However, according to global distribution of crop exposure to ozone presented by Avnery et al. (2011) the highest exposure of crops to ozone generally occurs in the northern hemisphere and Brazil due to greater ozone-precursor emissions and concentrations during the crops' growing season. Ozone exposure during the soybean and maize growing seasons is high in the northern hemisphere, whereas in the southern hemisphere the high ozone levels occur during the periods of high biomass burning (August and October), which are coincident with the maize growing season in the Democratic Republic of Congo and the wheat growing season in Brazil.

8.2 The Basis of Ozone-Detrimental Effects

Since the early studies on the effects of ozone on plant species, it was observed that this pollutant is by nature a strong oxidizing agent capable of being rapidly converted in the intracellular space to different reactive oxygen species (ROS) (Castagna and Ranieri, 2009; Iriti and Faoro, 2008). Ozone movement into the apoplastic space is largely controlled by stomatal gas exchange, and immediately after its entry in the sub-stomatal chamber it is spontaneously decomposed to ROS, such as hydrogen peroxide (H_2O_2), superoxide radicals ($O_2^{\bullet-}$), hydroxyl radicals ($OH^{\bullet-}$) and nitric oxide (NO), or it can react with a number of compounds present in cell wall, apoplastic fluid and plasma membrane (Laisk et al., 1989; Castagna and Ranieri, 2009; Sharma et al., 2012). Studies performed with different species have reported that following ozone exposure (100–150 ppb) both H_2O_2 and $O_2^{\bullet-}$ are extensively accumulated in the leaf tissue, especially in sensitive plants (Guidi et al., 2010; Caregnato et al., 2013), and Scebba et al. (2003) report that extracellular ROS accumulation is one of the earliest detectable responses to ozone. Mahalingam et al. (2006) observed that ozone elicits a biphasic ROS burst in Arabidopsis with a smaller peak at 4 h and a larger peak at 16 h, and $O_2^{\bullet-}$ was the major ROS generated in response to 150 ppb of ozone. The direct harmful effects of ozone on leaves thus depend on the stomatal ozone flux, which is largely dependent on the gradient of ozone from outside to inside the leaf (Tuzet et al., 2011). The reactions of ozone within the aqueous matrix of the cell wall (the apoplast) with extracellular antioxidants may control the actual amount of ozone that can reach the cell membrane, thereby changing the rate of ozone uptake via stomata (Tuzet et al., 2011), and apoplastic ROS quenching antioxidant capacity can be considered the first line of defence against ozone-harmful damages (Dizengremel et al., 2008).

Following transient exposure to high levels of ozone, the over-production of ROS can lead to oxidation of membrane lipids, proteins and enzymes, as well as a variety of organic metabolites localized in the cell. The initial signals produced by ozone can thus be later translated in responses at the tissue level, leading to hypersensitive response, accelerated senescence and programmed cell death (Mahalingam et al., 2006). Despite their destructive activity, ROS are well-described as second messengers in a variety of cellular processes, which includes tolerance to abiotic and biotic environmental stresses (Wohlgemuth et al., 2002; Guidi et al., 2010). Besides the antioxidant enzymatic systems found in different cellular compartments, plants possess a range of antioxidant metabolites and detoxifying apparatus responsible for scavenging ROS. Plants can limit ozone-induced damages by protective mechanisms that involve the accumulation of compounds with high reducing potentials like ascorbate (AsA), glutathione (GSH), carotenoids, tocopherols and secondary metabolites, such as phenolic compounds (Baier et al., 2005). Moreover, ozone response pathways overlap with the programmed cell death induced in response to plant pathogens, and both stresses can induce the oxidative burst that leads to excessive ROS production, which activates the biosynthesis of ethylene, salicylic acid and jasmonic acid (Sharma and Davis, 1997; Kangasjarvi et al., 2005). These plant hormones coordinate different metabolic pathways involved in cell defence, and current evidence suggests that ethylene promotes endogenous ROS formation and lesion propagation, salicylic acid is required for programmed cell death and jasmonic acid is involved in cell lesion containment (Rao, 2000; Baier et al., 2005).

A number of authors have pointed out that the main level of ozone defence relies both on the existing content of cellular antioxidants (e.g. AsA and GSH) and the intensity of the detoxifying pathways that are responsible for regenerating these metabolites (Luwe et al., 1993; Calatayud et al., 2001; Dizengremel et al., 2008, 2009). The protective role of AsA as ROS-scavenger was first supported by the enhanced ozone-sensitivity shown by *Arabidopsis thaliana* mutants deficient in AsA synthesis (Conklin et al., 1996). Even so, the relationship between ozone sensitivity and apoplastic AsA concentration remains controversial and some studies have postulated that elevated apoplastic AsA levels cannot always be sufficient to render a plant tolerant to ozone (Ranieri et al., 1999; D'Haese et al., 2005; Di Baccio et al., 2008). The apoplast can be easily and rapidly depleted of AsA, allowing the subsequent oxidative action of ROS in foliar cells (Van Hove et al., 2001), and thus an efficient protective mechanism requires the transfer of AsA from intracellular detoxifying systems to the cell wall. The antioxidant role played by AsA is known to be dependent on the cell's ability to maintain it in a reduced state, which occurs through the Halliwell–Asada cycle (AsA–GSH) (Smirnoff, 1996; Noctor and Foyer, 1998; Di Baccio et al., 2008). Using high ozone concentrations (300 ppb), Luwe et al. (1993) observed a time-dependent relationship between oxidation of both extracellular AsA and intracellular GSH pool, while the cellular AsA redox state was unaltered during fumigation. As reported by numerous studies, AsA regeneration is tightly coupled to GSH within the cell and transport activity was responsible for replenishing the reduced apoplastic AsA pool (for review see Smirnoff, 1996; Noctor and Foyer, 1998). In the symplasm, GSH and NAD(P)H are responsible for reducing the AsA molecule. The reduction of GSSG (oxidized GSH) into GSH occurs through the action of glutathione reductase (GR), and together with other enzymes, such as thioredoxins and peroxiredoxins, the GSH/GSSG couple plays a redox sensor role (Foyer and Noctor, 2005). In fact, regeneration of reduced AsA and GSH can be provided by enzymes that use the reducing power of NAD(P)H, which clearly appears as a key regulator in most regeneration processes (Noctor, 2006).

Thus, the capacity of cells to appropriately maintain the antioxidant levels depends on carbon metabolism changes concomitant with alteration in gene expression (Foyer and Noctor, 2005). In higher plants, chronic ozone fumigation impairs the photosynthetic process and the carbon dioxide (CO_2) assimilation due to a decrease in ribulose-1,5-bisphosphate carboxylase/oxygenase (Rubisco) activity and quantity, together with the destruction of photosynthetic pigments (Fontaine et al., 1999; Anderson et al., 2003; Iglesias et al., 2006; Calatayud et al., 2007). While photosynthesis is limited, the activity and quantity of PEPcase (phosphoenolpyruvate carboxylase) is strongly increased, allowing accumulation of four-carbon acids (Dizengremel et al., 2008). Several enzymes of glycolysis and the pentose phosphate pathway are also activated, providing precursors for the anapleurotic pathway (oxaloacetate and malate) that will produce higher amounts of reducing power (NADPH and NADH) to further help the detoxification process (Dizengremel et al., 2009). Ozone exposure lasting several days increases the levels of ROS, impairs the photosynthetic machinery and the Calvin cycle, causing the exhaustion of carbon availability as the demand for reducing power and energy are increased. In a meta-analysis study performed with data from 53 peer-reviewed works published between 1980 and 2007 that evaluated the responses of wheat (*Triticum aestivum*) to elevated ozone, Feng et al. (2008) demonstrated that ozone exposure to an average concentration of 43 ppm significantly decreased photosynthetic rates (20%), Rubisco activity (19%), stomatal conductance (22%) and chlorophyll content (40%), and such biochemical modifications affected the whole plant by inducing a larger decrease in below-ground (27%) biomass than in above-ground (18%) biomass.

Ozone-induced reductions in photosynthesis not only change carbon assimilation, but also affect carbon translocation and accumulation in different plant parts. This arises either from a reduction in carbon translocation from source leaves to distant sink, which, according to Grantz and Farrar (2000), occurs due to phloem inhibition transport, or from the effect of ozone on ethylene synthesis, a hormone that controls shoot and root growth, promotes senescence and abscission, and more recently, has been associated with the disruption of ABA-induced stomatal opening regulation (Wilkinson et al., 2012). The negative impacts on root biomass might lead to reductions in grain and fruit production, since the ability of the plant to take up the nutrients and water required to sustain growth and yield is compromised (Ashmore, 2005; Rai and Agrawal, 2012). Besides, under ozone stress the pool of non-structural carbohydrates essential for growth, including sugars and starch, is affected both due to reduction in the carbohydrate synthesis and by a shift of carbon compounds to repair processes and defence metabolites (Fuhrer and Booker, 2003; Booker et al., 2009; Wang and Frei, 2011). The synthesis of defence metabolites might divert resources away from the synthesis of other sets of metabolites, so analysis of the plant metabolite profiling could be assessed to identify the trade-offs between primary and secondary metabolism (Stitt et al., 2010).

8.3 Ozone and the Changes in Plant Defence Metabolism

Probably one of the most important adjustments made by plants to avoid environmental stress is to change the chemical composition of leaves, flowers, fruits, roots and stems. In certain varieties of wheat, rice, bean, soybean and sorghum, the physiological stress imposed by ozone modifies the chemical composition of crops, affecting not only the grain size and weight but also the nutritional composition of the final agricultural products (Biswas et al., 2008; Booker et al., 2009; Iriti et al., 2009; Betzelberger et al., 2012; Wang et al., 2012). Ozone exposure can activate the biosynthesis of plant secondary metabolites, a diverse group of organic compounds with important adaptive significance in protecting plants against predators and pathogens, in providing reproductive advantage as attractants of pollinators and seed-dispersing animals, and as allelopathic agents (Harborne, 1993; Croteau et al., 2000). Besides the importance for the plant itself, secondary metabolites determine a number of nutritional aspects of food, including colour, taste, smell and antioxidative, anticarcinogenic, anti-inflammatory and cholesterol-lowering properties (Hounsome et al., 2008). Thus, shifts in the chemical composition of important field crops can lead to loss of potentially beneficial components and have detrimental impacts on food safety and consumer's health.

Based on their biosynthetic origins, plant secondary compounds can be divided into three major groups: the phenylpropanoids, the terpenoids and the alkaloids. Phytochemicals arising from these pathways include compounds with a powerful antioxidant capacity, able to efficiently scavenge different ROS (Prior et al., 1998; Di Baccio et al., 2008; Iriti and Faoro, 2008). Many phenolic compounds, which are primarily derived from the phenylpropanoid pathway, are known to work as effective antioxidant molecules because the electron reduction potential of the phenolic radical is lower than the electron reduction potential of oxygen radicals, and also because phenoxyl radicals are generally less reactive than oxygen radicals (Rice-Evans et al., 1997). Phenolic compounds such as flavonoids are responsible for determining distinguishing traits of plant parts, establishing, for example, flower colours, and leaves and grain flavours (tastes and odours).

In plants, the phenylpropanoid metabolism is induced in response to stress, and enhancement of key enzyme activities and accumulation of secondary metabolites occur early after exposure, in order to improve the resistance against pathogen attack and/or tolerance to environmental pollutants (Iriti and Faoro, 2009). Ozone can elevate the level of flux through the phenylpropanoid pathway stimulating the production of phenolic compounds, including lignin, suberin, tannin, stilbenes and flavonoids (Eckey-Kaltenbach et al., 1994; Tuomainen et al., 1996; Saleem et al., 2001). According to some studies, the phenylpropanoid pathway is one of the most affected targets of ozone, inducing gene transcription and enzyme activities (Tosti et al., 2006; Di Baccio et al., 2008). Shikimate dehydrogenase (SKDH) is one of the key enzymes of the shikimate pathway, a metabolic route that produces aromatic amino acids and a large number of phenolic compounds. Increased accumulation of flavonoids, such as quercetin and chlorogenic acid, has been found in different natural and cultivated plant species exposed to elevated ozone levels (Saleem et al., 2001; Saviranta et al., 2010) and, as suggested by Appel (1993), these compounds further increase resistance against ozone damage by scavenging $OH^{\bullet-}$ and H_2O_2. Increased levels of transcription of genes involved in flavonoid biosynthesis were also found in ozone-resistant leguminous cultivars (Puckette et al., 2008), suggesting that a number of transcription factors and signalling genes differently enable resistant plants to adapt more rapidly to ozone stress. Furthermore, Booker and Miller (1998) in a greenhouse study with soybeans, observed that after 6 h of ozone fumigation a rapid and coordinated increase occurred in the activities of two phenylpropanoid pathway enzymes, phenylalanine ammonia lyase (PAL) and 4-coumarate:CoA ligase, and that the stimulation of these enzymes' activities remained elevated for several days.

Terpenoids are the most structurally diversified class of plant natural products, with functional roles in plants as structural components of membranes (sterols), electron carriers (quinones), photosynthetic pigments (phytol, carotenoids) and hormones (gibberelins, abscisic acid) (Croteau et al., 2000). Isoprenoids have several bioactive properties and thus have long been used in the pharmacological industry and in the human diet (Hounsome et al., 2008). In plants, isoprene may act as protecting thermal agents, but a more general antioxidant action has been recently hypothesized on the basis of protection against abiotic stress (Spinelli et al., 2011). However, the response of the isoprenoid biosynthetic pathway to ozone may vary considerably, and according to Calfapietra et al. (2009) it is especially dependent on the length and level of exposure to the pollutant. Measurements of isoprene emission carried out in *Populus tremuloides* chronically exposed to ozone (1.5-fold the ambient levels for several years) indicated that isoprene synthesis and emission were decreased, and such responses were associated with reductions in isoprene synthase messenger RNA and reduced levels of dimethylallyl diphosphate (DMADP), the main substrate for isoprene synthesis (Calfapietra et al., 2007, 2008). Puckette et al. (2008) reported that in a *Medicago truncatula* ozone-resistant accession exposed to acute ozone treatment (300 nl l^{-1} for 6 h), key genes related to isoprenoid biosynthesis pathway were strongly up-regulated at 12 h post-treatment.

Alkaloids are nitrogen-containing compounds mainly derived from amino acids, which possess great interest because of their pronounced toxicological, pharmacological, nutritional and medicinal properties (e.g. caffeine, nicotine, morphine, quinine). Most alkaloids are very toxic and, therefore, they are found to play an important role in plant chemical defence against herbivores and microorganisms (Harborne, 1993). Glycoalkaloids such as α-solanine and α-chaconine, for example, are naturally occurring phytotoxins in potato that may cause a bitter taste and gastroenteritis. For food safety purposes an upper limit for total glycoalkaloid content of 20 mg per 100 g of potato is generally accepted; if they occur in too high concentrations they can be considered lethal to humans (Sinden et al., 1976; Smith et al., 1996; Friedman and McDonald, 1997). Studies concerning the effects of ozone on the alkaloids biosynthetic pathway are still scarce, and most of them deal with the influence of the pollutant on the nitrogen metabolism (see Iriti and Faoro, 2009). In a study with tobacco plants (*Nicotiana tabacum*) grown under high ozone concentrations (80–100 ppb) the authors observed that treated plants had higher levels of total nitrogen (primarily reduced nitrogen) and lower levels of nicotine (a pyridine alkaloid), which increased the survival and growth response of tobacco hornworm larvae (*Manduca sexta*) once the plant chemical defence contents were modified by the pollutant (Jackson *et al.*, 2000). In addition, Langebartels *et al.* (1991) observed that a single ozone treatment (5 or 7 h) had a strong influence on the levels of polyamines (putrescine and spermidine), important alkaloid precursors, that can improve ozone tolerance either acting as ROS scavenger molecules or inhibiting ethylene biosynthesis and thus reducing the senescence.

As secondary metabolites are products of primary metabolism, an excessive activation of defence compound biosynthesis can have detrimental effects on plant fitness-relevant functions such as growth and reproduction (Bolton, 2009). If priority is given to the defence-related processes the availability of carbon and nitrogen resources may become limiting (Manderscheid *et al.*, 1992; Le Bot *et al.*, 2009). As observed by Saleem *et al.* (2001), long-term ozone exposure of a sensitive silver birch clone (*Betula pendula* Roth) increased total phenolic content (16.2%) at the expense of growth, suggesting that changes in carbon allocation towards chemical defence resulted in lower biomass production. Under chronic ozone exposure, shifts in the partitioning of photosynthates may severely influence the content of carbohydrates and minerals on roots and lower leaves mainly because carbon compound allocation to young leaves and seed production are necessary to maximize resource acquisition to survival. Such changes are known to be dependent on environmental factors such as temperature, soil nutrients, nitrogen availability and water stress (for review see Fuhrer and Booker, 2003; Bender and Weigel, 2011; Ainsworth *et al.*, 2012).

8.4 The Metabolic Shift and the Influences on Plant Quality

The negative impacts of ozone on yield have become a great threat to global food security, especially in developing countries, where food shortages are a risk in the face of rapidly increasing populations. However, elevated ozone levels not only threaten agriculture and food security by reducing the food quantity, but also by changing the food quality. While the ozone negative impacts on crop yields are obvious, their effects on crop quality are almost unknown (Ashmore, 2005).

Crop quality may be affected either by changes in primary metabolite production (e.g. carbohydrates, proteins), or as a consequence of increased secondary metabolism synthesis. Ozone-induced deviations of available resources from growth to defence metabolism might alter the chemical composition of crops and consequently the quality of harvested products of crops (Iriti and Faoro, 2009; Wang and Frei, 2011). Phytochemicals arising from defence (secondary) metabolism are important for human health and nutrition, especially because some of them are source of biologically active substances not only with health-benefits potential, such as chemopreventive, anti-inflammatory and antioxidants, but also health-harmful potential, such as carcinogenic and toxic compounds (Chen and Kong, 2004; Crozier *et al.*, 2006; Korkina, 2007).

Among the few studies that assess the quality of the marketable crop products (grains, tubers, fruits and vegetables), most of them investigate the content of proteins, carbohydrates and lipids, while very few evaluate the change on secondary compound content. Visible injuries induced by ozone are also of great importance especially when the marketable value of the crop depends on the appearance. For example, in leafy vegetables visible injuries may make the product unmarketable (Ashmore, 2005). In addition, the ozone may alter the quality of forages, making them less digestible and less nutritious for ruminants, allowing the emergence of ozone secondary effects, such as reduction in the milk and meat production from grazing animals (Vandermeiren and Pleije, 2011). Here we divided the ozone impacts on crop quality into seven categories of quality parameters, which according Wang and Frei (2011) are: protein, lipids, carbohydrates, secondary compounds, minerals, physical and sensory aspects, and nutritive value of forage for ruminant animals.

8.4.1 Proteins

Plant foodstuffs are a great source of dietary protein for humans and animals. On a global basis, plant proteins provide about 60% of the human daily protein intake, mainly from cereal grains. However, in developing countries this value may be even higher (FAO, 2010). Therefore, protein content plays a significant role in determining the nutritional quality of many crops, especially in developing countries.

Plants exposed to ozone are known to have protein concentration of harvested fractions altered. Generally, an increment in the amount of grain proteins is often associated with crops exposed to ozone, as seen in wheat, soybean, bean and rice (Table 8.1). However, this increase is not large enough to compensate the grain yield loss, so the grain protein yield is reduced (Mulchi et al., 1988; Pleijel et al., 1997; Feng et al., 2008; Piikki et al., 2008; Frei et al., 2012; Zheng et al., 2013). Differently from the grains, the protein concentration in leaves does not show a trend as seen in Table 8.1.

In wheat (*Triticum aestivum*), which is considered to be one of the most ozone-sensitive crops (Mills et al., 2007), changes in grain protein content are a very important effect elicited by ozone, especially because it is a major source of plant protein worldwide (FAO, 2010; Vandermeiren and Pleije, 2011). The protein concentration usually increases

Table 8.1. Effect of ozone stress on the protein content of major crops and forage.

Crop species	Organ	Ozone effect	References
Bahiagrass, *Paspalum notatum* Flugge	Leaf	−/↑	Muntifering et al. (2000)
Bean, *Phaseolus vulgaris* L.	Seed	↑	Iriti et al. (2009)
Broccoli, *Brassica oleracea* L.	Flower + stalk	↑	Vandermeiren et al. (2012)
Clover, *Trifolium subterraneum* L.	Leaf	↑	Sanz et al. (2005)
Corn, *Zea mays* L.	Seed	−	Garcia et al. (1983)
Grass, *Briza maxima*	Leaf	↓	Sanz et al. (2011)
Grassland species mixture	Leaf	↓	Gilliland et al. (2012)
Lespedeza, *Lespedeza cuneata*	Leaf	−	Powell et al. (2003)
Little bluestem, *Schizachyrium scoparium*	Leaf	↓	Powell et al. (2003)
Mustard, *Brassica campestris* L.	Seed	↓	Singh et al. (2009), Tripathi and Agrawal (2012)
Peanut, *Arachis hypogaea* L.	Seed	−	Burkey et al. (2007)
Rapeseed, *Brassica napus* L.	Seed	−/↓/↑	Bosac et al. (1998), Ollerenshaw et al. (1999), Vandermeiren et al. (2012)
Rice straw, *Oryza sativa* L.	Leaf	↑	Frei et al. (2011)
Rice, *Oryza sativa* L.	Seed	↓/↑	Rai et al. (2010), Frei et al. (2012), Wang et al. (2012), Zheng et al. (2013)
Soybean, *Glycine max* (L.) Merr.	Seed	−/↑	Howell and Rose (1980), Grunwald and Endress (1984), Mulchi et al. (1988)
Wheat, *Triticum aestivum* L.	Seed	−/↑	Fuhrer et al. (1990, 1992), Pleijel et al. (1991, 1997, 1998, 1999, 2006), Feng et al. (2008), Piikki et al. (2008), Zheng et al. (2013)
Wheat, *Triticum aestivum* L.	Leaf	↓	Feng et al. (2008)

Ozone stress increase (↑), decrease (↓) or does not show significant difference (−) on protein content.

in wheat grain of plants grown under ozone exposure, while in leaf it is reduced (Table 8.1). Zheng et al. (2013) discuss that the higher protein levels in grains are likely a consequence of reduced carbohydrate levels. In addition, there are indications that not only the amount but also the composition of the proteins is affected by ozone, for example, the dry gluten/protein ratio was increased in wheat grains from plants grown at ambient ozone levels (Vandermeiren et al., 1992). Moreover, Fuhrer et al. (1992) found a small but significant increase in Zeleny values with increasing ozone concentration, indicating a trend towards better protein quality.

Rice is listed as the grain crop with the second highest world production (FAO, 2010), providing over 21% of the caloric needs of the world's population and up to 76% of the caloric intake of the population of South-east Asia (Fitzgerald et al., 2009). Mills et al. (2007) identified rice as a moderately sensitive staple crop, and a number of studies with different cultivars found that not only yield but other major growth parameters are severely affected by ozone (Rai et al., 2010; Sarkar and Agrawal, 2012). Rai et al. (2010) observed a reduction in seed protein concentration in two rice cultivars grown in non-filtered chambers (NFC) when compared to filtered chambers (FC). However, these results contrast with those reported by Zheng et al. (2013), Frei et al. (2012) and Wang et al. (2012), who showed increments in seed protein levels of rice plants grown in NFCs.

Besides wheat and rice, legumes are great sources of protein, being up to three times richer in protein than cereal grains (Duranti and Gius, 1997). Analysing Table 8.1 it is possible to note that protein concentration in legume grains increases when plants are exposed to ozone, which is the case for soybean and bean. However, seeds from groundnut (*Arachis hypogaea* L.), which is considered to be sensitive to ozone, do not have their protein content modified by the pollutant (Burkey et al., 2007). In contrast, the ozone tends to decrease protein content in seeds of *Brassica* genus as shown by Bosac et al. (1998), Ollerenshaw et al. (1999), Singh et al. (2009) and Tripathi and Agrawal (2012), although Vandermeiren et al. (2012) have shown that ozone increases protein content in rapeseed.

8.4.2 Lipids

With some exceptions, in contrast to animal fats, vegetable oils contain predominantly unsaturated fatty acids which are very important to human health. Some unsaturated fatty acids like linoleic acid (omega-6 family) and α-linolenic acid (omega-3 family) are essential for humans because we are not able to completely synthesize them. However, plants have this ability and plant products are the major source of essential fatty acids in the human food chain. Thus, changes induced by ozone on plants' lipid content should be considered.

As seen in Table 8.2, ozone effects on seed lipid content does not show a clear trend, although in studies with mustard and rapeseed, which are two major world sources of vegetable oils, ozone decreases lipid content in most cases (Bosac et al., 1998; Ollerenshaw et al., 1999; Singh et al., 2009; Tripathi and Agrawal, 2012). Rapeseed oil is a valuable plant oil for human nutrition due to its high content of monounsaturated and polyunsaturated fatty acids combined with a very low proportion of saturated fatty acids (Vandermeiren et al., 2012). A study conducted by Vandermeiren et al. (2012) showed that ozone led to a shift in fatty acid composition of the vegetable oil derived from seeds of oilseed rape. The authors observed that the content of oleic acid (18:1) significantly declined, linoleic acid (18:2) increased and linolenic acid (18:3) showed no differences. Total monounsaturated fatty acids were decreased by ozone exposure, while total saturated fatty acids were increased, leading to oil quality decreases.

Singh et al. (2009) observed that in response to ambient ozone the contents of oil, protein and minerals (Ca, Mg, K, P, Zn) were significantly decreased in mustard seeds when compared to the plants grown in air filtered chambers at the recommended NPK (nitrogen, phosphorus and potassium) fertilization. However, these effects were suppressed when 1.5× recommended NPK was added to the soil. Tripathi and Agrawal (2012)

Table 8.2. Effect of ozone stress on the lipid content of major crops.

Crop species	Organ	Ozone effect	References
Bean, *Phaseolus vulgaris* L.	Seed	↑	Iriti *et al.* (2009)
Maize, *Zea mays* L.	Seed	–	Garcia *et al.* (1983)
Mustard, *Brassica campestris* L.	Seed	↓	Singh *et al.* (2009), Tripathi and Agrawal (2012)
Peanut, *Arachis hypogaea* L.	Seed	–	Burkey *et al.* (2007)
Rapeseed, *Brassica napus* L.	Seed	–/↓/↑	Bosac *et al.* (1998), Ollerenshaw *et al.* (1999), Vandermeiren *et al.* (2012)
Rice, *Oryza sativa* L.	Seed	↑	Frei (2012)
Soybean, *Glycine max* (L.)	Seed	–/↓	Howell and Rose (1980), Mulchi *et al.* (1988), Grunwald and Endress (1984)

Ozone stress increase (↑), decrease (↓) or does not show significant difference (–) on lipid content.

reported that in mustard seeds the fatty acid profile was altered by ozone, reporting that saturated fatty acid content was reduced after ozone exposure. However, monounsaturated fatty acid, polyunsaturated fatty acid and ω-6 fatty acid showed a gain after the treatment. Among the fatty acid components, linoleic acid was decreased whereas oleic, erucic and linolenic acids were enhanced in response to ozone. Lower levels of linolenic acid and higher contents of oleic acid are preferred for cooking and frying purpose (Nesi *et al.*, 2008). Some environmental factors like ozone are able to alter the seed oil:protein ratio. In soybean plants grown in ambient ozone, Howell and Rose (1980) and Grunwald and Endress (1984) found a significant lower oil:protein ratio in the seeds, which was associated with a decrease in seed total oil content.

8.4.3 Carbohydrates

To analyse the ozone impacts on the quality of carbohydrates in crop products they can be separated into three components: sugar, starch and fibre content. Ozone effects on fibre content of plant foodstuff for human consumption is the least studied, although their intake has important implications for health. For example, human consumption of soluble and insoluble dietary fibres has been related with weight loss, and some studies have found that a diet with higher insoluble fibre content can reduce the risk of bowel cancer and heart diseases (Ceyhan *et al.*, 2012).

Regarding the impacts on carbohydrate constituents, we observed that the majority of studies report that starch and reducing sugar (glucose and fructose) concentration decreases while fibre content is enhanced in many species exposed to ozone, despite experimental differences in ozone treatments (Table 8.3). Sucrose content showed no change except in the study conducted by Köllner and Krause (2000), who found that sucrose levels decreased after ozone exposure.

Potato (*Solanum tuberosum* L.) has a great importance for human nutrition and, in terms of production, is the fourth most important crop in global scale, coming after wheat, rice and maize (FAO, 2010). Potato tubers have several applications in the food industry, for which quality has a major importance. In this context, starch and reducing sugar content of the tuber plays an important role to determine the potato tuber quality. The starch content of potato tubers must be sufficiently high to avoid excessive absorption of fat during frying, whereas the reducing sugar content should be low to prevent the darkening of chips due to the Maillard reaction, which is unacceptable in fried potato products (Roe *et al.*, 1990; Vandermeiren *et al.*, 2005).

Sucrose content may also contribute with Maillard reaction through the by-products formed after sucrose hydrolysis induced by heat during frying (Leszkowiat *et al.*, 1990). According to Mills *et al.* (2007) potato is a moderately sensitive crop to ozone; even so, the pollutant is able to change tuber quality. Pell *et al.* (1980) and Pell and Pearson (1984)

Table 8.3. Effect of ozone stress on the carbohydrate content of major crops, forages and woody trees.

Carbohydrates	Crop species	Organ	Ozone effect	References
Starch	Maize, *Zea mays* L.	Seed	–	Garcia *et al.* (1983)
	Potato, *Solanum tuberosum* L.	Tuber	↓ / –	Pell and Pearson (1984), Köllner and Krause (2000), Vorne *et al.* (2002), Vandermeiren *et al.* (2005)
	Rice, *Oryza sativa* L.	Seed	↓	Rai *et al.* (2010), Frei *et al.* (2012)
	Sweet potato, *Ipomoea batatas* (L.) Lam.	Tuber	↓	Keutgen *et al.* (2008)
	Wheat, *Triticum aestivum* L.	Seed	↓	Fuhrer *et al.* (1990, 1992), Feng *et al.* (2008)
Reducing sugar (fructose and glucose)	Grape, *Vitis vinifera* L.	Fruit	– / ↓	Soja *et al.* (1997), Soja *et al.* (2004)
	Ladino clover, *Trifolium repens* L.	'Shoot'	↓	Blum *et al.* (1982)
	Mustard, *Brassica campestris* L.	Seed	↓	Tripathi and Agrawal (2012)
	Potato, *Solanum tuberosum* L.	Tuber	↓ / ↑	Pell *et al.* (1980, 1988), Pell and Pearson (1984), Vorne *et al.* (2002)
	Rapeseed, *Brassica napus* L.	Seed	– / ↓	Bosac *et al.* (1998)
	Rice, *Oryza sativa* L.	Seed	↑	Rai *et al.* (2010)
	Sweet potato, *Ipomoea batatas* (L.) Lam.	Tuber	– / ↓	Keutgen *et al.* (2008)
Sucrose	Potato, *Solanum tuberosum* L.	Tuber	– / ↓	Pell *et al.* (1980, 1988), Köllner and Krause (2000), Vorne *et al.* (2002), Vandermeiren *et al.* (2005)
	Strawberry, *Fragaria× ananassa* Duch.	Fruit	–	Keutgen and Pawelzik (2008)
	Sweet potato, *Ipomoea batatas* (L.) Lam.	Tuber	–	Keutgen *et al.* (2008)
Fibre (NDF, ADF and lignin)	Bahiagrass, *Paspalum notatum* Flugge	Leaf	– / ↑	Muntifering *et al.* (2000)
	Bean, *Phaseolus vulgaris* L.	Seed	↑	Iriti *et al.* (2009)
	Clover, *Trifolium* spp.	Leaf	– / ↑	Sanz *et al.* (2005), Muntifering *et al.* (2006), Gonzalez-Fernandez *et al.* (2008)
	Maize, *Zea mays* L.	Seed	–	Garcia *et al.* (1983)
	Grass, *Briza maxima*	Leaf	↑	Sanz *et al.* (2011)
	Grassland species mixture	Leaf	– / ↑	Gilliland *et al.* (2012)
	Poa pratensis L.	Leaf	– / ↑	Bender *et al.* (2006)
	Lespedeza, *Lespedeza cuneata*	Leaf	↑	Powell *et al.* (2003)
	Little bluestem, *Schizachyrium scoparium*	Leaf	– / ↑	Powell *et al.* (2003)
	Rice straw, *Oryza sativa* L.	Leaf	↑	Frei *et al.* (2011)

Ozone stress increase (↑), decrease (↓) or does not show significant difference (–) on carbohydrate content.

reported that ozone increases the reducing sugar content, while Pell *et al.* (1988) and Vorne *et al.* (2002) reported that ozone improves potato tuber quality by decreasing the content of reducing sugars. On the other hand, the reduction of the starch content observed in many studies might have a negative impact on tuber quality (Table 8.3). Similarly to potato, sweet potato starch and reducing sugar content decreased while sucrose did not change after ozone exposure (Table 8.3). Ozone decreased reducing sugar content in seeds of both *Brassica* species, unlike rice grain where reducing sugar content was enhanced (Table 8.3). Fuhrer *et al.* (1990, 1992) and Feng *et al.* (2008) reported lower starch content in wheat grains in response to ozone.

Quality is a major determinant of fruit crop value (Soja *et al.*, 2004), and in strawberry, for example, fruit carbohydrate and sugar compound profiles play an important role in flavour and quality (Keutgen and Pawelzik, 2007). Concentrations of sucrose and glucose in two different cultivars of strawberry (cv. Korona and cv. Elsanta) were not significantly influenced by ozone, while fructose content decreased in fruit of cv. Elsanta grown under ozone exposure. Although there were no changes in sucrose and glucose content, the authors found that ozone pollution during the growth phase tended to reduce the sweetness index in both cultivars (Keutgen and Pawelzik, 2008). In grape berries, the accumulation of reducing sugar showed a greater decrease than grape yield at the highest ozone exposure in most experimental replicates (Soja *et al.*, 1997, 2004).

8.4.4 Secondary compounds

In recent years, a number of studies have pointed out the benefits of phytochemicals to human health, which are considered to have the ability to act as anti-inflammatory, antioxidant, antiviral and anticancer agents. However, some secondary compounds, like alkaloids, might be very toxic for us, and many of the functions of secondary metabolites remain unknown and still need to be elucidated (Crozier *et al.*, 2006; Hounsome *et al.*, 2008). Thus, ozone-mediated changes in the composition of plants' secondary compounds may represent a risk for human consumption.

Changes in the composition of the secondary compounds elicited by ozone exposure are presented in Table 8.4. Phenolics have been the most extensively studied metabolites. These molecules are derived from the phenylpropanoid biosynthetic pathway, which is strongly responsive to diverse environmental stress (Korkina, 2007). The majority of studies tabulated herein reported an increase or no changes on phenolics in the edible fractions of a variety of crops grown under ozone exposure, except in leaves of little bluestem (primary growth) reported by Powell *et al.* (2003). In most cases, PAL is the main enzyme responsible for the accumulation of phenolics in agricultural products produced under stressful conditions. Booker and Miller (1998) observed in soybean leaves an increase in PAL activity after 6 h of ozone treatment, and these activities remained elevated for several days. In the same study they found a continuous increase throughout ozone exposure showing a relationship between phenolic content and PAL activity.

The majority of studies were performed in leaves while only three studies assessed other parts of plant, such as fruit and seed. Among these three works, Iriti *et al.* (2009) conducted the most complete work regarding secondary compounds. They evaluated the ozone effect in bean seeds, and found an increase in total phenolic content and changes in phenolics ratio. Separately, the majority of phenolics assessed (delphinidin-3-glucoside, cyanidin-3-glucoside, peonidin-3-glucoside, kaempferol, kaempferol-3-glucoside, caffeic acid, p-coumaric acid and sinapic acid) decreased while only two of them (petunidin-3-glucoside and pelargonidin-3-glucoside) increased. The authors also observed an increase in antioxidant potential; however, they suggested that this change was unrelated to the modification in the phenolic compounds. In tobacco leaves three phenolic compounds were shown to be elevated when plants were grown under ozone exposure. Caffeoylputrescine, which represents the major phenolic component of the apoplastic fluid of leaves, was increased four-fold after ozone treatment (Langebartels *et al.*, 1991).

Table 8.4. Effect of ozone stress on the secondary compounds and vitamins of major crops, forages and woody trees.

Secondary compounds and vitamins	Crop species	Organ	Ozone effect	References
Alkaloids	Potato, *Solanum tuberosum* L.	Tuber	– / ↑ / ↓	Speroni *et al.* (1981), Pell and Pearson (1984), Donnelly *et al.* (2001), Vorne *et al.* (2002), Vandermeiren *et al.* (2005)
Ascorbate (vitamin C)	Broccoli, *Brassica oleracea* L.	Flower + stalk	–	Vandermeiren *et al.* (2012)
	Lettuce, *Lactuca sativa* L.	Leaf	↓	Calatayud *et al.* (2002)
	Potato, *Solanum tuberosum* L.	Tuber	– / ↑	Vorne *et al.* (2002)
	Spinach, *Spinacia oleracea* L.	Leaf	↓	Calatayud *et al.* (2003)
	Strawberry, *Fragaria× ananassa* Duch.	Fruit	↓	Keutgen and Pawelzik (2008)
Carotenoids	Sweet potato, *Ipomoea batatas* (L.) Lam.	Leaf	↓ / –	Keutgen *et al.* (2008)
Glucosinolate	Broccoli, *Brassica oleracea* L.	Flower + stalk	– / ↑ / ↓	Vandermeiren *et al.* (2012)
	Rapeseed, *Brassica napus* L.	Seed	–	Vandermeiren *et al.* (2012)
	Rapeseed, *Brassica napus* L.	Leaf	– / ↑ / ↓	Gielen *et al.* (2006), Himanen *et al.* (2008)
Phenolic	Bahiagrass, *Paspalum notatum* Flugge	Leaf	–	Muntifering *et al.* (2000)
	Bean, *Phaseolus vulgaris* L.	Seed	↑	Iriti *et al.* (2009)
	Clover, *Trifolium* spp.	Leaf	– / ↑	Muntifering *et al.* (2006), Saviranta *et al.* (2010)
	Lespedeza, *Lespedeza cuneata*	Leaf	–	Powell *et al.* (2003)
	Little bluestem, *Schizachyrium scoparium*	Leaf	– / ↓	Powell *et al.* (2003)
	Rice straw, *Oryza sativa* L.	Leaf	↑	Frei *et al.* (2011)
	Silver birch, *Betula pendula*	Leaf	↑	Saleem *et al.* (2001)
	Soybean, *Glycine max* L.	Leaf	↑	Keen and Taylor (1975), Booker and Miller (1998)
	Strawberry, *Fragaria× ananassa* Duch.	Fruit	–	Keutgen and Pawelzik (2008)
	Rice, *Oryza sativa* L.	Seed	–	Frei *et al.* (2012)
	Tobacco, *Nicotiana tabacum* L.	Leaf	↑	Langebartels *et al.* (1991)
Tocopherol (vitamin E)	Broccoli, *Brassica oleracea* L.	Flower + stalk	–	Vandermeiren *et al.* (2012)
	Wheat, *Triticum aestivum* L.	Seed	–	Fuhrer *et al.* (1990)
	Rapeseed, *Brassica napus* L.	Seed	↓	Vandermeiren *et al.* (2012)

Ozone stress increase (↑), decrease (↓) or does not show significant difference (–) on secondary compounds and vitamin content.

Apoplastic ascorbate content, which is thought to be the first line of defence against ROS in leaves (Castagna and Ranieri, 2009), declined 15% in spinach leaves and 35% in lettuce leaves grown under ozone fumigation (Calatayud *et al.*, 2002, 2003). In potatoes, the amounts of ascorbic acid are moderate, but because of the high consumption in some regions in Europe it is one of most important sources of vitamin C (FAO, 2010; Lee and Kader, 2000). Increases in ascorbate content in tuber of potato plants exposed to ozone observed by Vorne *et al.* (2002) indicate an improvement in the tuber quality. In strawberry fruit, Keutgen and Pawelzik (2008) observed that the levels of total ascorbic acid significantly decreased in ozone-exposed plants of two different cultivars (cv. Korona and cv. Elsanta). Tocopherols (vitamin E) are powerful antioxidants, therefore a decrease in vitamin E could be considered as a negative effect on plants' nutritional value. Vitamin E in oilseed rape was significantly reduced at increasing ozone concentrations (Vandermeiren *et al.*, 2012). In contrast, broccoli and wheat did not show any difference in the vitamin E content in plants grown under high levels of ozone.

Glycoalkaloids are phytotoxins naturally found in potato and, in most cases, the total glycoalkaloids concentration in tuber is not significantly affected by ozone (Speroni *et al.*, 1981; Donnelly *et al.*, 2001; Vorne *et al.*, 2002). Although Donnelly *et al.* (2001) did not find any difference in total glycoalkaloids, the authors observed an increase in α-solanine content and no difference in α-chaconine content. Pell and Pearson (1984) observed that the response to the ozone depends on the cultivar, and total glycoalkaloid content decreased in tubers of cv. Norchip and increased in those of cv. Cherokee.

Glucosinolates, a group of nitrogen- and sulfur-containing secondary compounds involved in chemical protection against herbivores and stress, are characteristic of the Brassicaceae and some families of the Capparales order. These metabolites are toxic to some animals (including humans) in high concentrations, but in low concentrations seem to have benefits for humans health, such as anticancer properties (Tripathi and Mishra, 2007; Sarıkamış, 2009). Vandermeiren *et al.* (2012) observed that the total glucosinolate content of the rapeseed seeds and broccoli head was not significantly changed when plants were grown under ozone exposure, although in broccoli head, the ozone exposure increased the aliphatic glucosinolate and decreased the indolic glucosinolate. In leaves of rapeseed plants chronically exposed to ozone a clear change in the glucosinolate profile was found, although no changes in total glucosinolate could be observed (Himanen *et al.*, 2008). Gielen *et al.* (2006) in a study with two lines of *Brassica napus* L. subspecies *oleifera* with different concentrations of glucosinolates, observed that ozone exposure only changed the glucosinolate content in the line with high glucosinolate concentration, which diminished the leaf total glucosinolate content in the presence of ozone.

8.4.5 Minerals

Human adequate mineral intake is needed for good health and to prevent nutritional disorders. Agricultural products are rich sources of several essential minerals, hence a large number of studies have measured mineral concentration of the edible parts of different crops. It is known that mineral contents are influenced very much by surrounding environment, however, few studies have investigated the interactions between mineral concentration of the edible parts of the crop and ozone stress (Wang and Frei, 2011; Ceyhan *et al.*, 2012).

Although the effect of ozone on minerals is important, we do not find any pattern of influence, making it difficult to properly discuss the results found so far, and more studies are needed to fill the gaps and provide robustness to data. Even so, the results concerning the mineral content of plants exposed to ozone are presented in Table 8.5. Regarding mineral content, seeds of wheat and rice were the most studied agricultural products. In wheat grains ozone exposure increased seven of nine studied minerals, while in rice grains results agree that N and P amount decreased in response to ozone, while Mn and Cu content increased and Na was not changed.

Garcia *et al.* (1983) found that the seed of maize had increased micronutrients (Fe, Zn and Cu), whereas the macronutrients decreased or did not change due to ozone exposure. In mustard, Singh *et al.* (2009) observed that in exposed plants the levels of Ca, K and Mg decrease with normal application of NPK, but when 1.5× the normal NPK application was added the Ca, Mg and K concentrations were not modified in the presence of ozone.

8.4.6 Physical and sensory aspects

Physical aspects such as mass, size, shape, visual damage, texture and flavour are also important in determining the quality of marketable crop products, especially in horticultural products. Symptoms of visible injury appear

Table 8.5. Effect of ozone stress on the mineral content of major crops.

Crop species	Ozone effect											Organ	References
	N	P	K	Ca	Mg	Na	Fe	Mn	Zn	Cu	B		
Maize, *Zea mays* L.	nd	–	–	nd	↓	nd	↑	–	↑	↑	nd	Seed	Garcia *et al.* (1983)
Ladino clover, *Trifolium repens*	–	–↑	–↑	–	–	↓	↑	–↑	–	–↑	–	Stalk + leaf	Blum *et al.* (1982)
Mustard, *Brassica campestris* L.	nd	–	–↓	–↓	–↓	nd	nd	nd	–↓	nd	nd	Seed	Singh *et al.* (2009)
Potato, *Solanum tuberosum* L.	nd	–	–	↑	–↓	nd	–	–	–	nd	nd	Stalk + leaf	Fangmeier *et al.* (2002)
Potato, *Solanum tuberosum* L.	–↑	–	–	–	–	nd	↓	–↑	–	nd	nd	Tuber	Fangmeier *et al.* (2002)
Rice, *Oryza sativa* L.	↓	↓	↓↑	–↓↑	↓↑	–	–↓	↑	–↑	↑	nd	Seed	Rai *et al.* (2010), Frei *et al.* (2012), Wang *et al.* (2012), Zheng *et al.* (2013)
Sweet potato, *Ipomoea batatas* (L.) Lam.	↑	–	–	↑	↑	nd	nd	nd	nd	nd	nd	Tuber	Keutgen *et al.* (2008)
Wheat, *Triticum aestivum* L.	nd	↑	↑	–↑	↑	–	–	↑	↑	↑	nd	Seed	Fuhrer *et al.* (1990), Pleijel *et al.* (2006), Feng *et al.* (2008), Zheng *et al.* (2013)

Ozone stress increase (↑), decrease (↓) or does not show significant difference (–) on mineral content. Minerals not determined (nd).

typically in the leaves, consequently leafy vegetables become highly susceptible to toxic effects of ozone. In broad-leaved plants, the symptoms due to acute exposure include bleaching (small unpigmented necrotic spots), flecking (small brown necrotic areas fading to grey or white), stippling (small punctuate spots, which may be white, black or red) and bifacial necrosis (when the entire tissue through the leaf is killed developing a range of colour from white to dark orange-red). Whereas, injuries attributed to chronic exposure appear as chlorosis (yellowing due to the chlorophyll breakdown) and bronzing (red-brown pigmentation induced by phenylpropanoid accumulation) (Krupa *et al.*, 2001; Iriti and Faoro, 2008). In a study with spinach and lettuce exposed to elevated ozone (NFCs+ ozone), visible foliar injury symptoms in the form of blackish and necrotic bifacial lesions were mainly observed in the interveinal and marginal area of mature leaves (Calatayud *et al.*, 2002, 2003). Temple *et al.* (1990) reported that lettuce and onions chronically exposed to ozone exhibited severe leaf injury, while no injury symptoms could be observed in broccoli. These visible injuries are particularly undesirable when the marketable value of the crop depends on the appearance, especially because it can cause an obvious loss of economic crop value.

In some cases flavour may be also modified by changing compound profile. The soluble solids content (an indirect measurement of sugar content, i.e. sweetness) of watermelon fruit was decreased from 4 to 8% due to exposure to ambient levels of ozone, leading to decrease in fruit quality (Gimeno *et al.*, 1999). Ozone also tended to reduce the sweetness index in two different cultivars of strawberry (Keutgen and Pawelzik, 2008). In potato (cv. Cherokee) exposed to ozone, Pell and Pearson (1984) observed an increase in total glycoalkaloids, which may lead to bitterness. In addition, another study with potato plants exposed to ozone in open-top chambers found that paste from tubers was more viscous under elevated ozone in one year (1998) and starch granules were more resistant to swelling under elevated ozone in the following year (1999) (Donnelly *et al.*, 2001).

Chalk is an opaque area in the rice grain and is an important quality characteristic in rice. Chalk areas are undesirable because it alters rice cooking and appearance, which negatively affects rice quality. Moreover, chalky grains tend to be weaker and break easily, thus decreasing mill yield (Wang and Frei, 2011). Wang *et al.* (2012) showed that chalky grain percentage was higher due to ozone exposure, while chalkiness area and chalkiness degree remain unchanged. Furthermore, they observed that long-term ozone exposure increased surface firmness and reduced acceptability of cooked rice. The study suggested the starch in rice grain grown in high ozone levels exhibited lower viscosity and elasticity.

8.4.7 Feed value of forage for ruminant animals

Forage quality is determined by its digestibility, nutrient content (proteins, lipid, sugars, starch, minerals) and antinutrient content (Waghorn and Clark, 2004; Vandermeiren and Pleije, 2011). Generally, ozone decreases forage quality and it can lead to secondary effects such as lower milk and meat production from grazing animals (Vandermeiren and Pleije, 2011), and thus it is possible to link ozone indirectly with impairment of food security.

Digestibility and protein content are the most studied aspects regarding the impacts of effect on forage quality. Fibre fractions (neutral detergent fibre, NDF; acid detergent fibre, ADF; lignin) and phenolic content are important parameters used to determine the plant material digestibility and, commonly, these two parameters are inversely correlated with forage digestibility (Wang and Frei, 2011). Analysing Table 8.6 it is notable that ozone decreased leaf digestibility (Table 8.6), whereas fibre content increased (Table 8.3) in the vast majority of cases. The ADF fraction and lignin is inversely correlated to forage digestibility, while NDF is more closely associated with voluntary forage intake than with digestibility (Jung and Allen, 1995).

In clovers grown under ozone exposure, Muntifering *et al.* (2006) observed a decrease in IVDMD (*in vitro* dry-matter digestibility), in IVCWD (*in vitro* cell-wall digestibility) and lignin, but no differences in NDF and soluble phenolics concentration were reported. Similarly, González-Fernández *et al.* (2008) observed a negative impact on clover forage grown under ozone, and NDF and lignin enhanced while IVDMD decreased.

In a study performed by Gilliland *et al.* (2012), a mixture of grassland species (*Lolium arundinacea*, *Paspalum dilatatum*, *Cynodon dactylon* and *Trifolium repens*) exposed to twice (2×) ambient ozone concentration contained approximately 8% more NDF and 15% greater concentration of soluble phenolics than forage grown in non-filtered air (NF), but concentrations of ADF and lignin of forage were approximately equal. In addition, the authors fed white rabbits (*Oryctolagus cuniculus*) with these two

Table 8.6. Effect of ozone stress on the digestibility of forage crops for ruminant herbivores.

Crop species	Organ	Ozone effect	References
Clover, *Trifolium* spp.	Leaf	↓	Muntifering *et al.* (2006), González-Fernández *et al.* (2008)
Grassland species mixture	Leaf	↓	Gilliland *et al.* (2012)
Lespedeza, *Lespedeza cuneata*	Leaf	–/↓	Powell *et al.* (2003)
Little bluestem, *Schizachyrium scoparium*	Leaf	–/↓	Powell *et al.* (2003)
Poa pratensis L.	Leaf	↓	Bender *et al.* (2006)
Rice straw, *Ozyza sativa* L.	Leaf	↓	Frei *et al.* (2011)

Ozone stress increase (↑), decrease (↓) or does not show significant difference (–) on digestibility.

forages and observed that the digestibility was 5.5 g day^{-1} greater for rabbits that ingested the NF than the forages grown under 2× ambient ozone. The nutritive quality of little bluestem and *Sericea lespedeza* exposed to ambient and 2× ambient ozone concentration were decreased by 2% and 7%, respectively, and the authors explain that this is a result of increased levels of cell wall constituents and decreased *in vitro* digestibility (Powell *et al.*, 2003).

In *Poa pratensis*, a high-yielding perennial pasture grass in Europe, early-season ozone exposure caused a loss in the relative feed value of 8%, which is enough to have nutritional implications for herbivore utilization, with consequences in voluntary intake and digestibility (Bender *et al.*, 2006). Frei *et al.* (2011) studied the effect of ozone on the nutritive quality of rice straw, a by-product of rice grain with important feed value for ruminant livestock. The effects of ozone on the chemical composition of straw were clearly dependent on the ozone level, with significant changes even at ambient ozone concentrations. Increases in crude ash, lignin and phenolic concentration adversely affected the digestibility as demonstrated in incubation experiments simulating rumen digestion *in vitro*. Taking all these studies together, it is possible to note that ozone-induced changes in foliar chemistry can drive alterations in forage quality, which has severe economic and nutritional implications for their utilization by ruminant herbivores.

8.5 Conclusions

In a scenario where future ozone levels are predicted to increase, it is especially important to further understand the dynamic interactions between ozone, plant development and carbon allocation. Together with seasonal rising temperatures and CO_2 concentrations, ozone exposure changes the timing of carbon dynamics in plants with major detrimental impacts on crop growth rates and seed development, increasing stress among plants (Long *et al.*, 2005; Fuhrer, 2009). The metabolic switch elicited in plants exposed to medium to elevated ozone levels might lead to 'hidden' changes in qualitative and nutritional properties of natural products, with an overall risk and consequences on the food and feed chain. After analysing numerous works performed around the world with different plants and distinct crop varieties, we can conclude that the negative impacts of ozone need to be considered in a combination of yield and quality parameters. Changes associated with secondary metabolite biosynthesis can be detrimental for the plant's fitness, and when we consider the allocation costs, modifications of food value and composition could possibly be more significant than biomass yield reductions alone in the assessment of ozone effects (Fuhrer and Booker, 2003; Bender and Weigel, 2011; Vandermeiren and Pleije, 2011).

Besides, some quality aspects of crops such as enhanced seed protein content and secondary metabolites are apparently improved by ozone, and as suggested by Iriti and Faoro (2009), it may be favourable in crops and plants that provide foodstuffs and beverages enriched with bioactive phytochemicals. Even so, reports concerning the effects of chemically altered marketable crops are still lacking, and the consequences for human nutrition need to be studied in more detail. Future challenges thus include mitigation of ozone-induced changes and development of ozone-tolerant crops, especially in regions where agroecosystems are presented as the key strategy to sponsor communities' food supply in attempt to avoid further ecological impacts and to improve the quality of the agricultural products consumed by us.

References

Agrawal, M., Singh, B., Rajput, M., Marshall, F. and Bell, J.N. (2003) Effect of air pollution on peri-urban agriculture: a case study. *Environmental Pollution* 126, 323–329.

Ainsworth, E., Yendrek, C.R., Sitch, S., Collins, W.J. and Emberson, L.D. (2012) The effects of tropospheric ozone on net primary productivity and implications for climate change. *Annual Review of Plant Biology* 63, 637–661.

Anderson, P.D., Palmer, B., Houpis, J.L.J., Smith, M.K. and Pushnik, J.C. (2003) Chloroplastic responses of ponderosa pine (*Pinus ponderosa*) seedlings to ozone exposure. *Environment International* 29, 407–413.

Appel, H.M. (1993) Phenolics in ecological interactions: the importance of oxidation. *Journal of Chemical Ecology* 19, 1521–1552.

Ashmore, M.R. (2005) Assessing the future global impacts of ozone on vegetation. *Plant, Cell & Environment* 28, 949–964.

Avnery, S., Mauzerall, D.L., Liu, J. and Horowitz, L.W. (2011) Global crop yield reductions due to surface ozone exposure: 1. Year 2000 crop production losses and economic damage. *Atmospheric Environment* 45, 2284–2296.

Baier, M., Kandlbinder, A., Golldack, D. and Dietz, K.J. (2005) Oxidative stress and ozone: perception, signalling and response. *Plant, Cell & Environment* 28, 1012–1020.

Bender, J. and Weigel, H.J. (2011) Changes in atmospheric chemistry and crop health: a review. *Agronomy for Sustainable Development* 31, 81–89.

Bender, J., Muntifering, R.B., Lin, J.C. and Weigl, H.J. (2006) Growth and nutritive quality of *Poa pratensis* as influenced by ozone and competition. *Environmental Pollution* 142, 109–115.

Betzelberger, A.M., Yendrek, C.R., Sun, J., Leisner, C.P., Nelson, R.L., Ort, D.R. and Ainsworth, E.A. (2012) Ozone exposure response for U.S. soybean cultivars: linear reductions in photosynthetic potential, biomass, and yield. *Plant Physiology* 160, 1827–1839.

Biswas, D.K., Xu, H., Li, Y.G., Liu, M.Z., Chen, Y.H., Sun, J.Z. and Jiang, G.M. (2008) Assessing the genetic relatedness of higher ozone sensitivity of modern wheat to its wild and cultivated progenitors/relatives. *Journal of Experimental Botany* 59, 951–963.

Blum, U., Smith, G. and Fites, R. (1982) Effects of multiple O_3 exposures on carbohydrate and mineral contents of ladino clover. *Environmental and Experimental Botany* 22, 143–154.

Bolton, M.D. (2009) Primary metabolism and plant defense – fuel for the fire. *Molecular Plant-Microbe Interactions* 22, 487–497.

Booker, F.L. and Miller, J.E. (1998) Phenylpropanoid metabolism and phenolic composition of soybean [*Glycine max* (L.) Merr.] leaves following exposure to ozone. *Journal of Experimental Botany* 49, 1191–1202.

Booker, F.L., Muntifering, R., McGrath, M., Burkey, K., Decoteau, D., Fiscus, E., Manning, W., Krupa, S., Chappelka, A. and Grantz, D. (2009) The ozone component of global change: potential effects on agricultural and horticultural plant yield, product quality and interactions with invasive species. *Journal of Integrative Plant Biology* 51, 337–351.

Bosac, C., Black, V.J., Roberts, J.A. and Black, C.R. (1998) Impact of ozone on seed yield and quality and seedling vigour in oilseed rape (*Brassica napus* L.). *Journal of Plant Physiology* 153, 127–134.

Burkey, K.O., Booker, F.L., Pursley, W.A. and Heagle, A.S. (2007) Elevated carbon dioxide and ozone effects on peanut: II. Seed yield and quality. *Crop Science* 47, 1488–1497.

Calatayud, A., Alvarado, J.W. and Barreno, E. (2001) Changes in chlorophyll a fluorescence, lipid peroxidation, and detoxificant system in potato plants grown under filtered and non-filtered air in open-top chambers. *Photosynthetica* 39, 507–513.

Calatayud, A., Ramirez, J.W., Iglesias, D.J. and Barreno, E. (2002) Effects of ozone on photosynthetic CO_2 exchange, chlorophyll a fluorescence and antioxidant systems in lettuce leaves. *Physiologia Plantarum* 116, 308–316.

Calatayud, A., Iglesias, D.J., Talon, M. and Barreno, E. (2003) Effects of 2-month ozone exposure in spinach leaves on photosynthesis, antioxidant systems and lipid peroxidation. *Plant Physiology and Biochemistry* 41, 839–845.

Calatayud, V., Cerveró, J. and Sanz, M.J. (2007) Foliar, physiological and growth responses of four maple species exposed to ozone. *Water, Air, and Soil Pollution* 185, 239–254.

Calfapietra, C., Wiberley, A.E., Falbel, T.G., Linskey, A.R., Mugnozza, G.S., Karnosky, D.F., Loreto, F. and Sharkey, T.D. (2007) Isoprene synthase expression and protein levels are reduced under elevated O_3 but not under elevated CO_2 (FACE) in field-grown aspen trees. *Plant, Cell & Environment* 30, 654–661.

Calfapietra, C., Mugnozza, G.S., Karnosky, D.F., Loreto, F. and Sharkey, T.D. (2008) Isoprene emission rates under elevated CO_2 and O_3 in two field-grown aspen clones differing in their sensitivity to O_3. *The New Phytologist* 179, 55–61.

Calfapietra, C., Fares, S. and Loreto, F. (2009) Volatile organic compounds from Italian vegetation and their interaction with ozone. *Environmental Pollution* 157, 1478–1486.

Caregnato, F.F., Bortolin, R.C., Divan Junior, A.M. and Moreira, J.C.F. (2013) Exposure to elevated ozone levels differentially affects the antioxidant capacity and the redox homeostasis of two subtropical *Phaseolus vulgaris* L. varieties. *Chemosphere* 93, 320–330.

Castagna, A. and Ranieri, A. (2009) Detoxification and repair process of ozone injury: from O-3 uptake to gene expression adjustment. *Environmental Pollution* 157, 1461–1469.
Ceyhan, E., Kahraman, A. and Onder, M. (2012) The impacts of environment on plant products. *International Journal of Bioscience, Biochemistry and Bioinformatics* 2, 48–51.
Chen, C. and Kong, A.N.T. (2004) Dietary chemopreventive compounds and ARE/EpRE signaling. *Free Radical Biology & Medicine* 36, 1505–1516.
Collins, W.J., Stevenson, D., Johnson, C. and Derwent, R. (2000) The European regional ozone distribution and its links with the global scale for the years 1992 and 2015. *Atmospheric Environment* 34, 255–267.
Conklin, P.L., Williams, E.H. and Last, R.L. (1996) Environmental stress sensitivity of an ascorbic acid-deficient *Arabidopsis* mutant. *Proceedings of the National Academy of Sciences of the United States of America* 93, 9970–9974.
Croteau, R., Kutchan, T.M. and Lewis, N.G. (2000) Natural products (secondary metabolites). In: Buchanan, B.B., Gruissen, W. and Jones, R.L. (eds) *Biochemistry & Molecular Biology of Plants*. American Society of Plant Physiologists, Rockville, Maryland, pp. 1250–1318.
Crozier, A., Clifford, M.N. and Ashihara, H. (2006) *Plant Secondary Metabolites: Occurrence, Structure and Role in the Human Diet*, 1st edn. Blackwell Publishing, Oxford, UK.
D'Haese, D., Vandermeiren, K., Asard, H. and Horemans, N. (2005) Other factors than apoplastic ascorbate contribute to the differential ozone tolerance of two clones of *Trifolium repens* L. *Plant, Cell & Environment* 28, 623–632.
Dentener, F., Keating, T. and Akimoto, H. (2010) *Hemispheric Transport of 2010 part A: ozone and particulate matter*. Air Pollution Studies No.17. United Nations Publication, Geneva, Switzerland.
Di Baccio, D., Castagna, A., Paoletti, E., Sebastiani, L., Ranier, A. and Ranieri, A. (2008) Could the differences in O(3) sensitivity between two poplar clones be related to a difference in antioxidant defense and secondary metabolic response to O(3) influx? *Tree Physiology* 28, 1761–1772.
Dizengremel, P., Le Thiec, D., Bagard, M. and Jolivet, Y. (2008) Ozone risk assessment for plants: central role of metabolism-dependent changes in reducing power. *Environmental Pollution* 156, 11–15.
Dizengremel, P., Le Thiec, D., Hasenfratz-Sauder, M.-P., Vaultier, M.-N., Bagard, M. and Jolivet, Y. (2009) Metabolic-dependent changes in plant cell redox power after ozone exposure. *Plant Biology* 1, 35–42.
Donnelly, A., Lawson, T., Craigon, J., Black, C.R., Colls, J.J. and Landon, G. (2001) Effects of elevated CO_2 and O_3 on tuber quality in potato (*Solanum tuberosum* L.). *Agriculture, Ecosystems & Environment* 87, 273–285.
Duranti, M. and Gius, C. (1997) Legume seeds: protein content and nutritional value. *Field Crops Research* 53, 31–45.
Eckey-Kaltenbach, H., Ernst, D., Heller, W. and Sandermann, H. (1994) Biochemical plant responses to ozone. IV. Cross-induction of defensive pathways in parsley (*Petroselinum crispum* L.) plants. *Plant Physiology* 104, 67–74.
Emberson, L.D., Ashmore, M.R., Murray, F., Kuylenstierna, J.C.I., Percy, K.E., Izuta, T., Zheng, Y., Shimizu, H., Sheu, B.H., Liu, C.P., Agrawal, M., Wahid, A., Abdel-Latif, N.M., Tienhoven, M., van-Bauer, L.I. and Domingos, M. (2001) Impacts of air pollutants on vegetation in developing countries. *Water, Air, and Soil Pollution* 130, 107–118.
Emberson, L.D., Büker, P., Ashmore, M.R., Mills, G., Jackson, L.S., Agrawal, M., Atikuzzaman, M.D., Cinderby, S., Engardt, M., Jamir, C., Kobayashi, K., Oanh, N.T.K., Quadir, Q.F. and Wahid, A. (2009) A comparison of North American and Asian exposure–response data for ozone effects on crop yields. *Atmospheric Environment* 43, 1945–1953.
EPA Environmental Protection Agency (1996) *Air Quality Criteria for Ozone and Related Photochemical Oxidants*. United States Environmental Protection Agency, Washington, DC.
Fangmeier, A., Temmerman, L. De, Black, C., Persson, K. and Vorne, V. (2002) Effects of elevated CO_2 and/or ozone on nutrient concentrations and nutrient uptake of potatoes. *European Journal of Agronomy* 17, 353–368.
FAO (2010) Food and agricultural commodities production. Food and Agriculture Organization of the United Nations. Available at: http://faostat.fao.org/site/339/default.aspx (accessed November 2014).
Feng, Z., Kobayashi, K. and Ainsworth, E.A. (2008) Impact of elevated ozone concentration on growth, physiology, and yield of wheat (*Triticum aestivum* L.): a meta-analysis. *Global Change Biology* 14, 2696–2708.
Fitzgerald, M.A., McCouch, S.R. and Hall, R.D. (2009) Not just a grain of rice: the quest for quality. *Trends in Plant Science* 14, 133–139.
Fontaine, V., Pelloux, J., Podor, M., Afif, D., Ge, D., Poincare, H., Nancy, F.V., Gérant, D., Grieu, P. and Dizengremel, P. (1999) Carbon fixation in *Pinus halepensis* submitted to ozone: opposite response of

ribulose-1,5-bisphosphate carboxylase/oxygenase and phosphoenolpyruvate carboxylase. *Physiologia Plantarum* 105, 187–192.

Foyer, C.H. and Noctor, G. (2005) Redox homeostasis and antioxidant signaling: a metabolic interface between stress perception and physiological responses. *The Plant Cell* 17, 1866–1875.

Frei, M., Kohno, Y., Wissuwa, M., Makkar, H.P.S. and Becker, K. (2011) Negative effects of tropospheric ozone on the feed value of rice straw are mitigated by an ozone tolerance QTL. *Global Change Biology* 17, 2319–2329.

Frei, M., Kohno, Y., Tietze, S., Jekle, M., Hussein, M.A., Becker, T. and Becker, K. (2012) The response of rice grain quality to ozone exposure during growth depends on ozone level and genotype. *Environmental Pollution* 163, 199–206.

Friedman, M. and McDonald, G.M. (1997) Potato glycoalkaloids: chemistry, analysis, safety, and plant physiology. *Critical Reviews in Plant Sciences* 16, 55–132.

Fuhrer, J. (2009) Ozone risk for crops and pastures in present and future climates. *Naturwissenschaften* 96, 173–194.

Fuhrer, J. and Booker, F. (2003) Ecological issues related to ozone: agricultural issues. *Environment International* 29, 141–154.

Fuhrer, J., Lehnherr, B., Moeri, P.B., Tschannen, W. and Shariat-Madari, H. (1990) Effects of ozone on the grain composition of spring wheat grown in open-top field chambers. *Environmental Pollution* 65, 181–192.

Fuhrer, J., Grimm, A.G., Tschannen, W. and Shariat-Madari, H. (1992) The response of spring wheat (*Triticum aestivum* L.) to ozone at higher elevations. *New Phytologist* 121, 211–219.

Garcia, W., Cavins, J., Inglett, G., Heagle, A. and Kwolek, W.F. (1983) Quality of corn grain from plants exposed to chronic levels of ozone. *Cereal Chemistry* 60, 388–391.

Gielen, B., Vandermeiren, K., Horemans, N., D'Haese, D., Serneels, R. and Valcke, R. (2006) Chlorophyll a fluorescence imaging of ozone-stressed *Brassica napus* L. plants differing in glucosinolate concentrations. *Plant Biology* 8, 698–705.

Gilliland, N.J., Chappelka, A.H., Muntifering, R.B., Booker, F.L. and Ditchkoff, S.S. (2012). Digestive utilization of ozone-exposed forage by rabbits (*Oryctolagus cuniculus*). *Environmental Pollution* 163, 281–286.

Gimeno, B.S., Bermejo, V., Reinert, R., Zheng, Y. and Barnes, J. (1999) Adverse effects of ambient ozone on watermelon yield and physiology at a rural site in Eastern Spain. *New Phytologist* 144, 245–260.

González-Fernández, I., Bass, D., Muntifering, R., Mills, G. and Barnes, J. (2008) Impacts of ozone pollution on productivity and forage quality of grass/clover swards. *Atmospheric Environment* 42, 8755–8769.

Grantz, D.A. and Farrar, J.F. (2000) Ozone inhibits phloem loading from a transport pool: compartmental efflux analysis in Pima cotton. *Australian Journal of Plant Physiology* 27, 859–868.

Grunwald, C. and Endress, A.G. (1984) Fatty acids of soybean seeds harvested from plants exposed to air pollutants. *Journal of Agricultural and Food Chemistry* 32, 50–53.

Guidi, L., Degl'Innocenti, E., Giordano, C., Biricolti, S. and Tattini, M. (2010) Ozone tolerance in *Phaseolus vulgaris* depends on more than one mechanism. *Environmental Pollution* 158, 3164–3171.

Harborne, J.B. (1993) *Introduction to Ecological Biochemistry*, 3rd edn. Elsevier Academic Press, London.

Himanen, S.J., Nissinen, A., Auriola, S., Poppy, G.M., Stewart, C.N., Holopainen, J.K. and Nerg, A.M. (2008) Constitutive and herbivore-inducible glucosinolate concentrations in oilseed rape (*Brassica napus*) leaves are not affected by Bt Cry1Ac insertion but change under elevated atmospheric CO_2 and O_3. *Planta* 227, 427–437.

Hounsome, N., Hounsome, B., Tomos, D. and Edwards-Jones, G. (2008) Plant metabolites and nutritional quality of vegetables. *Journal of Food Science* 73, 48–65.

Howell, R.K. and Rose, L.P. (1980) Residual air pollution effects on soybean seed quality. *Plant Disease* 64, 385–386.

Iglesias, D.J., Calatayud, A., Barreno, E., Primo-Millo, E. and Talon, M. (2006) Responses of citrus plants to ozone: leaf biochemistry, antioxidant mechanisms and lipid peroxidation. *Plant Physiology and Biochemistry* 44, 125–131.

IPCC (Intergovernmental Panel on Climate Change) (2007) *The Physical Science Basis. Contribution of Working Group I to the Fourth Annual Assessment Report of the Intergovernmental Panel on Climate Change*. Cambridge University Press, Cambridge, UK.

Iriti, M. and Faoro, F. (2008) Oxidative stress, the paradigm of ozone toxicity in plants and animals. *Water, Air and Soil Pollution* 187, 285–301.

Iriti, M. and Faoro, F. (2009) Chemical diversity and defence metabolism: how plants cope with pathogens and ozone pollution. *International Journal of Molecular Sciences* 10, 3371–3399.

Iriti, M., Di Maro, A., Bernasconi, S., Burlini, N., Simonetti, P., Picchi, V., Panigada, C., Gerosa, G., Parente, A. and Faoro, F. (2009) Nutritional traits of bean (*Phaseolus vulgaris*) seeds from plants chronically exposed to ozone pollution. *Journal of Agricultural and Food Chemistry* 57, 201–208.

Jackson, D.M., Rufty, T.W., Heagle, A.S., Severson, R.F. and Eckel, R.V.W. (2000) Survival and development of tobacco hornworm larvae on tobacco plants grown under elevated levels of ozone. *Journal of Chemical Ecology* 26, 1–19.

Jung, H.G. and Allen, M.S. (1995) Characteristics of plant cell walls affecting intake and digestibility of forages by ruminants. *Journal of Animal Science* 73, 2774–2790.

Kangasjarvi, J., Jaspers, P. and Kollist, H. (2005) Signalling and cell death in ozone-exposed plants. *Plant, Cell & Environment* 28, 1021–1036.

Keen, N.T. and Taylor, O. (1975) Ozone injury in soybeans. *Plant Physiology* 55, 731–733.

Keutgen, A. and Pawelzik, E. (2007) Modifications of taste-relevant compounds in strawberry fruit under NaCl salinity. *Food Chemistry* 105, 1487–1494.

Keutgen, A.J. and Pawelzik, E. (2008) Influence of pre-harvest ozone exposure on quality of strawberry fruit under simulated retail conditions. *Postharvest Biology and Technology* 49, 10–18.

Keutgen, N., Keutgen, A.J. and Janssens, M.J.J. (2008) Sweet potato [*Ipomoea batatas* (L.) Lam.] cultivated as tuber or leafy vegetable supplier as affected by elevated tropospheric ozone. *Journal of Agricultural and Food Chemistry* 56, 6686–6690.

Köllner, B. and Krause, G.H.M. (2000) Changes in carbohydrates, leaf pigments and yield in potatoes induced by different ozone exposure regimes. *Agriculture, Ecosystems & Environment* 78, 149–158.

Korkina, L. (2007) Phenylpropanoids as naturally occurring antioxidants: from plant defense to human health. *Cellular and Molecular Biology* 53, 15–25.

Krupa, S.V., Nosal, M. and Legge, A.H. (1998) A numerical analysis of the combined open-top chamber data from the USA and Europe on ambient ozone and negative crop responses. *Environmental Pollution* 101, 157–160.

Krupa, S., McGrath, M.T., Andersen, C.P., Booker, F.L., Burkey, K.O., Chappelka, A.H., Chevone, B.I., Pell, E.J. and Zilinskas, B.A. (2001) Ambient ozone and plant health. *Plant Disease* 85, 4–12.

Laisk, A., Kull, O. and Moldau, H. (1989) Ozone concentration in leaf intercellular air spaces is close to zero. *Plant Physiology* 90, 1163–1167.

Langebartels, C., Kerner, K., Leonardi, S., Schraudner, M., Trost, M. and Heller, W. (1991) Biochemical plant responses to ozone. 1. Differential induction of polyamine and ethylene biosynthesis in tobacco. *Plant Physiology* 95, 882–889.

Le Bot, J., Bénard, C., Robin, C., Bourgaud, F. and Adamowicz, S. (2009) The 'trade-off' between synthesis of primary and secondary compounds in young tomato leaves is altered by nitrate nutrition: experimental evidence and model consistency. *Journal of Experimental Botany* 60, 4301–4314.

Lee, S.K. and Kader, A.A. (2000) Preharvest and postharvest factors influencing vitamin C content of horticultural crops. *Postharvest Biology and Technology* 20, 207–220.

Leszkowiat, M., Barichello, V., Yada, R., Coffin, R., Lougheed, E. and Stanley, D. (1990) Contribution of sucrose to nonenzymatic browning in potato chips. *Journal of Food Science* 55, 281–282.

Long, S.P., Ainsworth, E.A., Leakey, A.D.B. and Morgan, P.B. (2005) Global food insecurity. Treatment of major food crops with elevated carbon dioxide or ozone under large-scale fully open-air conditions suggests recent models may have overestimated future yields. *Philosophical Transactions of the Royal Society B - Biological Sciences* 360, 2011–2020.

Luwe, M., Takahama, U. and Heber, U. (1993) Role of ascorbate in detoxifying ozone in the apoplast of spinach (*Spinacia oleracea* L.) leaves. *Plant Physiology* 101, 969–976.

Mahalingam, R., Jambunathan, N., Gunjan, S.K., Faustin, E., Weng, H. and Ayoubi, P. (2006) Analysis of oxidative signalling induced by ozone in *Arabidopsis thaliana*. *Plant Cell & Environment* 29, 1357–1371.

Manderscheid, R., Jager, H.J. and Kress, L.W. (1992) Effects of ozone on foliar nitrogen metabolism of *Pinus taeda* L. and implications for carbohydrate metabolism. *New Phytologist* 121, 623–633.

Mills, G., Buse, A., Gimeno, B., Bermejo, V., Holland, M., Emberson, L. and Pleijel, H. (2007) A synthesis of AOT40-based response functions and critical levels of ozone for agricultural and horticultural crops. *Atmospheric Environment* 41, 2630–2643.

Mittal, M.L., Hess, P.G., Jain, S.L., Arya, B.C. and Sharma, C. (2007) Surface ozone in the Indian region. *Atmospheric Environment* 41, 6572–6584.

Mulchi, C.L., Lee, E., Tuthill, K. and Olinick, E.V. (1988) Influence of ozone stress on growth processes, yields and grain quality characteristics among soybean cultivars. *Environmental Pollution* 53, 151–169.

Muntifering, R., Crosby, D., Powell, M. and Chappelka, A. (2000) Yield and quality characteristics of bahiagrass (*Paspalum notatum*) exposed to ground-level ozone. *Animal Feed Science and Technology* 84, 243–256.

Muntifering, R.B., Chappelka, A.H., Lin, J.C., Karnosky, D.F. and Somers, G.L. (2006) Chemical composition and digestibility of *Trifolium* exposed to elevated ozone and carbon dioxide in a free-air (FACE) fumigation system. *Functional Ecology* 20, 269–275.

Nesi, N., Delourme, R., Brégeon, M., Falentin, C. and Renard, M. (2008) Genetic and molecular approaches to improve nutritional value of *Brassica napus* L. seed. *Comptes Rendus Biologies* 331, 763–771.

Noctor, G. (2006) Metabolic signalling in defence and stress: the central roles of soluble redox couples. *Plant, Cell & Environment* 29, 409–425.

Noctor, G. and Foyer, C.H. (1998) Ascorbate and glutathione: keeping active oxygen under control. *Annual Review of Plant Physiology and Plant Molecular Biology* 49, 249–279.

Oksanen, E., Pandey, V., Pandey, A.K., Keski-Saari, S., Kontunen-Soppela, S. and Sharma, C. (2013) Impacts of increasing ozone on Indian plants. *Environmental Pollution* 177, 189–200.

Ollerenshaw, J., Lyons, T. and Barnes, J. (1999) Impacts of ozone on the growth and yield of field-grown winter oilseed rape. *Environmental Pollution* 104, 53–59.

Pell, E.J. and Pearson, N.S. (1984) Ozone-induced reduction in quantity and quality of two potato cultivars. *Environmental Pollution Series A: Ecological and Biological* 35, 345–352.

Pell, E.J., Weissberger, W.C. and Speroni, J.J. (1980) Impact of ozone on quantity and quality of greenhouse-grown potato plants. *American Chemical Society* 14, 568–571.

Pell, E.J., Pearson, N.S. and Vinten-Johansen, C. (1988) Qualitative and quantitative effects of ozone and/or sulfur dioxide on field-grown potato plants. *Environmental Pollution* 53, 171–186.

Piikki, K., De Temmerman, L., Ojanperä, K., Danielsson, H. and Pleijel, H. (2008) The grain quality of spring wheat (*Triticum aestivum* L.) in relation to elevated ozone uptake and carbon dioxide exposure. *European Journal of Agronomy* 28, 245–254.

Pleijel, H., Skärby, L., Wallin, G. and Sellden, G. (1991) Yield and grain quality of spring wheat (*Triticum aestivum* L., cv. Drabant) exposed to different concentrations of ozone in open-top chambers. *Environmental Pollution* 69, 151–168.

Pleijel, H., Ojanperä, K. and Mortensen, L. (1997) Effects of tropospheric ozone on the yield and grain protein content of spring wheat (*Triticum aestivum* L.) in the Nordic countries. *Acta Agriculturae Scandinavica, Section B - Soil & Plant Science* 47, 20–25.

Pleijel, H., Danielsson, H., Gelang, J., Sild, E. and Sellde, G. (1998) Growth stage dependence of the grain yield response to ozone in spring wheat (*Triticum aestivum* L.). *Agriculture, Ecosystems & Environment* 70, 61–68.

Pleijel, H., Mortensen, L., Fuhrer, J., Ojanpera, K. and Danielsson, H. (1999) Grain protein accumulation in relation to grain yield of spring wheat (*Triticum aestivum* L.) grown in open-top chambers with different concentrations of ozone, carbon dioxide and water availability. *Agriculture, Ecosystems & Environment* 72, 265–270.

Pleijel, H., Eriksen, A.B., Danielsson, H., Bondesson, N. and Selldén, G. (2006) Differential ozone sensitivity in an old and a modern Swedish wheat cultivar – grain yield and quality, leaf chlorophyll and stomatal conductance. *Environmental and Experimental Botany* 56, 63–71.

Powell, M.C., Muntifering, R.B., Lin, J.C. and Chappelka, A.H. (2003) Yield and nutritive quality of sericea lespedeza (*Lespedeza cuneata*) and little bluestem (*Schizachyrium scoparium*) exposed to ground-level ozone. *Environmental Pollution* 122, 313–322.

Prior, R.L., Cao, G., Martin, A., Sofic, E., McEwen, J., O'Brien, C., Lischner, N., Ehlenfeldt, M., Kalt, W., Krewer, G. and Mainland, C.M. (1998) Antioxidant capacity as influenced by total phenolic and anthocyanin content, maturity, and variety of Vaccinium species. *Journal of Agricultural and Food Chemistry* 46, 2686–2693.

Puckette, M.C., Tang, Y. and Mahalingam, R. (2008) Transcriptomic changes induced by acute ozone in resistant and sensitive *Medicago truncatula* accessions. *BMC Plant Biology* 8, 46.

Rai, R. and Agrawal, M. (2012) Impact of tropospheric ozone on crop plants. *Proceedings of the National Academy of Sciences, India Section B: Biological Sciences* 82, 241–257.

Rai, R., Agrawal, M. and Agrawal, S.B. (2010) Threat to food security under current levels of ground level ozone: a case study for Indian cultivars of rice. *Atmospheric Environment* 44, 4272–4282.

Ranieri, A., Castagna, A., Padu, E., Moldau, H., Rahi, M. and Soldatini, G.F. (1999) The decay of O-3 through direct reaction with cell wall ascorbate is not sufficient to explain the different degrees of O-3-sensitivity in two poplar clones. *Journal of Plant Physiology* 154, 250–255.

Rao, M.V. (2000) Jasmonic acid signaling modulates ozone-induced hypersensitive cell death. *The Plant Cell Online* 12, 1633–1646.

Rice-Evans, C., Miller, N. and Paganga, G. (1997) Antioxidant properties of phenolic compounds. *Trends in Plant Science* 2, 152–159.

Roe, M.A., Faulks, R.M. and Belsten, J.L. (1990) Role of reducing sugars and amino acids in fry colour of chips from potatoes grown under different nitrogen regimes. *Journal of the Science of Food and Agriculture* 52, 207–214.

Saleem, A., Loponen, J., Pihlaja, K. and Oksanen, E. (2001) Effects of long-term open-field ozone exposure on leaf phenolics of European silver birch (*Betula pendula* Roth). *Journal of Chemical Ecology* 27, 1049–1062.

Sanz, J., Muntifering, R.B., Bermejo, V., Gimeno, B.S. and Elvira, S. (2005) Ozone and increased nitrogen supply effects on the yield and nutritive quality of *Trifolium subterraneum*. *Atmospheric Environment* 39, 5899–5907.

Sanz, J., Bermejo, V., Muntifering, R., González-Fernández, I., Gimeno, B.S., Elvira, S. and Alonso, R. (2011) Plant phenology, growth and nutritive quality of *Briza maxima*: responses induced by enhanced ozone atmospheric levels and nitrogen enrichment. *Environmental Pollution* 159, 423–430.

Sarıkamış, G. (2009) Glucosinolates in crucifers and their potential effects against cancer: Review. *Canadian Journal of Plant Science* 89, 953–959.

Sarkar, A. and Agrawal, S.B. (2012) Evaluating the response of two high yielding Indian rice cultivars against ambient and elevated levels of ozone by using open top chambers. *Journal of Environmental Management* 95, S19–24.

Saviranta, N.M.M., Julkunen-Tiitto, R., Oksanen, E. and Karjalainen, R.O. (2010) Leaf phenolic compounds in red clover (*Trifolium pratense* L.) induced by exposure to moderately elevated ozone. *Environmental Pollution* 158, 440–446.

Scebba, F., Pucciarelli, I., Soldatini, G.F. and Ranieri, A. (2003) O_3-induced changes in the antioxidant systems and their relationship to different degrees of susceptibility of two clover species. *Plant Science* 165, 583–593.

Sharma, P., Ambuj, B.J., Dubey, R.S. and Pessarakli, M. (2012) Reactive oxygen species, oxidative damage, and antioxidative defense mechanism in plants under stressful conditions. *Journal of Botany* 2012, 1–26.

Sharma, Y.K. and Davis, K.R. (1997) The effects of ozone on antioxidant responses in plants. *Free Radical Biology and Medicine* 23, 480–488.

Sinden, S., Deahl, K. and Aulenbach, B. (1976) Effect of glycoalkaloids and phenolics on potato flavor. *Journal of Food Science* 41, 520–523.

Singh, P., Agrawal, M. and Agrawal, S.B. (2009) Evaluation of physiological, growth and yield responses of a tropical oil crop (*Brassica campestris* L. var. Kranti) under ambient ozone pollution at varying NPK levels. *Environmental Pollution* 157, 871–880.

Smirnoff, N. (1996) The function and metabolism of ascorbic acid in plants. *Annals of Botany* 78, 661–669.

Smith, D., Roodick, J. and Jones, J. (1996) Potato glycoalkaloids: some unaswered questions. *Trends in Food Science & Technology* 7, 126–131.

Soja, G., Eid, M., Gangl, H. and Redl, H. (1997) Ozone sensitivity of grapevine (*Vitis vinifera* L.): evidence for a memory effect in a perennial crop plant? *Phyton (Austria)* 37, 265–270.

Soja, G., Reichenauer, T.G., Eid, M., Soja, A.M., Schaber, R. and Gangl, H. (2004) Long-term ozone exposure and ozone uptake of grapevines in open-top chambers. *Atmospheric Environment* 38, 2313–2321.

Speroni, J., Pell, E. and Weissberger, W. (1981) Glycoalkaloid levels in potato tubers and leaves after intermittent plant exposure to ozone. *American Potato Journal* 58, 407–414.

Spinelli, F., Cellini, A., Marchetti, L., Nagesh, K.M. and Piovene, C. (2011) Emission and function of volatile organic compounds in response to abiotic stress. In: Shanker, A. and Venkateswarlu, B. (eds) *Abiotic Stress in Plants - Mechanisms and Adaptations*. InTech, Rijeka, Croatia, pp. 367–394.

Stitt, M., Lunn, J. and Usadel, B. (2010) *Arabidopsis* and primary photosynthetic metabolism - more than the icing on the cake. *The Plant Journal for Cell and Molecular Biology* 61, 1067–1091.

Temple, P.J., Jones, T.E. and Lennox, R.W. (1990) Yield loss assessments for cultivars of broccoli, lettuce, and onion exposed to ozone. *Environmental Pollution* 66, 289–299.

Tosti, N., Pasqualini, S., Borgogni, A., Ederli, L., Falistocco, E., Crispi, S. and Paolocci, F. (2006) Gene expression profiles of O_3-treated *Arabidopsis* plants. *Plant, Cell & Environment* 29, 1686–1702.

Tripathi, M.K. and Mishra, A.S. (2007) Glucosinolates in animal nutrition: a review. *Animal Feed Science and Technology* 132, 1–27.

Tripathi, R. and Agrawal, S.B. (2012) Effects of ambient and elevated level of ozone on *Brassica campestris* L. with special reference to yield and oil quality parameters. *Ecotoxicology and Environmental Safety* 85, 1–12.

Tuomainen, J., Pellinen, R., Roy, S., Kiiskinen, M., Eloranta, T., Karjalainen, R. and Kangasjärvi, J. (1996) Ozone affects birch (*Betula pendula* Roth) phenylpropanoid, polyamine and active oxygen detoxifying pathways at biochemical and gene expression level. *Journal of Plant Physiology* 148, 179–188.

Tuzet, A., Perrier, A., Loubet, B. and Cellier, P. (2011) Modelling ozone deposition fluxes: the relative roles of deposition and detoxification processes. *Agricultural and Forest Meteorology* 151, 480–492.

UNEP (United Nations Environment Programme) (1999) *Global Environment Outlook*. Earthscan, London.

Van Hove, L.W., Bossen, M.E., San Gabino, B.G. and Sgreva, C. (2001) The ability of apoplastic ascorbate to protect poplar leaves against ambient ozone concentrations: a quantitative approach. *Environmental Pollution* 114, 371–382.

Vandermeiren, K. and Pleije, H. (2011) Effects of ozone on food and feed quality. In: Mills, G. and Harmens, H. (eds) *Ozone Pollution: A Hidden Threat to Food Security*. International Cooperative Programme on Effects of Air Pollution on Natural Vegetation and Crops, Environment Centre Wales, UK, pp. 53–60.

Vandermeiren, K., De Temmerman, L., Staquet, A. and Baeten, H. (1992) Effects of air filtration on spring wheat grown in open-top field chambers at a rural site. II. Effects on mineral partitioning, sulphur and nitrogen metabolism and on grain quality. *Environmental Pollution* 77, 7–14.

Vandermeiren, K., Black, C., Pleijel, H. and De Temmerman, L. (2005) Impact of rising tropospheric ozone on potato: effects on photosynthesis, growth, productivity and yield quality. *Plant, Cell & Environment* 28, 982–996.

Vandermeiren, K., De Bock, M., Horemans, N., Guisez, Y., Ceulemans, R. and De Temmerman, L. (2012) Ozone effects on yield quality of spring oilseed rape and broccoli. *Atmospheric Environment* 47, 76–83.

Vingarzan, R. (2004) A review of surface ozone background levels and trends. *Atmospheric Environment* 38, 3431–3442.

Vorne, V., Ojanperä, K., De Temmerman, L., Bindi, M., Högy, P., Jones, M., Lawson, T. and Persson, K. (2002) Effects of elevated carbon dioxide and ozone on potato tuber quality in the European multiple-site experiment 'CHIP-project'. *European Journal of Agronomy* 17, 369–381.

Waghorn, G.C. and Clark, D.A. (2004) Feeding value of pastures for ruminants. *New Zealand Veterinary Journal* 52, 320–331.

Wang, Y. and Frei, M. (2011) Stressed food – The impact of abiotic environmental stresses on crop quality. *Agriculture, Ecosystems & Environment* 141, 271–286.

Wang, Y., Yang, L., Han, Y., Zhu, J., Kobayashi, K., Tang, H. and Wang, Y. (2012) The impact of elevated tropospheric ozone on grain quality of hybrid rice: A free-air gas concentration enrichment (FACE) experiment. *Field Crops Research* 129, 81–89.

WHO (World Health Organization) (2005) *Air Quality Guidelines For Particulate Matter, Ozone, Nitrogen Dioxide and Sulfur Dioxide: Global update 2005 - Summary of risk assessment*. World Health Organization Press, Geneva, Switzerland.

Wilkinson, S., Mills, G., Illidge, R. and Davies, W.J. (2012) How is ozone pollution reducing our food supply? *Journal of Experimental Botany* 63, 527–536.

Wohlgemuth, H., Mittelstrass, K., Kschieschan, S., Bender, J., Weigel, HJ., Overmyer, K., Kangasjärvi, J., Sandermann, H. and Langebartels, C. (2002) Activation of an oxidative burst is a general feature of sensitive plants exposed to the air pollutant ozone. *Plant, Cell & Environment* 25, 717–726.

Zheng, F., Wang, X., Zhang, W., Hou, P., Lu, F., Du, K. and Sun, Z. (2013) Effects of elevated O_3 exposure on nutrient elements and quality of winter wheat and rice grain in Yangtze River Delta, China. *Environmental Pollution* 179, 19–26.

9 Potentiality of Ethylene in Sulfur-Mediated Counteracting Adverse Effects of Cadmium in Plants

Mohd Asgher, M. Iqbal R. Khan, Mehar Fatma and Nafees A. Khan*
Department of Botany, Aligarh Muslim University, India

Abstract

Plants are exposed to different kinds of stresses including both biotic and abiotic during the course of their lifetime. Among abiotic stresses heavy metal stress is a serious issue reducing crop productivity. Cadmium (Cd) is a highly toxic heavy metal, and occupies seventh place among the top 20 toxins mainly due to its negative influence on the biochemical systems of cells. This is considered as an extremely significant pollutant because of its higher toxicity and solubility in water. It is dispersed in the natural and agricultural environments mainly through anthropogenic activities and has a long biological half-life. It is a toxic pollutant for humans, animals and plants even at low doses. Cadmium gains entry into the environment as components of phosphate fertilizers and industrial waste disposal. Sulfur (S) plays a significant role in detoxification of Cd since it is a constituent of most of the defence compounds involved in Cd detoxification. Optimum S nutrition is helpful in reducing Cd translocation within the plants. Plants synthesize cysteine (Cys)-rich, metal-binding peptides, which include phytochelatins and metallothioneins, on exposure to the toxic doses of heavy metals. Detoxification of the heavy metal occurs through chelation and sequestration in the vacuole. In fact, Cd exposure induces the activity of enzymes involved in the sulfate reductive assimilation pathway and glutathione (GSH) biosynthesis. Glutathione has been considered as a marker for various stresses. Sulfur assimilation led to the synthesis of Cys and methionine (Met). Met is the precursor of ethylene, with 1-aminocyclopropane-1-carboxylic acid (ACC) being an intermediate in the conversion of Met to ethylene. Ethylene is the gaseous plant hormone and is now considered to regulate many plant developmental processes throughout the plants' life from germination to senescence but also mediate plants' responses to stresses. This chapter focuses on the interactive role of ethylene, S, antioxidants system and tolerance of cadmium in plants.

9.1 Introduction

Cadmium (Cd), considered as one of the most important environmental pollutants, is a toxic heavy metal. It can bind to free sulfhydryl (SH) residues and interfere with homoeostasis leading to the displacement of essential elements such as zinc (Zn), iron (Fe) and calcium (Ca) from proteins, all of which in turn might trigger oxidative injuries and finally inhibit growth and development of plants (DalCorso et al., 2008). Since Cd can also be a potential health risk to humans, it becomes necessary to limit the entry of Cd from soil into the food chain. In plants, Cd toxicity stimulates the formation of reactive oxygen

*E-mail: naf9@lycos.com

species (ROS), which in excessive amounts damage photosynthetic machinery and inhibit photosynthesis (Rodríguez-Serrano et al., 2009). In order to protect plants from Cd toxicity plants are endowed with antioxidant machinery, that minimizes the ROS produced during oxidative stress. Sulfur (S), one of the least abundant essential macronutrients in plants, is taken up from the soil mainly as sulfate by roots. It is crucial for the survival of plants under different types of stresses. Sulfur is a component of certain amino acids, metal clusters, and a diverse range of primary and secondary metabolites, such as glutathione (GSH), phytochelatins (PCs) and glucosinolates that have a role in protection of plants from oxidative as well as other environmental stresses (Rausch and Wachter, 2005). Ethylene as a plant hormone influences many aspects of plant growth and photosynthesis (Pierik et al., 2006; Acharya and Assmann, 2009). The ethylene response also depends on the availability of mineral nutrients. One of the most common and well-studied responses of plants to various environmental stresses is the enhanced production of ethylene. Ethylene production of plants was shown to be raised by exposure to various physico-chemical stresses (Abeles et al., 1992; Morgan and Drew, 1997). The present chapter focuses on the interactive role of ethylene, S and the antioxidants system in minimizing Cd-induced oxidative stress.

9.2 Cadmium Stress in Plants: a General Aspect

9.2.1 Availability of cadmium in soil

Cadmium availability to plants is greater in acid soils and its solubility is increased by root exudates (Zhu et al., 1999). Cadmium is present in the soil solution predominantly as Cd but also as Cd-chelates (Tudoreanu and Phillips, 2004). In acid soils mobility and availability of Cd is much higher than in calcareous, neutral and slightly alkaline soil; it was demonstrated that increase in redox potential (Eh) leads to a decrease in exchangeable Cd along with an increase in its reducible form. Cadmium availability to plants is dependent on the pH and ionic strength of the soil medium (Sajidu et al., 2006).

The rhizosphere is an extremely important region, which acts as an interface connecting plant roots and soil. During plant growth and metabolism, roots release organic substances to the rhizosphere which also controls the entry of water, nutrients and other chemical compounds that may be beneficial or harmful to plants. Root exudates of plants can change the chemical and some physical properties of the rhizosphere, and this in turn affects Cd absorption. This influence of root exudates on bioavailability and toxicity of Cd may include modification of the pH and Eh of the rhizosphere, formation of chelates with Cd ions, and also alteration of community construction as well as of rhizospheric microbe population and activities. Soil pH, organic matter and clay content, presence of other ions, root exudates, types and cultivars of crop plants affect bioavailability of Cd in soil, and hence crop uptake, of which pH is perhaps the most important. The bioavailability in Cd-contaminated acidic soils is substantially higher in comparison to neutral and alkaline soils (Sarwar et al., 2010). Cadmium availability can be manipulated through the use of various amendments, such as lime (Tlustos et al., 2006) in acidic soil, the reason for this being the increase of pH is expected to reduce Cd availability.

9.2.2 Uptake, accumulation and transportation of cadmium

The process of metal uptake by plant roots from soil is extremely complex and involves transfer of metals from the soil to the surface of roots and further to the inside of root cells. As soon as Cd enters the roots, it can reach xylem through apoplastic and or symplastic pathways. Plants have developed a range of mechanisms to absorb metal from the soil and transport it to the aerial parts (Benavides et al., 2005).

The uptake rate of heavy metals depends on the pH value of the soil solution, the organic matter content in the soil and the concentrations of other ions. At higher pH values, the solubility of Cd salts in the soil solution is reduced due to the formation of low soluble compounds and, as a result, the biological availability of soil Cd decreases (Salt et al., 1995). At lower concentration (2.5–90 nM), it transports across the membrane in an active, energy requiring H$^+$ ATPase mediated process whereas at high concentration of Cd the uptake is a non-metabolic (passive) process, involving diffusion coupled with sequestration. After uptake by the plant roots, Cd becomes accumulated in cytosol, cytosolic organelles and vacuoles, which is a major compartment of Cd^{2+} accumulation in plants. At low level of exposure it forms a complex in the cytosol with GSH, whereas at high Cd exposure levels, it is transported into the vacuoles, where it forms complexes with organic acids and PCs (Grant et al., 1998). In the presence of a substantial concentration of Cd salts in the soil, Cd uptake by the plant increases proportionally to increasing soil Cd. In addition, Cd might enter root cells as Cd-chelates through YSL (Yellow-Stripe 1-Like) proteins (Curie et al., 2009). Cadmium can then reach the stele through a symplasmic pathway formed by the cytoplasms of individual root cells connected by plasmodesmata. The Cd species transported through the symplasm are unknown, but could include Cd or Cd-chelates (Verbruggen et al., 2009). Root to shoot translocation of Cd is generally driven by transpiration and occurs through the xylem (Ueno et al., 2008). This indicates that Cd transfer from the root medium to the xylem in the hyperaccumulator *Arabidopsis halleri* was an energy-dependent process. However, the relative contribution of the symplastic and apoplastic pathways to the delivery of cations to xylem is still little known (White, 2001). It was observed that in the cytoplasm, Cd-rich vesicular structures were formed during its accumulation; further, uptake rates into epidermal storage cells were higher than into standardized epidermal and mesophyll cells (Leitenmaier and Küpper, 2011).

Uptake of Cd by roots and its transport to the different organs of plants, even at low concentration exposure, result in an inhibitory effect on content of mineral nutrients and homoeostasis, shoot and root growth and development (Macek et al., 2002; Farinati et al., 2010). In higher plants Cd uptake depends on several factors, including availability and concentration in soil, or water, and to some extent in the atmosphere (Clemens, 2006). Cadmium absorption can be influenced by the soil concentration of other mineral elements such as Ca, Zn and Fe. Addition of Ca or Zn reduces uptake of Cd (Cosio et al., 2005). The apoplastic transport of heavy metal occurs through the intercellular spaces and cell walls, and symplastic through protoplasts and via plasmodesmata without participation of any organelles (Redjala et al., 2009).

Metal cation homoeostasis is necessary for plant nutrition and resistance to toxic heavy metals. Metal cation transporters are essential in various steps of plant nutrition since these transport proteins mediate uptake of metal by root cells and transfer of metals between cells and organs. They are also involved in metal detoxification by mediating the transport of metal chelates or metal cations from the cytosol to the vacuolar compartment (Rea et al., 1998). Eight family members have been characterized and identified for P-type ATPase. In *Arabidopsis*, HMA1–HMA4 and HMA5–HMA8 were predicted to transport Zn/Cd/lead (Pb)/cobalt (Co) and copper (Cu)/silver (Ag), respectively. HMA2 and HMA4 are plasma membrane localized heavy metal exporters of Zn and Cd, which are involved in heavy metal tolerance and metal hyperaccumulation (Mills et al., 2003; Hussain et al., 2004; Courbot et al., 2007).

Cadmium is probably exported from the vacuole by NRAMP (natural resistance-associated macrophage protein) transporters, such as orthologues of AtNRAMP3 and AtNRAMP4 responsible for Cd efflux from the vacuole (Verbruggen et al., 2009). In *A. halleri*, *HMA4* is a key gene conferring a remarkable ability in this species to hyperaccumulate and hypertolerate Cd; to date, it is the only gene for which there is genetic evidence for a role in both Zn and Cd tolerance (Courbot et al., 2007; Hanikenne et al., 2008). Another member of this gene family, *AtHMA3*, is required for the sequestration of Cd into vacuoles so as to limit transport to xylem (Morel et al., 2009). Zrt-Irt-like proteins (ZIP) transporters probably control the influx of metal ions into cells, and may also be involved in Zn homoeostasis (Colangelo and Guerinot, 2006). Miyadate et al. (2011)

showed that in rice, *OsHMA3*, located at qCdT7, was responsible for controlling rate of translocation of Cd from roots to shoots, and the mechanism was shown to be by sequestering Cd into the vacuoles of root cells. Bhuiyan *et al.* (2011) observed that a *yeast cadmium factor 1* (*YCF1*) gene over-expression in *Brassica juncea* conferred enhanced tolerance to Cd and Pb stress in all transgenic lines, indicating that *YCF1* plays an active role in translocation of heavy metals conjugated to glutamine synthetase (GS) from cytoplasm to vacuoles to sequester Cd and Pb as Cd-GS or Pb-GS in the vacuole (Song *et al.*, 2003).

Transporters play significant roles in selective import or removal of molecules across biological membranes. Involvement of several families of membrane transport proteins in metal homoeostasis has been established to date, which includes cation diffusion facilitators (CDF), ZIP, cation exchangers (CAX), Cu transporters (COPT), heavy-metal P-type ATPases (HMA), NRAMP and the ATP-binding cassette (ABC) transporters (Williams *et al.*, 2000; Maser *et al.*, 2001; Cobbett *et al.*, 2003; Hall and Williams, 2003). The *Arabidopsis thaliana* genome is capable of encoding more than 120 ABC proteins. ABC proteins were originally identified in plants as transporters involved in the final detoxification process, i.e. vacuolar deposition. These proteins have multiple functions, since they are involved in different plant processes such as organ growth, nutrition and development, and also in abiotic stress responses, as well as during plant–environment interaction. Kim *et al.* (2006) reported that over-expression of AtATM3 (*Arabidopsis thaliana* ABC transporter) leads to greater Cd and Pb tolerance, whereas the gene knockout make the plants more susceptible to heavy metals. Similarly, in yeast, heavy metal ATPase of *Thlaspi* (TcHMA4) increased their efflux and thereby increased tolerance to these metals (Papoyan and Kochian, 2004). Cadmium tolerance is conferred in plants by members of the ABC transporter family, which have been shown in *Arabidopsis* to include MRP3 (multidrug-resistance related protein, Kolukisaoglu *et al.*, 2002), ATM3 (ABC transporter of the mitochondria, Kim *et al.*, 2006) and PDR8 (pleiotropic drug resistance, Kim *et al.*, 2007). It has also been shown that in *Arabidopsis*, AtOS1, a member of the Abc1 family which is localized in the chloroplasts, is involved in Cd-induced oxidative stress signalling. AtOS1 does not transport Cd but seems to have a crucial role in tolerance, possibly through a putative kinase activity (Jasinski *et al.*, 2008). YCF1 is an ABC transporter that confers Cd tolerance into the vacuole through the Mn and Ca transport of Cd conjugates (Szczypka *et al.*, 1994; Li *et al.*, 1997). Oda *et al.* (2011) reported that *OsABCG43* is a Cd inducible-transporter gene capable of conferring Cd tolerance in rice. Enhanced Cd tolerance was also induced in *Arabidopsis* by over-expression of AtABCC1 and could be a good candidate for increasing the storage capacity of vacuoles (Park *et al.*, 2012). Since many transporters are involved in conferring heavy metal tolerance, the identified genes can be potential targets for genetic engineering of plants with increased tolerance to heavy metals and capacity of accumulation, which would be desirable traits in phytoremediation. A list of genes involved in Cd tolerance has been given in Table 9.1.

9.2.3 Cadmium toxicity in plants

Effect on plants: morphological aspects

Though Cd at high concentrations inhibits growth and development of plants, it may stimulate growth and development at low concentrations depending on the specific plant species (Wahid and Ghani, 2008). Plants grown in high Cd contaminated soil show visible symptoms of injury as chlorosis, caused by reduction in chloroplast number per cell as well as a change in cell size, growth inhibition, browning of root tips and finally death (Baryla *et al.*, 2001; Dai *et al.*, 2006). Cadmium causes various phytotoxic symptoms including inhibition of growth, leaf chlorosis and root putrescence (Skorzynska-Polit *et al.*, 2010; Valentovicova *et al.*, 2010). Increasing concentration of Cd significantly reduced number of leaves as well as leaf area, negatively affected photosynthetic carbon fixation and consequently reduced dry matter accumulation in plants (Sharma *et al.*, 2010).

Table 9.1. Genes involved in Cd stress tolerance.

Gene for Cd tolerance	Plants	Role	References
PCS	Brassica juncea	Increased phytochelation protects plant from heavy metal toxicity	Shanmugaraj et al. (2013)
HvPCS	Hordeum vulgare	Expression of gene increased in the Cd presence	Kaznina et al. (2012)
OsPDR5/ABCG43	Oryza sativa	Cellular Cd tolerance	Oda et al. (2011)
YCF1	B. juncea	Enhanced tolerance to Cd over-expression conferred and Pb stress in all transgenic lines	Bhuiyan et al. (2011)
GmOASTL4	Nicotiana tabacum	Increased cysteine levels and enhanced Cd tolerance	Ning et al. (2010)
PrP_4A	Pisum sativum	PRs and defence-related proteins could protect against Cd toxicity	Rodríguez-Serrano et al. (2009)
TaTM20	Saccharomyces cerevisiae	Decrease in intracellular content of Cd	Kim et al. (2008)
AtPCS1	B. juncea	Phytochelation synthesis	Gasic and Korban (2007)
ERF1 and ERF2	Arabidopsis thaliana	Able to bind several pathogenesis-related promoters and dehydration responses element and regulates ERF protein expression	Weber et al. (2006)
CAD/RMl1	A. thaliana	glu-cys synthatase/GSH biosynthesis	Vernoux et al. (2001)
AtCys-3A	A. thaliana	Over-expression of gene provides Cd tolerance	Domínguez-Solís et al. (2001)
TaPCS1	S. cerevisiae	Expression makes yeast cells more Cd tolerant	Clemens et al. (1999)
Glutamyl cysteine synthase (gsh1)	B. juncea	Increased Cd tolerance and higher concentration of PCs, Glu, Cys, GSH	Zhu et al. (1999)

Excess Cd in cowpea led to development of marginal chlorosis on young and middle leaves; later, affected leaves turned yellow and necrotic, dried and collapsed, and no pods were produced at highest level of Cd due to toxicity of the elements (Dube et al., 2003). It has been shown that heavy metals affect several parameters of growth in higher plants including biomass (Arun et al., 2005). The Cd toxicity effects are evident in terms of injury symptoms on the above-ground parts, reduced growth and yield. Plant will show visual Cd toxicity symptoms on exposure to Cd stress; the most common symptoms are characterized by brown and short roots, chlorosis, few tillers and reduced biomass and senescence (Cosio et al., 2006). Since changes in morphological traits are correlated to heavy metals these may be considered as suitable bio-indicators of heavy metal pollution and may also be used for classifying the species as sensitive or tolerant to the different heavy metals.

Effect on plants: physiological and biochemical aspects

In higher plants, heavy metals inhibit biosynthesis of chlorophylls and accessory pigments, which in turn lead to inhibition of photosynthesis. Activities of carbonic anhydrase and ribulose 1,5 bisphosphate carboxylase oxygenase (Rubisco) was shown to be reduced by exposure to Cd, and the mechanism of photosynthetic response involved was both stomatal and non-stomatal (Mobin and Khan, 2007). Decrease in Rubisco activity is accompanied by a declining level of soluble sugars, suggesting that under Cd stress sugar synthesis is reduced, relative to CO_2 fixation capacity of Rubisco (Leitao et al., 2003; Afef et al., 2011). Cadmium can also inhibit efficiency of PSII activity and may induce a photoinhibitory effect in in vitro conditions in isolated thylakoid membranes (Pagliano et al., 2006). Cadmium inhibition of electron transport at PSI may induce inhibition of PSI electron transport rate besides its effect on PSII activity (Neelam and Rai, 2003).

Cadmium concentrations gradually inhibit electron transport altogether from water-splitting system up to PSI, which indicates that Cd has multiple inhibitory sites on the photosynthetic apparatus, affecting both PSII and PSI activities (Perreault et al., 2011). Chlorophyll content in the leaves was significantly decreased by Cd in the soil, which could be mediated through the reduced uptake of magnesium (Mg), an integral part of the chlorophyll molecule due to Cd toxicity (Shamsi et al., 2007). In mitochondria, Cd inhibits transport of electrons and protons and disorganizes the electron transport chain and disrupts activities of glycolytic and pentose phosphate pathway enzymes (Seregin and Ivanov, 2001). Faller et al. (2005) showed that photoactivation of PSII was inhibited by Cd^{2+} due to competitive binding to the essential Ca binding site. Chlorophyll fluorescence also decreases due to excess Cd in the environment affecting chloroplast function or CO_2 fixation (Iqbal et al., 2010). High amounts of Cd can be taken up and accumulated by plants leading to disturbed physiological metabolisms in plants such as in transpiration, photosynthesis, respiration and nitrogen (N) assimilation (Zhou et al., 2006; Wang et al., 2008; Gill et al., 2012). Further, it has been reported that high concentrations of heavy metals can inhibit activities of photosynthetic enzymes and block the photosynthetic electron transport chain with a reduction in chlorophyll and carotenoid content (Thapar et al., 2008).

Exposure of plants to Cd decreased the total chlorophyll, amino acid and increased the proline and ascorbic acid slightly. Increased Cd concentrations significantly enhanced proline accumulation and decreased N metabolism in Brassica juncea (Asgher et al., 2013). Accumulation of free amino acids is one of the common responses of plants to many abiotic stresses. Environmental stresses accumulate proline in large amounts; this is followed by other amino acids such as those derived from aspartic acid, including asparagine, isoleucine, leucine, valine and methionine (Vassilev and Lidon, 2011). In Cd-stressed plants, the significance of proline lies in its contribution to water balance maintenance, protection of enzymes and biomolecules and also, in many cases, detoxification of ROS. Proline-mediated alleviation of drought could contribute to Cd tolerance. Formation of ROS such as superoxide anion and hydrogen peroxide in mitochondria, chloroplast and peroxisomes in plants under Cd exposure is an indicator that Cd induces oxidative stress. In plants, mitochondria produce ROS at high rates, a more reduced electron transport chain produces more ROS with the respiratory complexes I and III being the main sites of the production (Bartoli et al., 2004). Heavy metal exposure causes lipid peroxidation; excessive amounts of Cd may also cause decreased uptake of nutrient elements, inhibition of various enzyme activities and induction of oxidative stress including changes in the activities of enzymes involved in antioxidant defence system (Sandalio et al., 2001).

Nitrogen is extremely important in plant growth and yield as it is a component of amino acids, proteins, nucleic acids and other cell constituents which are essential for the plant. Accumulation of toxic Cd^{2+} in seedlings was shown to significantly modify activities of enzymes involved in primary N assimilation such as GS, glutamate synthase (GOGAT), nitrate reductase (NR) and nitrite reductase (NiR) in both root and shoot of the seedlings (El-Shora and Ali, 2011). Cadmium treatment influences the activity behaviour of key enzymes of various metabolic pathways, which in most cases was due to interaction of Cd with enzyme (-SH) groups (Seregin and Ivanov, 2001). Plant roots often serve as storage sites preventing toxic dosages from reaching the stem and grain (Grifferty and Barrington, 2000). Cadmium significantly reduces the normal H:K exchange and the activity of plasma membrane ATPase (Obata et al., 1996), and also strongly affects the activity of several enzymes, such as glucose-6-phosphate dehydrogenase, glutamate dehydrogenase, malic enzyme, isocitrate dehydrogenase (Van Assche and Clijsters, 1990; Mattioni et al., 1997), rubisco and carbonic anhydrase (Mobin and Khan, 2007). Cadmium accumulated in vacuoles and apoplasts also plays an important role in scavenging of free radicals produced in plant cells (Sridhar et al., 2005).

Effect of cadmium on plants: molecular aspects

The Cd-induced cell death may involve apoptotic features, such as condensation of chromatin, nuclear DNA cleavage into oligonucleosomes and formation of apoptotic bodies. Berboodi and Samadi (2004) reported that a necrotic process was started when plants were exposed to high Cd concentrations leading to cell lysis and leakage, condensation of chromatin, marginal and condensed nucleus, fragmentation of DNA and formation of apoptotic bodies, all of which are indicators of apoptotic death. Cadmium can also disrupt assembly–disassembly of microtubule, leading to an alteration in cell cycle and division (Fusconi et al., 2006). In root-tip cells of the plant, Cd damages nucleoli and, in growing plants, it alters the synthesis of RNA and inhibits ribonuclease activity (Shah and Dubey, 1995). While a very high level of DNA damage was induced in roots by Cd, in the leaves the damage was less; however, continuous Cd treatments finally caused an increase in leaf DNA damage through ROS production. This increase in DNA damage of leaves could be associated with necrotic and apoptotic DNA fragmentation, since in these plants growth was inhibited with leaves becoming distorted and yellowish (Gichner et al., 2008). Cai and Cherian (2003) reported that Cd negatively affects structural integrity of DNA.

9.2.4 Reactive oxygen species generation and signalling under cadmium stress

Cadmium, like many other abiotic and biotic stress factors, evokes oxidative stress in plants, which occurs when there is an imbalance in any cell compartment between production of ROS and antioxidant defence, leading to damage. ROS are produced as by-products of some biochemical processes and some of them play a role in signal transduction. Plants under normal conditions have adapted antioxidative defence to maintain the redox equilibrium. The main ROS producer in green plant parts in the light are chloroplasts and peroxisomes while in non-green plant parts ROS production takes place in mitochondria. Asada (2006) demonstrated in chloroplast thylakoids PSI and PSII are the main sites of ROS production.

Involvement of mitogen-activated protein kinases (MAPKs) in abiotic responses in plants have been well established. This signalling module involved transduction of extracellular signals to the nucleus for appropriate cellular adjustment. The pathway is evolutionarily conserved among eukaryotic organisms. In plants, a novel MAPK gene *OsMSRMK2* from japonica-type rice was activated by higher amounts of mercury (Hg), Cu and Cd ions (Agrawal et al., 2002). It has been recently reported that stress due to excessive Cd and Cu activates different kinase enzymes belonging to the MAPK family (Jonak et al., 2004). Cadmium-induced MAPKinase activity was dependent on NADPH oxidase activity and functional state of mitochondria and Cd ion may cause mitochondrial dysfunction (Yeh et al., 2007). In *A. thaliana* MPK3 and MPK6 was activated by Cd in a ROS-dependent manner. Interestingly, Cd elicited much higher levels of MPK3 and MPK6 activation in roots than in leaves, resulting in a cellular response to overcome the heavy metal toxicity (Liu et al., 2010). Further, it is demonstrated in *Zea mays* that ZmMPK3 transcript levels are induced after exposure to high concentrations of Cd (Wang et al., 2010). Once it enters the plant cell, Cd induces ROS, which activates the MAPK cascade. After activation, the MAPK module is translocated into the nucleus or cytoplasm to trigger the cellular responses through phosphorylation of downstream proteins (Fiil et al., 2009; Nadarajah and Sidek, 2010).

Exposure to excess Cu or Cd ions resulted in a complex activation pattern of four distinct MAPKs: SIMK, MMK2, MMK3 and SAMK (stress-activated MAPK) (Jonak et al., 2004). This indicates that MAPK cascades involved in signalling are activated by different heavy metals. The presences of an elevated level of heavy metal ions initiates a broad range of cellular responses.

9.3 Regulation of Sulfur Uptake and Assimilation

Sulfur is a macronutrient which is essential for growth and development of plants. In nature, sulfate is the most abundant form of S and is also the main source of S for plants; within the plant, sulfate is assimilated into cysteine. This amino acid (Cys) is an essential component of proteins, besides which it also plays important roles as precursor for several essential biomolecules such as vitamins and co-factors (Droux, 2004; Wirtz and Droux, 2005), many defence compounds and antioxidants like GSH, which is an essential determinant of cellular redox homoeostasis (Rausch and Wachter, 2005). In general, Cd has a high affinity to metabolic processes of the S metabolism, and its first effects are on ATP-sulfurylase (ATPS; De Knecht et al., 1995) and adenosine 5'-phosphosulfate sulfotransferase (Nussbaum et al., 1988). Sulfate is taken up by transport systems in the root and distributed by sulfate transporters in the whole plant (Buchner et al., 2004). Sulfur is obtained by plants from the soil in the form of sulfate in addition to sulfur dioxide and hydrogen sulfide (H_2S) from the atmosphere and to a minor extent by leaves via stomata through a step in which S gets converted into sulfide and Cys. Sulfate transporters facilitate the uptake of sulfate from soil during the S assimilation pathway (Davidian and Kopriva, 2010).

Assimilation of sulfate is highly regulated in a demand-driven manner (Kopriva and Rennenberg, 2004; Kopriva, 2006). ATPS has been assumed as the rate-limiting step enabling and initiating S metabolism. A high ATPS activity is expected to provide tolerance to plants against stress. ATP-sulfurylase expression and activity is weakly induced upon S depletion but repressed through GSH (Teuveny and Filner, 1997). In higher plants, uptake and assimilation of S are crucial factors in determining crop yield, quality, and also during resistance to different types of stresses. Plastids are the sites of sulfate reduction and Cys synthesis occur in the plastids as also in the mitochondria and cytosol. ATPS catalyses sulfate activation by ATP to adenosine 5'-phosphosulfate (APS), which is the first step in the pathway. In plants and the majority of bacteria, APS reductase (APR) reduces sulfate to sulfite (Kopriva et al., 2002). Sulfite is further reduced to sulfide by a ferredoxin-dependent sulfite reductase and sulfide incorporated into the amino acid skeleton of O-acetyl-l-Ser (OAS) by OAS (thiol) lyase, forming Cys (Leustek et al., 2000). Serine acetyl transferase (SAT), which catalyses the formation of O-acetyl-L Ser (OAS) from L-Ser and acetyl-CoA, links the Ser metabolism to Cys biosynthesis (Saito, 2004). Subsequently, Cys is formed by the condensation of sulfide and OAS, catalysed by Cys synthase (OAS thiol lyase). An overview of assimilation of S in the organelle involved and detoxification of Cd in the vacuole has been depicted in Fig. 9.1. In *Arabidopsis* five genes encoding SAT are localized in plastid (*SAT1*), mitochondria (*SAT3*) and cytosol (*SAT2*, *SAT4*, *SAT5*) (Kawashima et al., 2005). Serine acetyl transferase (*SAT2* and *SAT4*) differ from other SATS in their amino acid sequence composition and are less expressed (Kawashima et al., 2005). Exposure of plants to Cd induces the activity of enzymes involved in the sulfate assimilation pathway (Herbette et al., 2006; Khan et al., 2007). Expression of APSR, SiR and OASS is increased in Cd-treated Indian mustard (*Brassica juncea*) (Minglin et al., 2005; Alvarez et al., 2009). ATP-sulfurylase and SAT play important roles in heavy metal tolerance and accumulation (Hawkesford, 2003; Freeman et al., 2004).

9.4 Detoxification Mechanisms for Cadmium Tolerance in Plants

Different plant species have developed suitable and specific mechanisms of heavy metal detoxification which help them to survive. Some of these are: exclusion, chelation and/or compartmentalization of the metal ions; expression of more general stress response mechanisms such as ethylene and stress proteins; enhancing antioxidants systems; and mineral nutrients. A brief account of the

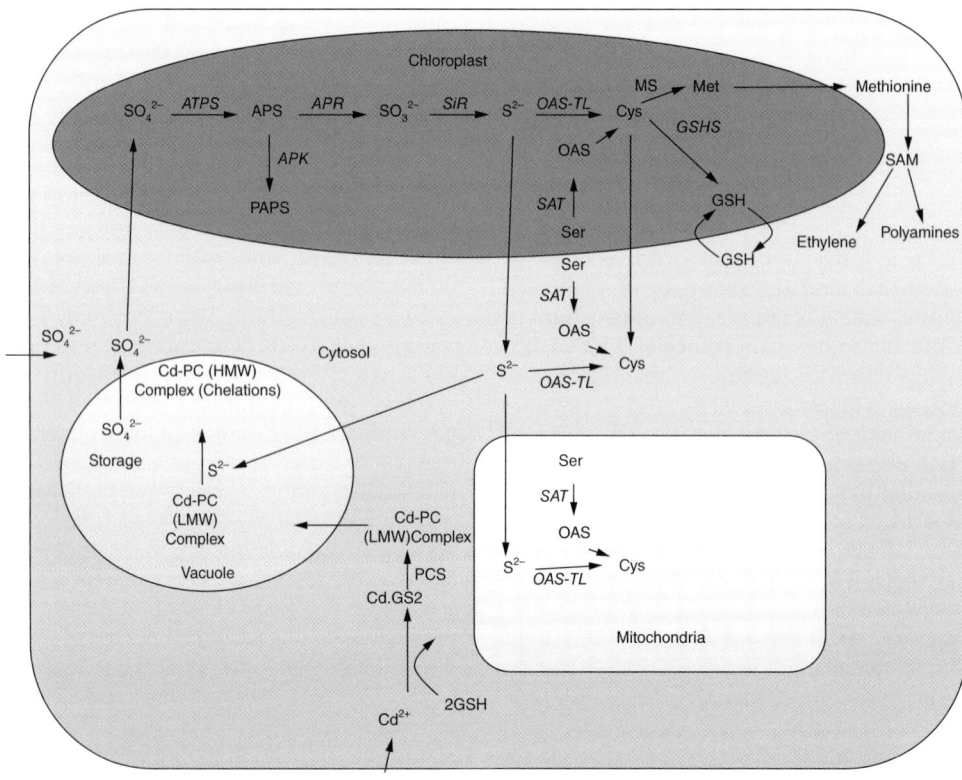

Fig. 9.1. An overview of sulfur assimilatory pathway compartmentalization and detoxification of cadmium in plants. APK, APS kinase; APR, APS reductase; APS, adenosine 5-phosphosulfate; ATPS, ATP sulfurylase; Cd, cadmium; Cys, cysteine; GSH, reduced glutathione; GSHS, glutathione synthetase; HMW, high molecular weight; LMW, low molecular weight; MS, methionine synthase; OAS, O-acetyl serine; OAS-TL, OAS (thiol)lyase; PAPS, 3′-phospho-5′adenylylsulfate; PC, phytochelatin; PCS, phytochelatin synthase; S^{2-}, sulfide; SAM, S-adenosyl methionine; SAT, serine acetyltransferase; Ser, serine; SiR, sulfite reductase; SO_3^{2-}, sulfite; SO_4^{2-}, sulfate; Cd-GS, glutathionatocadmium

detoxifying substances involved against Cd toxicity is explained below.

9.4.1 Role of antioxidants

Antioxidative enzymes play important roles in adaptation and survival of plants during periods of stress. Plants have adapted various strategies to overcome stresses by the activation of various enzymatic and non-enzymatic scavenging systems to mitigate ROS, thus protecting cells from oxidative damage (Sairam and Tyagi, 2004). Enzymatic antioxidants include superoxide dismutase (SOD), catalase (CAT), ascorbate peroxidase (APX) and glutathione reductase (GR), which have a potential to alleviate deleterious effects resulting from cellular oxidative stress by these oxygen radical detoxifying enzymes (Mishra et al., 2006). The coordinate functioning of the different enzymes is essential for maintenance of steady-state level of ROS.

Superoxide dismutase is a key enzyme involved in the regulation of intracellular concentrations of ROS. Superoxide is rapidly converted to H_2O_2 by the activity of SOD in the different compartments of plant cells (Bowler et al., 1992). Increase in SOD activity during oxidative stress has been observed by several researchers (Scebba et al., 2006; Mobin and Khan, 2007). Thus, increased SOD activity in plant cells showed that it plays a positive role in controlling the cellular level of these ROS and repairing oxidative damage (Miller et al., 2008). Lee et al. (2007) reported that Cu/Zn-SOD and APX gene expressions in transgenic fescue plants showed tolerance to methyl viologen as well as heavy metal stress.

Catalase is another antioxidant enzyme that eliminates H_2O_2 by converting it into O_2 and H_2O and is indispensable for ROS detoxification during stress (Miller et al., 2008). Hence, CAT activity probably plays an important role in protection against Cd-induced oxidative damage (Mobin and Khan, 2007; Zhang et al., 2009). Over-expressing of *BjCAT3*, a CAT gene in tobacco, could enhance the tolerance under Cd stress as the seedlings of tobacco plants respond positively by showing longer root length (Guan et al., 2009). Cadmium at 50 µM causes a rapid inactivation of both iso-enzymes of APX, one of which is thylakoid-bound and the other stromal. Inactivation of APX could be due to decreased ascorbate concentration; this has been supported by *in vitro* treatment with exogenous ascorbate and study of APX kinetic properties under Cd stress (Liu et al., 2008). Activity of APX by an H_2O_2-scavenger that belongs to the ascorbate-glutathione (AsA-GSH) cycle was inhibited at all Cd concentrations tested. The reduction in APX activity may be due to GSH depletion and a subsequent reduction in the AsA-GSH cycle (Goncalves et al., 2007).

Glutathione reductase is extremely important in a cell's defence against ROS under various abiotic stresses including heavy metals. Glutathione reductase maintains homoeostasis of GSH and GSSG crucial for signalling stress response proteins and regulating oxidative stress (Szalai et al., 2009). GR appears to play a significant role in detoxification of Cd-induced ROS, possibly via the AsA-GSH cycle. Nouairi et al. (2009) reported that in *Brassica napus* leaves after 15 days of treatment with Cd, GR activity increased significantly up to 10 µM Cd and then declined at higher concentrations. Gill et al. (2011) reported that in *B. juncea* significant increase in activities of SOD, CAT, APX and GR was noted on exposure to Cd stress, irrespective of the cultivars. Higher increase of SOD, CAT, APX and GR activity resulted in efficient scavenging of Cd-induced ROS. Interestingly, the increased activity of GR protected the plants from ROS by maintaining the high ratio of GSH/GSSG, which is necessary for ascorbate regeneration as well as for the activation of several CO_2-fixing enzymes (Gill and Tuteja, 2010).

Non-enzymatic antioxidants are essentially composed of relatively high concentrations of GSH, ascorbate and α-tocopherol (Bowler et al., 1992). Other important non-enzymatic antioxidants include Cys, mannitol, vitamin E, anthocyanin, some alkaloids and carotene.

Glutathione, an important store of reduced S, is a major form of transported reduced S, and is involved in resistance to different stresses. It has a specific role in maintaining a cellular redox status (Kopriva and Koprivova, 2005). Glutathione contains one free–SH group (also known as thiol), which is received from Cys, whereas in GSSG this group is connected by a disulfide bridge (Wonisch and Schaur, 2001). The tripeptide GSH (α-glutamylcysteinylglycine) is one of the important antioxidants in plants and is involved in plant metabolism and plant defence, primarily through stress signalling and defence gene expression. It also plays a role in the detoxification of ROS, xenobiotics, herbicides and heavy metals such as Cd (Nakamura et al., 2013) and protects proteins from oxidation through a process called glutathionylation. Its main role in the antioxidative defence is its ability to produce ascorbic acid, a powerful water-soluble antioxidant through the AsA-GSH cycle. A significant increase in the messenger RNA level of genes involved in GSH synthesis (*GSH1* and *GSH2*) during Cd stress was noticed in *Arabidopsis* (Semane et al., 2007). Other functions for GSH include the formation of PCs, which have an affinity for heavy metals and are transported as complexes into the vacuole, thus allowing the plants to have some level of resistance to heavy metals (Sharma and Dietz, 2006). Environmental stresses cause increased ROS levels, and GSH response can be crucial for adaptive responses. Glutathione S-transferases have also been reported to be essential in the detoxification of xenobiotics, toxins and catabolic products (Marrs, 1996; Leustek et al., 2000). The reduction of GSSG to GSH by using NADPH as an electron donor thus regenerates the GSH and is mediated by GR (Chalapathi and Reddy, 2008). In transgenic plants, it was shown that GR plays an important protective role to oxidative stress caused by photoinhibition and GR activity increased dramatically in response to ethylene or salicylic acid (SA), and was reduced by H_2O_2 or paraquat (Choi et al., 2004).

In plants, biosynthesis of Cys plays a vital role in the S cycle, as the inorganic S taken up from the soil is reduced to Cys, the first reduced S-containing compounds. In *A. thaliana* plants, several OASTLs and SAT-encoding genes have been identified by sequence homology and functionality (Alvarez et al., 2010). Most of the accumulation of cysteine occurs in cytosol by the action of the major cytosolic OASTL, encoded by OAS-A1 (Lopez-Martin et al., 2008). Cysteine synthesis depends therefore on OAS-TL activity, S availability and OAS supplied by SAT (Leustek et al., 2000). Plants contain a large number of nuclear genes that encode cytosolic, plastid and mitochondrial isoforms of SAT and OAS-TL. At least three cDNAs for each enzyme have been isolated from *A. thaliana*, the most extensively studied plant in relation to S metabolism. Cysteine is synthesized at the final step of sulfate assimilation pathway. Various detoxification processes activated following heavy metal exposure result in increased synthesis of S-containing compounds such as Cys, which play significant roles in tolerance and survival of plants (Rausch and Wachter, 2005). Domínguez-Solís et al. (2004) reported that increased rate of Cys biosynthesis could be correlated to enhanced Cd tolerance.

Cysteine can function directly during protein synthesis and that of GSH, or serve as a donor of reduced S for synthesis of Met and various co-enzymes (Davidian and Kopriva, 2010; Takahashi et al., 2011). Over-expression of GmOASTL4 in tobacco increased the Cys levels and elevated Cd stress tolerance (Ning et al., 2010). It has been found in various plants that over-expression of genes conferred enhanced tolerance to Cd stress (Table 9.1). Transgenic tobacco plants which showed significantly enhanced tolerance for Cd, selenium and nickel supplemented in an agar medium also over-expressed Cys synthase in both the cytosol and the chloroplast (Kawashima et al., 2004).

In plants tocopherols have numerous functions such as preventing oxidative stress and protecting fatty acids in membranes from oxidation, and are one of the most effective single-oxygen quenchers. *VTE1* gene-encoded tocopherol cyclase (VTE1) acts as a catalyst in the synthesis of tocopherol (Liu et al., 2008). Over-expressing VTE1 from *Arabidopsis* in transgenic lines of tobacco showed decreased H_2O_2 content lipid peroxidation and electrolyte leakage in comparison with the wild type. Thus, they concluded that increase in vitamin E is due to expression of VTE1 in plants and this also leads to increased tolerance to environmental stresses (Siefermann-Harms, 1987). It also contributes to the defence against heavy metals. In *Arabidopsis* plants enzymes involved in vitamin E biosynthesis are up-regulated in response to Cu and Cd, and vitamin E-deficient mutants (vte1) showed enhanced oxidative stress and sensitivity to both metals (Collin et al., 2008). Manipulation of genes through genetic approaches which can enhance stress tolerance, that protect and maintain cellular functions or structure of cellular components, has been targeted to produce transgenic plants. Molecular analysis of plants over-expressing different enzymes of the Halliwell-Asada cycle can provide greater insights into the oxidative stress-tolerance mechanism. Studies have revealed that enhancing ROS protection by the constitutive overexpression of antioxidant defence enzymes in transgenic plants is beneficial.

9.4.2 Role of phytochelates

After Hg, Cd stands second in terms of its potential to induce PCs synthesis. Once Cd has entered the cytosol, another system strictly related to S metabolism is promptly activated and finally results in the production of important complexing agents, termed phytochelatins, which may contribute decisively towards rendering the metal ineffective. The binding of heavy metals by a family of peptide ligands, the PCs, is dependent on the synthesis of the precursor thiol compounds Cys and GSH (Cobbett, 2000).

Cd, when added to the reaction medium becomes associated with GSH and forms bis(glutathionato)cadmium (Cd·GS2), which in presence of phytochelatin synthase (PCS) forms low molecular weight Cd-PC complex (LMW) in the cytosol and subsequently transports this into the vacuole, and combines there with more Cd and sulfide to form high molecular weight Cd-PC complex (HMW), thus limiting Cd circulation inside the cytosol (Fig. 9.1).

Hence, PCs protect plants from the deleterious effect of heavy metals. These are thiolate peptides with a primary structure (γ-Glu-Cys)n-Gly, and are non-translationally synthesized from GSH. During Cd exposure, phytochelatin synthase, a dipeptidyl transpeptidase catalyses synthesis of PC peptides from GSH. This enzyme is present constitutively in the cell, but requires heavy metal ions for activity. Cadmium exerts its effect by forming a metal-GSH thiolate which acts as the acceptor molecule in the transpeptidation reaction (Vatamaniuk et al., 2000). This is due to the fact that Cd has a high affinity for different functional groups in biological molecules, specially for thiols. In the cell, they inactivate proteins by uncontrolled binding or cause oxidative stress by depleting GSH pools (Clemens, 2006).

In certain plants and in yeasts (*Saccharomyces pombe*) as well as *Candida glabrata*, sulfide ions also play important roles in the efficacy of Cd detoxification by PCs. High molecular weight PC–Cd complexes contain both Cd and acid-labile sulfide. The amount of Cd per molecule and the stability of the complex is increased by the incorporation of sulfide into the HMW complexes. Sulfide ions (S^{2-}) are also present in metal-MtIII complexes. These ions improve the stabilization of metal-MtIII compounds, as a result of which detoxification is also enhanced (Dameron et al., 1989). Glutamine synthetase (GS) and gamma-glutamylcysteine synthase (γ-ECS) synthesis genes over-expression elevates Cd resistance and accumulation in Indian mustard (Bennett et al., 2003). In *Arabidopsis*, over-expression of *PCS* gene (*AtPCS1*) itself caused sensitivity to Cd in tobacco over-expressing AtPCS1 increased absorption and Cd tolerance (Pomponi et al., 2006). Recent studies indicate that plants possess various classes of metal transporters which are involved in metal uptake and homoeostasis, and also in tolerance. These include heavy metal CPx-ATPase, the Nramp and CDF family and the ZIP family. Wheat TaPCS1 has been cloned independently and found to confer enhanced tolerance and accumulation of Cd, which when expressed in yeast also mediates GSH-dependent PC biosynthesis (Clemens et al., 1999). Kaznina et al., 2012 studied the amount of PCs in the barley root cells and expression of *HvPCS* gene increased in the Cd presence independently of plant age.

9.4.3 Synthesis of protein

Cadmium exposure induces the synthesis of considerable number of stress proteins, probably heat shock proteins (HSPs), with molecular weight ranging between 10 kDa and 70 kDa. Synthesis of a number of stress peptides including HSPs and chaperones was found to be associated with Cd stress in various plants (Clemens, 2001). While HSPs act as molecular chaperones in normal protein folding and assembly, they may also be involved in the repair of proteins and protection under stress conditions. It is reported that antibody localization showed that HSP70 was present not only in the nucleus and cytoplasm but also at the plasma membrane. This was an indication that HSP70 could be involved in the protection of membranes against Cd damage. Ultrastructural studies have revealed that a short heat stress given prior to heavy-metal stress induces a tolerance effect by preventing membrane damage (Neumann et al., 1994).

9.4.4 Phytohormones and nutrients status in cadmium tolerance

Phytohormones are also assumed to be involved in the adaptation and survival of plants by modifying heavy metal toxicity and Cd tolerance (Piotrowska-Niczyporuk et al., 2012; Asgher et al., 2014). Application of SA was shown to decrease both the uptake and transport of Cd, alleviate Cd-induced inhibition of nutrient absorption and also led to significant increases in contents of chlorophyll and carotenoid (Saidia et al., 2013). Pre-treatment of maize plants with SA resulted in an increase of lipid content by low and mild Cd-stress. It is quite clear that SA plays a protective role on the lipid membranes of the Cd-treated maize plants (Ivanova et al., 2008). *Haem oxygenase-1* (*HO-1*) gene has been found to be involved in SA-induced alleviation of oxidative stress caused by Cd stress in lucerne by decreasing distribution of ROS in the root tips (Cui et al., 2012).

Nitric oxide (NO), an important signalling molecule, also plays a role in the cellular responses of plants to heavy metal toxicity tolerance and improved crop productivity (Gill et al., 2013). Nitric oxide donors (N-tert-butylphenylnitrone, 3-morpholinosydonimine, sodium nitroprusside and ASC + NaNO$_2$) were effective in reducing CdCl$_2$-induced toxicity and CdCl$_2$-increased malondialdehyde (MDA) content. It was shown that in rice leaves, NO played a protective role against CdCl$_2$-induced toxicity (Hsu and Kao, 2004). Cadmium toxicity in rice leaves could also be alleviated by polyamines, mainly spermidine and spermine, which reduce Cd uptake and production of MDA and H$_2$O$_2$ (Hsu and Kao, 2007). However, there is a strong possibility that they can effectively stabilize and protect the membrane systems against the toxic effects of metal ions, particularly the redox active metals.

Lipid peroxidation induced by Cd was observed to be reduced with the supplementation of brassinosteroids (BRs). Seed germination enhancement and seedling growth by BRs could be one of the mechanisms responsible for their ameliorative influence on the inhibitory effect of Cd toxicity. Out of the two BRs, 28-homobrassinolide was more effective than 24-epibrassinolide in stress alleviation (Anuradha and Rao, 2007). Application of methyljasmonates with Cd caused alleviation of Cd damages by a reduction of MDA content (lipid peroxidation) and H$_2$O$_2$ content along with increase in activities of antioxidant enzymes in soybean plants. The observed increase in enzyme activities could be due to increased Cd tolerance (Keramat et al., 2009). Jasmonic acid (JA) regulates genes involved in GSH and PCS in *Arabidopsis* under Cd treatment that help in minimizing the stress (Xiang and Oliver, 1998). Ethylene has been shown to detoxify H$_2$O$_2$ by inducing the activity of APX (Melhorn, 1990) and to regulate expression of genes encoding metallothioneins and defence proteins. Cellular response may be regulated by the synthesis of JA and SA to minimize Cd damage (Rodríguez-Serrano et al., 2006).

Epibrassinolide (EBR) application to Cd-stressed plants remarkably decreased the Cd content in both the leaves and roots compared with Cd alone. These findings give a positive role for EBR in reducing pollutant residues for food safety and also strengthening the role of EBR in phytoremediation (Ahammed et al., 2013).

Cadmium treatments increased the uptake of Cu, Fe, Zn, manganese (Mn) and phosphorus (P) in roots; in leaves, Zn and P increased but Mn decreased. The increase of Fe and Cu might enhance the damage due to oxidative stress in leaves and roots because they are redox-metals that catalyse ROS production through the Haber-Weiss reaction (Candan and Tarhan, 2003). Magnesium is one of the essential macronutrients for plants (Karley and White, 2009). Besides being the central atom of the chlorophyll molecule, it is also essential as a co-factor for several enzymes and chelation to nucleotidyl phosphate forms (Shaul, 2002). Mg content decreased in leaves exposed to higher Cd; however, decreased Mg suggests there might be an antagonism effect between Cd and Mg (Hermans et al., 2011). The increased P content may contribute to the alleviation of Cd toxicity because P can react with heavy metal in forming insoluble compounds like P-Zn (Wang et al., 2009). Pankovica et al. (2000) suggested that in sunflower, optimal N supply may reduce the inhibitory effects of Cd on photosynthesis by increasing RUBP activity. The ameliorative effect of potassium against Cd-toxicity-induced oxidative damage may be because Cd availability to the plants becomes reduced along with an increase in the activities of antioxidative enzymes leading to a reduction in the ROS, as observed in mustard (Umar et al., 2008). Many researchers showed that addition of Zn to soil reduced crop Cd concentrations by increasing the germination index and vigour index at low metal concentration (25 and 50 ppm) (Patel et al., 2013). However, others reported that addition of Zn to soils led to increased Cd uptake (Moraghan, 1993). Thus, it is clear that S, along with other macronutrients and micronutrients, plays a significant role in decreasing Cd uptake and accumulation in crop plants (Sarwar et al., 2010). Hassan et al. (2005) speculated that S in nutrient solution reduces Cd availability by forming an insoluble Cd–S complex. Sulfur is a fundamental nutrient required for synthesis of Cd-binding proteins. Adequate S availability seems to play a favourable role in alleviation of heavy metal stress. Finally, accumulation of thiol pools in roots to cope with toxic heavy metals is suggested. Plants require an enhanced supply of reduced S (Cys or GSH) to compensate increased S demand for PCs synthesis (Astolfi et al., 2005). Phenotypically, the first visible S depletion symptoms appear on young leaves, which constitute the main sink organ for newly acquired sulfate (Anderson, 2005). In *Arabidopsis*, a high supply of Ca alleviates Cd toxicity (Suzuki, 2005). Similarly, high Mg alleviates toxicity in *Brassica* (Kashem and Kawai, 2007), which is also alleviated by low Mg conditions in *Arabidopsis* (Hermans et al., 2011) where a reduction in the uptake and accumulation of Cd have been reported. Addition of $CaCl_2$ to cadmium-stressed common bean plants improved growth (Suzuki, 2005; Cakmak, 2008). It demonstrated that Ca reduced Cd uptake and caused a modest reduction in Cd toxicity (Shahrtash et al., 2011).

Cd application led to a significant increase in root IAA levels but a decrease in indole acetic acid (IAA) levels in shoot when compared to controls. This may be due to the up-regulation of the transcription of gene encoding nitrilase (*AtNIT*) in root with no effect in shoots. Expression of AtAAO encoding for the enzyme aldehyde oxidase either in roots or shoots is not influenced by Cd. From these studies, it is suggested that AtNIT up-regulation was mainly involved in the increase of IAA levels in roots which promoted lateral root formation and growth, thereby increasing root branching (Vitti et al., 2013).

9.4.5 Involvement of sulfur in cadmium stress alleviation

Sulfur is an essential macro-element necessary for normal plant growth and development, and also plays a pivotal role in the protection of plants against environmental stresses, including heavy metal toxicity (Alscher et al., 1997). The ability of S in alleviating stress may be due to its participation in the synthesis of GSH and PCs. Cadmium causes a transient depletion of GSH pool which is a critical step in Cd sensitivity and an inhibition of activity of some antioxidative enzymes, especially of GR (Schutzendubel and Polle, 2002). Sulfur is involved in -SH and disulfide bonds (S–S) formation. Thiol groups have redox properties and hence are important in the stress response of plants. Thiol groups can be easily oxidized, forming S–S groups. Stabilization of protein structure is governed by these bonds, and in many enzymes active centres comprise of thiol groups.

The involvement of S in alleviation of Cd toxicity via enhancement of GSH is confirmed from the studies showing that alleviation of Cd toxicity by S is S-level dependent. Addition of S into medium having a lower S level increased GSH content, resulting in marked alleviation of Cd toxicity, while further addition of S no longer enhanced GSH content (Hassan et al., 2005). Masood et al. (2012) reported that S alleviates oxidative stress induced by Cd through enhancement in ATPS activity, Cys content, GSH and GR activity. Sulfur, when applied to the soil, alleviated Cd toxicity and lowered the reduction caused by Cd by restoration of growth and by increasing AsA and GSH contents (Anjum et al., 2008). Increased Cys availability is a major factor responsible for imparting Cd tolerance in *Arabidopsis*, which can be achieved by the exogenous application of thiol-related compounds. Interestingly, Cd hypersensitivity is due to enhanced synthesis of PCs in *Arabidopsis* (Lee et al., 2003), suggesting that the manipulation of Cys biosynthesis rather than the production of PCs may be more important for phytoremediation purposes. *Atcys-3A* gene over-expression in *A. thaliana* resulted in increased Cd tolerance (Domínguez-Solís et al., 2004). The activity of ATPS and levels of Cys and GSH were increased in both plants, when exposed to sub-lethal Cd levels; these findings also provide an idea that increasing ATPS activity can enhance S metabolism and hence heavy metal tolerance (Khan et al., 2009).

Application of S mitigated the adverse effects of Cd stress by enhancing total soluble carbohydrates, photosynthetic pigments and antioxidant enzymes; the application of 1 mM of S was more effective in alleviating the adverse effect of Cd stress (Gaafar *et al.*, 2012). Hassan *et al.* (2005) demonstrated that the higher S levels (0.4 and 0.6 mmol) helped alleviate Cd toxicity, with a concomitant increase in growth, and a decrease in Cd and MDA content in both roots and shoots in comparison to lower S level (0.2 mmol). As a consequence, higher S levels alleviated the oxidative stress, leading to a reduced MDA content and less growth inhibition by Cd toxicity. Chen and Huerta (1997) evaluated the effect of S on the growth of barley seedling under Cd stress and reported that though plant biomass is reduced in both S treatments under Cd-stress, the reduction is less with the high S treatment. From this study, they postulated that S is a critical nutritional factor in plants to counter Cd toxicity in barley seedling. Leaf dry weight, total chlorophyll and sugar contents and NR activity and protein content increased with the treatment of S and decreased under Cd stress. Thus, S could alleviate the Cd-induced impairment of biochemical features of the plant and the enhancement of nitrate accumulation in leaves (Anjana *et al.*, 2006). Sun *et al.* (2013) further reported that sodium hydrosulfide, a H_2S donor, mitigated syndromes associated with Cd toxicity via the enhancement of antioxidant system and cellular Cd homoeostasis by decreasing the amount of H_2O_2 accumulation and lipid peroxidation.

9.5 Ethylene, the Gaseous Hormone

Phytohormones play central roles in plants, which enable them to adapt to different abiotic stresses by mediating a broad range of adaptive responses. The 'classical' phytohormones, identified during the first half of the 20th century, are the auxins, gibberellins (GA), cytokinins, abscisic acid (ABA) and ethylene. More recently, several hormones, including BRs, JA, SA, NO and strigolactones have been reported. Environmental stresses induce ethylene production in large amounts (Wang *et al.*, 2006). The phytohormones that have an important role against abiotic and biotic stress are ABA, JA, SA and ethylene. Phytohormones SA, JA and ethylene are important in defence against pathogen and pest attack (Bari and Jones, 2009).

Ethylene is a gaseous plant hormone, which regulates diverse aspects of plants' cellular and developmental processes, including cell expansion, senescence, leaf abscission, seed germination and fruit ripening. Ethylene production is predominantly enhanced by exogenous application of auxins at high concentrations (Vandenbussche *et al.*, 2003). Ethylene in the gaseous form is released into the environment from the intercellular space where its concentration is in equilibrium with that dissolved in the cytoplasm. 1-aminocyclopropane-1-carboxylic acid, which originates from the amino acid L-Met, is known to be the immediate precursor of ethylene (Adams and Yang, 1979).

9.5.1 Ethylene biosynthesis: a relationship between sulfur and ethylene

Ethylene is synthesized from carbons C-3 and C-4 of methionine via two intermediates: S-adenosyl-L-Met (SAM) and ACC, while C_1 and C_2 were incorporated into CO_2 and formic acid, respectively. The conversion of SAM to ACC by ACS releases 5'-methylthioadenosine (MTA) and 5-methylthioribose (MTR). In the ethylene biosynthetic pathway, the production of ACC by ACC synthase (ACS) is the rate-limiting step; this is followed by the conversion of ACC to ethylene by ACC oxidase (Bleecker and Kende, 2000). 5'-methylthioadenosine is recycled to Met by the Yang cycle, utilizing ATP. 1-aminocyclopropane-1-carboxylicacid can be either converted to ethylene by ACC oxidase (ACO) or inactivated by conjugation to form malonyl-ACC or glutamyl-ACC. Oxidation of ACC generates CO_2 and cyanide as by-products; cyanide is then converted to β-cyanoalanine to prevent its accumulation at toxic levels (Wang *et al.*, 2002). Biosynthesis of ethylene is regulated both by developmental processes as well as by external stresses. The pathway of S-assimilation leads to cysteine biosynthesis, the first organic compound synthesized in the sulfate assimilatory pathway which leads to Met and SAM (Kopriva, 2006; Takahashi *et al.*, 2011). Methionine, the precursor for ethylene biosynthesis, is a fundamental metabolite in plant cells. It is first converted to SAM, then ACC, and finally ethylene in three consecutive reactions catalysed by the activities of enzymes SAM synthetase, ACS and ACO, respectively (Bleecker and Kende, 2000). SAM is a substrate for the biosynthesis of ethylene, polyamines and nicotianamine.

9.5.2 Ethylene signalling pathway

Ethylene can bind to receptors, which are similar to two-component regulators (Chang *et al.*, 1993). Unlike these receptors present in the bacterial plasma membrane the ethylene receptors in plants reside predominantly at the endoplasmic reticulum (Grefen *et al.*, 2008). All ethylene receptors have a sensor domain that can be categorized into a transmembrane domain and a GAF domain (found in cGMP phosphodiesterases, adenylate cyclases and Fh1a transcription factors), a histidine kinase domain and a response domain. The GAF domain binds cyclic nucleotides in a number of bacterial proteins, and the chromophore in the plant photoreceptor phytochrome (Aravind and Ponting, 1997).

Molecular studies performed on *Arabidopsis* have revealed that ethylene perception in plants is mediated through a family of ethylene receptors, including the (ethylene response) ETR1and ETR2, (ethylene response sensor) ERS1 and ERS2 and (ethylene insensitive) EIN4 gene products (Hua and Meyerowitz, 1998). Ethylene response factor (ERF1) and ethylene response sensor (ERS1) that comprise group I harbour three hydrophobic transmembrane segments and possess a conserved histidine kinase domain. Conversely, ETR2, EIN4 and ERS2 that belong to group II contain four transmembrane domains, but lack several of the key features of the kinase (Hua *et al.*, 1998). These proteins present on the cell membrane typically consist of a sensor protein and a response regulator protein, which function together to regulate adaptive responses to a broad range of environmental signals (Sakakibara *et al.*, 2000). Ethylene binding sites are located in the amino terminal half of ETR1 proteins (Schaller and Bleecker, 1995) and carboxyl terminal half of polypeptide with homology to histidine kinases and response regulators (Chang *et al.*, 1993). Ethylene receptor can change the interaction between the receptor and constitutive triple response 1 (CTR1), a Raf-like kinase (Kieber *et al.*, 1993; Clark *et al.*, 1998). The inactivated CTR1 will inhibit ethylene response repression and in turn alter gene expression. Binding of ethylene to its receptors is through a Cu co-factor, which is possibly delivered by the Cu transporter RAN1. Genetic studies showed that hormone binding results in the inactivation of receptor function. In the absence of ethylene, therefore, the receptors are hypothesized to be in a functionally active form that constitutively activates a Raf-like serine/threonine (Ser/Thr) kinase, CTR1, which is also a negative regulator of the pathway (Kieber *et al.*, 1993). Constitutive triple response 1 is a Raf-like Ser/Thr kinase with similarity to a MAPK kinase kinase (MAPKKK), suggesting the involvement of a MAP-kinase-like signalling cascade in the regulation of ethylene signalling. EIN2, EIN3, EIN5, and EIN6 are positive regulators of ethylene responses, which act downstream of CTR1. Inactivation of CTR1 potentiates signalling mediated by the C-terminal cytoplasmic domain of ethylene insensitive 2 (EIN2) (Alonso *et al.*, 1999). When EIN2 is activated a transcriptional cascade involving the EIN3/EIL and ERF transcription factors is initiated. EIN2 signalling leads to activation of the EIN3 transcription factor in the nucleus. Ethylene insensitive 4 is a direct activator of the ethylene response factor (ERF) genes. Ethylene response transcription factors in turn bind to ethylene response elements of downstream ethylene-induced genes. When the dominant *etr1-1* mutation was introduced into the binding pocket of the ETR1 receptor, ethylene binding was suppressed (O'Malley *et al.*, 2005). EIN3, another important transcription factor in ethylene signalling, accumulates in the presence of ethylene through the control of E3 ligase EIN3 binding F-Box protein 1 (EBF1) and EBF2 (Yanagisawa *et al.*, 2003; Gagne *et al.*, 2004).

By changing the mere concentration of a signalling compound, signalling cannot be induced since the activation of signalling paths, receptor concentrations and the presence of a downstream signal transmission apparatus are important factors. Klee (2004) has shown that less receptor would make the plants more sensitive because larger percentage receptor inactivations are induced by a small amount of ethylene.

9.5.3 Ethylene in sulfur-mediated alleviation of cadmium stress

Formation of ethylene in plants is related to S assimilation via Cys. Sulfur influences ethylene sensitivity and involvement of ethylene in GSH synthesis control and alleviation of Cd stress. Due to the central role of ethylene in abiotic stress and oxidative damage the ACC oxidase is a suitable indicator for oxidative damage. Expression of ERF proteins that belong to the APETALA2 (AP2)/ET-responsive-element-binding protein (EREBP) family is affected by Cd. Over-expression of an EREBP increases ROS detoxification and reduction in ROS-induced cell death (Ogawa et al., 2005). Therefore, over-expression of plant ERF genes enhanced tolerance to stresses, indicating that these genes could be used as candidate genes to improve crop resistance. Studies on *A. thaliana* by Weber et al. (2006) showed that ERF1 and ERF2 are induced in the roots after 2 h of Cd treatment. Vassilev et al. (2004) reported Cd stress increased ethylene production at 14 and 28 mg Cd kg^{-1} sand concentration by decreasing linolenic acid content regarded as a monitor of lipid peroxidation whereas at a concentration of 42 mg Cd kg^{-1} sand, ethylene production was decreased; this was probably the disruption of the chloroplast membrane leading to a loss of enzyme activities in the ethylene biosynthetic pathway. Iqbal et al. (2011) reported that application of ethephon (an ethylene source) enhanced the activities of NR and ATPS leading to enhanced N and S assimilation, which in turn resulted in increased photosynthetic responses in mustard (*B. juncea* L.) cultivars that differ in photosynthetic capacity.

Application of ethephon and N increased ethylene and influenced photosynthetic and growth response of plants, ethylene increased the stomatal conductance of plants, thereby increasing the diffusion of CO_2 and thus photosynthesis (Iqbal et al., 2011). The ethylene signalling pathways also seem to be involved in the early phase of response to Cd. Genes encoding the ethylene responsive factors ERF2 (*At5g47220*) and ERF5 (*At5g47230*) were up-regulated in all conditions analysed in both root and shoot tissues, whereas ACO and ACS genes encoding were up-regulated in both root and shoot tissues, but only after 30 h with 50 µM Cd (Herbette et al., 2006). Exogenous ethylene greatly stimulated Cd-induced cell death and Cd treatment enhanced endogenous ethylene production in tomato (Lakimova et al., 2005). Sulfur alleviates Cd-induced oxidative stress via ethylene as reported in Indian mustard (Masood et al., 2012). Lakimova et al. (2008) reported that Cd treatment induced a transient increase in ethylene production during the first 24 h. Suppression of ethylene production with aminoethoxyvinyl glycine (AVG) inhibited while addition of ethylene stimulated Cd-induced cell death.

The Cd-dependent oxidative damage to membranes induces JA and ethylene production, which, in turn, could modulate the defensive response in conjunction with ROS (Rodríguez-Serrano et al., 2009). Ethylene may also affect the photosynthesis process since higher activity of ACS with increasing ethephon concentrations stimulate photosynthetic rate, which might be due to increased stomatal conductance (Khan, 2004). On the other hand, application of AVG resulted in a decrease in ACC synthase activity, ethylene, photosynthesis and growth (Khan, 2005). It is thus quite evident that heavy metal stress in plants produces ROS, triggering signalling molecules such as JA and SAM, and activates detoxification process through the involvement of GSH/PCs biosynthesis (Ahsan et al., 2008). Most of the up-regulated proteins were related to S and GSH metabolism.

9.6 Conclusion and Future Prospects

As a whole, these studies provide the framework to understand better the mechanisms that govern plant cell response to Cd-induced stress and defence strategies adopted. Heavy metal toxicity issues in plants as well as in soils are a significant problem throughout the world. The adverse effects of heavy metal stress on different metabolic processes such as photosynthesis, respiration, water relations and also membrane stability leads to decreased productivity. Various genetic approaches can be used for mitigating the stress effects by developing crop plants with enhanced tolerance. Cadmium stress can be alleviated by using S compounds and enzymes involved in S assimilation which detoxify Cd-induced stress. Plants with higher S accumulation capacity are expected to show more tolerance to Cd stress. Sulfur application to plants thus may provide a novel strategy to reduce Cd toxicity. Since antioxidants have a positive role in controlling Cd stress defence, increasing their level through genetic approaches may provide protection against this stress. Over-expression of certain genes can also confer tolerance. The mechanisms of alleviation of Cd stress also include the biosynthesis of phytohormones. It has been shown that S alleviates Cd-induced stress via ethylene. Though it is known that ethylene production provides protection to various biotic and abiotic stresses in plants, little is known about the interaction of ethylene with S, and this reveals new potential targets for the development of mechanisms against stress. So it is interesting to study ethylene and S in relation to Cd toxicity. It is explained in this chapter that sulfur can induce tolerance to Cd stress and mitigate photosynthetic inhibition through ethylene by maintaining high GSH levels. Further research will focus on details of the functions of ethylene in the S-mediated regulation of GSH synthesis and antioxidant system under Cd stress.

Acknowledgements

Financial assistance to authors by University Grants Commission and Department of Science and Technology, New Delhi is gratefully acknowledged.

References

Abeles, F.B., Morgan, P.W. and Saltveit, M.E. (1992) *Ethylene in Plant Biology*, 2nd edn. San Diego, Academic Press, New York.

Acharya, B.R. and Assmann, S.M. (2009) Hormone interactions in stomatal function. *Plant Molecular Biology* 69, 451–462.

Adams, D.O. and Yang, S.F. (1979) Ethylene biosynthesis. Identification of 1-aminocyclopropane-1-carboxylic acid as an intermediate in the conversion of methionine to ethylene. *Proceedings of National Academy of Sciences USA* 76, 170–174.

Afef, N.H., Leila, S., Donia, B., Houda, G. and Chiraz, C.H. (2011) Relationship between physiological and biochemical effects of cadmium toxicity in *Nicotiana rustica*. *American Journal of Plant Physiology* 6, 294–303.

Agrawal, G.K., Rakwal, R., Jwa, N.S., Han, K.S. and Agrawal, V.P. (2002) Molecular cloning and mRNA expression analysis of the first rice jasmonate biosynthetic pathway gene allene oxide synthase. *Plant Physiology and Biochemistry* 40, 771–782.

Ahammed, G.J., Choudhary, S.P., Chen, S., Xia, X., Shi, K., Zhou, Y. and Yu, J. (2013) Role of brassinosteroids in alleviation of phenanthrene–cadmium co-contamination-induced N photosynthetic inhibition and oxidative stress in tomato. *Journal of Experimental Botany* 64, 199–213.

Ahsan, N., Lee, D.G., Alam, I., Kim, P.J., Lee, J.J., Ahn, Y.O., Kwak, S.S., Lee, I.J., Bahk, J.D., Kang, K.Y., Renaut, J., Komatsu, S. and Lee, B.H. (2008) Comparative proteomic study of arsenic-induced differentially expressed proteins in rice roots reveals glutathione plays a central role during As stress. *Proteomics* 8, 3561–3576.

Alonso, J.M., Hirayama, T., Roman, G., Nourizadeh, S. and Ecker, J.R. (1999) EIN2, a bifunctional transducer of ethylene and stress responses in *Arabidopsis*. *Science* 284, 2148–2152.

Alscher, R.G., Donahue, J.L. and Cramer, C.L. (1997) Reactive oxygen species and antioxidants: relationships in green cells. *Physiologia Plantarum* 100, 224–233.

Alvarez, C., Calo, L., Romero, L.C., García, I. and Gotor, C. (2010) An O-acetylserine (thiol) lyase homolog with L-cysteine desulfhydrase activity regulates cysteine homeostasis in *Arabidopsis*. *Plant Physiology* 152, 656–669.

Alvarez, S., Berla, B.M., Sheffield, J., Cahoon, R.E., Jez, J.M. and Hicks, L.M. (2009) Comprehensive analysis of the *Brassica juncea* root proteome in response to cadmium exposure by complementary proteomic approaches. *Proteomics* 9, 2419–2431.

Anderson, J.W. (2005) Regulation of sulphur distribution and redistribution in grain plants. In: Saito, K., De Kok, L.J., Stulen, I., Hawkesford, M.J., Schnug, E., Sirko, A. and Rennenberg, H. (eds) *Sulphur Transport and Assimilation in Plants in the Postgenomic Era*. Backhys Publishers, Leiden, the Netherlands, pp. 23–31.

Anjana, Umar, S. and Iqbal, M. (2006) Functional and structural changes associated with cadmium in mustard plant: effect of applied sulphur. *Communications in Soil Science and Plant Analysis* 37, 1205–1217.

Anjum, N.A., Umar, S., Ahmad, A., Iqbal, M. and Khan, N.A. (2008) Sulphur protects mustard (*Brassica campestris* L.) from cadmium toxicity by improving leaf ascorbate and glutathione. *Plant Growth and Regulation* 54, 271–279.

Anuradha, S. and Rao, S.S.R. (2007) The effect of brassinosteroids on radish (*Raphanus sativus* L.) seedlings growing under cadmium stress. *Plant, Soil and Environment* 53, 465–472.

Aravind, L. and Ponting, C.P. (1997) The GAF domain: an evolutionary link between diverse phototransducing proteins. *Trends in Biochemical Science* 22, 458–459.

Arun, K.S., Carlos, C., Herminia, L.Z. and Avudainayagam, S. (2005) Chromium toxicity in plants. *Environment International* 31, 739–775.

Asada, K. (2006) Production and scavenging of reactive oxygen species in chloroplasts and their functions. *Plant Physiology* 141, 391–396.

Asgher, M., Khan, M.I.R., Iqbal, N., Masood, A. and Khan, N.A. (2013) Cadmium tolerance in mustard cultivars: dependence on proline accumulation and nitrogen assimilation. *Journal of Functional and Environmental Botany* 3, 30–42.

Asgher, M., Khan, M.I.R., Anjum, N.A. and Khan, N.A. (2014) Minimizing cadmium toxicity in plants: role of plant growth regulators. *Protoplasma* doi: 10.1007/s00709-014-0710-4.

Astolfi, S., Zuchi, S. and Passera, C. (2005) Effect of cadmium on H+ATPase activity of plasma membrane vesicles isolated from roots of different S-supplied maize (*Zea mays* L.) plants. *Plant Science* 169, 361–368.

Bari, R. and Jones, J. (2009) Role of plant hormones in plant defence responses. *Plant Molecular Biology* 69, 473–488.

Bartoli, C.G., Gomez, F., Martinez, D.E. and Guiamet, J.J. (2004) Mitochondria are the main target for oxidative damage in leaves of wheat. *Journal of Experimental Botany* 55, 1663–1669.

Baryla, A., Carrier, P., Frank, F., Coulomb, C., Sahut, C. and Havaud, M. (2001) Leaf chlorosis in oiseed rape plants (*Brassica napus*) grown in cadmium polluted soil: causes and consequences of photosynthesis and growth. *Planta* 212, 699–709.

Benavides, M.P., Gallego, S.M. and Tomaro, M.L. (2005) Cadmium toxicity in plants. *Brazilian Journal of Plant Physiology* 17, 21–34.

Bennett, L.E., Burkhead, J.L., Hale, K.L., Terry, N., Pilon, M. and Pilon-Smits, E.A.H. (2003) Analysis of transgenic Indian mustard plants for phytoremediation of metal-contaminated mine tailings. *Journal of Environmental Quality* 32, 432–440.

Berboodi, B.S.H. and Samadi, L. (2004) Detection of apoptotic bodies and oligonucleosomal DNA fragments in cadmium treated root apical cells of *Allium cepa* L. *Plant Science* 167, 411–416.

Bhuiyan, M.S., Min, S.R., Jeong, W.J., Sultana, S., Choi, K.S., Song, W.Y., Lee, Y., Lim, Y.P. and Liu, J.R. (2011) Overexpression of a yeast cadmium factor 1 (YCF1) enhances heavy metal tolerance and accumulation in *Brassica juncea*. *Plant Cell Tissue and Organ Culture* 105, 85–91.

Bleecker, A.B. and Kende, H. (2000) Ethylene: a gaseous signal molecule in plants. *Annual Review of Cellular Development Biology* 16, 1–18.

Bowler, C.H., Van Montagu, M. and Inzé, D. (1992) Superoxide dismutase and stress tolerance. *Annual Review of Plant Physiology and Plant Molecular Biology* 43, 83–116.

Buchner, P., Elisabeth, C., Stuiver, E., Westerman, S., Wirtz, M., Hell, R., Hawkesford, M.J. and De Kok, L.J. (2004) Regulation of sulphate uptake and expression of sulphate transporter genes in *Brassica oleracea* as affected by atmospheric H_2S and pedospheric sulphate nutrition. *Plant Physiology* 136, 3396–3408.

Cai, L. and Cherian, G. (2003) Zinc-metallothionein protects from DNA damage induced by radiation better than glutathione and Cu or Cd metallothioneins. *Toxicology Letters* 136, 193–198.

Cakmak, I. (2008) Zinc crops 2007: improving crop production and human health. *Plant and Soil* 306, 1–2.

Candan, N. and Tarhan, L. (2003) Relationship among chlorophyll-carotenoid content, antioxidant enzyme activities and lipid peroxidation levels by Mg^{2+} deficiency in the *Mentha pulegium* leaves. *Plant Physiology and Biochemistry* 41, 35–40.

Chalapathi, R.A.S.V. and Reddy, A.R. (2008) Glutathione reductase: a putative redox regulatory system in plant cells. In: Khan, N.A., Singh, S. and Umar, S. (eds) *Sulphur Assimilation and Abiotic Stress in Plants*. Springer, Berlin, pp. 111–147.

Chang, C., Kwok, S.F., Bleecker, A.B. and Meyerowitz, E.M. (1993) *Arabidopsis* ethylene-response gene ETR1: similarity of product to two-component regulators. *Science* 262, 539–544.

Chen, Y. and Huerta, A.J. (1997) Effects of sulphur nutrition on photosynthesisin cadmium-treated barley seedlings. *Journal of Plant Nutrition* 20, 845–856.

Choi, D.G., Yoo, N.H., Yu, C.Y., de los Reyes, B. and Yun, S.J. (2004) The activities of antioxidant enzymes in response to oxidative stresses and hormones in paraquat-tolerant *Rehmannia glutinosa* plants. *Journal of Biochemistry and Molecular Biology* 37, 618–624.

Clark, K.L., Larsen, P.B., Wang, X. and Chang, C. (1998) Association of the *Arabidopsis* CTR1 Raf-like kinase with the ETR1 and ERS1 ethylene receptors. *Proceedings of the National Academy of Sciences USA* 95, 5401–5406.

Clemens, S. (2001) Molecular mechanisms of plant metal tolerance and homeostasis. *Planta* 212, 475–486.

Clemens, S. (2006) Toxic metal accumulation, responses to exposure and mechanisms of tolerance in plants. *Biochimie* 88, 1707–1719.

Clemens, S., Kim, E.J., Neumann, D. and Schroeder, J.I. (1999) Tolerance to toxic metals by a gene family of phytochelatin synthases from plants and yeast. *EMBO Journal* 18, 3325–3333.

Cobbett, C.S. (2000) Phytochelatin biosynthesis and function in heavy metal detoxification. *Current Opinion in Plant Biology* 32, 211–216.

Cobbett, C.S., Hussain, D. and Haydon, M.J. (2003) Structural and functional relationships between type 1B heavy metal-transporting P-type ATPases in *Arabidopsis*. *New Phytologist* 159, 315–321.

Colangelo, E.P. and Guerinot, M.L. (2006) Put the metal to the petal: metal uptake and transport throughout plants. *Current Opinion in Plant Biology* 9, 322–330.

Collin, V.C., Eymery, F., Genty, B., Rey, P. and Havaux, M. (2008) Vitamin E is essential for the tolerance of *Arabidopsis thaliana* to metal-induced oxidative stress. *Plant Cell and Environment* 31, 244–257.

Cosio, C., De Santis, L., Frey, B., Diallo, S. and Keller, C. (2005) Distribution of cadmium in leaves of *Thlaspi caerulescens*. *Journal of Experimental Botany* 56, 765–775.

Cosio, C., Vollenweider, P. and Keller, C. (2006) Localization and effects of cadmium in leaves of a cadmium-tolerant willow (*Salix viminalis* L.) Macrolocalization and phytotoxic effects of cadmium. *Environmental and Experimental Botany* 58, 64–74.

Courbot, M., Willems, G., Motte, P., Arvidsson, S., Roosens, N., Saumitou-Laprade, P. and Verbruggen, N. (2007) A major quantitative trait locus for cadmium tolerance in *Arabidopsis halleri* co localizes with HMA4, a gene encoding a heavy metal ATPase. *Plant Physiology* 144, 1052–1065.

Cui, W., Li, L., Gao, Z., Wu, H., Xie, Y. and Shen, W. (2012) Haem oxygenase-1 is involved in salicylic acid-induced alleviation of oxidative stress due to cadmium stress in *Medicago sativa*. *Journal of Experimental Botany* 63, 5521–5534.

Curie, C., Cassin, G., Couch, D., Divol, F., Higuchi, K., Le Jean, M., Misson, J., Schikora, A., Czernic, P. and Mari, S. (2009) Metal movement within the plant: contribution of nicotianamine and yellow stripe 1-like transporters. *Annals of Botany* 103, 1–11.

Dai, L.P., Xiong, Z.T., Huang, Y. and Li, M.J. (2006) Cadmium-induced changes in pigments, total phenolics, and phenylalanine ammonia-lyase activity in fronds of *Azolla imbricate*. *Environmental Toxicology* 21, 505–512.

DalCorso, G., Farinati, S., Maistri, S. and Furini, A. (2008) How plants cope with cadmium: staking all on metabolism and gene expression. *Journal of Integrated Plant Biology* 50, 1268–1280.

Dameron, C.T., Reese, R.N., Mehra, R.K., Kortan, A.R., Caroll, P.J., Steigerwald, M.L., Brus, L.E. and Winge, D.R. (1989) Biosynthesis of cadmium sulfide quantum semiconductor crystallites. *Nature* 338, 596–597.

Davidian, J.C. and Kopriva, S. (2010) Regulation of sulphate uptake and assimilation – the same or not the same? *Molecular Plant* 3, 314–325.

De Knecht, J.A., Van Baren, N., Ten Bookum, W.T., Wong Fong Song, H.W., Koevoets, P.L.M., Schat, H. and Verkleij, J.A.C. (1995) Synthesis and degradation of phytochelatins in cadmium-sensitive and cadmium-tolerant *Silene vulgaris*. *Plant Science* 106, 9–18.

Dominguez-Solis, J.R., Gutiérrez-Alcalá, G., Romero, L.C. and Gotor, C. (2001) The cytosolic O-acetylserine (thiol) lyase gene is regulated by heavy metals and can function in cadmium tolerance. *Journal of Biological Chemistry* 276, 9297–9302.

Dominguez-Solis, J.R., Lopez-Martin, M.C., Ager, F.J., Ynsa, M.D., Romero, L.C. and Gotor, C. (2004) Increased cysteine availability is essential for cadmium tolerance and accumulation in *Arabidopsis thaliana*. *Plant Biotechnology* 2, 469–476.

Droux, M. (2004) Sulphur assimilation and the role of sulphur in plant metabolism: a survey. *Photosynthesis Research* 79, 331–348.

Dube, B.K., Sinha, P., Gopal, R. and Chatterjee, C. (2003) Modulation of radish metabolism by zinc phytotoxicity. *Indian Journal of Plant Physiology* 8, 302–306.

El-Shora, H.E. and Ali, A.S. (2011) Changes in the activities of nitrogen metabolism enzymes in cadmium stressed marrow seedling. *Asian Journal of Plant Science* 10, 117–124.

Faller, P., Kienzler, K. and Liszlay, A.K. (2005) Mechanism of Cd^{2+} toxicity: Cd^{2+} inhibited photoactivation of photosystem by competitive binding to the essential Ca^{2+} site. *Biochimica et Biophysica Acta* 1706, 158–164.

Farinati, S., DalCorso, G., Varotto, S. and Furini, A. (2010) The *Brassica juncea* BjCdR15, an ortholog of *Arabidopsis* TGA3, is a regulator of cadmium uptake, transport and accumulation in shoots and confers cadmium tolerance in transgenic plants. *New Phytologist* 185, 964–978.

Fiil, B.K., Petersen, K., Petersen, M. and Mundy, J. (2009) Gene regulation by MAP kinase cascades. *Current Opinion in Plant Biology* 12, 615–621.

Freeman, J.L., Persans, M.W., Nieman, K., Albrecht, C., Peer, W., Pickering, I.J. and Salt, D.E. (2004) Increased glutathione biosynthesis plays a role in nickel tolerance in *Thlaspi* nickel hyperaccumulators. *Plant Cell* 16, 2176–2191.

Fusconi, A., Repetto, O., Bonab, E., Massab, N., Galloa, C., Dumas-Gaudot, E. and Bertab, G. (2006) Effects of cadmium on meristem activity and nucleus ploidy in roots of *Pisum sativum* L. cv. Frisson seedlings. *Environmental and Experimental Botany* 58, 253–260.

Gaafar, A.R.Z., Ghdan, A.A., Siddiqui, M.H., Al-Whaibi, M.H., Basalah, M.O., Ali, H.M. and Sakran, A.M. (2012) Influence of sulphur on cadmium (Cd) stress tolerance in *Triticum aestivum* L. *African Journal of Biotechnology* 43, 10108–10114.

Gagne, J.M., Smalle, J., Gingerich, D.J., Walker, J.M., Yoo, S.D., Yanagisawa, S. and Vierstra, R.D. (2004) *Arabidopsis* EIN3-binding F-box 1 and 2 form ubiquitin-protein ligases that repress ethylene action and promote growth by directing EIN3 degradation. *Proceedings of National Academy of Sciences USA* 101, 6803–6808.

Gasic, K. and Korban, S. (2007) Transgenic Indian mustard plants expressing an *Arabodopsis* phytochelatin synthase (AtPCS1) exhibit enhanced As and Cd tolerance. *Plant Molecular Biology* 64, 361–369.

Gichner, T., Znidar, I. and Szakova, J. (2008) Evaluation of DNA damage and mutagenicity induced by lead in tobacco plants. *Mutation Research* 652, 186–190.

Gill, S.S. and Tuteja, N. (2010) Reactive oxygen species and antioxidant machinery in abiotic stress tolerance in crop plants. *Plant Physiology and Biochemistry* 48, 909–930.

Gill, S.S., Khan, N.A., Anjum, N.A. and Tuteja, N. (2011) Amelioration of cadmium stress in crop plants by nutrients management: morphological, physiological and biochemical aspects. *Plant Stress* 5, 1–23.

Gill, S.S., Khan, N.A. and Tuteja, N. (2012) Cadmium at high dose perturbs growth, photosynthesis and nitrogen metabolism while at low dose it up regulates sulphur assimilation and antioxidant machinery in garden cress. *Plant Science* 182, 112–120.

Gill, S.S., Hasanuzzaman, M., Nahar, K., Macovei, A. and Tuteja, N. (2013) Importance of nitric oxide in cadmium stress tolerance in crop plants. *Plant Physiology and Biochemistry* 63, 254–261.

Goncalves, J.F., Becker, G.A., Cargnelutti, D., Tabaldi, L.A., Pereira, L.B., Battisti, V., Spanevello, M.R., Morsch, M.V., Nicoloso, F.T. and Schetinger, M.R.C. (2007) Cadmium toxicity causes oxidative stress and induces response of the antioxidant system in cucumber seedlings. *Brazilian Journal of Plant Physiology* 19, 223–232.

Grant, C., Buckley, W., Bailey, L. and Selles, F. (1998) Cadmium accumulation in crops. *Canadian Journal of Plant Science* 78, 1–17.

Grefen, C., Stadele, K., Ruzicka, K., Obrdlik, P., Harter, K. and Horak, J. (2008) Subcellular localization and *in vivo* interactions of the *Arabidopsis thaliana* ethylene receptor family members. *Molecular Plant* 1, 308–320.

Grifferty, A. and Barrington, S. (2000) Zinc uptake by young wheat plants under two transpiration regimes. *Journal of Environmental Quality* 2, 443–446.

Guan, Z.Q., Chai, T.Y., Zhang, Y.X., Xu, J. and Wei, W. (2009) Enhancement of Cd tolerance in transgenic tobacco overexpressing a Cd-induced catalase cDNA. *Chemosphere* 76, 623–630.

Hall, J.L. and Williams, L.E. (2003) Transition metal transporters in plants. *Journal of Experimental Botany* 54, 2601–2613.

Hanikenne, M., Talke, I.N., Haydon, M.J., Lanz, C., Nolte, A., Motte, P., Kroymann, J., Weigel, D. and Kramer, U. (2008) Evolution of metal hyperaccumulation required cis-regulatory changes and triplication of HMA4. *Nature* 453, 391–395.

Hassan, M.J., Wang, Z. and Zhang, G. (2005) Sulphur alleviates growth inhibition and oxidative stress caused by cadmium toxicity in rice. *Journal of Plant Nutrition* 28, 1785–1800.

Hawkesford, M.J. (2003) Transporter gene families in plants: the sulphate transporter gene family redundancy or specialization? *Physiologia Plantarum* 117, 155–163.

Herbette, S., Taconnat, L., Hugouvieux, V., Piette, L., Magniette, M.L., Cuine, S., Auroy, P., Richaud, P., Forestier, C., Bourguignon, J., Renou, J.P., Vavasseur, A. and Leonhardt, N. (2006) Genome-wide transcriptome profiling of the early cadmium response of *Arabidopsis* roots and shoots. *Biochimie* 88, 1751–1765.

Hermans, C., Chen, J., Coppens, F., Inze, D. and Verbruggen, N. (2011) Low magnesium status in plants enhances tolerance to cadmium exposure. *New Phytologist* 192, 428–436.

Hsu, Y.T. and Kao, C.H. (2004) Cadmium toxicity is reduced by nitric oxide in rice leaves. *Plant Growth Regulation* 42, 227–238.

Hsu, Y.T. and Kao, C.H. (2007) Cadmium-induced oxidative damage in rice leaves in reduced by polyamines. *Plant and Soil* 291, 27–37.

Hua, J. and Meyerowitz, E.M. (1998) Ethylene responses are negatively regulated by a receptor gene family in *Arabidopsis thaliana*. *Cell* 94, 261–271.

Hua, J., Sakai, H., Nourizadeh, S., Chen, Q.G., Bleecker, A.B., Ecker, J.R. and Meyerowitz, E.M. (1998) Ein4 and ERS2 are members of the putative ethylene receptor family in *Arabidopsis*. *Plant Cell* 10, 1321–1332.

Hussain, D., Haydon, M., Wang, Y., Wong, E., Sherson, S., Young, J., Camakaris, J., Harper, J. and Cobbett, C. (2004) P-type ATPase heavy metal transporters with roles in essential zinc homeostasis in *Arabidopsis*. *Plant Cell* 16, 1327–1339.

Iqbal, N., Masood, A., Nazar, R., Syeed, S. and Khan, N.A. (2010) Photosynthesis, growth and antioxidant metabolism in mustard (*Brassica juncea* L.) cultivars differing in cadmium tolerance. *Agricultural Sciences in China* 9, 519–527.

Iqbal, N., Nazar, R., Syeed, S., Masood, A. and Khan, N.A. (2011) Exogenously-sourced ethylene increases stomatal conductance, photosynthesis, and growth under optimal and deficient nitrogen fertilization in mustard. *Journal of Experimental Botany* 62, 4955–4963.

Ivanova, E., Staikova, T. and Velcheva, I. (2008) Cytotoxicity and genotoxicity of heavy metal- and cyanide-contaminated waters in some regions for production and processing of ore in Bulgaria. *Bulgarian Journal of Agricultural Science* 14, 262–268.

Jasinski, M., Sudre, D., Schansker, G., Schellenberg, M., Constant, S., Martinoia, E. and Bovet, L. (2008) AtOSA1, a member of the Abc1-like family, as a new factor in cadmium and oxidative stress response. *Plant Physiology* 147, 719–731.

Jonak, C., Nakagami, H. and Hirt, H. (2004) Heavy metal stress. Activation of distinct mitogen-activated protein kinase pathways by copper and cadmium. *Plant Physiology* 136, 3276–3283.

Karley, A.J. and White, P.J. (2009) Moving cationic minerals to edible tissues: potassium, magnesium, calcium. *Current Opinion in Plant Biology* 12, 291–298.

Kashem, A. and Kawai, S. (2007) Alleviation of cadmium phytotoxicity by magnesium in Japanese mustard spinach. *Soil Science and Plant Nutrition* 53, 246–251.

Kawashima, C.G., Noji, M., Nakamura, M., Ogra, Y., Suzuki, K.T. and Saito, K. (2004) Heavy metal tolerance of transgenic tobacco plants over-expressing cysteine synthase. *Biotechnology Letters* 26, 153–157.

Kawashima, C.G., Berkowitz, O., Hell, R., Noji, M. and Saito, K. (2005) Characterization and expression analysis of a serine acetyltransferase gene family involved in a key step of the sulphur assimilation pathway in *Arabidopsis*. *Plant Physiology* 137, 220–230.

Kaznina, N.M., Titov, A.F., Topchieva, L.V., Laidinen, G.F. and Batova, Y.V. (2012) Barley plant response to cadmium action as dependent on plant age. *Russian Journal of Plant Physiology* 59, 65–70.

Keramat, B., Kalantari, K.M. and Arvin, M.J. (2009) Effects of methyl jasmonate in regulating cadmium induced oxidative stress in soybean plant (*Glycine max* L.) *African Journal of Microbiology Research* 3, 240–244.

Khan, N.A. (2004) Activity of 1-aminocyclopropane carboxylic acid synthase in two mustard cultivars differing in photosynthetic capacity. *Photosynthetica* 42, 477–480.

Khan, N.A. (2005) The influence of exogenous ethylene on growth and photosynthesis of mustard following defoliation. *Scientia Horticulturae* 105, 499–505.

Khan, N.A., Samiullah, Singh S. and Nazar, R. (2007) Activities of antioxidative enzymes, sulphur assimilation, photosynthetic activity and growth of wheat (*Triticum aestivum*) cultivars differing in yield potential under cadmium stress. *Journal of Agronomical Crop Science* 193, 435–444.

Khan, N.A., Anjum, N.A., Nazar, R. and Iqbal, N. (2009) Increased activity of ATP-sulphurylase, contents of cysteine and glutathione reduce high cadmium-induced oxidative stress in high photosynthetic potential mustard cultivar. *Russian Journal of Plant Physiology* 56, 670–677.

Kieber, J.J., Rothenberg, M., Roman, G., Feldmann, K.A. and Ecker, J.R. (1993) CTR1, a negative regulator of the ethylene response pathway in *Arabidopsis*, encodes a member of the raf family of protein kinases. *Cell* 72, 427–441.

Kim, D.Y., Bovet, L., Kushnir, S., Noh, E.W., Martinoia, E. and Lee, Y. (2006) AtATM3 is involved in heavy metal resistance in *Arabidopsis*. *Plant Physiology* 140, 922–932.

Kim, D.Y., Bovet, L., Maeshima, M., Martinoia, E. and Lee, Y. (2007) The ABC transporter AtPDR8 is a cadmium extrusion pump conferring heavy metal resistance. *The Plant Journal* 50, 207–218.

Kim, Y.Y., Kim, D.Y., Shim, D., Song, W.Y., Lee, J., Julian, I., Schroeder, Kim, S., Moran, N. and Lee, Y. (2008) Expression of the novel wheat gene TM20 confers enhanced cadmium tolerance to bakers' yeast. *The Journal of Biological Chemistry* 283, 15893–15902.

Klee, H.J. (2004) Ethylene signal transduction. Moving beyond *Arabidopsis*. *Plant Physiology* 135, 660–667.

Kolukisaoglu, H.U., Bovet, L., Klein, M., Eggmann, T., Geisler, M., Wanke, D., Martinoia, E. and Schulz, B. (2002) Family business: the multidrug-resistance related protein (MRP) ABC transporter genes in *Arabidopsis thaliana*. *Planta* 216, 107–119.

Kopriva, S. (2006) Regulation of sulphate assimilation in *Arabidopsis* and beyond. *Annals of Botany* 97, 479–495.

Kopriva, S. and Koprivova, S.A. (2005) Sulphate assimilation and glutathione synthesis in C4 plants. *Photosynthesis Research* 86, 363–372.

Kopriva, S. and Rennenberg, H. (2004) Control of sulphate assimilation and glutathione synthesis: interaction with N and C metabolism. *Journal of Experimental Botany* 55, 1831–1842.

Kopriva, S., Buchert, T., Fritz, G., Suter, M., Benda, R., Schunemann, V., Koprivova, A., Schurmann, P., Trautwein, A.X., Kroneck, P.M.H. and Brunold, C. (2002) The presence of an iron-sulphur cluster in adenosine 5′-phosphosulphate reductase separates organisms utilizing adenosine 5′-phosphosulphate and phosphoadenosine 5′-phosphosulphate for sulphate assimilation. *The Journal of Biological Chemistry* 277, 21786–21791.

Lakimova, E., Kapchina-Toteva, V., de Jong, A., Atanassov, A. and Woltering, E. (2005) Involvement of ethylene, oxidative stress and lipid-derived signals in cadmium induced programmed cell death in tomato suspension cells. *BMC Plant Biology* 5, 1–2.

Lakimova, E.T., Woltering, E.J., Kapchina-Toteva, V.M., Harren, F.J. and Cristescu, S.M. (2008) Cadmium toxicity in cultured tomato cells, role of ethylene, proteases and oxidative stress in cell death signaling. *Cell Biology International* 32, 1521–1529.

Lee, S., Moon, J.S., Ko, T.S., Petros, D., Goldsbrough, P.B. and Korban, S.S. (2003) Overexpression of *Arabidopsis* phytochelatin synthase paradoxically leads to hypersensitivity to cadmium stress. *Plant Physiology* 131, 656–663.

Lee, S.H., Ahsan, N., Lee, K.W., Kim, D.H., Lee, D.G., Kwak, S.S., Kwon, S.Y., Kim, T.H. and Lee, B.H. (2007) Simultaneous overexpression of both Cu-Zn superoxide dismutase and ascorbate peroxidise in transgenic tall fescue plants confers increased tolerance to a wide range of abiotic stresses. *Plant Physiology* 164, 1626–1638.

Leitao, L., Goulas, P. and Biolley, J.P. (2003) Timecourse of Rubisco oxidation in beans (*Phaseolus vulgaris* L.) subjected to a long-term ozone stress. *Plant Science* 165, 613–620.

Leitenmaier, B. and Küpper, H. (2011) Cadmium uptake and sequestration kinetics in individual leaf cell protoplasts of the Cd/Zn hyperaccumulator *Thlaspi caerulescens*. *Plant Cell and Environment* 34, 208–219.

Leustek, T., Martin, M.N., Bick, J.A. and Davies, J.P. (2000) Pathways and regulation of sulphur metabolism revealed through molecular and genetic studies. *Annual Review of Plant Physiology and Plant Molecular Biology* 51, 141–165.

Li, Z.S., Lu, Y.P., Zhen, R.G., Szczypka, M., Thiele, D.J. and Rea, P.A. (1997) A new pathway for vacuolar cadmium sequestration in *Saccharomyces cerevisiae*: YCF1-catalyzed transport of bis (glutathionato) cadmium. *Proceedings of National Academy of Sciences USA* 94, 42–47.

Liu, X., Hua, X., Guo, J.Q.D., Wang, L., Liu, Z., Jin, Z., Chen, S. and Liu, G. (2008) Enhanced tolerance to drought stress in transgenic tobacco plants overexpressing VTE1 for increased tocopherol production from *Arabidopsis thaliana*. *Biotechnology Letters* 30, 1275–1280.

Liu, X.M., Kim, K.E., Kim, K.C., Nguyen, X.C., Han, H.J., Jung, M.S. and Chung, W.S. (2010) Cadmium activates *Arabidopsis* MPK3 and MPK6 via accumulation of reactive oxygen species. *Phytochemistry* 71, 614–618.

Lopez-Martin, M.C., Romero, L.C. and Gotor, C. (2008) Cytosolic cysteine in redox signaling. *Plant Signaling and Behaviour* 3, 880–881.

Macek, T., Mackova, M., Pavlıkova, D., Szakova, J., Truksa, M., Singh-Cundy, A., Kotrba, P., Yancey, N. and Scouten, W.H. (2002) Accumulation of cadmium by transgenic tobacco. *Acta Biotechnology* 22, 101–106.

Marrs, K.A. (1996) The functions and regulation of glutathione S-transferases in plants. *Annual Review of Plant Physiology and Plant Molecular Biology* 47, 127–158.

Maser, P., Thomine, S., Schroeder, J.I., Ward, J.M., Hirschi, K., Sze, H., Talke, I.N., Amtmann, A., Maathuis, F.J., Sanders, D., Harper, J.F., Tchieu, J., Gribskov, M., Persans, M.W., Salt, D.E., Kim, S.A. and Guerinot, M.L. (2001) Phylogenetic relationships within cation transporter families of *Arabidopsis*. *Plant Physiology* 126, 1646–1667.

Masood, A., Iqbal, N. and Khan, N.A. (2012) Role of ethylene in alleviation of cadmium-induced photosynthetic capacity inhibition by sulphur in mustard. *Plant Cell and Environment* 35, 524–533.

Mattioni, C., Lacerenza, N.G., Troccoli, A., De Leonardis, A.M. and Di Fonzo, N. (1997) Water and salt stress-induced alterations in proline metabolism of *Triticum durum* seedlings. *Physiologia Plantarum* 101, 787–792.

Melhorn, H. (1990) Ethylene-promoted ascorbate peroxidase activity protects plants against hydrogen peroxide, ozone and paraquat. *Plant Cell and Environment* 13, 971–976.

Miller, G., Shulaev, V. and Mitter, R. (2008) Reactive oxygen signaling and abiotic stress. *Physiologia Plantarum* 133, 481–489.

Mills, R., Krijger, G., Baccarini, P., Hall, J.L. and Williams, L. (2003) Functional expression of AtHMA4, a P1B-type ATPase of the Zn/Co/Cd/Pb subclass. *The Plant Journal* 35, 164–176.

Minglin, L., Yuxiu, Z. and Tuanyao, C. (2005) Identification of genes up-regulated in response to Cd exposure in *Brassica juncea* L. *Gene* 363, 151–158.

Mishra, S., Srivastava, S., Tripathi, R.D., Govindarajan, R., Kuriakose, S.V. and Prasad, M.N.V. (2006) Phytochelatin synthesis and response of antioxidants during cadmium stress in *Bacopa monnieri* L. *Plant Physiology and Biochemistry* 44, 25–37.

Miyadate, H., Adachi, S., Hiraizumi, A., Tezuka, K., Nakazawa, N., Kawamoto, T., Katou, K., Kodama, I., Sakurai, K., Takahashi, H., Nagasawa, N.S., Watanabe, A., Fujimura, T. and Akagi, H. (2011) OsHMA, a P1B-type of ATPase affects root-to-shoot cadmium translocation in rice by mediating efflux into vacuoles. *New Phytologist* 189, 190–199.

Mobin, M. and Khan, N.A. (2007) Photosynthetic activity, pigment composition and antioxidative response of two mustard (*Brassica juncea*) cultivars differing in photosynthetic capacity subjected to cadmium stress. *Journal of Plant Physiology* 164, 601–610.

Moraghan, J.T. (1993) Accumulation of cadmium and selected elements in flax seed grown on a calcareous soil. *Plant and Soil* 150, 61–68.

Morel, M., Crouzet, J., Gravot, A., Auroy, P., Leonhardt, N., Vavasseur, A. and Richaud, P. (2009) AtHMA3, a P1B-ATPase allowing Cd/Zn/Co/Pb vacuolar storage in *Arabidopsis*. *Plant Physiology* 149, 894–904.

Morgan, P.W. and Drew, M.C. (1997) Ethylene and plant response to stress. *Physiologia Plantarum* 100, 620–630.

Nadarajah, K. and Sidek, H.M. (2010) The green MAPKs. *Australian Journal of Plant Science* 9, 1–10.

Nakamura, S., Suzui, N., Nagasaka, T., Komatsu, F., Ishioka, N.S., Ito-Tanabata, S., Kawachi, N., Rai, H., Hattori, H., Chino, M. and Fujimaki, S. (2013) Application of glutathione to roots selectively inhibits cadmium transport from roots to shoots in oilseed rape. *Journal of Experimental Botany* 64, 1073–1081.

Neelam, A. and Rai, L.C. (2003) Differential responses of three cyanobacteria to UV-B and Cd. *Journal of Microbiology and Biotechnology* 13, 544–551.

Neumann, D., Lichtenberger, O., Günther, D., Tschiersch, K. and Nover, L. (1994) Heat-shock proteins induce heavy-metal tolerance in higher plants. *Planta* 194, 360–367.

Ning, H., Zhang, C., Yao, Y. and Yu, D. (2010) Overexpression of a soybean O-acetylserine (thiol) lyase encoding gene GmOASTL4 in tobacco increases cysteine levels and enhances tolerance to cadmium stress. *Biotechnology Letters* 32, 564–557.

Nouairi, I., Ammar, W.B., Youssef, N.B., BenMiled, D.D., Ghorbal, M.H. and Zarrouk, M. (2009) Antioxidant defense system in leaves of Indian mustard (*Brassica juncea*) and rape (*Brassica napus*) under cadmium stress. *Acta Physiologiae Plantarum* 31, 237–247.

Nussbaum, S., Schmutz, D. and Brunold, C. (1988) Regulation of assimilatory sulphate reduction by cadmium in *Zea mays* L. *Plant Physiology* 88, 1407–1410.

Obata, H., Inoue, N. and Umebayashi, M. (1996) Effect of cadmium on plasma membrane ATPase from plant roots differing in tolerance to cadmium. *Soil Science and Plant Nutrition* 42, 361–366.

Oda, K., Otani, M., Uraguchi, S., Akihiro, T. and Fujiwara, T. (2011) Rice ABCG43 is Cd inducible and confers Cd tolerance on yeast. *Bioscience, Biotechnology and Biochemistry* 75, 1211–1213.

Ogawa, T., Pan, L., Kawai-Yamada, M., Yu, L.H., Yamamura, S., Loyama, T., Kitajima, S., Ohme-Takagi, M., Sato, F. and Uchimiya, H. (2005) Functional analysis of *Arabidopsis thaliana* ethylene-responsive element binding protein, AtEBP, conferring resistance to Bax and abiotic stress induced plant cell death. *Plant Physiology* 138, 1436–1445.

O'Malley, R.C., Rodriguez, F.I., Esch, J.J., Binder, B.M., O'Donnell, P., Klee, H.J. and Bleecker, A.B. (2005) Ethylene-binding activity, gene expression levels, and receptor system output for ethylene receptor family members from *Arabidopsis* and tomato. *The Plant Journal* 41, 651–659.

Pagliano, C., Raviolo, M., Vecchia, F.D., Gabbrielli, R., Gonneli, C., Rascio, N., Barbato, R. and Rocca, N. (2006) Evidence for PSII donor-side damage and photoinhibition induced by cadmium treatment on rice (*Oryza sativa* L.) *Photochemistry and Photobiology* 84, 70–78.

Pankovica, D., Plesniciar, M., Arsenijevica-Maksimovica, I., Petrovic, N., Sakac, Z. and Kastori, R. (2000) Effects of nitrogen nutrition on photosynthesis in Cd treated sunflower plants. *Annals of Botany* 86, 841–847.

Papoyan, A. and Kochian, L.V. (2004) Identification of *Thlaspi caerulescens* genes that may be involved in heavy metal hyperaccumulation and tolerance. Characterization of a novel heavy metal transporting ATPase. *Plant Physiology* 136, 3814–3823.

Park, J., Song, W.Y., Ko, D., Eom, Y., Hansen, T.H., Schiller, M., Lee, T.G., Martinoia, E. and Lee, Y. (2012) The phytochelatin transporters AtABCC1 and AtABCC2 mediate tolerance to cadmium and mercury. *The Plant Journal* 69, 278–288.

Patel, H.V., Parmar, S.R., Chudasama, C.J. and Mangrola, A.V. (2013) Interactive studies of zinc with cadmium and arsenic on seed germination and antioxidants properties of *Phaseolus aureus* Roxb. *International Journal of Plant, Animal and Environmental Sciences* 3, 166–174.

Perreault, F., Dionne, J., Didur, O., Juneau, P. and Popovic, R. (2011) Effect of cadmium on photosystem II activity in *Chlamydomonas reinhardtii*: alteration of O–J–I–P fluorescence transients indicating the change of apparent activation energies within photosystem II. *Photosynthesis Research* 107, 151–157.

Pierik, R., Tholen, D., Poorter, H., Visser, E.J.W. and Voesenek, L.A.C.J. (2006) The Janus face of ethylene: growth inhibition and stimulation. *Trends in Plant Science* 11, 176–183.

Piotrowska-Niczyporuk, A., Bajguz, A.B., Zambrzycka, E. and Godlewska-Żyłkiewicz, B. (2012) Phytohormones as regulators of heavy metal biosorption and toxicity in green alga *Chlorella vulgaris* (Chlorophyceae). *Plant Physiology and Biochemistry* 52, 52–65.

Pomponi, M., Censi, V., Di Girolamo, V., De Paolis, A., di Toppi, L.S., Aromolo, R., Costantino, P. and Cardarelli, M. (2006) Overexpression of *Arabidopsis* phytochelatin synthase in tobacco plants enhances Cd^{2+} tolerance and accumulation but not translocation to the shoot. *Planta* 223, 180–190.

Rausch, T. and Wachter, A. (2005) Sulphur metabolism: a versatile platform for launching defence operations. *Trends in Plant Science* 10, 503–509.

Rea, P.A., Li, Z.S., Lu, Y.P., Drozdowicz, Y.M. and Martinoia, E. (1998) From vacuolar GS-X pumps to multispecific ABC transporters. *Annual Review of Plant Physiology and Plant Molecular Biology* 49, 727–760.

Redjala, T., Sterckeman, T. and Morel, J.L. (2009) Cadmium uptake by roots: contribution of apoplast and of high and low-affinity membrane transport systems. *Environmental and Experimental Botany* 67, 235–242.

Rodríguez-Serrano, M., Romero-Puertas, M.C., Zabalza, A., Corpas, F.J., Gómez, M., del Río, L.A. and Sandalio, L.M. (2006) Cadmium effect on oxidative metabolism of pea (*Pisum sativum* L.) roots: imaging of reactive oxygen species and nitric oxide accumulation *in vivo*. *Plant Cell and Environment* 29, 1532–1544.

Rodríguez-Serrano, M., Romero-Puertas, M.C., Pazmiño, D.M., Testillano, P.S., Risueño, M.C., del RíoL, A. and Sandalio, L.M. (2009) Cellular response of pea plants to cadmium toxicity: cross talk between reactive oxygen species, nitric oxide, and calcium. *Plant Physiology* 150, 229–243.

Saidia, I., Ayounia, M., Dhiebb, A., Chtouroub, Y., Chaïbia, W. and Djebali, W. (2013) Oxidative damages induced by short-term exposure to cadmium in bean plants: protective role of salicylic acid. *South African Journal of Botany* 85, 32–38.

Sairam, R.K. and Tyagi, A. (2004) Physiology and molecular biology of salinity stress tolerance in plants. *Current Science* 86, 407–412.

Saito, K. (2004) Sulphur assimilatory metabolism. The long and smelling road. *Plant Physiology* 136, 2443–2450.

Sajidu, S.M.I., Henry, E.M.T., Persson, I., Masamba, W.R.L. and Kayambazinthu, D. (2006) pH dependence of sorption of Cd^{2+} Zn^{2+} Cu^{2+} and Cr^{3+} on crude water and sodium chloride extracts of *Moringa stenopetala* and *Moringa oleifera*. *African Journal of Biotechnology* 5, 2397–2401.

Sakakibara, H., Taniguchi, M. and Sugiyama, T. (2000) His-Asp phosphorelay signaling: a communication avenue between plants and their environment. *Plant Molecular Biology* 42, 273–278.

Salt, D.E., Prince, R.C., Pickering, I.J. and Raskin, I. (1995) Mechanisms of cadmium mobility and accumulation in Indian mustard. *Plant Physiology* 109, 1427–1433.

Sandalio, L.M., Dalurzo, H.C., Gomez, M., Romero-Puertas, M.C. and del Rio, L.A. (2001) Cadmium-induced changes in the growth and oxidative metabolism of pea plants. *Journal of Experimental Botany* 52, 2115–2126.

Sarwar, N., Naeem, A., Saifullah, Malhi, B.S., Farid, S.S. and Zia, M.H. (2010) Role of mineral nutrition in minimizing cadmium accumulation by plants. *Journal of Science, Food and Agriculture* 90, 925–937.

Scebba, F., Arduini, I., Ercoli, L. and Sebastiani, L. (2006) Cadmium effects on growth and antioxidant enzymes activities in *Miscanthus sinensis*. *Biologia Plantarum* 50, 688–692.

Schaller, G.E. and Bleecker, A.B. (1995) Ethylene-binding sites generated in yeast expressing the *Arabidopsis* ETR1 gene. *Science* 270, 1809–1811.

Schutzendubel, A. and Polle, A. (2002) Plant responses to abiotic stresses: heavy metal induced oxidative stress and protection by mycorrhization. *Journal of Experimental Botany* 52, 1351–1365.

Semane, B., Cuypers, A., Smeets, K., Belleghem, F., Horemans, N. and Schat, H. (2007) Cadmium responses in *Arabidopsis thaliana*: glutathione metabolism and antioxidative defence system. *Physiologia Plantarum* 129, 519–528.

Seregin, I.V. and Ivanov, V.B. (2001) Physiological aspects of cadmium and lead toxic effects on higher plants. *Russian Journal of Plant Physiology* 48, 523–544.

Shah, K. and Dubey, R.S. (1995) Effect of cadmium on RNA level as well as activity and molecular forms of ribonuclease in growing rice seedlings. *Plant Physiology and Biochemistry* 33, 577–584.

Shahrtash, M., Mohsenzadeh, S., Zare, H. and Mohabatkar, H. (2011) Role of calcium in alleviation of cadmium toxicity in maize seedling. *Electronic Journal of Environmental, Agricultural and Food Chemistry* 10, 2404–2412.

Shamsi, I.H., Wei, K., Jilani, G. and Zhang, G.P. (2007) Interactions of cadmium and aluminum toxicity in their effect on growth and physiological parameters in soybean. *Journal of Zhejiang University Science* 8, 181–188.

Shanmugaraj, M.B., Chandra, H.M, Srinivasan, B. and Ramalingam, S.K. (2013) Cadmium induced physio-biochemical and molecular response in *Brassica juncea*. *International Journal of Phytoremediation* 15, 206–218.

Sharma, R.K., Agrawal, M. and Agrawal, S.B. (2010) Physiological, biochemical and growth responses of lady's finger (*Abelmoschus esculentus* L.) plants as affected by Cd contaminated soil. *Bulletein of Environmental and Contaminated Toxicology* 84, 765–770.

Sharma, S.S. and Dietz, K.J. (2006) The significance of amino acids and amino acid-derived molecules in plant responses and adaptation to heavy metal stress. *Journal of Experimental Botany* 57, 711–726.

Shaul, O. (2002) Magnesium transport and function in plants: the tip of the iceberg. *BioMetals* 15, 309–323.

Siefermann-Harms, D. (1987) The light-harvesting and protective functions of carotenoids in photosynthetic membranes. *Physiologia Plantarum* 69, 561–568.

Skorzynska-Polit, E., Drazkiewicz, M. and Krupa, Z. (2010) Lipid peroxidation and antioxidative response in *Arabidopsis thaliana* exposed to cadmium and copper. *Acta Physiologiae Plantarum* 32, 169–175.

Song, W.Y., Sohn, E.J., Martinoia, E., Lee, Y.J., Yang, Y.Y., Jasinski, M., Forestier, C., Hwang, I. and Lee, Y. (2003) Engineering tolerance and accumulation of lead and cadmium in transgenic plants. *Nature Biotechnology* 21, 914–919.

Sridhar, B.B.M., Diehl, S.V., Han, F.X., Monts, D.L. and Su, Y. (2005) Anatomical changes due to uptake and accumulation of Zn and Cd in Indian mustard (*Brassica juncea*). *Environmental and Experimental Botany* 54, 131–141.

Sun, J., Wang, R., Zhang, X., Yu, Y., Zhao, R., Li, Z. and Chen, S. (2013) Hydrogen sulfide alleviates cadmium toxicity through regulations of cadmium transport across the plasma and vacuolar membranes in *Populus euphratica* cells. *Plant Physiology and Biochemistry* 65, 67–74.

Suzuki, N. (2005) Alleviation by calcium of cadmium-induced root growth inhibition in *Arabidopsis* seedlings. *Plant Biotechnology* 22, 19–25.

Szalai, G., Kellos, T., Galiba, G. and Kocsy, G. (2009) Glutathione as an antioxidant and regulatory molecule in plants under abiotic stress conditions. *Plant Growth Regulation* 28, 66–80.

Szczypka, M.S., Wemmie, J.A., Moye-Rowley, W.S. and Thiele, D.J. (1994) A yeast metal resistance protein similar to human cystic fibrosis transmembrane conductance regulator (CFTR) and multidrug resistance-associated protein. *The Journal of Biological Chemistry* 269, 22853–22857.

Takahashi, H., Kopriva, S., Giordano, M., Saito, K. and Hell, R. (2011) Sulphur assimilation in photosynthetic organisms: molecular functions and regulations of transporters and assimilatory enzymes. *Annual Review of Plant Biology* 62, 157–184.

Teuveny, Z. and Filner, P. (1997) Regulation of adenosine triphosphate sulphurylase in cultured tobacco cells. *The Journal of Biology and Chemistry* 252, 1858–1864.

Thapar, R., Srivastava, A.K., Bhargava, P., Mishra, Y. and Rai, L.C. (2008) Impact of different abiotic stress on growth, photosynthetic electron transport chain, nutrient uptake and enzyme activities of Cu-acclimated *Anabaena doliolum*. *Journal of Plant Physiology* 165, 306–316.

Tlustos, P., Szakova, J., Hruby, J., Hartman, I., Najmanova, J., Nedelnik, J., Pavlikova, D. and Batysta, M. (2006) Removal of As, Cd, Pb and Zn from contaminated soil by high biomass producing plants. *Plant, Soil and Environment* 52, 413–423.

Tudoreanu, L. and Phillips, C.J.C. (2004) Modeling cadmium uptake and accumulation in plants. *Advances in Agronomy* 84, 121–157.

Ueno, D., Iwashita, T., Zhao, F. and Ma, J.F. (2008) Characterization of Cd translocation and identification of the Cd form in xylem sap of the Cd-hyperaccumulator *Arabidopsis* halleri. *Plant Cell and Physiology* 49, 540–548.

Umar, S., Diva, I., Anjum, N.A. and Iqbal, M. (2008) Research Findings: II Potassium nutrition reduces cadmium accumulation and oxidative burst in mustard (*Brassica campestris* L.). *Electronic International Fertilizer Correspondent* no. 16.

Valentovicova, K., Haluskova, L., Huttova, J., Mistrık, I. and Tamas, L. (2010) Effect of cadmium on diaphorase activity and nitric oxide production in barley root tips. *Journal of Plant Physiology* 167, 10–14.

Van Assche, F. and Clijsters, H. (1990) Effects of heavy metals on enzyme activity in plants. *Plant Cell and Environment* 13, 195–206.

Vandenbussche, F., Vriezen, W.H., Smalle, J., Laarhoven, L.J., Harren, F.J. and Van Der Straeten, D. (2003) Ethylene and auxin control the *Arabidopsis* response to decreased light intensity. *Plant Physiology* 133, 517–527.

Vassilev, A. and Lidon, F. (2011) Cd-induced membrane damages and changes in soluble protein and free amino acid contents in young barley plants. *Emirates Journal of Food and Agriculture* 23, 130–136.

Vassilev, A., Lidon, F., Scotti, P., Da Graca, M. and Yordanov, I. (2004) Cadmium-induced changes in chloroplast lipids and photosystem activities in barley plants. *Biologia Plantarum* 48, 153–156.

Vatamaniuk, O.K., Mari, S., Lu, Y.P. and Rea, P.A. (2000) Mechanism of heavy metal ion activation of phytochelatin (PC) synthase. *The Journal of Biology and Chemistry* 275, 31451–31459.

Verbruggen, N., Hermans, C. and Schat, H. (2009) Mechanisms to cope with arsenic or cadmium excess in plants. *Current Opinion in Plant Biology* 12, 364–372.

Vernoux, T., Wilson, R.C., Seeley, K.A., Reichheld, J.P., Muroy, S., Brown, S., Maughan, S.C., Cobbett, C.S., Van Montagu, M., Inze, D., May, M.J. and Sung, Z.R. (2001) The Root Meristem less1/Cadmium Sensitive 2 gene defines a glutathione-dependent pathway involved in initiation and maintenance of cell division during postembyronic root development. *Plant Cell* 12, 97–109.

Vitti, A., Nuzzaci, M., Scopa, A., Tataranni, G., Remans, T., Jaco, V. and Sofo, A. (2013) Auxin and cytokinin metabolism and root morphological modifications in *Arabidopsis* thaliana seedlings infected with cucumber mosaic virus (CMV) or exposed to cadmium. *International Journal of Molecular Science* 14, 6889–6902.

Wahid, A. and Ghani, A. (2008) Varietal differences in mungbean (*Vigna radiata*) for growth, yield, toxicity symptoms and cadmium accumulation. *Annals of Applied Biology* 152, 59–69.

Wang, C., Zhang, S.H., Wang, P.F., Hou, J., Zhang, W.J., Li, W. and Lin, Z.P. (2009) The effect of excess Zn on mineral nutrition and antioxidative response in rapeseed seedlings. *Chemosphere* 11, 1468–1476.

Wang, J., Ding, H., Zhang, A., Ma, F., Cao, J. and Jiang, M. (2010) A novel mitogen-activated protein kinase gene in maize (*Zea mays*), ZmMPK3, is involved in response to diverse environmental cues. *Journal of Integrative Plant Biology* 52, 442–452.

Wang, K., Li, H. and Ecker, J. (2002) Ethylene biosynthesis and signaling networks. *Plant Cell* 14, 131–151.

Wang, L., Zhou, Q.X., Ding, L.L. and Sun, Y.B. (2008) Effect of cadmium toxicity on nitrogen metabolism in leaves of *Solanum nigrum* L. as a newly found cadmium hyperaccumulator. *Journal of Hazardous Material* 154, 818–825.

Wang, Y., Feng, H., Qu, Y., Cheng, J., Zhao, Z., Zhang, M., Wang, X. and An, L. (2006) The relationships between reactive oxygen species and nitric oxide in ultraviolet-B-induced ethylene production in leaves of maize seedlings. *Environmental and Experimental Botany* 57, 51–61.

Weber, M., Trampczynska, A. and Clemens, S. (2006) Comparative transcriptome analysis of toxic metal responses in *Arabidopsis thaliana* and the Cd-hypertolerant facultative metallophyte *Arabidopsis halleri*. *Plant Cell and Environment* 29, 950–963.

White, P.J. (2001) The pathways of calcium movement to the xylem. *Journal of Experimental Botany* 52, 891–899.

Williams, L.E., Pittman, J.K. and Hall, J.L. (2000) Emerging mechanisms for heavy metal transport in plants. *Biochimica Biophysica et Acta* 1465, 104–126.

Wirtz, M. and Droux, M. (2005) Synthesis of the sulphur amino acids: cysteine and methionine. *Photosynthesis Research* 86, 345–362.

Wonisch, W. and Schaur, R.J. (2001) Chemistry of glutathione. In: Grill, D., Tausz, M. and De Kok, L.J. (eds) *Significance of Glutathione to Plant Adaptation to the Environment*. Kluwer Academic Publishers, Dordrecht, the Netherlands, pp. 13–26.

Xiang, C. and Oliver, D.J. (1998) Glutathione metabolic genes coordinately respond to heavy metals and jasmonic acid in *Arabidopsis*. *Plant Cell* 10, 1539–1550.

Yanagisawa, S., Yoo, S.D. and Sheen, J. (2003) Differential regulation of EIN3 stability by glucose and ethylene signaling in plants. *Nature* 425, 521–525.

Yeh, T.C., Marsh, V., Bernat, B.A., Ballard, J., Colwell, H., Evans, R.J. and Wallace, E. (2007) Biological characterization of ARRY-142886 (AZD6244), a potent, highly selective mitogen-activated protein kinase kinase 1/2 inhibitor. *Clinical Cancer Research* 13, 1576–1583.

Zhang, S., Zhang, H., Qin, R., Jiang, W. and Liu, D. (2009) Cadmium induction of lipid peroxidation and effects on root tip cells and antioxidant enzyme activities in *Vicia faba* L. *Ecotoxicology* 18, 814–823.

Zhou, W.B., Philippe, J. and Qiu, B.S. (2006) Growth and photosynthetic responses of the bloom-forming cyanobacterium *Microcystis aeruginosa* to elevated levels of cadmium. *Chemosphere* 65, 1738–1746.

Zhu, Y.L., Pilon-Smits, E.A., Tarun, A.S., Weber, S.U., Jouanin, L. and Terry, N. (1999) Cadmium tolerance and accumulation in Indian mustard is enhanced by overexpressing gamma-glutamylcysteine synthetase. *Plant Physiology* 121, 1169–1178.

10 Heavy Metal and Metalloid Stress in Plants: The Genomics Perspective

Piyalee Panda,[1] Lingaraj Sahoo[2] and Sanjib Kumar Panda[1]*

[1]*Plant Molecular Biotechnology Laboratory, Department of Life Science and Bioinformatics, Assam University, Silchar;* [2]*Department of Biotechnology, Indian Institute of Technology, Guwahati, India*

Abstract

Heavy metal and metalloid stress are major abiotic stress factors that limit crop production and reduce agricultural yield. Beside natural factors, human activities have contributed to the enormous increase in heavy metal and/or metalloid pollution in the environment. Both heavy metals and metalloids exert deleterious impacts on plant growth and development. High production of reactive oxygen species (ROS) results in oxidative damage to important cellular components. Our understanding of these factors and the key mechanisms involving a wide array of genes and their expression is far from complete. In the past few decades, several molecular mechanisms were identified, which are particularly related to heavy metal transport and hypertolerance. Moreover, expression profiles of several major genes associated with heavy metal/metalloid stress were elucidated and provided vital insights to our understanding of heavy metal stress in plants. This review focuses on the molecular physiology of heavy metal and metalloid stress in plants. We discuss here the aspects related to the physiology of heavy metal stress, ROS and several molecular events that are associated with metal tolerance.

10.1 Introduction

Environmental pollution imposes a significant impact on global climate. The change in global climatic conditions has intensified the frequency of many abiotic stress factors, which imposes a severe threat to agricultural productivity. It is estimated that approximately 70% of the total reduction in crop yield worldwide is due to various abiotic stresses. This has also threatened global food security. Heavy metal contamination of the environment occurs mainly due to extensive industrial pollution and other anthropogenic activities. They pose a severe threat to human health and their hazardous effects are being constantly monitored and reviewed by several international bodies (Jarup, 2003). Heavy metal exposure is usually chronic and causes a large number of diseases such as mental lapses, kidney damage, hepatic damage, skin poisoning, cancer and neurological disorders (Jarup, 2003). Plants require metals for their normal metabolism. Heavy metals like cadmium (Cd), chromium (Cr), arsenic (As), lead (Pb) and mercury (Hg) are regarded as non-essential elements and do not have any role in physiology and metabolism. On the other hand copper (Cu), zinc (Zn), manganese (Mn), magnesium (Mg), cobalt (Co), etc.

---

*E-mail: drskpanda@gmail.com

are essential elements required by the plants. Exceeding the threshold level of these essential elements causes alterations in normal physiological functions.

In the past few decades, extensive industrialization and anthropogenic activities have released these toxic components into the environment. The concentration of heavy metals in soil is increasing constantly. Cadmium is a toxic heavy metal, whose permissible limit is 100 mg kg^{-1} soil (Salt et al., 1995; Yadav, 2010). High concentration of Cd in soil causes loss of photosynthetic efficiency, disturbs water uptake by roots, causes nutrient imbalance and produces a high concentration of reactive oxygen species (ROS) in cells (Sandalio et al., 2001; Choudhury and Panda, 2004; Wojcik and Tukiendrof, 2004; Mohanpuria et al., 2007). Chromium is at third place amongst the top six toxic threats of the world, affecting approximately 7.3 million people around the world. In plants, Cr has no biological significance. It is used in the paint industry, leather tanning and metal welding. In nature, Cr exists as trivalent (Cr III) and hexavalent (Cr VI) forms. The bioavailability of Cr (VI) is considerably high, which makes it more toxic than other forms (Panda and Choudhury, 2005). Cr is associated with inhibition of seed germination, degradation of chlorophyll and pigments, and produces ROS in cells (Panda, 2007; Ali et al., 2011). Lead is another toxic heavy metal that does not have any role in plants' physiology and metabolism (Yadav, 2010). Pb is usually released into the environment due to disposal of industrial wastes and burning of fossil fuels. In plants, Pb causes inhibition of antioxidant enzymes and produces a high concentration of ROS causing oxidative damage (Choudhury and Panda, 2004; Sharma and Dubey, 2005).

Nickel is released into the environment due to widespread mining, smelting waste, pesticides and phosphate fertilizers (Gimeno-Gracía et al., 1996). Ni causes changes in physiological processes leading to chlorosis and necrosis of leaves (Zornoza et al., 1999; Pandey and Sharma, 2002; Rahman et al., 2005). Impaired growth, nutrient imbalance and loss of cell membrane function is associated with Ni toxicity in plants (Yadav, 2010). Ni causes inhibition of H-ATPase activity and initiates peroxidation of membrane lipids (Ros et al., 1992). Cobalt concentration in the environment increases due to burning of sewage, burning of fossil fuels and degradation of Co alloys (Barceloux, 1999). High concentration of Co is also associated with oxidative stress and disruption of important physiological process in plants (Chatterjee and Chatterjee, 2000). Copper and Zn are toxic to plants beyond their threshold levels. Cu toxicity is strongly associated with production of ROS and oxidative damage in plants (Stadtman and Oliver, 1991; Thounaojam et al., 2013) and also causes loss of photosynthetic efficiency in plants along with growth inhibition (Hegedüs et al., 2001). Zn in excess causes growth retardation, imbalance in nutrient levels and causes oxidative damage (Choi et al., 1996; Ebbs and Kochian, 1997; Fontes and Cox, 1998; Panda and Choudhury, 2005; Choudhury et al., 2013). Understanding the molecular physiology of heavy metal stress in plants is far from complete. With advancement in molecular biology and functional genomics, several mechanisms and regulatory networks involved in heavy metal stress and tolerance have been elucidated. Such findings have opened new avenues to explore and understand heavy metal stress in plants. In this chapter, we highlight certain aspects of heavy metal stress in plants and the molecular mechanisms underlying stress response and tolerance. Sources of heavy metal contamination in soil and plants and their overall effect on plant metabolism are highlighted in Fig. 10.1.

10.2 Reactive Oxygen Species and Oxidative Stress

ROS are inevitable entities of aerobic life. Abiotic stresses such as drought, salt, heat and heavy metals are known to produce ROS, which are also considered as by-products of aerobic metabolism, and cellular compartments that have strong electron flow are associated with high ROS production. ROS include entities like hydrogen peroxide (H_2O_2), superoxide radical ($O_2^{\bullet-}$), hydroxyl radical ($OH^{\bullet}$) and singlet oxygen ($^1O_2$) (Halliwell and Gutteridge, 1989). During heavy metal stress in plants,

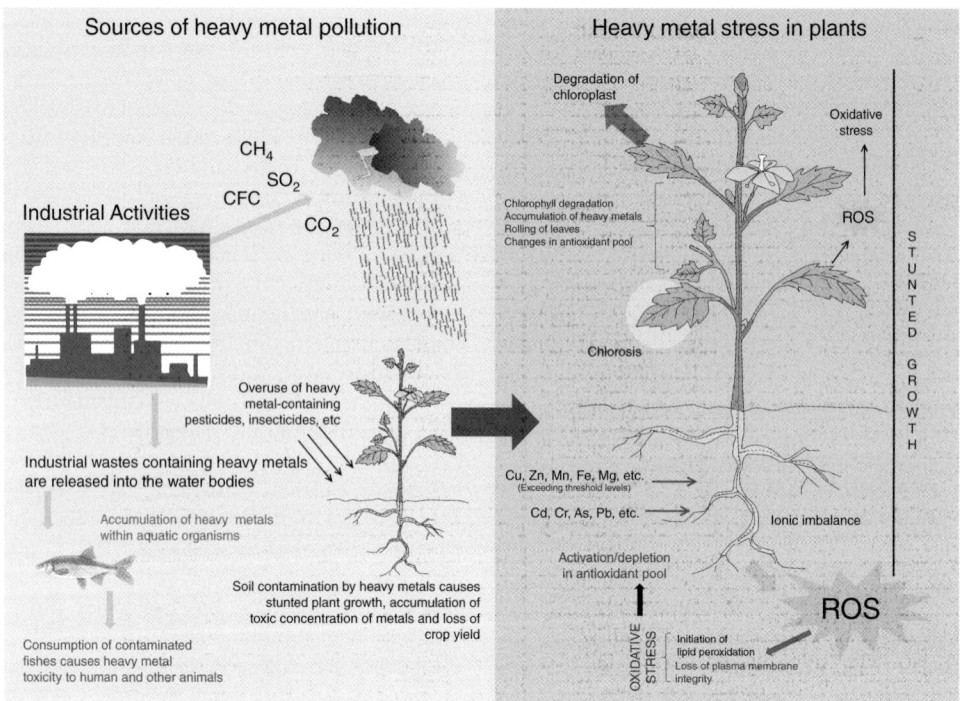

Fig. 10.1. Sources of heavy metal pollution and their effects on plants.

ROS are primarily produced and exert severe oxidative load. Studies have demonstrated that high ROS production during stress can be one of the major factors for loss of crop yield. Further, during abiotic stresses, ROS metabolism and related signalling mechanisms are common features (Munne-Bosch et al., 2013). Plants have intrinsic antioxidant enzymes and other defence metabolites, which scavenge deleterious effects of ROS in cells. These antioxidants protect the cell from oxidative damage and support plant growth (Foyer and Noctor, 2005). The production of ROS in cells is genetically programmed (Shao et al., 2008). ROS, like H_2O_2 and $O_2^{\bullet-}$, besides causing oxidative stress also act as important signalling molecules (Choudhury et al., 2013). The production of ROS in plants is shown in Fig. 10.2.

Heavy metal toxicity in plants causes displacement of essential ions in the cells. In case of Cd toxicity, studies have shown that calcium (Ca) ions are replaced in the photosystem II (PSII), which results in inactivation of PSII photoactivation (Faller et al., 2005; Sharma and Dietz, 2008). The main source of ROS in cells results due to electron transfer in mitochondria, chloroplast and oxidative metabolism in the peroxisomes (Sharma and Dietz, 2008). In *Pinus sylvestris*, Cd treatment for 6 h generates ROS (Schutzendubel et al., 2001). Studies have also shown that in *Medicago sativa* and *Nicotiana tabacum* Cd treatment causes high production of ROS (Garnier et al., 2006; Matsui, 2006). ROS like $OH^{\bullet}$ can cause peroxidation of polyunsaturated fatty acids (PUFA) in the membrane (Mithofer et al., 2004). In plants, the enzymic lipid peroxidation is catalysed by α-dioxygenases and lipooxygenase, which can convert unsaturated fatty acids to lipid peroxides. Oxygen is produced during photosynthesis. In PSII, $^1O_2$ are continuously produced. H_2O_2 can be reduced to $OH^{\bullet}$ by $O_2^{\bullet-}$ in presence of transition metals (Apel and Hirt, 2004). In comparison to other ROS, $OH^{\bullet}$ is most reactive and cells do not possess a proper scavenging system for it (Apel and Hirt, 2004). The production of $OH^{\bullet}$ can only be restricted in plants by controlling the reactions leading to its formation (Apel and Hirt, 2004). Plant defence mechanisms include different antioxidant

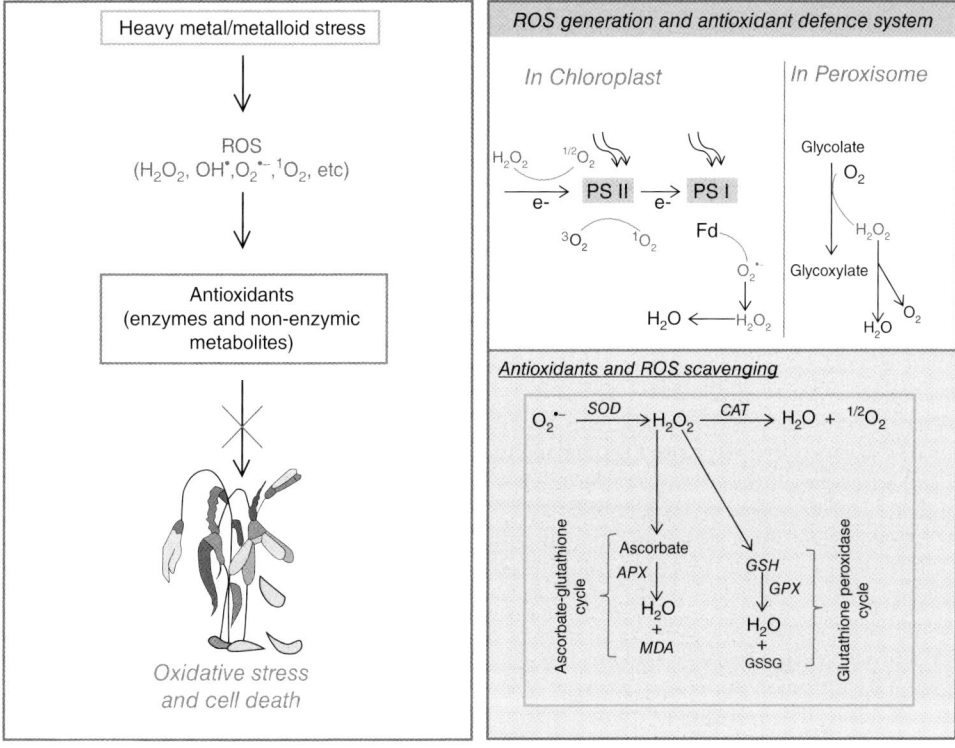

Fig. 10.2. Heavy metal/metalloid-induced ROS generation and antioxidant defence system in plants.

enzymes and non-enzymic antioxidants. The enzymic ROS-scavenging antioxidants include catalase (CAT), superoxide dismutase (SOD), ascorbate peroxidase (APX), glutathione peroxidase (GPX), etc. (Halliwell and Gutteridge, 1989). SOD converts $O_2^{\bullet-}$ to H_2O_2, which is further acted upon by CAT, APX and GPX to form H_2O (Halliwell and Gutteridge, 1989; Apel and Hirt, 2004). The APX requires the ascorbate-glutathione cycle, and detoxification of H_2O_2 to H_2O takes place by oxidation of ascorbate to mono-dehydroascorbate (MDA) (reviewed by Apel and Hirt, 2004). Non-enzymic antioxidant metabolites are also key players in ROS detoxification. These include ascorbate, glutathione, α-tocopherol, etc. (reviewed by Choudhury et al., 2013). Studies have shown that a high ratio of ascorbate and glutathione (GSH:GSSG) plays a significant role in ROS detoxification (Apel and Hirt, 2004). High concentration of ROS in cells can alter gene expression levels and impose oxidative stress.

The recognition of diverse signalling pathways behind Al-induced oxidative stress has opened the way to understand the role of oxidative damage and whether it can be a possible biomarker to understand Al stress and tolerance in plants. Earlier, through molecular approaches, the expression of the levels of typical ROS-responsive genes were found to be altered by Al, for example, oxidative stress-responsive genes such as catalase was induced in wheat (Snowden and Gardner, 1993) and *Arabidopsis* (Richards et al., 1994). Studies on *Arabidopsis* have shown that long-time exposure of Al resulted in progressive increase of peroxidase mRNA levels along with induction of SOD mRNA and subsequent decline of catalase. Ezaki et al. (2001) demonstrated that ectopic expression of *NtPox* (tobacco peroxidase) and *parB* (tobacco glutathione S-transferase) enhanced the activities of glutathione S-transferase and peroxidase in transgenic *Arabidopsis* during Al stress. The over-expression of *parB* and *NtPox* conferred

Al tolerance, which suggests that these genes are responsible for alleviation of oxidative damage generated by ROS. Genes responding to H_2O_2 such as PR-proteins were up-regulated in wheat under Al stress (Cruz-Ortega et al., 1997). It was also shown that the transcript level of alanine aminotransferase was reduced under Al stress, which suggested crisis in the metabolism of sugar/amino acids that ultimately disturbs redox homoeostasis. Later, using microarray analysis, a similar set of Al-inducible genes was identified in *Arabidopsis* (Kumari et al., 2008; Zhao et al., 2009) and maize (Maron et al., 2008). Thus, ROS-induced responses are critically important for Al tolerance. The expressed sequence tag (EST) analysis of gene expression in rye under Al stress showed the involvement of novel oxidative stress genes, which included glutathione peroxidase, glucose-6-phosphate dehydrogenase and ascorbate peroxidase. Al induces the organ-specific expression of a cell wall-associated receptor kinase 1 (*WAK1*) gene (Sivaguru et al., 2003). It was also shown that the transgenic plant, which over-expresses *WAK1*, confers significant tolerance to Al in comparison to the wild type in terms of root growth. Mitochondrial function was altered significantly by Al (Panda et al., 2008). Al enhanced production of mitochondrial H_2O_2 and $O_2^{\bullet -}$, caused the opening of mitochondrial permeability transition (MPT) pore and depolarized the mitochondrial inner membrane potential (Panda et al., 2008). Subsequent *in vitro* studies on respiratory changes and ROS production in isolated mitochondria and whole cell of tobacco have shown that AOX pathway is severely affected by Al stress and over-expression of AOX conferred Al tolerance (Panda et al., 2013).

10.3 Molecular Physiology of Heavy Metal and Metalloid Stress Tolerance

In the past few decades our understanding on heavy metal/metalloid stress tolerance in plants has improved considerably. Several intrinsic factors associated with heavy metal stress and tolerance are now known. Transcription factors and gene expression analysis have helped in categorizing key factors associated with heavy metal stress in plants. Heavy metal stress is related to several important traits and these traits are controlled by diverse and complex mechanisms involving several genes, which are either up- or down-regulated. In this section, we summarize some important aspects of different molecular mechanisms associated with heavy metal stress and tolerance in plants. Studies on yeast have significantly contributed to our understanding of heavy metal stress responses in higher eukaryotes. Several metal transporters were identified that mediate uptake of heavy metals from soil into the root system. For example, COPT1 was identified as Cu transporter, while ZIP family transporters facilitates uptake of Zn (*ZRT*) and Fe (*IRT*) (Kampfenkel et al., 1995; Fox and Guerinot, 1998; Saier, 2000; Clemens, 2001). In *Arabidopsis*, several members of ZIP were identified. In *Noccaea caerulescens* (*Thlaspi caerulescens*), a ZIP transporter ZNT1 was cloned, which mediates both Zn and Cd uptake (Pence et al., 2000). Lombi et al. (2002) later cloned the orthologue of the *Arabidopsis IRT1* from *N. caerlescens*.

In plants, Zn transporters are located in the plasma membrane and control the entry of Zn into the cells. Several gene families were identified to be involved in metal transport. One of the important families is the cation diffusion family (CDF) (Kramer et al., 2007). Although the function of CDF still remains largely unknown, it is believed that CDF plays some role in conferring heavy metal tolerance (Kramer et al., 2007). Lang et al. (2005) reported four CDF genes, *BjCET1–4* from *Brassica juncea*, and later evaluated their role in tolerance to Zn and Cd stress (Xu et al., 2009). It was demonstrated that the heterologous expression of *BjCET2* in double mutant yeast Δ*zrc1* and Δ*cot1* improved metal tolerance and reduced the uptake rate of Zn and Cd (Xu et al., 2009). The over-expression of *BjCET2* in transgenic *B. juncea* conferred high tolerance to Zn and Cd (Xu et al., 2009). In *BjCET2*-deficient lines, both Zn and Cd sensitivity was considerably high. The results suggest that *BjCET2* plays a crucial role in conferring tolerance to Zn and Cd in *B. juncea* (Xu et al., 2009). The *mer* (mercuric ion resistance) are bacterial heavy metal resistance determinants that represent influx-type transporters (Silver and Phung, 1996). MerC is a bacterial heavy metal transporter from *mer* operon that senses and transports Hg and Cd (Kusano *et al.*, 1990; Sasaki *et al.*, 2005; Kiyono *et al.*, 2012). The intracellular membrane transport in eukaryotes is associated with vesicle formation, transport and fusion with the target membrane (Kiyono *et al.*, 2012). In *Arabidopsis*, soluble N-ethylmaleimide-sensitive factor attachment protein receptor (SNARE) molecules, SYP111 and SYP121, are associated with transport of secretory vesicles at the plasma membrane (Kiyono *et al.*, 2012). In *Arabidopsis*, studies have shown that SNAREs can be used as a marker for organelle targeting, which can direct MerC to specific membranes in yeast models (Kiyono *et al.*, 2010). In yeast cells, expression of MerC resulted in high accumulation of Hg (Kiyono *et al.*, 2010). Thus, SNARE targeting of MerC to the plasma membrane represents an important approach for efficient phytoaccumulation of Hg and Cd (Kiyono *et al.*, 2011, 2012).

Metal hyper-accumulation and hypertolerance are naturally selected and hypertolerance is considered as intense abiotic stress-resistance traits (Kramer, 2010). Researches on metal hyper-accumulator model plants like *Arabidopsis halleri* have provided immense information on metal hyper-accumulation and hypertolerance. *A. halleri* is known to be a Zn/Cd hyper-accumulator species. Studies on root elongation tolerance test have revealed that *A. halleri* can tolerate 76-fold higher Zn and eight-fold higher Cd (Bert *et al.*, 2003; Willems *et al.*, 2007). Based on transcriptome comparison between *A. halleri* and metal non-accumulator *A. lyrata*, candidate genes were identified (Becher *et al.*, 2004; Weber *et al.*, 2004; Talke *et al.*, 2006) along with heterologous screening of cDNA libraries and functional genomics approaches (Gortz *et al.*, 1998; van der Zaal *et al.*, 1999; Hussain *et al.*, 2004). The results revealed a great level of synteny of *A. halleri* and *A. lyrata* with *A. thaliana*. It was further demonstrated that *AhHMA4* (heavy metal ATPase 4) is required for high degree Cd hypertolerance in *A. halleri* (Hanikenne *et al.*, 2008). For Zn hyper-accumulation and normal Cd uptake (in non-hyper-accumulating species of *A. halleri*), *AhHMA4* is required (Hanikenne *et al.*, 2008). Studies on metal-sensitive yeast mutants revealed that *AhHMA4* encodes plasma membrane protein belonging to P-type ATPase (heavy metal pump family) that is capable of conferring Zn and Cd tolerance (Talke *et al.*, 2006; Courbot *et al.*, 2007). The protein function of HMA4 in both *A. halleri* and *A. thaliana* has no major functional differences, with major difference of 6- and 53-fold higher transcript abundance of *AhHMA4* in *A. halleri* (Talke *et al.*, 2006; Kramer, 2010). As compared to *A. halleri*, metal hyper-accumulation and hypertolerance are usually non-metal specific in *N. caerulescens* (Kramer, 2010). Analogous to *AhHMA4*, the transcript level of metal transport protein 1 (*MTP1*) is higher in Zn and Cd hyper-accumulating *N. caerulescens* and nickel (Ni) hyper-accumulating *N. goesingense* as compared to non-hyper-accumulating types (Kramer, 2010). The Zn transporter 1 (*NcZNT1*) in *N. caerulescens* encodes for Zn and Cd transporter that is homologous to AtZIP4 (ZIP, Zn regulated transporter/iron regulated transporter protein) family protein of *A. thaliana* (Pence *et al.*, 2000). Another important aspect that is concerned with hyper-accumulators (e.g. Ni hyper-accumulators) is histidine level. In Ni hyper-accumulating species like *Alyssum*, steady-state concentrations of histidine were found to be higher than that of Ni-non-accumulators (Kerkeb and Kramer, 2003). The high level of histidine was due to elevated transcript levels of two genes, which encode ATP-phosphoribosyl transferase (Ingle *et al.*, 2005). Hyper-accumulation of heavy metals is commonly found in approximately 0.2% of all angiosperms (Kramer, 2010). Using model species like *A. halleri* and *N. caerulescens*, a further understanding of heavy metal transport and hypertolerance in plants can be obtained. In future, it will be significant to map genes and to establish phenotyping methods for understanding metal tolerance.

While discussing the molecular physiology of metalloid stress in plants, we focus our attention on Al and the molecular physiology of its stress and tolerance. Al stress tolerance is less understood. Al is considered as a toxic metalloid that causes severe loss of crop productivity and yield. Under normal conditions, Al exists as non-toxic Al-oxide and alumino-silicates (Ma *et al.*, 2001) but turns to toxic trivalent cations under acid soil conditions. More than 40% of world arable soil is acidic and Al is the major toxic compound affecting crop productivity (Panda and Matsumoto, 2007). In plants, Al tolerance occurs in two ways: (i) exclusion of Al from roots; and (ii) activation of intercellular tolerance (Kochian, 1995). Al exclusion is related to the increase in the pH gradient in the rhizosphere (Taylor and Foy, 1985), organic acid excretion from roots (Miyasaka et al., 1991) and formation of a barrier in the rhizosphere or in the roots (Horst et al., 1982). Organic acid (OA) excretion is one of the important aspects that has been considerably explored. Miyasaka et al. (1991) reported OA secretion in *Phaseolus vulgaris* under Al stress. Using different varieties having contrasting Al tolerance, they showed that Al tolerance is related to high secretion of citric acid. The secretion of OA varies greatly amongst different species of plants and plays a crucial role in conferring Al tolerance (Pellet et al., 1995; Ma et al., 1997; Yang et al., 2001). Transgenic approaches to understanding the role of OA in Al tolerance has also provided significant understanding to OA secretion and metabolism. Ectopic expression of bacterial citrate synthase (CA) conferred considerable Al tolerance in tobacco, which was followed by over-expression of plant mitochondrial orthologues that confer Al-tolerance in *Arabidopsis* (del la Fuente et al., 1997; Koyama et al., 2000; Anoop et al., 2003). In transgenic tobacco, the co-transformation of phosphenolpyruvate (PEP) carboxylase enhanced Al tolerance (Wang et al., 2012). Further, studies on membrane proteins lead to identification of genes that encode for OA transporters. *TaALMT1* (*Triticum aestivum aluminium activated malate transporter 1*) was cloned from wheat and is responsible for malate efflux (Sasaki et al., 2004). The electrophysiological studies have revealed that Al activates *TaALMT1* and the gene was later successfully cloned into Al-sensitive barley cultivars that conferred Al tolerance in hydroponic solutions and acid soils (Delhaize et al., 2004). Later, *TaALMT1* was transformed in wheat, showing high expression and was followed by a high malate efflux (Pereira et al., 2010). In sorghum and barley, Al-activated citrate transporter was identified (Furukawa et al., 2007; Magalhaes et al., 2007). These genes were highly homologous to *Arabidopsis* ferric reductase defective 3 (*FRD3*) citrate transporter (Green and Rogers, 2004), belonging to multi-drug and toxic compound extrusion protein (*MATE*) super family. The ectopic expression of *SbMATE* conferred Al tolerance in *Arabidopsis*, which indicated that Al-responsive citrate transporter can improve Al tolerance in crop plants using transgenic approaches (Magalhaes et al., 2007). Maron et al. (2013) reported that higher copy number of *MATE1* is effectively linked with Al tolerance in maize. The high expression of *MATE1* and increase in its copy number were directly correlated and the maize inbred lines with higher copy numbers were Al tolerant with high *MATE1* expression (Maron et al., 2013). Other studies have also uncovered certain molecular mechanisms related to OA excretion. In *Arabidopsis*, *AtALMT1*, a homologue of wheat *ALMT1*, is a critical factor for Al tolerance by affecting malate efflux from roots (Hoekenga et al., 2006). Kobayashi et al. (2007) reported that *AtALMT1* is indeed expressed under Al stress and the efflux of malate was highly induced. The specificity of *AtALMT1* expression is dependent strictly on the availability of Al in the medium; malate efflux is blocked when Al is removed from the medium (Kobayashi et al., 2007). Ligaba et al. (2009) reported a role of protein phosphorylation and dephosphorylation in *TaALMT1*-mediated malate excretion in wheat.

Other factors that could be responsible for intrinsic Al tolerance were identified in plants. Mutant analysis and comparative genomics in *Oryza sativa* and *Arabidopsis* led to the identification of various other genes critical for Al tolerance, but would have no role in OA excretion and Al exclusion. *Arabidopsis ALS3* (encoding aluminium sensitive 3) and encoding bacterial-type ATPase, and their homologous genes in rice *STAR2* (sensitive to Al rhizotoxicity 2; ALS3 homologue) were identified from the Al-sensitive mutants (Larsen et al., 2005). Further research identified that *STAR2* interacts with *STAR1*, both have similar structure of half-type bacterial ATPase and are critical for Al tolerance in rice (Huang et al., 2009). The *STAR1/STAR2* complex transports UDP-glucose, which is critical for Al tolerance. ALS3 may also interact with *Arabidopsis* homologue of STAR1 (*AtSTAR1*) and contribute Al-tolerance in a similar manner in rice (Huang *et al*., 2010a). Other transporters such as vacuole localizing ATPase, *ALS1* (Larsen *et al*., 2007) and a type of rice Nramp (natural resistance-associated macrophage protein) (Xia *et al*., 2010) were isolated. Several Al-tolerant genes and transcription factors have been identified in the past few decades. *STOP1* (sensitive to proton rhizotoxicity 1) is a type of Cys2/His2 zinc finger protein transcription factor, which plays a crucial role in Al tolerance in *Arabidopsis* by regulating function of multiple genes (Iuchi *et al*., 2007). At least three Al tolerance genes, *AtALMT1*, *ALS3* and *AtMATE*, are co-regulated in *stop1* mutants (Liu *et al*., 2009; Sawaki *et al*., 2009). This would account for greater Al sensitivity of *stop1* mutants than knockout (KO) mutants of single Al tolerance gene. *ART1* (aluminium resistance transcription factor 1) homologous to *STOP1* was identified from rice (Yamaji *et al*., 2009) and similarly co-regulated multiple Al-tolerant genes such as *STAR1* and *2* (ALS3 homologue). These reports established that the Al tolerance system is operating in a similar manner in some plant species, while the major Al tolerance mechanisms are different (Fig. 10.3).

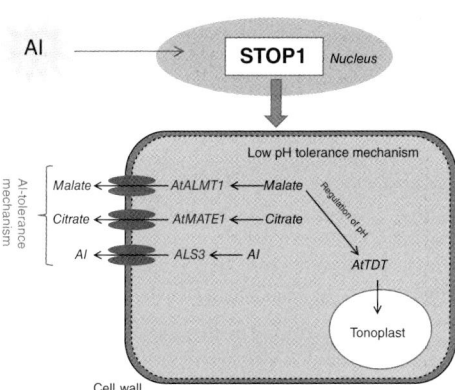

Fig. 10.3. Mechanism of organic excretion regulated by *STOP1* under aluminium stress in plants.

10.4 Role of miRNA in Regulation of Heavy Metal Stress

In plants, abiotic stress responses depend on appropriate regulation of gene expression. Large numbers of genes are regulated under abiotic stresses including heavy metals. Several approaches including comparative genomics have been used to understand transcriptome changes during such stresses. With the discovery of microRNAs (miRNA), the post-transcriptional regulation of gene expression began to provide more evident understanding of stress tolerance in plants. miRNAs regulate gene expression post-transcriptionally and play a significant role in regulation of plant growth and development (Yang and Chen, 2013). They are encoded by the *MIR* genes that are transcribed by RNA poly II forming stem loop structures (Lee *et al*., 2004). The miRNAs and their targets respond to a wide range of heavy metals stress in plants (Huang *et al*., 2009; Yang and Chen, 2013). Use of advanced tools like high-throughput sequencing and genome wide analysis in the characterization of miRNAs in stress responses has now become a powerful tool. Studies on the effect of heavy metals (As, Cd, Hg and Mn) and metalloids (Al) on plants such as *Oryza sativa* L., *Brassica napus*, *Brassica juncea*, *Medicago truncatula* and *Phaseolus vulgaris* revealed that miRNAs are major players in regulation of stress response (Huang *et al*., 2010b; Valdes-Lopez *et al*., 2010; Chen *et al*., 2012; Liu and Zhang, 2012; Zhou *et al*., 2012a, b; Srivastava *et al*., 2013). It was further demonstrated that during Al and Mn stress several miRNAs were induced, while some were depressed under As, Cd and Hg stress (Yang and Chen, 2013). miRNAs such as miR159, miR166, miR162, miR171, miR390 and miR396 are down-regulated and the expression of miR156, miR393 and miR395 are up-regulated during severe heavy metal stress (Yang and Chen, 2013).

In plants, $O_2^{\bullet-}$ is produced during stressed conditions and SOD represents the first line of defence. SOD in plants is classified into three types: FeSOD, MnSOD and Cu/ZnSOD (Fridovich, 1995). In *Arabidopsis* and *O. sativa* miR398 was found to target Cu/ZnSOD genes, *CSD1* and *CSD2* (Bonnet *et al*., 2004;

Jones-Rhoades and Bartel, 2004). Studies have shown that during oxidative stress conditions miR398 is down-regulated. The mRNA levels of *CSD1* and *CSD2* increased in response to Cu and Fe treatment (reviewed by Lu and Huang, 2008). In transgenic *Arabidopsis*, the over-expression of miR398 resistant type of *CSD2* resulted in higher accumulation of *CSD2* mRNA than those plants that over-expresses regular *CSD2*, which reflects high tolerance to stresses including heavy metals (Lu and Huang, 2008; Sunkar, 2010). In *B. napus*, Cd treatment resulted in differential expression of several miRNAs such as miR158, miR161, miR162, miR164, miR171, miR319, miR395 and miR398 in leaves and roots (Zhou *et al.*, 2012a). In *P. vulgaris*, Mn treatment also showed differences in miRNA expression including miR170, miR172, mir1508, miR1511, miR1526 and miR2118 in roots and nodules (Valdes-Lopez *et al.*, 2010). During sulfate deficiency, miR395-mediated gene expression is shown to be critical for sulfur assimilation (Huang *et al.*, 2010a).

For normal growth and development plants require sulfur, which is an essential component for biosynthesis of important metabolites like glutathione and phytochelatins (Na and Salt, 2011). The sulfate transporter 2;1 (*SULTR2;1*) gene encodes for a sulfate transporter (*LAST*) for sulfate translocation from roots to leaves (Takahashi *et al.*, 2011). Studies have shown that miR395 targets *SULTR2;1* along with three ATP sulfurylases (*APS*) gene: *APS1*, *APS3* and *APS4* (Khraiwesh *et al.*, 2012). Using 5′ RACE, the role of miR395 during Cd stress in *B. napus* was evaluated. It was observed that plants that over-express miR395 can tolerate a higher concentration of Cd than the wild types (Huang *et al.*, 2010b) without showing any symptoms of stress. This clearly indicated that miR395 is critically involved in Cd detoxification (Zhang *et al.*, 2013). A similar trait for miR395 was previously reported in plants under Al and Mn stress (Valdes-Lopez *et al.*, 2010; Chen *et al.*, 2012).

10.5 Future Directions

The molecular physiology of heavy metal (and metalloid) stress and tolerance in higher plants has progressed in last few decades. Several critical genes associated with tolerance were identified and most of them have been tested successfully in transgenic plant systems. Plants have evolved their intrinsic tolerance mechanisms to encounter stress. It has been demonstrated that an array of genes are involved that regulate changes to stresses. Studies have indicated that plants also evolved an active tolerance system that is induced by stress. Our understanding on ROS-mediated oxidative stress and functions of the antioxidant system and the related gene expression has also improved approaches in understanding heavy metal tolerance in plants. Identification of putative metal transporter(s) enriched the basis of metal hyper-accumulation and hypertolerance. Our future perspective is directing us to develop new heavy metal (and metalloid)-tolerant crop plants and their introduction into breeding programmes. Besides this, understanding of redox signalling events and components and cross-talk mechanisms under stress can help the understanding of metal stress in crop plants. With the integration of physiology to genomics and developing system biological approaches, better understanding of metal stress perception and tolerance in crops and its subsequent application to breeding programmes in future can be expected.

References

Ali, S., Bai, P., Zeng, F., Cai, S., Shamsi, I.H., Qui, B., Wu, F. and Zhang, G. (2011) The ecotoxicological and interactive effects of chromium and aluminium on growth, oxidative damage and antioxidant enzymes on two barley genotypes differing in Al tolerance. *Environmental Experimental Botany* 70, 185–191.

Anoop, V.A., Basu, U., McCammon, M.T., McAlister-Henn, L. and Taylor, G.J. (2003) Modulation of citrate metabolism alters aluminium tolerance in yeast and transgenic canola overexpressing a mitochondrial citrate synthase. *Plant Physiology* 132, 2205–2217.

Apel, K. and Hirt, H. (2004) Reactive oxygen species: metabolism, oxidative stress and signal transduction. *Annual Review of Plant Biology* 55, 373–399.

Barceloux, D.G. (1999) Cobalt. *Journal of Toxicology and Clinical Toxicology* 37, 201–206.

Becher, M., Talke, I.N., Krall, L. and Kramer, U. (2004) Cross-species microarray transcript profiling reveals high constitutive expression of metal homeostasis genes in shoots of the zinc hyperaccumulator *Arabidopsis halleri*. *Plant Journal* 37, 251–268.

Bert, V., Meerts, P., Saumitou-Laprade, P., de Laguerie, P. and Petit, P. (2003) Genetic basis of Cd tolerance and hyperaccumulation in *Arabidopsis halleri*. *Plant and Soil* 249, 9–18.

Bonnet, E., Wuyts, J., Rouze, P. and de Peer, Y.V. (2004) Detection of 91 potential conserved plant microRNAs in *Arabidopsis thaliana* and *Oryza sativa* identifies important target genes. *Proceedings of National Academy of Sciences USA* 101, 11511–11516.

Chatterjee, J. and Chatterjee, C. (2000) Phytotoxicity of cobalt, chromium and copper in cauliflower. *Environmental Pollution* 109, 69–74.

Chen, L., Wang, T., Zhao, M., Tian, Q. and Zhang, W.H. (2012) Identification of aluminium responsive microRNAs in *Medicago truncatula* by genome wide high throughput sequencing. *Planta* 235, 375–386.

Choi, J.M., Pak, C.H. and Lee, C.W. (1996) Micronutrient toxicity in French marigold. *Journal of Plant Nutrition* 19, 901–916.

Choudhury, S. and Panda, S.K. (2004) Induction of oxidative stress and ultrastructural changes in moss *Taxithelium nepalense* (Schwager.) Broth under lead and arsenic phytotoxicity. *Current Science* 87(3), 342–348.

Choudhury, S., Panda, P., Sahoo, L. and Panda, S.K. (2013) Reactive oxygen species signaling in plants under abiotic stress. *Plant Signaling and Behavior* 8:e23681. doi: 10.4161/psb 26381.

Clemens, S. (2001) Molecular mechanisms of plant metal tolerance and homeostasis. *Planta* 212, 475–486.

Courbot, M., Willems, G., Motte, P., Arvidsson, S., Roosens, N., Saumitou-Laprade, P. and Verbruggen, N. (2007) A major quantitative trait locus for cadmium tolerance in *Arabidopsis halleri* colocalizes with HMA4, a gene encoding a heavy metal ATPase. *Plant Physiology* 144, 1052–1065.

Cruz-Ortega, R., Cushman, J.C. and Ownby, J.D. (1997) cDNA clones encoding 1,3-β-glucanase and fimbrin-like cytoskeletal protein are induced by Al toxicity in wheat roots. *Plant Physiology* 114, 1453–1460.

del la Fuente, J.M., Ramirez-Rodriguez, V., Cabera-Ponce, J.L. and Herrea-Estrella, L. (1997) Aluminium tolerance in transgenic plants by alteration of citrate synthesis. *Science* 276, 1566–1567.

Delhaize, E., Ryan, P.R., Yamamoto, Y., Sasaki, T. and Matsumoto, H. (2004) Engineering high-level aluminium tolerance in barley with the *ALMT1* gene. *Proceedings of National Academy of Sciences USA* 101, 15249–15254.

Ebbs, S.D. and Kochian, L.V. (1997) Toxicity of zinc and copper to *Brassica* species: implications for phytoremediation. *Journal of Environmental Quality* 26, 776–781.

Ezaki, B., Katsuhara, M., Kawamura, M. and Matsumoto, H. (2001) Different mechanism of four aluminium (Al)-resistant transgenes for Al toxicity in *Arabidopsis*. *Plant Physiology* 127, 918–927.

Faller, P., Kienzler, K. and Krieger-Liszkay, A. (2005) Mechanism of Cd^{2+} toxicity: Cd^{2+} inhibits photoactivation of photosystem II by competitive binding to the essential Ca^{2+} site. *Biochemistry Biophysics Acta* 1706, 158–164.

Fontes, R.L.S. and Cox, F.R. (1998) Zinc toxicity in soybean grown at high iron concentration in nutrient solution. *Journal of Plant Nutrition* 21, 1723–1730.

Fox, T.C. and Guerinot, M.L. (1998) Molecular biology of cation transport in plant. *Annual Review of Plant Physiology and Plant Molecular Biology* 49, 583–591.

Foyer, C.H. and Noctor, G. (2005) Oxidant and antioxidant signaling in plants: a re-evaluation of the concept of oxidative stress in a physiological context. *Plant Cell and Environment* 28, 1056–1071.

Fridovich, I. (1995) Superoxide radical and superoxide dismutase. *Annual Reviews of Biochemistry* 64, 97–112.

Furukawa, J., Yamaji, N., Wang, H., Mitani, N., Murata, Y., Sato, K., Katsuhara, M., Takeda, K. and Ma, J.F. (2007) An aluminium activated citrate transporter in barley. *Plant and Cell Physiology* 48, 1081–1091.

Garnier, L., Simon-Plas, F., Thuleau, P., Agnel, J.P., Blein, J.P., Ranjeva, R. and Montillet, J.L. (2006) Cadmium affects tobacco cells by a series of three waves of reactive oxygen species that contribute to cytotoxicity. *Plant Cell and Environment* 29, 1956–1969.

Gimeno-Gracia, E., Andreu, V. and Boluda, R. (1996) Heavy metal incidence in the application of inorganic fertilizers and pesticides to rice farming soils. *Environmental Pollution* 92, 19–25.

Gortz, N., Fox, T., Connolly, E., Park, W., Guerinot, M.L. and Eide, D. (1998) Identification of zinc transporter genes from *Arabidopsis* that responds to zinc deficiency. *Proceedings of National Academy of Sciences USA* 95(12), 7220–7224.

Green, L.S. and Rogers, E.E. (2004) *FRD3* controls iron localization in *Arabidopsis*. *Plant Physiology* 136, 2523–2531.

Halliwell, B. and Gutteridge, J.M.C. (1989) *Free Radicals in Biology and Medicine*. Clarendon Press, Oxford, UK.

Hanikenne, M., Talke, I.N., Haydon, M.J., Lanz, C., Nolte, A., Motte, P., Kroymann, J., Weigel, D. and Kramer, U. (2008) Evolution of metal hyperaccumulation required *cis*-regulatory changes and triplication of HMA4. *Nature* 453(7193), 391–395.

Hegedüs, A., Erdei, S. and Horvath, G. (2001) Comparative study of H_2O_2 detoxifying enzymes in green and greening barley under cadmium stress. *Plant Science* 160, 1085–1093.

Hoekenga, O.A., Maron, L.G., Piñeros, M.A., Cancado, G.M., Shaff, J., Kobayashi, Y., Ryan, P.R., Dong, B., Delhaize, E., Sasaki, T., Matsumoto, H., Yamamoto, Y., Koyama, H. and Kochian, L.V. (2006) *AtALMT1*, which encodes a malate transporter, is identified as one of the several genes critical for aluminium tolerance in *Arabidopsis*. *Proceedings of the National Academy of Sciences USA* 103, 9738–9743.

Horst, W.J., Wagner, A. and Marschner, H. (1982) Mucilage protects roots from aluminium injury. *Zeitschrift für Pflanzenphysiolgie* 105, 435–444.

Huang, C.F., Yamaji, N., Mitani, N., Yano, M., Nagamura, Y. and Ma J.F. (2009) A bacterial-type ABC transporter is involved in aluminium tolerance in rice. *Plant Cell* 21, 655–667.

Huang, C.F., Yamaji, N. and Ma, J.F. (2010a) Knockout of a bacterial-type ATP-binding cassette transporter gene, *AtSTAR1*, results in increased aluminium sensitivity in *Arabidopsis*. *Plant Physiology* 153, 1669–1677.

Huang, S.Q., Xiang, A.L., Che, L.L., Chen, S., Li, H., Song, J.B. and Yang, Z.M. (2010b) A set of miRNA from *Brassica napus* in response to sulphate deficiency and cadmium stress. *Plant Biotechnology Journal* 8, 887–899.

Hussain, D., Haydon, M.J., Wang, Y., Wong, E., Sherson, S.M., Young, J., Camakaris, J., Harper, J.F. and Cobbett, C.S. (2004) P-type ATPase heavy metal transporters with role in essential zinc homeostasis in *Arabidopsis*. *Plant Cell* 16(5), 1327–1339.

Ingle, R.A., Mugford, S.T., Ress, J.D., Campbell, M.M. and Smith, J.A.C. (2005) Constitutively high expression of histidine biosynthetic pathway contributes to nickel tolerance in hyperaccumulator plants. *Plant Cell* 17, 2089–2106.

Iuchi, S., Koyama, H., Iuchi, A., Kobayashi, Y., Kitabayashi, S., Kobayashi, Y., Ikka, T., Hirayama, T., Shinozaki, K. and Kobayashi, M. (2007) Zinc finger protein STOP1 is critical for proton tolerance in *Arabidopsis* and coregulates a key gene in aluminium tolerance. *Proceeding of National Academy of Sciences USA* 104, 9900–9905.

Jarup, L. (2003) Hazards of heavy metal contamination. *British Medical Bulletin* 68, 167–182.

Jones-Rhoades, M.W. and Bartel, D.P. (2004) Computational identification of plant microRNAs and their targets, including a stress-induced miRNA. *Molecular Cell* 14, 787–799.

Kampfenkel, K., Van Montagu, M. and Inźe, D. (1995) Effect of iron excess on *Nicotiana plumbaginifolia* plants. *Plant Physiology* 107, 725–735.

Kerkeb, L. and Kramer, U. (2003) The role of histidine in xylem loading of nickel in *Alyssum lesbiacum* and *Brassica juncea*. *Plant Physiology* 131, 716–724.

Khraiwesh, B., Zhu, J.K. and Zhu, J. (2012) Role of miRNAs and siRNAs in biotic and abiotic stress responses of plants. *Biochemistry Biophysics Acta* 1819, 137–148.

Kiyono, M., Miyahara, K., Sone, Y., Pan-Hou, H., Uraguchi, S., Nakamura, R. and Sakabe, K. (2010) Engineering and expression of the heavy metal transporter MerC in *Saccharomyces cerevisiae* for increased cadmium accumulation. *Applied Microbiology and Biotechnology* 86(2), 753–759.

Kiyono, M., Sone, Y., Miyahara, K., Oka, Y., Nakamura, M., Nakamura, R., Sato, M.H., Pan-Hou, H., Sakabe, K. and Inoue, K. (2011) Genetic expression of bacterial *merC* fused with plant SNARE in *Saccharomyces cerevisiae* increased mercury accumulation. *Biochemical and Engineering Journal* 56, 137–141.

Kiyono, M., Oka, Y., Sone, Y., Tanaka, M., Nakamura, R., Sato, M.H., Pan-Hou, H., Sakabe, K. and Inoue, K. (2012) Expression of bacterial heavy metal transporter MerC fused with a plant SNARE, SYP121, in *Arabidopsis thaliana* increases cadmium accumulation and tolerance. *Planta* 235, 841–850.

Kobayashi, Y., Hoekenga, O.A., Itoh, H., Nakashima, M., Saito, S., Shaff, J.E., Maron, L.G., Piñeros, M.A., Kochian, L.V. and Koyama, H. (2007) Characterization of *AtALMT1* expression in aluminium-inducible malate release and its role for rhizotoxic stress tolerance in *Arabidopsis*. *Plant Physiology* 145, 843–852.

Kochian, L.V. (1995) Cellular mechanism of aluminium toxicity and resistance in plants. *Annual Review of Plant Physiology and Plant Molecular Biology* 46, 237–260.

Koyama, H., Kawamura, A., Kihara, T., Hara, T., Takita, E. and Shibata, D. (2000) Overexpression of mitochondrial citrate synthase in *Arabidopsis thaliana* improved growth on phosphorus-limited soil. *Plant and Cell Physiology* 41(9), 1030–1037.

Kramer, U. (2010) Metal hyperaccumulation in plants. *Annual Review of Plant Biology* 61, 517–534.

Kramer, U., Talke, I.N. and Hanikenne, M. (2007) Transition metal transporter. *FEBS Letters* 581, 2263–2272.

Kumari, M., Taylor, G.J. and Deyholos, M.K. (2008) Transcriptome response to aluminium stress in root of *Arabidopsis thaliana*. *Molecular Genetics and Genomics* 279, 339–357.

Kusano, T., Ji, G.Y., Inoue, C. and Silver, S. (1990) Constitutive synthesis of a transport function encoded by *Thiobacillus ferrooxidans merC* gene encoded in *Escherichia coli*. *Journal of Bacteriology* 172, 2688–2692.

Lang, M.L., Zhang, Y.X. and Chai, T.Y. (2005) Identification of genes upregulated in response to Cd exposure in *Brassica juncea* L. *Gene* 363, 151–158.

Larsen, P.B., Geisler, M.J., Jones, C.A., Williams, K.M. and Cancel, J.D. (2005) *ALS3* encodes a phloem-localized ABC transporter like protein that is required for aluminium tolerance in *Arabidopsis*. *Plant Journal* 41, 353–363.

Larsen, P.B., Cancel, J., Rounds, M. and Ochoa, V. (2007) *Arabidopsis ALS1* encodes a root tip and stele localized half type ABC transporter required for root growth in an aluminium toxic environment. *Planta* 225, 1447–1458.

Lee, Y., Kim, M., Yeom, K.H., Lee, S., Baek, S.H. and Kim, V.N. (2004) MicroRNA genes are transcribed by RNA polymerase II. *EMBO Journal* 23, 4051–4060.

Ligaba, A., Kochian, L.V. and Piñeros, M. (2009) Phosphorylation at S384 regulates the activity of the TaALMT1 malate transporter that underlies aluminium resistance in wheat. *Plant Journal* 60, 411–423.

Liu, J.P., Magalhaes, J.V., Shaff, J. and Kochian, L.V. (2009) Aluminium-activated citrate and malate transporters from the MATE and ALMT families function independently to confer *Arabidopsis* aluminum tolerance. *Plant Journal* 57, 389–399.

Liu, Q. and Zhang, H. (2012) Molecular identification and analysis of arsenite stress responsive miRNAs in rice. *Journal of Agricultural and Food Chemistry* 60, 6524–6536.

Lombi, E., Tearall, K.L., Howarth, J.R., Zhao, F.J., Hawkesford, M.J. and McGrath, S.P. (2002) Influence of iron status on cadmium and zinc uptake by different ecotypes of hyperaccumulator *Thalspi caerulescens*. *Plant Physiology* 128, 1359–1367.

Lu, X.-Y. and Huang, X.-L. (2008) Plant miRNA and abiotic stress responses. *Biochemical and Biophysical Research Communications* 368, 458–462.

Ma, J.F., Zheng, S.J. and Matsumoto, H. (1997) Specific secretion of citric acid induced by aluminium stress in *Cassia tora* L. *Plant and Cell Physiology* 38, 1019–1025.

Ma, J.F., Ryan, P.R. and Delhaize, E. (2001) Aluminium tolerance in plants and complexing role of organic acids. *Trends in Plant Sciences* 6(6), 273–278.

Magalhaes, J.V., Liu, J., Guimarães, C.T., Lana, U.G., Alves, V.M., Wang, Y.H., Schaffert, R.E., Hoekenga, O.A., Piñeros, M.A., Shaff, J.E., Klein, P.E., Carnerio, N.P., Coelho, C.M., Trick, H.N. and Kochian, L.V. (2007) A gene in the multidrug and toxic compound extrusion (MATE) family confers aluminium tolerance in sorghum. *Nature Genetics* 39, 1156–1161.

Maron, L.G., Kirst, M., Mao, C., Milner, M.J., Menossi, M. and Kochian, L.V. (2008) Transcriptional profiling of aluminium toxicity and tolerance responses in maize roots. *New Phytologist* 179, 116–128.

Maron, L.G., Guimarães, C.T., Krist, M., Albert, P.S., Brichler, J.A., Bradbury, P.J., Buckler, E.S., Coluccio, A.E., Danilova, T.V., Kudrna, D., Magalhaes, J.V., Piñeros, M.A., Schatz, M.C., Wing, R.A. and Kochian, L.V. (2013) Aluminium tolerance in maize is associated with higher *MATE1* gene copy number. *Proceedings of National Academy of Sciences USA* (published online before print) doi: 10.1073/pnas.1220766110.

Matsui, K. (2006) Green leaf volatiles: hydroperoxide lyase pathway of oxylipin metabolism. *Current Opinion in Plant Biology* 9, 274–280.

Mithofer, A., Schulze, B. and Boland, W. (2004) Biotic and heavy metal stress response in plants: evidence of common signals. *FEBS Letters* 566, 1–5.

Miyasaka, S.C., Buta, J.C., Howell, R.K. and Foy, C.D. (1991) Mechanism of aluminium tolerance in snapbeans. Root exudation of citric acid. *Plant Physiology* 96, 737–743.

Mohanpuria, P., Rana, N.K. and Yadav, S.K. (2007) Cadmium induced oxidative stress influence on glutathione metabolic genes of *Camellia sinensis* (L.) O. Kuntze. *Environmental Toxicology* 22, 368–374.

Munne-Bosch, S., Queval, G. and Foyer, C.H. (2013) The impact of global change factors on redox signaling underpinning stress tolerance. *Plant Physiology* 161, 5–19.

Na, G.N. and Salt, D.E. (2011) The role of sulfur assimilation and sulfur containing compounds in trace element homeostasis in plants. *Environmental and Experimental Botany* 72, 18–25.

Panda, S.K. (2007) Chromium-mediated oxidative stress and ultrastructural changes in root cells of developing rice seedlings. *Journal of Plant Physiology* 164, 1419–1428.

Panda, S.K and Choudhury, S. (2005) Chromium stress in plants. *Brazilian Journal of Plant Physiology* 17(1), 131–136.

Panda, S.K. and Matsumoto, H. (2007) Molecular physiology of heavy metal toxicity and tolerance in plants. *Botanical Review* 73(4), 326–347.

Panda, S.K., Yamamoto, Y., Kondo, H. and Matsumoto, H. (2008) Mitochondrial alterations related to programmed cell death in tobacco cells under aluminium stress. *Comptes Rendus Biologies* 331, 597–610.

Panda, S.K., Sahoo, L., Katsuhara, M. and Matsumoto, H. (2013) Overexpression of alternative oxidase gene confers aluminium tolerance by altering the respiratory capacity and the response to oxidative stress in tobacco cells. *Molecular Biotechnology* doi: 10.1007/s12033-012-9595-7.

Pandey, N. and Sharma, C.P. (2002) Effect of heavy metal Co^{2+}, Ni^{2+} and Cd^{2+} on growth and metabolism of cabbage. *Plant Science* 163, 753–758.

Pellet, D.M., Grunes, D.L. and Kochian, L.V. (1995) Organic acid exudation as an aluminium-tolerance mechanism in maize (*Zea mays*). *Planta* 196, 788–795.

Pence, N.S., Larsen, P.B., Ebbs, S.D., Letham, D.L., Lasat, M.M., Gravin, D.F., Eide, D. and Kochian, L.V. (2000) The molecular physiology of heavy metal transport in the Zn/Cd hyperaccumulator *Thlaspi caerulescens*. *Proceedings of National Academy of Sciences USA* 97(9), 4956–4960.

Pereira, J.F., Zhou, G., Delhaize, E., Richardson, T., Zhou, M. and Ryan, P.R. (2010) Engineering greater aluminium resistance in wheat by over-expressing *TaALMT1*. *Annals of Botany* 106, 205–214.

Rahman, H., Sabreen, S., Alam, S. and Kawai, S. (2005) Effect of nickel on growth and composition of metal micronutrients in barley plants grown in nutrient solution. *Journal of Plant Nutrition* 28, 393–404.

Richards, K.D., Snowden, K.C. and Gardner, R.C. (1994) wali6 and wali7 genes induced by aluminium in wheat (*Triticum aestivum* L.) roots. *Plant Physiology* 105, 1455–1456.

Ros, R., Cooke, D.T., Martinez-Cortina, C. and Picazo, I. (1992) Nickel and cadmium related changes in growth, plasma membrane and lipid composition, ATPase hydrolytic activity and proton pumping of rice (*Oryza sativa* L. cv. Bahia) shoots. *Journal of Experimental Botany* 43, 1475–1481.

Saier, M.H. Jr (2000) A functional phylogenetic classification system for transmembrane solute transporter. *Microbiology and Molecular Biology Reviews* 64, 354–411.

Salt, D.E., Prince, R.C., Pickering, I.J. and Raskin, I. (1995) Mechanism of cadmium mobility and accumulation in Indian mustard. *Plant Physiology* 109, 1427–1433.

Sandalio, L.M., Dalurzo, H.C., Gomez, M., Romero-Puertas, M.C. and del Rio, L.A. (2001) Cadmium induced changes in growth and oxidative metabolism of pea plants. *Journal of Experimental Botany* 52(364), 20015–20126.

Sasaki, T., Yamamoto, Y., Ezaki, B., Katsuhara, M., Ahn, S.J., Ryan, P.R., Delhaize, E. and Matsumoto, H. (2004) A wheat gene encoding an aluminium-activated malate transporter. *Plant Journal* 37, 645–653.

Sasaki, Y., Minakawa, T., Miyazaki, A., Silver, S. and Kusano, T. (2005) Functional dissection of mercuric ion transporter, MerC, from *Acidithiobacillus ferrooxidans*. *Bioscience Biotechnology and Biochemistry* 69, 1394–1402.

Sawaki, Y., Iuchi, S., Kobayashi, Y., Ikka, T., Sakurai, N., Fujita, M., Shinozaki, K., Shibata, D., Kobayashi, M. and Koyama, H. (2009) *STOP1* regulates multiple genes that protect *Arabidopsis* from proton and aluminium toxicities. *Plant Physiology* 150, 281–294.

Schutzendubel, A., Schwanz, P., Teichmann, T., Gross, K., Langenfeld-Heyser, R., Goldbold, D.L. and Polle, A. (2001) Cadmium induced changes in antioxidative systems, hydrogen peroxide content, and differentiation in Scots pine roots. *Plant Physiology* 127, 887–898.

Shao, H.B., Chu, L.Y., Lu, Z.H. and Kang, C.M. (2008) Primary antioxidant free radical scavenging and redox signaling pathways in higher plant cells. *International Journal of Biological Sciences* 4, 8–14.

Sharma, P. and Dubey, R.S. (2005) Lead toxicity in plants. *Brazilian Journal of Plant Physiology* 17, 35–52.

Sharma, S.S. and Dietz, K.-J. (2008) The relationship between metal toxicity and cellular redox imbalance. *Trends in Plant Science* 14(1), 43–49.

Silver, S. and Phung, L.T. (1996) Bacterial heavy metal resistance: new surprise. *Annual Reviews of Microbiology* 50, 753–789.

Sivaguru, M., Ezaki, B., He, Z.-H., Tong, H., Osawa, H., Baluska, F., Volkmann, D. and Matsumoto, H. (2003) Aluminium-induced gene expression and protein localization of cell wall associated receptor kinase in *Arabidopsis*. *Plant Physiology* 132, 2256–2266.

Snowden, K.C. and Gardner, R.C. (1993) Five genes induced by aluminium in wheat (*Triticum aestivum*) roots. *Plant Physiology* 103, 855–861.

Srivastava, S., Srivastava, A.K., Suprasanna, P. and D'Souza, S.F. (2013) Identification and profiling of arsenic stress-induced microRNAs in *Brassica juncea*. *Journal of Experimental Botany* 64, 303–315.

Stadtman, E.R. and Oliver, C.N. (1991) Metal catalyzed oxidation of proteins. *Journal of Biological Chemistry* 266, 2005–2008.

Sunkar, R. (2010) MicroRNAs with macro-effects on plant stress response. *Seminars in Cell and Developmental Biology* 21, 805–811.

Takahashi, H., Kopriva, S., Giordono, M., Saito, K. and Hell, R. (2011) Sulfur assimilation in photosynthetic organisms: molecular functions and regulations of transporters and assimilatory enzymes. *Annual Review of Plant Biology* DOI: 10.1146/annurev-arplant-042110-103921

Talke, I.N., Hanikenne, M. and Kramer, U. (2006) Zinc dependant global transcriptional control, transcriptional de-regulation and higher gene copy number for gene in metal homeostasis of the hyperaccumulator *Arabidopsis halleri*. *Plant Physiology* 142, 148–167.

Taylor, G.J. and Foy, C.D. (1985) Mechanism of aluminium tolerance in *Triticum aestivum* L. (wheat). II. Differential pH induced by spring cultivars in nutrient solutions. *American Journal of Botany* 72(5), 702–706.

Thounaojam, T.C., Panda, P., Choudhury, S., Patra, H.K. and Panda, S.K. (2013) Zinc ameliorates copper-induced oxidative stress in developing rice (*Oryza sativa* L.) seedlings. *Protoplasma*. DOI: 10.1007/s00709-013-0525-8.

Valdes-Lopez, O., Yang, S.S., Aparicio-Fabre, R., Graham, P.H., Reyes, J.L., Vance, C.P. and Harnandez, G. (2010) MicroRNA expression profile in common bean (*Phaseolus vulgaris*) under nutrient deficiency stresses and manganese toxicity. *New Phytologist* 187, 805–818.

van der Zaal, B.J., Nuteboom, L.W., Pinas, J.E., Chardonnens, A.N., Schat, H., Verkleij, J.A. and Hooykaas, P.J. (1999) Overexpression of a novel *Arabidopsis* gene related to putative zinc transporter genes from animals can lead to enhanced resistance and accumulation. *Plant Physiology* 119, 1047–1055.

Wang, Q., Yi, Q., Hu, Q., Zhao, Y., Nian, H., Kunzhi, L., Yu, Y., Izui, K. and Chen, L. (2012) Simultaneous overexpression of citrate synthase and phosphoenolpyruvate carboxylase in leaves augments citrate exclusion and Al resistance in transgenic tobacco. *Plant Molecular Biology Reporter* 30, 992–1005.

Weber, M., Harada, E., Vess, C., Roepenack-Lahaye, E. and Clemens, S. (2004) Comparative microarray analysis of *Arabidopsis thaliana* and *Arabidopsis halleri* roots identifies nicotinamine synthase, a ZIP transporter and other genes as potential metal hyperaccumulating factors. *Plant Journal* 37, 269–281.

Willems, G., Drager, D.B., Courbot, M., Gode, C., Verbruggen, N. and Saumitou-Laprade, P. (2007) The genetic basis of Zn tolerance in the metallophyte *Arabidopsis halleri* sp. *halleri* (Brassicaceae): an analysis of quantitative trait loci. *Genetics* 176, 659–674.

Wojcik, M. and Tukiendrof, A. (2004) Phytochelatin synthesis and cadmium localization in wild type of *Arabidopsis thaliana*. *Plant Growth Regulation* 44, 71–80.

Xia, J., Yamaji, N. and Ma, J.F. (2010) Plasma membrane-localized transporter for aluminium in rice. *Proceedings of National Academy of Sciences USA* 107(43), 18381–18385.

Xu, J., Chai, T., Zhang, Y., Lang, M. and Han, L. (2009) The cation-efflux transporter BjCET2 mediates zinc and cadmium accumulation in *Brassica juncea* L. leaves. *Plant Cell Reports* 28, 1235–1242.

Yadav, S.K. (2010) Heavy metal toxicity in plants: An overview on the role of glutathione and phytochelatins in heavy metal stress tolerance of plants. *South African Journal of Botany* 76, 167–179.

Yamaji, N., Huang, C.F., Nagao, S., Yano, M., Sato, Y., Nagamura, Y. and Ma, J.F. (2009) A zinc finger transcription factor ART1 regulates multiple genes implicated in aluminum tolerance in rice. *Plant Cell* 21(10), 3339–3349.

Yang, Z.M. and Chen, J. (2013) A potential role of microRNAs in plant response to metal toxicity. *Metallomics* 5, 1184–1190.

Yang, Z.M., Nian, H., Sivaguru, M., Tanakamaru, S. and Matsumoto, H. (2001) Characterization of aluminium induced citrate secretion in aluminium-tolerant soybean (*Glycine max*) plants. *Physiologia Plantarum* 113, 164–171.

Zhang, L.W., Song, J.B., Shu, X.X., Zhang, Y. and Yang, Z.M. (2013) miR395 is involved in detoxification of cadmium in *Brassica napus*. *Journal of Hazardous Material* 251, 204–211.

Zhao, C.R., Ikka, T., Sawaki, Y., Kobayashi, Y., Suzuki, Y., Hibino, T., Sato, S., Sakurai, N., Shibata, D. and Koyama, H. (2009) Comparative transcriptomic characterization of aluminium, sodium chloride, cadmium and copper rhizotoxicities in *Arabidopsis thaliana*. *BMC Plant Biology* 9, 32.

Zhou, Z.S., Song, J.B. and Yang, Z.M. (2012a) Genome wide identification of *Brassica napus* microRNAs and their targets in response to cadmium. *Journal of Experimental Botany* 63, 4597–4613.

Zhou, Z.S., Zeng, H.Q., Lui, Z.P. and Yang, Z.M. (2012b) Genome wide identification of *Medicago truncatula* microRNAs and their targets reveals their differential regulation by heavy metal. *Plant Cell Environment* 35, 86–99.

Zornoza, P., Robles, S. and Martin, N. (1999) Alleviation of nickel toxicity by ammonium supply to sunflower plants. *Plant and Soil* 208, 221–226.

11 Influence of Arsenic and Phosphate on the Growth and Metabolism of Cultivated Plants

Asok K. Biswas*
Plant Physiology and Biochemistry Laboratory, Centre for Advanced Study, Department of Botany, University of Calcutta, India

Abstract

In growing cereal and legume seedlings arsenate is more toxic for root growth than shoot growth. Arsenate treatment affects the activity of antioxidant scavenging enzymes, level of oxidative stress markers, sugar and starch contents as well as carbohydrate metabolizing enzyme activities causing adverse effects on growth and metabolism of seedlings of cereals and legumes. In wheat and rice arsenate treatment decreases total and soluble N_2 contents, nitrate and nitrite contents and activities of nitrate and nitrite reductase that lead to limitation of NO_3 uptake by root. The activities of ammonium assimilating enzymes, i.e. glutamine synthetase and glutamate synthase, are decreased whereas deaminating activity of glutamate dehydrogenase is increased leading to an accumulation of toxic NH_3. Arsenic treatment results in an alteration of shape of chloroplast and disorganizes the membrane structure in bean plants. All such alterations inhibited growth and development of test seedlings. On arsenate exposure, levels of various Krebs cycle intermediates and also activities of different respiratory enzymes are decreased causing changes in growth pattern in rice and wheat seedlings. In presence of arsenate, the glutathione level and the activities of the synthesizing enzymes that are required for normal growth and metabolism of rice seedlings are suppressed. The production of phytochelatins on arsenate exposure enhanced the detoxifying mechanism against arsenate in the test seedlings. Arsenate forms complex with the thiol group of phytochelatin and gets sequestered in the plant vacuole enabling researchers to design a process of detoxification of arsenic. This knowledge is helpful to produce plant cultivars that are more resistant to arsenic or that have reduced arsenic uptake. Combined application of phosphate with arsenate can ameliorate the damaging effects caused by arsenate treatment alone in cereal and legume seedlings. Hence, the use of phosphate-enriched fertilizers in arsenic-contaminated soil may help normal growth of cultivated plants.

11.1 Introduction

Arsenic is a naturally occurring highly toxic metalloid element. Arsenic poisoning has gained at present a major global importance as it affects millions of people worldwide. It is a group A carcinogen, most commonly found in two inorganic arsenic forms as oxidized states, trivalent meta-arsenite (As^{3+}) and pentavalent arsenate (As^{5+}) (Smith et al., 1992). Arsenic contamination is widespread due to geologic and anthropogenic activities such as smelting operations, fossil fuel combustion (Nriagu and Pacyna, 1988) and arsenic-based agro-chemicals, fertilizers and disposal of municipal and industrial wastes (Requejo and Tena, 2005).

---

*E-mail: drbiswasak@yahoo.co.in

Groundwater contamination by arsenic has been reported from many countries. The largest population at risk is in Bangladesh followed by West Bengal in India owing to extensive withdrawal of water for irrigation and potable purposes. Presence of arsenic in irrigation water or in soil at an elevated level has the ability to hamper normal growth and development of plants and tends to develop toxicity symptoms in them (Huang et al., 1992; Mandal et al., 1996; Dhar et al., 1997; Biswas et al., 1998; Nickson et al., 1998; Chowdhury et al., 1999).

Phosphorus is an essential nutrient for plants and an important component in cell metabolism. It has a vital functional role in energy transfer and acts as modulator of enzyme activity and gene transcription; hence its assimilation, storage and metabolism are of major importance to plant growth and development. Phosphorus has been used to ameliorate arsenic toxicity (Benson, 1953). Arsenic is analogous to phosphate in the periodic table; both have similar electron configuration and chemical properties. From physiological and electrophysiological studies it is evident that arsenate and phosphate share the same transport pathway in higher plants with the transporters having a higher affinity for phosphate than for arsenate (Ullrich-Eberius et al., 1989; Meharg et al., 1994).

The toxicity of arsenic is an important growth-limiting factor for crops. Studies on toxic effects of arsenic on plants have been of great interest to scientists because this might help to develop methods to combat this threat. Significant advances have been achieved in our understanding of the physiological and electrophysiological processes of plants that are induced by arsenic and phosphate exposure. In this chapter, a review is presented about growth and metabolic consequences exhibited by plants due to arsenic and phosphate exposure.

11.2 Uptake and Transport of Arsenic in Plants

In most plant species the uptake of arsenic from soil to plant is low due to various reasons, i.e. low bioavailability of arsenic in soil, restricted uptake by plant roots, limited translocation from roots to shoots, along with arsenic phytotoxicity in plant tissues at low concentrations.

Inorganic forms of arsenic are highly phytotoxic and it is a non-essential element for plants. Smith et al. (1998) demonstrated that in aerobic soils arsenate is the predominant arsenic species, while arsenite dominates under anaerobic conditions. Flooding of paddy soils may cause mobilization of arsenite into the soil solution and increase arsenic bioavailability to rice plants (Xu et al., 2008).

Arsenate (H_2AsO_4)$^-$, being structurally similar to phosphate (H_2PO_4)$^-$, can compete with phosphate for its uptake by the high affinity plasma membrane phosphate transporters (Asher and Reay, 1979) like Pht1 in *Arabidopsis* (Shin et al., 2004), Pho84 in yeast (Bun-Ya et al., 1991) and LePT1 in tomato (Daram et al., 1999). Arsenite (H_2AsO_3)$^-$ uptake occurs by a glycerol transporting channel (Meharg and Jardine, 2003). In yeast, *fps1* gene encodes a glycerol channel protein, the product Fps1 mediating the influx of arsenite into the cell (Wysocki et al., 2001). Physiological studies on rice have shown that transport of arsenite may be through aquaporins although the exact mechanism of uptake is not yet clear (Meharg and Jardine, 2003). It has further been demonstrated in rice that there are two types of arsenite transporters belonging to the NIP subfamily of aquaporins and their role in arsenic accumulation in shoot and grain has been established (Ma et al., 2001).

Some plant species are arsenic hyperaccumulators and are naturally arsenic tolerant. These include *Pteris vittata* and some other members of the Pteridaceae (Ma et al., 2001; Visoottiviseth et al., 2002; Ellis et al., 2006; Pickering et al., 2006; Zhao et al., 2009). The growth of these plants is not affected when they accumulate significantly high arsenic. In contrast to arsenic non-hyper-accumulating plants, in hyper-accumulators, arsenic is not restricted to the roots but is immediately transferred after uptake to the shoots. This would be an important aspect of the hyper-accumulation phenotype. It is not yet clear how these plants avoid arsenic toxicity when arsenic accumulates to high levels in the leaves and the mechanism for hyper-accumulation is also still being elucidated.

Availability and toxicity of arsenic in tomato (*Lycopersicum esculentum*) plants is determined by the arsenic species present in soil. After uptake, arsenic is mainly restricted to the root system and very little is transported to the shoot. Reduction of plant growth and fruit yield occur when monomethyl and dimethyl arsenic is absorbed by the plants (Carbonell-Barrachina et al., 1997; Burlo et al., 1999). In bean (*Phaseolus vulgaris*) plants, unlike tomato, most of the arsenic absorbed by roots is transported to the shoots resulting in chlorosis and necrosis (Carbonell-Barrachina et al., 1997). Understanding the molecular and genetic basis for uptake and metabolism of arsenic will be an important step for the development of plants as agents for the phytoremediation of contaminated sites (Salt et al., 1995). The effect of arsenic on two diverse plants *Pennisetum typhoides* (monocotyledenous) and *Pisum sativum* (dicotyledenous) shows arsenic accumulation increasing progressively with increasing concentrations of arsenic in the nutrient media.

In plants phosphorus is important for energy transfer and protein metabolism. Both arsenate (AsV) and inorganic phosphate (Pi) are analogous and can be transported by Pi transporter proteins located on plasma-membrane (Ullrich-Eberius et al., 1989; Meharg and Macnair, 1990, 1991, 1992; Wu et al., 2011). Both compete for uptake through the same transport system in arsenic-tolerant non-hyperaccumulators (Meharg and Macnair, 1992; Bleeker et al., 2003), arsenic-hyper-accumulators (Wang, 2002; Tu and Ma, 2003) and arsenic sensitive non-accumulators (Abedin et al., 2002; Esteban et al., 2003). Under low concentration of Pi, AsV may compete with Pi to enter into the plant and amplify Pi-deficient symptoms, but Pi fertilization can protect plants, including arsenic hyper-accumulator *P. vittata*, from arsenic toxicity (Tu and Ma, 2003). Arsenic may exert toxicity to plants by interfering with many physiological functions performed by phosphorus, since arsenate is chemically analogous to phosphate (Meharg and Hartley-Whitacker, 2002).

11.3 Influence on Seedling Growth

Arsenate exposure significantly alters the normal growth and development of rice (Choudhury et al., 2010) and wheat (Ghosh et al., 2013) seedlings. There is reduction of seedling growth with increasing concentrations of arsenate treatments. The rate of root growth inhibition is stronger than shoot growth inhibition. High concentration of arsenate (100 µM) treatment significantly damaged the roots with browning of tissues accompanied with anatomical changes. The anatomical damage of root tissues caused by arsenate treatment could be repaired by phosphate application in these seedlings (Choudhury et al., 2011). Several workers have reported the phenomenon of reduction of root growth in different plant species in response to arsenate exposure (Kapustka et al., 1995; Sneller et al., 1999, 2000; Hartley-Whitaker et al., 2001a; Harminder et al., 2007). It has been reported that arsenate is more toxic than arsenite for rice root growth (Abedin et al., 2002). Observed reduction of shoot and root growth in rice seedlings by arsenate application was found to be ameliorated by extraneous application of phosphate. Singh et al. (2007) demonstrated that arsenic-induced root-growth inhibition in mung bean (*Phaseolus aureus* Roxb.) was due to oxidative stress caused by enhanced lipid peroxidation. It was demonstrated that increasing concentrations of arsenic reduced seed germination, root and shoot growth in clover (*Trifolium incarnatum* L.), but not in black nightshade (*Solanum nigrum* L.). For the latter, low concentrations (3 mg kg^{-1} dry sand) of arsenic seemed to have stimulatory effects on germination (Marques et al., 2007). Arsenic stress inhibits growth of spinach seedlings by causing water deficiency and hindering the nutrient balance. Elevated levels of arsenic in soil may limit the growth and yield of canola (*Brassica napus*) (Cox and Kovar, 2001). In radish (*Raphanus sativa*) arsenic affects germination of seeds while in lettuce (*Lactuca sativa*) arsenic does not affect germination but inhibits, to some extent, the radicle development in seedlings (Grafe et al., 2001). When tomato plants were grown in soils contaminated with arsenic, growth of both vegetative and root systems declined. Tissue chlorosis and necrosis were absent in arsenic-contaminated tomato plants. *Anadenanthera peregrine*, found in Brazil, is resistant to arsenic and grows naturally in arsenic-contaminated areas; its arsenic resistance has been associated with arbuscular mycorrhizal fungal symbiosis. However, a strong negative influence of arsenic on the growth and nutritional status of *Anadenanthera peregrina* seedlings has been recorded by Gomes *et al.* (2012).

Plant growth inhibition by arsenic has been proposed to be due to slowing or arresting expansion and biomass accumulation, as well as negatively affecting plant reproductive capacity through losses in fertility, yield and fruit production (Garg and Singla, 2011). In wheat (*T. aestivum*), decreases in plant height, grain yield, number of filled grains, grain weight and root biomass was seen while arsenic concentration in root, straw and husk increased significantly (Abedin *et al.*, 2002; Therapong *et al.*, 2004).

Stimulation of plant growth by arsenic at low concentrations is one of the many paradoxes related to arsenic toxicity (Woolson *et al.*, 1971a, b; Carbonell-Barrachina *et al.*, 1997, 1998; Miteva and Merakchiyska, 2002; Garg and Singla, 2011). This has been demonstrated under arsenic exposure in cultured plants, such as *Arabidopsis thaliana* (Chen *et al.*, 2010), which indicates that the trait is not based on arsenic disrupting plant–biotic interactions. It probably results either from a direct interaction of arsenic with plant metabolism, or from an interaction of arsenic with plant nutrients. Although the mechanism is unknown, arsenic stimulation of Pi uptake may play a major role in this (Tu and Ma, 2003).

11.4 Influence on Oxidative Stress Markers

Reactive oxygen species (ROS) such as superoxide ($O_2^{\bullet-}$), hydroxyl radical ($\bullet OH$) and H_2O_2 are produced on exposure of plants to arsenate AsV and arsenite (AsIII) (Hartley-Whitaker *et al.*, 2001a; Requejo and Tena, 2005; Singh *et al.*, 2006; Ahsan *et al.*, 2008; Mallick *et al.*, 2011). Although arsenic itself is not a redox metal, ROS production in plants on exposure to inorganic arsenic is associated with valence change of arsenic (Mittler, 2002). ROS have a damaging effect on plant biomolecules such as proteins, amino acids, purine nucleotides and nucleic acids and also cause membrane lipid peroxidation (Møller *et al.*, 2007). This lipid peroxidation not only damages cellular function, but also leads to the production of lipid-derived radicals (Van Breusegem and Dat, 2006; Møller *et al.*, 2007). Induction of lipid peroxidation by AsV was demonstrated in the arsenic hyper-accumulator *P. vittata* (Srivastava *et al.*, 2005; Singh *et al.*, 2006). This indicates that ROS is produced as an outcome of plant arsenic response and the degree of redox imbalance in the cell may be an important determinant of ROS-induced toxicity. The molecular targets sensitive to the ROS produced by arsenic exposure are not yet clear, although several examples exist (Møller *et al.*, 2007).

On arsenate exposure, an increasing tendency of H_2O_2 level has been observed in rice (Choudhury *et al.*, 2011) and mung bean (Harminder *et al.*, 2007) seedlings. It is now established that during abiotic stress, H_2O_2 also functions as a signal molecule thus playing a dual role in plant defence (Stone and Yang, 2006). Proline, an amino acid, acts as a cytoplasmic osmoticum and also protects the protein against denaturation (Kavi-Kishor *et al.*, 2005). Application of AsV alleviates proline contents in rice seedlings. In contrast, application of phosphate with AsV induced significant reduction of proline content (Choudhury *et al.*, 2011).

Similarly, malondialdehyde (MDA), which is often used as an indicator of oxidative damage, is produced during peroxidation of membrane lipid by decomposition of polyunsaturated fatty acid (Mittler, 2002). It was found that MDA contents increase with the application of arsenate in rice seedlings (Choudhury *et al.*, 2011). This is also confirmed in other species where severe lipid peroxidation has been observed: *Holcus lanatus* (Hartley-Whitaker *et al.*, 2001a), an arsenic-sensitive fern species (Srivastava *et al.*, 2005), red clover (*Trifolium pretense*) (Mascher *et al.*, 2005) and bean (*P. vulgaris*) plant (Stoeva *et al.*, 2005). The increased level of MDA contents significantly decline in rice seedlings with the application of phosphate (Choudhury *et al.*, 2011).

11.5 Influence on the Activities of Antioxidant Enzymes

Under normal cellular conditions the ROS homoeostasis is delicately balanced. ROS imbalances are caused by even small fluctuations in environmental conditions such as temperature, light or nutrient availability, which in turn act as signals of cellular syndrome and are generally easily managed by pre-existing antioxidant defence mechanisms (Foyer et al., 2001; Van Breusegem and Dat, 2006; Møller et al., 2007). However, with increased ROS generation during different stresses, including arsenic exposure, these defence mechanisms may be compromised, resulting in cellular damage and finally cell death (Van Breusegem and Dat, 2006).

Plants exhibit a great variation in their response to arsenic toxicity. Arsenic is absorbed mainly by the plant root-system as arsenate (Macnair and Cumbes, 1978). Hyper-accumulator species, on exposure to arsenic, have been reported to increase their antioxidant mechanism, both enzymatic and non-enzymatic, leading to its detoxification and subsequent hyper-accumulation (Srivastava et al., 2005; Singh et al., 2006), whereas in non-hyperaccumulators, arsenic induces an oxidative stress leading to cellular damage in terms of H_2O_2 accumulation, enhanced lipid peroxidation and up-regulation of many of the scavenging enzymes (Hartley-Whitaker et al., 2001a). Heavy metal toxicity of plants brings about complex biochemical responses and several defensive mechanisms, which include production of enzymatic and non-enzymatic antioxidants meant to detoxify ROS that readily occur in plants caused by metal contamination (Meharg, 1994; Stoeva and Bineva, 2003).

Many enzymes are involved in defence strategies regulated by ROS. Superoxide dismutase (SOD) converts highly reactive superoxide to less active but longer-lasting H_2O_2. SOD (E.C 1.15.1.1) is the most important superoxide ($O_2\bullet^-$) scavenger and provides a first line of defence against cellular injury caused by environmental stresses (Gratao et al., 2005). In plants, SOD activity varies significantly with arsenic treatment. In Zea mays, Oryza sativa, As-sensitive clones of H. lanatus, and the As-hyper-accumulator P. vittata, the enzyme activity induced by low arsenic exposure is not further increased on exposure to higher concentrations of arsenic, rather it may decline (Mylona et al., 1998; Hartley-Whitaker et al., 2001a; Cao et al., 2004). This may be because SOD is a metallo-enzyme and different classes of isoforms are known to occur (Meharg and Hartley-Whitaker, 2002). In Arabidopsis genes encoding three classes of SOD (FeSOD, MnSOD, Cu/ZnSOD) respond differentially to arsenate at the transcript level. Transcripts for genes encoding a chloroplastic and a cytosolic Cu/ZnSOD are induced more than two-fold by arsenate exposure, while those for a FeSOD are down-regulated about five-fold (Abercrombie et al., 2008). From these findings questions may arise about the effects of these changes in the SOD transcript pool on the characteristics of SOD activity. It would be also of interest to determine the adaptive advantages, if characteristics and mechanisms of SOD activity changes.

Excess H_2O_2, which itself is a highly reactive oxidizing agent, is scavenged by catalase (CAT, EC 1.11.1.6). Both CAT and catechol peroxidase (CPX, EC 1.11.1.7) activities decline and H_2O_2 level increases in rice seedlings on exposure to arsenate (Choudhury et al., 2011). CPX is an active oxygen scavenging enzyme by which degradation of H_2O_2 is done with the oxidation of a reducing co-substrate. Due to low enzyme activities, the level of H_2O_2 increases in the arsenate-treated seedlings to generate toxicity, leading to reduction in growth and metabolism in the seedlings. Ascorbic acid oxidase (AOX, EC 1.10.3.3) is a copper-containing enzyme induced by stress that oxidizes ascorbic acid in presence of oxygen producing dehydroascorbic acid and water. During oxidative stress, this enzyme becomes active to protect plant cells. The activity of AOX is altered in rice seedlings with arsenate treatments. At the highest concentration applied, AOX activity declines as opposed to the situation under lower concentrations of arsenate with concomitant rise in activity (Choudhury et al., 2011). Joint application of phosphate with arsenate in rice seedlings altered the activities of these enzymes from that observed by arsenate treatment alone. Both SOD and AOX activities decreased while CAT and CPX activities increased, leading to better growth and metabolism in the test seedlings (Choudhury et al., 2011). It has been reported in chickpea (Cicer arietinum) that the high rate of phosphate supply serves to limit the activities of those antioxidant enzymes which were enhanced by arsenic (Gunes *et al.*, 2008).

In plants, the balance of H_2O_2 is maintained by a two-component system within cells for regulating their production and therefore of ROS. The first component having a group of non-enzymatic antioxidants includes reduced glutathione (GSH), phytochelatin (PC), ascorbate, carotenoids and anthocyanin. The second component is composed of monodehydroascorbate reductase, didehydroascorbate reductase and GSH reductase. Exposure to arsenic generally induces the accumulation of the non-enzymatic antioxidants (Schmöger *et al.*, 2000; Hartley-Whitaker *et al.*, 2001a; Bleeker *et al.*, 2003, 2006; Khan *et al.*, 2009; Song *et al.*, 2010; Choudhury *et al.*, 2011). The production of these molecules requires metabolic acclimations, coupled with diversion of carbon, nitrogen, sulfur and metabolic energy from normal growth and development. Efficient recycling of oxidized GSH and ascorbate to allow further cycles of H_2O_2 reduction is carried out by the enzymatic antioxidants. Reducing power in the form of NAD(P)H is also required for the reduction of H_2O_2 through the interdependent ascorbate-GSH cycle, which then diverts this energy from other metabolic processes. Ahsan *et al.* (2008) and Khan *et al.* (2009) have demonstrated that the enzymes involved in the recycling of oxidized GSH and ascorbate are often induced in plants by arsenic exposure. Thus during arsenic exposure the interdependent ascorbate-GSH cycle plays an important role in maintaining ROS balance in plants (Foyer and Noctor, 2011).

11.6 Influence on Respiratory Cycle

Respiration not only provides large quantities of adenosine triphosphate (ATP) and other metabolites required for seed germination, together with formation of new tissues and organs, but also contributes to other physiological and biochemical processes (Wang *et al.*, 2001). Cellular respiration is catalysed by many enzymes that require metals as co-factors, whereas higher concentrations of these metals inhibit enzyme activities. Heavy metal affects mitochondrial respiration by altering oxygen consumption rates, carbon dioxide release, citric acid cycle intermediates and ATP production.

The inhibition of succinic oxidase system by arsenate can be accounted for by the inhibition of succinic dehydrogenase (Slater, 1949). Arsenate may combine with two SH groups, either on the same succinic dehydrogenase molecule or on different molecules. Arsenate can disrupt the pyruvate and succinate oxidation pathway. This inhibition effectively blocks the TCA cycle, which results in marked depletion of ATP. In addition to succinic dehydrogenase, the oxidizing agents affect the components of the succinic oxidase system, which links the dehydrogenase to cytochrome C (Slater, 1949). It is thus probable that different heavy metals have a non-specific effect on the entire succinic oxidase system, and they may also combine with SH-groups in the dehydrogenase. Arsenate binds to SH groups, disrupts SH-containing enzymes, inhibits pyruvate and succinate oxidation pathways and the Krebs cycle, causing impaired oxidative phosphorylation. Replacing the stable phosphorus anion in phosphate with the less stable arsenate (V) anion leads to rapid hydrolysis of high-energy bonds in ATP. This leads to loss of high-energy phosphate bonds and effectively 'uncouples' oxidative phosphorylation. A detailed knowledge about the mechanism of interactions between arsenate and citrate cycle, cell macromolecules and the structures of the organelles is necessary to explain the limits of resistance of the cell, tissue and the whole plant to arsenate-induced stress.

In wheat, the respiratory rate of roots has been shown to be induced at lower concentrations of arsenic, cadmium and lead but inhibited at higher concentrations. Cytochrome oxidase (COD), a terminal oxidase in the respiratory chain of eukaryotic cells, transfers electrons between cytochrome a_3 and oxygen. It displays nine iso-enzymes in the leaves and 13 iso-enzymes in the roots. The expressions of these iso-enzymes displayed a decreasing trend with increasing arsenic concentrations (Shao *et al.*, 2011). Isocitrate dehydrogenase (IDH) participates in the Krebs cycle and catalyses the oxidative decarboxylation of isocitrate to α-ketoglutarate reducing NAD^+ to NADH. It is a rate-limiting enzyme in the

Krebs cycle. Malate dehydrogenase (MDH) also plays a very important role in the Krebs cycle, catalysing the transformation of malic acid to oxaloacetate. The expression of COD, MDH and IDH iso-enzymes in leaves and roots of wheat seedlings are induced by lower concentration of arsenic whereas it is reduced at higher concentrations of arsenic. Therefore it is likely that heavy metals usually affect the respiration rate by influencing the expression of iso-enzymes involved in the respiratory mechanism (Shao et al., 2011).

In rice seedlings, succinate and malate dehydrogenase activities declined considerably due to arsenic treatment, but could be recovered by phosphate treatment (Choudhury et al., unpublished data). Under comparable in vitro conditions, cadmium, lead, zinc, copper and nickel reduce the activities of both of these dehydrogenases (Mathys, 1975; Wu et al., 1999). Pyruvate dehydrogenase is considered as a primary target for the toxic action of arsenic. Any disruption of the action of this enzyme can undermine the ability of the cell to meet its energy requirements and therefore result in cellular damage and also death (Thangavel et al., 2003). However, the activity of pyruvate dehydrogenase was found to be enhanced in the shoots of arsenate-treated rice seedlings (Choudhury et al., unpublished data). According to Van-Assche and Clijsters (1990), induction of enzyme activity by high concentrations of heavy metals is a typical in vivo response of the plant coping with environmental stresses.

Meeta, 1997) and in red clover (Mascher et al., 2002). The chlorophyll concentration in the leaves of Pisum sativum increased, but the ratio of chlorophyll a/b decreased after exposure to arsenic (Mascher et al., 2002). However, in lettuce (Lactuca sativa), accumulation of arsenic does not adversely affect the photosynthetic apparatus and level of chloroplastic pigments, probably through the activation of tolerance mechanism (Gusman et al., 2013). It has been reported that in rice leaf arsenic-induced decrease in chloroplastic pigment contents, these are increased by the joint application of phosphate with arsenic (Choudhury et al., 2011). Stoeva and Bineva (2003) demonstrated that oat plants grown in arsenic-contaminated soil are subjected to stress; as a consequence, leaf-gas-exchange is suppressed, chlorophyll, protein content and fluorescence ratio of chlorophyll decrease. In Chinese brake (P. vittata L.) addition of arsenic does not affect the chloroplast ultrastructure of young pinna while chloroplasts in mature pinna are severely damaged (Li et al., 2006).

Arsenic is a competitive inhibitor of photosynthetic phosphorylation. Arsenic may create conditions in the thylakoids where the energy level exceeds the amount that can be dissipated by the metabolic pathways of the chloroplasts (Dat et al., 2000). As a consequence, the electron transport processes in the thylakoid membranes are impeded and toxic symptoms develop. These damage the chloroplast membrane and disorganize the membrane structure.

11.7 Effect on Photosynthetic Apparatus

Arsenic treatment resulted in an alteration of the chloroplast shape with concaving membrane bending and partial destruction along with changes in the accumulation and flow of assimilates leading to decreased chlorophyll contents in rice leaf. The damage to the chloroplast structure implies functional changes of the integral photosynthetic process resulting in reduction of photosynthesis rate (Miteva and Merakchiyska, 2002).

Chlorophyll biosynthesis was found to be inhibited by arsenate in maize (Gadre and

11.8 Influence on Nitrogen Metabolism

The assimilation of nitrate into amino acids involves three major reactions in plants. Nitrate is first reduced to ammonium by activities of nitrate reductase (NR, EC 1.6.6.1) and nitrite reductase (NiR, EC 1.6.6.4), which is a key regulatory step of $N-NO_3$ conversion to organic nitrogen (Campbell, 1999; Kaiser et al., 1999). Next the ammonium is incorporated into glutamine and glutamate primarily by the glutamine synthase–glutamate synthase cycle (GS/GOGAT cycle) (Miflin and Lea, 1982) and then assimilated into amino acids, nucleic acids, proteins, chlorophylls and other metabolites (Marschner, 1995; Stitt et al., 2002). Both NR and NiR extracted from arsenic-exposed seedlings show higher Km values, whereas glutamate dehydrogenases (GDHs) show a decrease in the Km value compared to normal plants. This suggests that inhibition in the activities of nitrogen assimilatory enzymes accompanied with decreased affinity of the enzymes towards their substrates eventually leads to a marked suppression of nitrogen assimilation and impaired growth of crop plants in arsenic-polluted environments (Jha and Dubey, 2004).

In *Silene vulgaris*, nitrogen uptake decreases with increasing absorption of arsenate by the plant. But when a high concentration of arsenite is present, the nitrogen uptake by the plant increases substantially indicating the divergent behaviour of the nitrogen metabolism caused by the two arsenic species (Schimdt et al., 2004). In crop plants, increasing levels of arsenic cause a marked reduction in the activities of the nitrate assimilatory enzymes, nitrate reductase (NR), nitrite reductase (NiR) and glutamine synthetase (GS), whereas it caused an increase in the activities of alanine and aspartate aminotransferases. The activities of aminating (NADH-GDH) and de-aminating (NAD$^+$-GDH) glutamate dehydrogenases increase at moderately toxic level of arsenic whereas a higher arsenic level is inhibitory to the enzymes. In *Tropaeolum majus*, nitrogen concentration decreases linearly with increasing concentrations of dimethylarsinate (DMA). This indicates that DMA prevents the uptake of nitrogen and hence interferes with formation of amino acids and proteins. Plants growing in arsenite-rich soil exhibit an elevated concentration of non-protein nitrogen, which can be an indication either for a stimulated uptake of nitrate or for an interrupted amino acid/protein synthesis (Schimdt et al., 2004).

In nitrogen assimilation, the rate-limiting enzyme nitrate reductase is known to be sensitive to metal stress (Campbell, 1999; Vajpayee et al., 1999, 2000; Rai et al., 2004; Kumar and Joshi, 2008). In *P. vittata* more NH_4 assimilation is generally associated with higher activity of NR and NiR. In the presence of a substantial level of NH_4 in tissues, the aminating glutamate dehydrogenase is directly involved in the formation of glutamate, which, besides being considered as the principal precursor of proline biosynthesis (Oaks, 1994), probably also participates in transamination to provide substrate for the formation of many other amino acids. This would be a biochemical adaptive feature of *P. vittata* that possibly plays a protective role under arsenic stress conditions. Sulfhydryl (SH) groups are required for NADH binding and catalytic activity of NR (Solomonson and Barber, 1990). Arsenic might affect the NR activity by binding to functional –SH groups present in the active sites of this enzyme (Sharma and Dubey, 2005; Xiong et al., 2006). In other studies, arsenic has also been shown to interact with functional sulfhydryl (SH) groups (Schmöger et al., 2000; Hartley-Whitaker et al., 2002; Schat et al., 2002).

In *Zea mays* L., Boussama et al. (1999) studied the effect of cadmium and demonstrated a significant amount of production of organic acids (mainly malate, citrate, oxalate and pectate) as a consequence of high nitrate reduction, which was also demonstrated in *P. vittata* (Singh et al., 2009). Subsequently, these organic acids may minimize the inhibitory effect of arsenic on NR activity in *P. vittata* because of arsenic inactivation in the symplast as organic acid complexes (Verkleij et al., 1991; Wagner, 1993). Arsenate application in wheat (*Triticum aestivum* L.) seedlings decreases total and soluble nitrogen contents as well as nitrate and nitrite contents. In arsenic-treated seedlings, amino acid contents are found to be almost equally reduced both in root and shoot (Ghosh et al., 2013).

11.9 Influence on Glutathione

Plants develop antioxidants that are crucial for their defence against oxidative stress (Aravind and Prasad, 2005; Nemat-Alla and Hassan, 2006). GSH is the most abundant non-enzymatic antioxidant in plant cells which participates in scavenging of ROS through the GSH cycle (Foyer et al., 2001). GSH maintains the reductive nature of the cell and regulates the binding of xenobiotics with cellular thiol. There has been a direct link between the thiol status of membranes and cellular GSH. Evidence in the literature suggests a protective role of GSH against arsenic-induced oxidative damage (Ito et al., 1998). Maiti and Chatterjee (2000) have reported that GSH is cytoprotective against arsenic and an increased GSH concentration can presumably protect the organ from arsenic-induced lipid peroxidation. GSH also stimulates the arsenic detoxification process by modulating arsenic methylation and glutathione S-transferase activity (Vahter et al., 1984; Lee et al., 1989). The oxidative damage induced by arsenate would probably be due to disturbance in the pro-oxidant–antioxidant balance (Yamanaka et al., 1991). Because of its high affinity for sulfhydryl groups in GSH, there might be implications in the maintenance of the thiol–disulfide balance.

GSH contents both in shoot and root of rice seedlings increase following arsenate treatment (Choudhury et al., unpublished data). Different forms of thiol have important roles to play as antioxidants and detoxicants for efficient defence mechanism in plants. Increasing contents of GSH have been regarded a limiting factor for the detoxification process (Aravind and Prasad, 2005). GSH being an essential component of thiol pool, it plays several roles in protection against oxidative stress and against heavy metals and xenobiotics (Mendoza-Cozatl and Moreno-Sanchez, 2006). In previous work by May et al. (1998) it was shown that GSH is an abundant and ubiquitous thiol with important roles in the storage and transport of reduced sulfur, synthesis of proteins and nucleic acids and a modulator of enzyme activity. It has been observed that joint application of phosphate along with arsenate in rice seedlings reduces GSH level (Choudhury et al., unpublished data). It is quite clear from these results that the enhanced GSH contents play important roles in overcoming the effects of arsenate toxicity by enhancing the rate of detoxification.

An arsenate-induced increase in the oxidation state of the redox pools of GSH and ascorbate leads to the formation of oxidized glutathione (GSSG) dimers as well as dehydroascorbate (Singh et al., 2006). This observed shift in redox state can be at least in two levels (Foyer et al., 2001). In the first instance, superoxide and the hydroxyl radical may directly oxidize both GSH and ascorbate, GSH and ascorbate acting as nucleophilic scavengers. H_2O_2 may further oxidize GSH and ascorbate through the action of specific peroxidases, or in the case of GSH, also through the action of glutaredoxins (GRXs) and GSH-S-transferases (GST, EC 2.5.1.18). Similar to antioxidative enzymes such as SOD and catalase, activities of GST, GRXs and/or peroxidase are often enhanced following arsenic exposure (Mylona et al., 1998; Stoeva and Bineva, 2003; Srivastava et al., 2005; Geng et al., 2006; Abercrombie et al., 2008; Ahsan et al., 2008; Norton et al., 2008; Chakrabarty et al., 2009). In response to AsV exposure in rice at least ten GST genes are up-regulated and two are down-regulated (Norton et al., 2008; Chakrabarty et al., 2009). However, changes in GST gene expression do not seem to have a role in the AsIII response, as fewer transcripts change in abundance (Chakrabarty et al., 2009), thus highlighting the differential effects of the two inorganic forms of arsenic on cellular metabolism.

GSTs catalyse the addition of reduced GSH to electrophilic substrates, tagging them for vacuolar sequestration (Edwards et al., 2000). These enzymes have cytoprotective activities and are essential for protecting the plants against environmental and biotic stresses (Marrs, 1996). Their ability to catalyse the conjugation of GSH points to their role in protecting cells against oxidative stress. The physiological selectivity of plants in arsenic toxicity may be due to increase in GST activity following arsenic treatment as shown in growing rice seedlings (Choudhury et al., unpublished data). Enhanced GST activities, concomitant with increasing levels of GSH in arsenic-treated rice seedlings, may suggest rapid and easy conjugation of GSH and subsequent detoxification of arsenic toxicity. From previous reports, it is confirmed that the GST-mediated conjugation of GSH is enhanced under stress conditions which increases the plant defence against several types of stresses (Jablonkai and Hatzios, 1993). The amelioration of arsenic toxicity in rice seedlings by joint application of phosphate with arsenic can be related to the levels of reduced GSH along with decreasing GST activities (Choudhury et al., unpublished data).

The oxidized form of glutathione (GSSG) may be converted back to the reduced form (GSH) by glutathione reductase (GR, EC 1.8.1.7). In growing rice seedlings, treatment with arsenate resulted in significantly increased activities of GR. Application of phosphate jointly with arsenate showed a reduction in GR activity when compared to arsenate treatment alone. A high reduction state of glutathione pool is maintained by glutathione, which plays an important role in cell defence against ROS and xenobiotics by sustaining the reduced status of GSH (Yoon et al., 2005). Therefore, enhanced GR activity coupled with an increase in level of GSH would accelerate the ability of rice seedlings to detoxify the toxic effect of arsenic (Choudhury et al., unpublished data). Anderson and Davis (2004) reported that the enzymes GR, GST and glutathione peroxidase (GPx) utilize GSH to play an important role in the plant defence mechanism.

Protection of the organism from oxidative damage has been perceived as the major biological role of GPx (EC 1.11.1.9). In the biological system, GPx reduces lipid hydroperoxides to their corresponding alcohols and reduces free hydrogen peroxide to water. The activity of GPx was found to be reduced in rice seedlings treated with arsenic. On the other hand, joint application of phosphate along with arsenate increased the level of GPx activity, which could protect rice seedlings from oxidative damage caused by arsenate (Choudhury et al., 2011). Herbette et al. (2002) and Tanaka et al. (2005) have shown that in some plants, GPx activity can reduce peroxides, much more efficiently or sometimes exclusively, by using the GSH as a reductant. It is reported that the level of GPx in rice is affected by different stress conditions (Agrawal et al., 2002). Such findings correlate with the studies in *Hordeum vulgare*, where two GPx isoforms are increased under osmotic stress but a third GPx isoform is down-regulated under the same conditions (Churin et al., 1999).

11.10 Influence on Arsenic Binding Proteins

Chelation of metal by a ligand followed by compartmentalization of the ligand–metal complex is a recurrent general mechanism for heavy metal detoxification in plants. Arsenic has also been found to form complexes with the thiol peptides known as phytochelatins, and also with glutathione, the precursor of phytochelatin (PC). PC synthesis is strongly induced by arsenic (Grill et al., 1987; Sneller et al., 1999; Schmöger et al., 2000). PCs consists of the amino acids glutamate (Glu), cysteine (Cys) and glycine (Gly) with the Glu and Cys residues linked through a γ-carboxylamide bond. PCs form a chain of structures with increasing repetitions of the γ-Glu-Cys dipeptide followed by Gly; $(\gamma\text{-Glu-Cys})_n$-Gly where n has been found to be as high as 11, but is generally in the range of 2 to 4, although PC_2 and PC_3 are most common (Cobbett, 2000). PCs occur in a wide variety of plant species and in some microorganisms. PCs are structurally related to glutathione (γ-Glu-Cys-Gly) and are presumed to be the products of a biosynthetic pathway (Rauser, 1995). The tripeptide glutathione (GSH) is synthesized from the amino acids Glu, Cys and Gly by the enzymes γ-glutamylcysteine synthetase and glutathione synthetase.

PC synthases catalyse the synthesis of PCs, which are peptides competent in the high-affinity binding of heavy metals, from glutathione (GSH) and/or from previously synthesized PCs (Schmöger et al., 2000; Vatamaniuk et al., 2000). PC synthase is a dipeptidyl transferase that undergoes acylation at two sites. This enzyme catalyses a dipeptidyl transfer reaction in which some of the energy liberated upon cleavage of the Cys-Gly bonds of the γ-Glu-Cys donors in the first phase of the catalytic cycle is conserved through the formation of a two-site substituted enzyme γ-Glu-Cys acyl intermediate subsequently hydrolysed in the second phase of the cycle. This provides energy for the formation of the new peptide bond required for PC chain extension.

Arsenate can be readily reduced to arsenite via arsenate reductase in a glutathione-dependent reaction (Mukhopadhyay et al., 2000), which is then complexed with thiols, particularly PCs. Different chain lengths of PCs (n = 2–4) are formed in plants under arsenate exposure, which are involved in the arsenic bindings, and arsenic preferentially binds to PC_3 forming the As-PC_3 complex rather than to PC_2 and GSH (Raab et al., 2004). Since As-PC complexes are unstable at pH > 7.2, they are preferentially stored in the vacuole (pH 4.5–5.9) rather than in the cytoplasm (pH 7.2–7.4) (Sneller et al., 1999). The As-PC_3 complex is the dominant complex formed in the arsenic-tolerant *H. lanatus*. In contrast, PC_2 was the predominant species induced by arsenic in *Rauvolfia serpentina* (Schmöger et al., 2000), *S. vulgaris* (Sneller et al., 1999) and in the arsenic-tolerant clone of *H. lanatus* (Hartley-Whitaker et al., 2001b).

PCs can increase the ability of rice plants to detoxify the toxicity by binding with arsenic. Arsenate exposure to rice seedlings leads to the production of PCs with different chain lengths (n = 2–4). However, PC_2 and PC_3 were identified as the predominant species in rice (Choudhury et al., unpublished data). The role of PCs in the detoxification of arsenic has been established for a number of higher plant species (Sneller et al., 1999; Schmöger et al., 2000; Hartley-Whitaker et al., 2001b; Schat et al., 2002; Bleeker et al., 2003). The PC_2 synthesis in *Lolium perenne* and *Agropyron repens* appears to be significantly higher than the production of PC_3 and PC_4. In *Urtica dioica*, *Glecoma hederacea*, *Leonurus marrubiastrum* and *Zea mays* the PC_3 is the dominant form, while the level of PC_4 is consistently moderate (Schulz et al., 2008). An exception to this is the arsenic hyper-accumulator *P. vittata*, in which PCs play only a limited role in the tolerance of arsenic (Zhao et al., 2003).

It has been shown that arsenic preferentially binds to PC_3 forming the As-PC_3 complex rather than to PC_2 and GSH in *H. lanatus* and *Pteris cretica* (Raab et al., 2004). In contrast, PC_2 is the predominant species induced by arsenic in *Rauvolfia serpentina* (Schmoger et al., 2000), *S. vulgaris* (Sneller et al., 1999) and in the As-tolerant clone of *H. lanatus* (Hartley-Whitaker et al., 2001b). PC_2-As-PC_2 is the major type of As-PC complex formed in plants. In the arsenic-tolerant *Cytisus striatus*, PC_4 is the major species (Bleeker et al., 2003). However, in the arsenic-sensitive clone of *H. lanatus*, PC_3 is dominant and PC_4 remained at a low concentration (Hartley-Whitaker et al., 2001b). The production of individual PCs or PC-related peptides is responsible for enhancing the ability of rice seedlings to detoxify arsenate toxicity. The ratio of PC_2 and PC_3 synthesis in roots of the growing rice seedlings suggest that the detoxification potential of root is sufficient to prevent the development of damage symptoms and changes in growth pattern in the shoot of rice plants (Choudhury et al., unpublished data).

11.11 Concluding Remarks

To date, though there has been substantial progress in understanding the interactions between plant cells and arsenic, there is no clear understanding about the exact nature of arsenic toxicity specifically on plant metabolism. It is known that arsenite is more toxic compared to arsenate towards plant growth and metabolism, and, following absorption by the roots, plants can transform arsenate to more reactive arsenite species. Though this reduction involves multi-step enzymatic reactions, only one enzyme has been identified so far. Although significant advances have been made on the understanding of physiological processes affected by arsenic and phosphate, further information on physiological, biochemical and molecular studies of arsenic-stressed plants will give us more insight into the mode of arsenic action in plants.

Various studies have shown that rice plants accumulate considerable amounts of arsenic in their roots. So, there remains a possibility of arsenic entering the food chain. This may cause concern for mankind since rice is the staple food for a large group of the world population. Therefore, arsenic resistance in economic crops, especially rice, is the most desired achievement that will stop arsenic from getting into the ecological food chain and hence will strive for the welfare of mankind.

In plants arsenic-induced oxidative stress generates various ROS, resulting in widespread responses in plants, i.e. readjustment of transport and metabolic processes along with growth inhibition. Plants have several mechanisms to withstand this toxic effect. To understand how plants overcome the effects of any heavy metal stress it is necessary to characterize the physiological as well as biochemical aspects. This paves the path for further molecular characterization, which is very much needed for designing the processes of detoxification and also genetic engineering in order to confer arsenic resistance in plants. The transporters involved in the transport of arsenite into vacuole and towards the xylem have not yet been identified. Molecular understanding of hyper-accumulation of arsenic is still rudimentary. Future research by taking advantage of recently developed analytical tools may help to address these issues.

This review will open up a new arena in metalloid pollution research and will provide the necessary background to the environmental scientists interested in arsenic toxicity. The present endeavour is to enhance the knowledge base of the biochemical mechanism concerning arsenic toxicity, which in the long run may help to devise new ways and means to reduce arsenic toxicity in plants. The application of phosphate fertilizers in arsenic-contaminated crop fields can be practised as a suitable remedial procedure for limiting arsenic uptake and its consequent influence on the growth and metabolism of crop plants.

Acknowledgements

This work was supported by funding from the UGC, New Delhi. The author acknowledges the research scholars of the Plant Physiology and Biochemistry laboratory for their technical help during preparation of the manuscript.

References

Abedin, M.J., Feldmann, J. and Meharg, A.A. (2002) Uptake kinetics of arsenic species in rice plants. *Plant Physiology* 128, 1120–1128.

Abercrombie, J.M., Halfhill, M.D., Ranjan, P., Rao, M.R., Saxton, A.M., Yuan, J.S. and Stewart C.N. Jr (2008) Transcriptional responses of *Arabidopsis thaliana* plants to As (V) stress. *BMC Plant Biology* 8, 87.

Agrawal, G.K., Rakwal, R., Jwa, N.S. and Agrawal, V.P. (2002) Effects of signaling molecules, protein phosphatase inhibitors and blast pathogen (*Magnaporthe grisea*) on the mRNA level of a rice (*Oryza sativa* L.) phospholipid hydroperoxide glutathione peroxidase (OsPHGPx) gene in seedling leaves. *Gene* 283, 227–236.

Ahsan, N., Lee, D.G., Alam, I., Kim, P.J., Lee, J.J., Ahn, Y.O., Kwak, S.S., Lee, I.J., Bahk, J.D., Kang, K.Y., Renaut, J., Komatsu, S. and Lee, B.H. (2008) Comparative proteomic study of arsenic-induced differentially expressed proteins in rice roots reveals glutathione plays a central role during As stress. *Proteomics* 8, 3561–3576.

Anderson, J.V. and Davis, D.G. (2004). Abiotic stress alters transcript profiles and activity of glutathione S-transferase, glutathione peroxidase, and glutathione reductase in Euphorbia esula. *Physiologia Plantarum* 120, 421–433.

Aravind, P. and Prasad, M.N.V. (2005) Modulation of cadmium-induced oxidative stress in *Ceratophyllum demersum* by zinc involves ascorbate-glutathione cycle and glutathione metabolism. *Plant Physiology and Biochemistry* 43, 107–116.

Asher, D.J. and Reay, P.F. (1979) Arsenic uptake by barley seedlings. *Journal of Plant Physiology* 6, 495–496.

Benson, N.R. (1953) Effect of season, phosphate and acidity on plant growth in arsenic toxic soils. *Soil Science* 76, 215–224.

Biswas, B.K., Dhar, R.K., Samanta, G., Mandal, B.K., Chakraborti, D., Faruk, I., Islam, K.S., Chowdhury, M.M., Islam, A. and Roy, S. (1998) Detailed study report of Samta, one of the arsenic-affected villages of Jessore District, Bangladesh. *Current Science* 74, 134–145.

Bleeker, P.M., Schat, H., Vooijs, R., Verkleij, J.A.C. and Ernst, W.H.O. (2003) Mechanisms of arsenate tolerance in *Cytisus striatus*. *New Phytologist* 157, 33–38.

Bleeker, P.M., Hakvoort, H.W.J., Bliek, M., Souer, E. and Schat, H. (2006) Enhanced arsenate reduction by a CDC25-like tyrosine phosphatase explains increased phytochelatin accumulation in arsenate-tolerant *Holcus lanatus*. *Plant Journal* 45, 917–929.

Boussama, N., Ouariti, O., Suzuki, A. and Ghorbal, M.H. (1999) Cd-stress on nitrogen assimilation. *Journal of Plant Physiology* 155, 310–317.

Bun-Ya, M., Nishimura, M., Harashima, S. and Oshima, Y. (1991) ThePHO84 gene of *Saccharomyces cerevisiae* encodes an inorganic phosphate transporter. *Molecular and Cellular Biology* 11, 3229–3238.

Burlo, F., Guijarro, I. and Carbonell-Barrachina, A.A. (1999) Arsenic species: effects on and accumulation by tomato plants. *Journal of Agricultural Food Chemistry* 47, 1247–1253.

Campbell, W.H. (1999) Nitrate reductase structure, function and regulation: bridging the gap between biochemistry and physiology. *Annual Review of Plant Physiology and Plant Molecular Biology* 50, 277–303.

Cao, X., Ma, L.Q. and Tub, C. (2004) Antioxidative responses to arsenic in the arsenic-hyperaccumulator Chinese brake fern (*Pteris vittata* L.). *Environmental Pollution* 128, 317–325.

Carbonell-Barrachina, A.A., Burlo, F., Burgos-Hernandez, A. and Mataix, J. (1997) The influence of arsenite concentration on arsenic accumulation in tomato and bean plants. *Scientia Horticulturae* 71, 167–176.

Carbonell-Barrachina, A.A., Burló, F. and Mataix, J. (1998) Response of bean micronutrient nutrition to arsenic and salinity. *Journal of Plant Nutrition* 21, 1287–1299.

Chakrabarty, D., Trivedi, P.K., Misra, P., Tiwari, M., Shri, M., Shukla, D., Kumar, S., Rai, A., Pandey, A., Nigam, D., Tripathi, R.D. and Tuli, R. (2009) Comparative transcriptome analysis of arsenate and arsenite stresses in rice seedlings. *Chemosphere* 74, 688–702.

Chen, W., Chi, Y., Taylor, N.L., Lambers, H. and Finnegan, P.M. (2010) Disruption of ptLPD1 or ptLPD2, genes that encode isoforms of the plastidial lipoamide dehydrogenase, confers arsenate hypersensitivity in *Arabidopsis*. *Plant Physiology* 153, 1385–1397.

Choudhury, B. (2012) Arsenic induced physio-chemical changes in rice (*Oryza sativa* L.) and its amelioration by phosphate. PhD thesis from the University of Calcutta under the supervision of A.K. Biswas.

Choudhury, B., Chowdhury, S. and Biswas, A.K. (2011) Regulation of growth and metabolism in rice (*Oryza sativa* L.) by arsenic and its possible reversal by phosphate. *Journal of Plant Interactions* 6, 15–24.

Choudhury, T.R., Pathan, K.M., Amin, M.N., Ali, M., Quraishi, S.B. and Mustafa, A.I. (2010) Adsorption of Cr (III) from aqueous solution by groundnut shell. *Journal of Environmental Science and Water Resources* 1, 144–150.

Chowdhury, T.R., Basu, G.K., Mandal, B.K., Biswas, B.K., Samanta, G., Chowdhury, U.K., et al. (1999) Arsenic poisoning in the Ganges delta. *Nature* 401, 545–547.

Churin, Y., Schilling, S. and Borner, T. (1999) A gene family encoding glutathione peroxidase homologues in *Hordeum vulgare* (barley). *FEBS Letters* 459, 33–38.

Cobbett, C.S. (2000) Phytochelatins and their roles in heavy metal detoxification. *Plant Physiology* 123, 825–832.

Cox, M.S. and Kovar, J.L. (2001) Soil arsenic effects on canola seedling and growth and ion uptake. *Communications Soil Science and Plant Analysis* 31, 107–117.

Daram, P., Brunner, S., Rausch, C., Steiner, C., Amrhein, N. and Bucher, M. (1999) *Pht2;1* encodes a low-affinity phosphate transporter from *Arabidopsis*. *Plant Cell* 11, 2153–2166.

Dat, J., Vandenabeele, S., Vranova, E., Van Montagu, M., Inz, D. and VanBreusegem, F. (2000) Dual action of the active oxygen species during plant stress responses. *Cell and Molecular Life Science* 57, 779–795.

Dhar, R.K., Biswas, B.K., Samanta, G., Mandal, B.K., Chakraborti, D., Roy, S., Jafar, A., Islam, A., Ara, G., Kabir, S., Khan, A.W., Ahmed, S.A. and Hadi, S.A. (1997) Groundwater arsenic calamity in Bangladesh. *Current Science* 73, 48–59.

Edwards, R., Dixon, D.P. and Walbot, V. (2000) Plant glutathione-S-transferases: enzymes with multiple functions in sickness and in health. *Trends in Plant Science* 5, 193–198.

Ellis, D.R., Gumaelius, L., Indirolo, E., Pickering, I.J., Banks, J.A. and Salt, D.E. (2006) A novel arsenate reductase from the arsenic hyperaccumulating fern *Pteris vittata*. *Plant Physiology* 141, 1544–1554.

Esteban, E., Carpena, R.O. and Meharg, A.A. (2003) High-affinity phosphate/arsenate transport in white lupin (*Lupinus albus*) is relatively insensitive to phosphate status. *New Phytologist* 158, 165–173.

Foyer, C.H. and Noctor, G. (2011) Ascorbate and glutathione: the heart of the redox hub. *Plant Physiology* 155, 2–18.

Foyer, C.H., Theodoulou, F. and Delrot, S. (2001) The functions of inter- and intracellular glutathione transport systems in plants. *Trends in Plant Science* 6, 486–492.

Gadre, R.P. and Meeta, J. (1997) Effect of As on chlorophyll and protein concentrations and enzymic activities in greening maize leaves. *Water, Air and Soil Pollution* 93, 109–115.

Garg, N. and Singla, P. (2011) Arsenic toxicity in crop plants: physiological effects and tolerance mechanisms. *Environmental Chemistry Letters* 9, 303–321.

Geng, C.N., Zhu, Y.G., Tong, Y.P., Smith, S.E. and Smith, F.A. (2006) Arsenate uptake by and distribution in two cultivars of winter wheat (*Triticum aestivum* L.). *Chemosphere* 62, 608–615.

Ghosh, S., Saha, J. and Biswas, A.K. (2013) Interactive influence of arsenate and selenate on growth and nitrogen metabolism in wheat (*Triticum aestivum* L.) seedlings. *Acta Physiologiae Plantarum* 35, 1873–1885.

Gomes, M.P., Duarte, D.M., Miranda, P.L.S., Barreto, L.C., Matheus, M.T. and Garcia, Q.S. (2012) The effects of arsenic on the growth and nutritional status of *Anadenanthera peregrine*, a Brazilian savanna tree. *Journal of Plant Nutrition and Soil Science* 175, 466–473.

Grafe, M., Eick, M.J. and Grossl, P.R. (2001) Adsorption of arsenate (V) and arsenite (III) on goethite in the presence and absence of dissolved organic carbon. *Soil Science Society of America Journal* 65, 1680–1687.

Gratao, P.L., Polle, A., Lea, P.J. and Azevedo, R.A. (2005) Making the life of heavy metal-stressed plants a little easier. *Functional Plant Biology* 32, 481–494.

Grill, E., Winnacker, E.L. and Zenk, M.H. (1987) Phytochelatins, a class of heavy-metal-binding peptides from plants, are functionally analogous to metallothioneins. *Proceedings of the National Academy of Sciences USA* 84, 439–443.

Gunes, A., Inal, A., Bagci, E.G. and Pilbeam, D.J. (2008) Silicon mediated changes of some physiological and enzymatic parameters symptomatic for oxidative stress in spinach and tomato grown in sodic-B toxic soil. *Plant and Soil* 290, 103–114.

Gusman, G.S., Oliviera J.A., Farnese, F.S. and Cambraia, J. (2013) Arsenate and arsenite: the toxic effects on photosynthesis and growth of lettuce plants. *Acta Physiologiae Plantarum* 35, 1201–1209.

Harminder, P.S., Daizy, R.B., Ravinder, K.K. and Komal, A. (2007) Arsenic-induced root growth inhibition in mungbean (*Phaseolus aureus* Roxb.) is due to oxidative stress resulting from enhanced lipid peroxidation. *Plant Growth Regulators* 53, 65–73.

Hartley-Whitaker, J., Ainsworth, G. and Meharg, A.A. (2001a) Copper- and arsenate-induced oxidative stress in *Holcus lanatus* L. clones with differential sensitivity. *Plant Cell and Environment* 24, 713–722.

Hartley-Whitaker, J., Ainsworth, G., Voojis, R., Ten-Bookum, W., Schat, H. and Meharg, A.A. (2001b) Phytochelatins are involved in differential arsenate tolerance in *Holcus lanatus*. *Plant Physiology* 126, 299–306.

Hartley-Whitaker, J., Woods, C. and Meharg, A.A. (2002) Is differential phytochelatin production related to decreased arsenate influx in arsenate tolerant *Holcus lanatus*. *New Phytologist* 155, 219–225.

Herbette, S., Lenne, C., Leblanc, N., Julien, J.L., Drevet, J.R. and Roeckel-Drevet, P. (2002) Two GPX-like proteins from *Lycopersicon esculentum* and *Helianthus annuus* are antioxidant enzymes with phospholipid hydroperoxide glutathione peroxidase and thioredoxin peroxidase activities. *European Journal of Biochemistry* 269, 2414–2420.

Huang, Y.Z., Qian, X.C., Wang, G.Q., Gu, Y.L., Wang, S.Z., Cheng, Z.H., Xiao, B.Y., Gang, J.M., Wu, Y.K., Kan, M.Y., *et al.* (1992) Syndrome of endemic arsenism and fluorosis: a clinical study. *Chinese Medical Journal* 105, 586–590.

Ito, H., Okamoto, K. and Kato, K. (1998) Enhancement of expression of stress proteins by agents that lower the levels of glutathione in cells. *Biochimica et Biophysica Acta – Gene Structure and Expression* 1397, 223–230.

Jablonkai, I. and Hatzios, K.K. (1993) In-vitro conjugation of chloroacetanilide herbicides and atrazine with thiols and contribution of nonenzymatic conjugation to their glutathione-mediated metabolism in corn. *Journal of Agricultural and Food Chemistry* 41, 1736–1742.

Jha, A.B. and Dubey, R.S. (2004) Arsenic exposure alters activity behaviour of key nitrogen assimilatory enzymes in growing rice plants. *Plant Growth Regulators* 43, 259–268.

Kaiser, W.M., Weiner, H. and Huber, S.C. (1999) Nitrate reductase in higher plants: a case study for transduction of environmental stimuli into control of catalytic activity. *Physiologia Plantarum* 105, 385–390.

Kapustka, L.A., Lipton, J., Galbraith, H., Cacela, D. and Lejeune, K. (1995) Metal and arsenic impacts to soil, vegetation communities and wildlife habitat in southwest Montana uplands contaminated by smelter emissions: II. Laboratory phytotoxicity studies. *Toxicological and Environmental Chemistry* 14, 1905–1912.

Kavi-Kishor, P.B., Sangam, S., Amrutha, R.N., Sri Laxmi, P., Naidu, K.R., Rao, K.R., Rao, S., Reddy, K.J., Theriappan, P. and Sreeniv, N. (2005) Regulation of proline biosynthesis, degradation, uptake and transport in higher plants: its Implications in plant growth and abiotic stress tolerance. *Current Science* 88, 424–438.

Khan, I., Ahmad, A. and Iqbal, M. (2009) Modulation of antioxidant defence system for arsenic detoxification in Indian mustard. *Ecotoxicology and Environmental Safety* 72, 626–634.

Kumar, S. and Joshi, U.N. (2008) Nitrogen metabolism as affected by hexavalent chromium in sorghum (*Sorghum bicolor* L.). *Environmental and Experimental Botany* 64, 135–144.

Lee, T.-C., Ko, J. and Jan, K.Y. (1989) Differential cytotoxocity of sodium arsenite in human fibroblasts and Chinese hamster ovary cells. *Toxicology* 56, 289–300.

Li, W.X., Chen, T.B., Huang, Z.C., Lei, M. and Liao, X.Y. (2006) Effect of arsenic on chloroplast ultrastructure and calcium distribution in arsenic hyperaccumulator *Pteris vittata* L. *Chemosphere* 62, 803–809.

Ma, L.Q., Komar, K.M.M., Tu, C., Zhang, W., Cai, Y. and Kennelley, E.D. (2001) A fern that hyperaccumulates arsenic. *Nature (London)* 409, 579.

Macnair, M.R. and Cumbes, Q. (1978) Evidence that arsenic tolerance in *Holcus lanatus* L. is caused by an altered phosphate system. *New Phytologist* 107, 387–394.

Maiti, S. and Chatterjee, A.K. (2000) Differential response of cellular antioxidant mechanism of liver and kidney to arsenic exposure and its relation to dietary protein deficiency. *Environmental Toxicology and Pharmacology* 8, 227–235.

Mallick, S., Sinam, G. and Sinha, S. (2011) Study on arsenate tolerant and sensitive cultivars of *Zea mays* L.: Differential detoxification mechanism and effect on nutrients status. *Ecotoxicology and Environmental Safety* 74, 1316–1324.

Mandal, B.K., Roy Chowdhury, T., Samanta, G., Basu, G.K., Chowdhury, P.P., Chanda, C.R., Lodh, D., Karan, N.K., Dhar, R.K., et al. (1996) Arsenic in groundwater in seven districts of West Bengal, India: the biggest arsenic calamity in the world. *Current Science* 70, 976–986.

Marques, A.P.G.C., Rangel, A.O.S.S. and Castro, P.M.L. (2007) Effect of arsenic, lead and zinc on seed germination and plant growth in black nightshade (*Solanum nigrum* L.) vs. clover (*Trifolium incarnatum* L.). *Fresenius Environmental Bulletin* 16, 896–903.

Marrs, K.A. (1996) The function and regulation of glutathione-*S*-transferases in plants. *Annual Review of Plant Physiology and Plant Molecular Biology* 47, 127–158.

Marschner, H. (1995) *Mineral Nutrition of Higher Plants*, 2nd edn. Special Publications of the Society for General Microbiology, Academic Press, USA.

Mascher, R., Lippmann, B., Holzinger, S. and Bergmann, H. (2002) Arsenate toxicity: effects on oxidative stress response molecules and enzymes in red clover plants. *Plant Science* 163, 961–969.

Mathys, W. (1975) The role of malate, oxalate, and mustard oil glucosides in the evolution of zinc-resistance in herbage plants. *Plant Physiology* 40(2), 130–136.

May, M.J., Vernoux, T., Sanchez-Fernandez, R. and VanMontagu, M.I.D. (1998) Evidence for post transcriptional activation of γ-glutamylcysteine synthetase during plant stress responses. *Proceedings of the National Academy of Sciences* 95, 12049–12054.

Meharg, A.A. (1994) Ecological impact of major industrial-chemical accident. *Reviews of Environmental Contamination and Toxicology* 138, 21–48.

Meharg, A.A. and Hartley-Whitaker, J. (2002) Arsenic uptake and metabolism in arsenic resistant and nonresistant plant species. *New Phytologist* 154, 29–43.

Meharg, A.A. and Jardine, L. (2003) Arsenite transport into paddy rice (*Oryza sativa*) roots. *New Phytologist* 157, 39–44.

Meharg, A.A. and Macnair, M.R. (1990) An altered phosphate uptake system in arsenate-tolerant *Holcus lanatus* L. *New Phytologist* 116, 29–35.

Meharg, A.A. and Macnair, M.R. (1991) Uptake, accumulation and translocation of arsenate in arsenate-tolerant and non-tolerant *Holcus lanatus* L. *New Phytologist* 117, 225–231.

Meharg, A.A. and Macnair, M.R. (1992) Suppression of the high affinity phosphate uptake system: a mechanism of arsenate tolerance in *Holcus lanatus* L. *Journal of Experimental Botany* 43, 519–524.

Meharg, A.A., Naylor, J. and Macnair, M.R. (1994) Phosphorus nutrition of arsenate-tolerant and nontolerant phenotypes of velvetgrass. *Journal of Environmental Quality* 23, 234–238.

Mendoza-Cozatl, D.G. and Moreno-Sanchez, R. (2006) Control of glutathione and phytochelatin synthesis under cadmium stress. Pathway modeling for plants. *Journal of Theoretical Biology* 238, 919–936.

Miflin, B.J. and Lea, P.J. (1982) Ammonium assimilation and amino acid metabolism. In: Boulter, D. and Parthier, B. (eds) *Encyclopedia of Plant Physiology, New Series*, 14, 5–64.

Miteva, E. and Merakchiyska, M. (2002) Response of chloroplasts and photosynthetic mechanism of bean plants to excess arsenic in soil. *Bulgarian Journal of Agricultural Science* 151–156.

Mittler, R. (2002) Oxidative stress, antioxidants and stress tolerance. *Trends in Plant Science* 7, 405–410.

Møller, I.M., Jensen, P.E. and Hansson, A. (2007) Oxidative modifications to cellular components in plants. *Annual Review of Plant Physiology* 58, 459–481.

Mukhopadhyay, R., Shi, J. and Rosen, B.P. (2000) Purification and characterization of Acr2p, the *Saccharomyces cerevisiae* arsenate reductase. *The Journal of Biological Chemistry* 275, 21149–21157.

Mylona, P.V., Polidoros, A.N. and Scandalios, J.G. (1998) Modulation of antioxidant responses by arsenic in maize. *Free Radical Biology and Medicine* 25, 576–585.

Nemat-Alla, M.M. and Hassan, N.M. (2006) Changes of antioxidants levels in two maize lines following atrazine treatments. *Plant Physiology and Biochemistry* 44, 202–210.

Nickson, R.T., McArthur, J.M., Burgess, W.G., Ahmed, K.M., Ravenscroft, P. and Rahman, M. (1998) Arsenic poisoning of Bangladesh groundwater. *Nature* 395, 338.

Norton, G.J., Lou-Hing, D.E., Meharg, A.A. and Price, A.H. (2008) Rice–arsenate interactions in hydroponics: whole genome transcriptional analysis. *Journal of Experimental Botany* 59, 2267–2276.

Nriagu, J.O. and Pacyna, J.M. (1988) Quantitative assessment of worldwide contamination of air, water and soils by trace metals. *Nature* 333, 134–139.

Oaks, A. (1994) Primary nitrogen assimilation in higher plants and its regulation. *Canadian Journal of Botany* 72(6), 739–750.

Pickering, I.J., Gumaelius, L., Harris, H.H., Prince, R.C., Hirsch, G., Banks, J.A., Salt, D.E. and George, G.N. (2006) Localizing the biochemical transformations of arsenate in hyperaccumulating fern. *Environmental Science and Technology* 40, 5010–5014.

Raab, A., Feldmann, J. and Meharg, A.A. (2004) The nature of arsenic-phytochelatin complexes in *Holcus lanatus* and *Pteris cretica*. *Plant Physiology* 134, 1113–1122.

Rai, V., Vajpayee, P., Singh, S.N. and Mehrotra, S. (2004) Effect of chromium accumulation on photosynthetic pigments, oxidative stress defense system, nitrate reduction, praline level and eugenol content of *Ocimum tenuiflorum* L. *Plant Science* 167, 1159–1169.

Rauser, W.E. (1995) Phytochelatins and related peptides. Structure, biosynthesis and function. *Plant Physiology* 109, 1141–1149.

Requejo, R. and Tena, M. (2005) Proteome analysis of maize roots reveals that oxidative stress is a main contributing factor to plant arsenic toxicity. *Phytochemistry* 66, 1519–1528.

Salt, D.E., Blaylock, M., Kumar, P.B.A.N., Dushenkov, V., Ensley, B.D., Chet, I. and Raskin, I. (1995a) Phytoremediation: a novel strategy for the removal of toxic metals from the environment using plants. *Biotechnology* 13, 468–475.

Schat, H., Llugany, M., Vooijs, R., Hartley-Whitaker, J. and Bleeker, P.M. (2002) The role of phytochelatins in constitutive and adaptive heavy metal tolerances in hyperaccumulator and non-hyperaccumulator metallophytes. *Journal of Experimental Botany* 53, 2381–2392.

Schmidt, A.C., Reisser, W., Mattusch, J., Wennrich, R. and Jung, K. (2004) Analysis of arsenic species accumulation by plants and the influence on their nitrogen uptake. *Journal of Analytical Atomic Spectrometry* 19, 172–177.

Schmöger, M.E.V., Oven, M. and Grill, E. (2000) Detoxification of arsenic by phytochelatins in plants. *Plant Physiology* 122, 793–801.

Schulz, H., Haertling, S. and Tanneberg, H. (2008) The identification and quantification of arsenic-induced phytochelatins: comparison between plants with varying As sensitivities. *Plant and Soil* 303, 275–287.

Shao, Y.L., Jiang, D., Zhang, L. and Ma, C.L. (2011) Effects of arsenic, cadmium and lead on growth and respiratory enzymes activity in wheat seedlings. *African Journal of Agricultural Research* 6(19), 4505–4512.

Sharma, P. and Dubey, R.S. (2005) Lead toxicity in plants. *Brazilian Journal of Plant Physiology* 17, 35–52.

Shin, H., Shin, H.S., Dewbre, G.R. and Harrison, M.J. (2004) Phosphate transport in *Arabidopsis*: Pht1;1 and Pht1;4 play a major role in phosphate acquisition from both low- and high-phosphate environments. *Plant Journal* 39, 629–642.

Singh, H.P., Batish, D.R., Kohli, R.K. and Arora, K. (2007) Arsenic-induced root growth inhibition in mung bean (*Phaseolus aureus* Roxb.) is due to oxidative stress resulting from enhanced lipid peroxidation. *Plant Growth Regulation* 53, 65–73.

Singh, N., Ma, L.Q., Joseph, C.V. and Raj, A. (2009) Effects of arsenic on nitrate metabolism in arsenic hyperaccumulating and non-hyperaccumulating ferns. *Environmental Pollution* 157, 2300–2305.

Singh, N., Ma, L.Q., Srivastava, M. and Rathinasabapathi, B. (2006) Metabolic adaptations to arsenic-induced oxidative stress in *Pteris vittata* L. and *Pteris ensiformis* L. *Plant Science* 170, 274–282.

Slater, E.C. (1949) The dihydrocozymase-cytochrome c-reductase activity of heart-muscle preparation. *Biochemical Journal* 46, 499–503.

Smith, A.H., Hopenhayn-Rich, C., Bates, M.N., Goeden, H.M., Hertz-Picciotto, P., Duggan, H.M., Wood, R., Kornett, M.J. and Smith, M.T. (1992) Cancer risks from arsenic in drinking water. *Environmental Health Perspective* 97, 259–267.

Smith, E., Naidu, R. and Alston, A.M. (1998) Arsenic in the soil environment: a review. *Advances in Agronomy* 64, 149–195.

Sneller, F.E.C., Van-Heerwaarden, L.M., Kraaijeveld-Smit, F.J.L., Ten-Bookum, W.M., Koevoets, P.L.M., Schat, H. and Verkleij, J.A.C. (1999) Toxicity of arsenate in *Silene vulgaris*, accumulation and degradation of arsenate-induced phytochelatins. *New Phytologist* 144, 223–232.

Sneller, F.E.C., Heerwaarden, L.M., Koevoets, P.L.M., Vooijs, R., Schat, H. and Verkleij, J.A.C. (2000) Derivatization of phytochelatins from *Silene vulgaris* induced upon exposure to arsenate and cadmium: comparison of derivatization with Ellman's reagent and monobromobimane. *Journal of Agricultural and Food Chemistry* 48, 4014–4019.

Solomonson, L.P. and Barber, M.J. (1990) Assimilatory nitrate reductase: functional properties and regulation. *Annual Review of Plant Physiology and Plant Molecular Biology* 41, 225–253.

Song, W.Y., Park, J., Mendoza-Cózatl, D.G., Suter-Grotemeyer, M., Shim, D., Hortensteiner, S., Geisler, M., Weder, B., Rea, P.A., Rentsch, D., Schroeder, J.I., Lee, Y. and Martinoia, E. (2010) Arsenic tolerance in *Arabidopsis* is mediated by two ABCC-type phytochelatin transporters. *Proceedings of National Academy of Sciences USA* 107, 21187–21192.

Srivastava, M., Ma, L.Q., Singh, N. and Singh, S. (2005) Antioxidant responses of hyper-accumulator and sensitive fern species to arsenic. *Journal of Experimental Botany* 56, 1332–1342.

Stitt, M., Muller, C., Matt, P., Gibon, Y., Carillo, P., Morcuende, R., Scheible, W.R. and Krapp, A. (2002) Steps towards an integrated view of nitrogen metabolism. *Journal of Experimental Botany* 53, 959–970.

Stoeva, N. and Bineva, T. (2003) Oxidative changes and photosynthesis in oat plants grown in As-contaminated soil. *Bulgarian Journal of Plant Physiology* 29, 87–95.

Stoeva, N., Berova, M. and Zlatev, Z. (2005) Effect of arsenic on some physiological parameters in bean plants. *Plant Biology* 49, 293–296.

Stone, J.R. and Yang, S. (2006) Hydrogen peroxide: a signaling messenger. *Antioxidants and Redox Signalling* 8, 243–270.

Tanaka, Y., Toshio, S., Masanori, T., Nobuyoshi, N., Noriaki, K. and Hasezawa, S. (2005) Ethylene inhibits abscisic acid-induced stomatal closure in *Arabidopsis*. *Plant Physiology* 138, 2337–2343.

Thangavel, S., Chien-Hung, C., Ling-Huei, Y., Alexander, S.S., Wang, S.-Y. L., Tsen-Chien, C. and Kun-Yan, J. (2003) Reactive oxygen species are involved in arsenic trioxide inhibition of pyruvate dehydrogenase activity. *Chemical Research in Toxicology* 16(3), 409–414.

Therapong, C., Datta, R. and Sarkar, D. (2004) Comparative arsenic stress response in monocot seedlings. *American Association of Petroleum Geologists* 88.

Tu, S. and Ma, L.Q. (2003) Interactive effects of pH, arsenic and phosphorus on uptake of As and P and growth of the arsenic hyperaccumulator *Pteris vittata* L. under hydroponic conditions. *Environmental and Experimental Botany* 50, 243–251.

Ullrich-Eberius, C.I., Sanz, A. and Novacky, A.J. (1989) Evaluation of arsenate- and vanadate-associated changes of electrical membrane potential and phosphate transport in *Lemna gibba* G1. *Journal of Experimental Botany* 40, 119–128.

Vahter, M., Marafante, E. and Dencker, L. (1984) Tissue distribution and retention of 74As-dimethylarsinic acid in mice and rats. *Archives of Environmental Contamination* and *Toxicology* 13, 259–264.

Vajpayee, P., Sharma, S.C., Rai, U.N., Tripati, R.D. and Yunus, M. (1999) Bioaccumulation of chromium and toxicity to photosynthetic pigments, nitrate reductase activity and protein content of *Nelumbo nucifera* Gaetrn. *Chemosphere* 39, 2159–2169.

Vajpayee, P., Tripati, R.D., Rai, U.N., Ali, M.B. and Singh, S.N. (2000) Chromium accumulation reduces chlorophyll biosynthesis, nitrate reductase activity and protein content of *Nymphaea alba*. *Chemosphere* 41, 1075–1082.

Van Assche, F. and Clijsters, H. (1990) Effects of heavy metals on enzyme activity in plants. *Plant Cell and Environment* 13, 195–206.

Van Breusegem, F. and Dat, J.F. (2006) Reactive oxygen species in plant cell death. *Plant Physiology* 141, 384–390.

Vatamaniuk, O.K., Mari, S., Lu, Y.P. and Rea, P.A. (2000) Mechanism of heavy metal ion activation of phytochelatin (PC) synthase: blocked thiols are sufficient for PC synthase-catalyzed transpeptidation of glutathione and related thiol peptides. *Journal of Biological Chemistry* 275, 31451–31459.

Verkleij, J.A.C., Lolkema, P.C., De Neeling, A.L. and Harmens, H. (1991) Heavy metal resistance in higher plants: biochemical and genetic aspects. In: Rozema, J. and Verkleij, J.C. (eds) *Ecological Responses to Environmental Stresses*. Kluwer Academic, Amsterdam, the Netherlands, pp. 8–19.

Visoottiviseth, P., Francesconi, K. and Sridokchan, W. (2002) The potential of Thai indigenous plant species for the phytoremediation of arsenic contaminated land. *Environmental Pollution* 118, 453–461.

Wagner, G.J. (1993) Accumulation of cadmium in crop plants and its consequences to human health. *Advances in Agronomy* 51, 173–212.

Wang, J. (2002) Mechanisms of arsenic hyperaccumulation in *Pteris vittata*: uptake kinetics, interactions with phosphate, and arsenic speciation. *Plant Physiology* 130, 1552–1561.

Wang, Y.H., Garvin, D.F. and Kochian, L.V. (2001) Nitrate-induced genes in tomato roots. Array analysis reveals novel genes that may play a role in nitrogen nutrition. *Plant Physiology* 127, 345–359.

Woolson, E.A., Axley, J.H. and Kearney, P.C. (1971a) Correlation between available soil arsenic, estimated by six methods, and response of corn (*Zea mays* L.). *Soil Science Society of America Proceedings* 35, 101–105.

Woolson, E.A., Axley, J.H. and Kearney, P.C. (1971b) The chemistry and phytotoxicity of arsenic in soils. I. Contaminated field soils. *Soil Science Society of America Proceedings* 35, 938–943.

Wu, J., Hsu, F.C. and Cunningham, S.D. (1999) Chelate-assisted Pb phytoextraction, Pb availability, uptake and translocation constraints. *Environmental Science & Technology* 33, 1898–1904.

Wu, J., van Geen, A., Ahmed, K.M., Alam, Y.A.J., Culligan, P.J., *et al.* (2011) Increase in diarrheal disease associated with arsenic mitigation in Bangladesh. *PLoS One* 6, e29593.

Wysocki, R., Chery, C.C., Wawrzycka, D., Van Hulle, M., Cornelis, R., Thevelein, J.M. and Tamas, M.J. (2001) The glycerol channel Fps1p mediates the uptake of arsenite and antimonite in *Saccharomyces cerevisiae*. *Molecular Microbiology* 40, 1391–1401.

Xiong, Z.T., Liu, C. and Geng, B. (2006) Phytotoxic effects of copper on nitrogen metabolism and plant growth in *Brassica pekinensis* Rupr. *Ecotoxicology and Environmental Safety* 64, 273–280.

Xu, X.Y., McGrath, S.P, Meharg, A. and Zhao, F.J. (2008) Growing rice aerobically markedly decreases arsenic accumulation. *Environmental Science & Technology* 42, 5574–5579.

Yamanaka, K., Hasegawa, A., Sawamura, R. and Okada, S. (1991) Cellular response to oxidative damage in lung induced by the administration of dimethylarsinic acid, a major metabolite of inorganic arsenics, in mice. *Toxicology and Applied Pharmacology* 108, 205–213.

Yoon, H.S., Lee, I.A., Lee H., Lee, B.H. and Jo, J. (2005) Over expression of a eukaryotic glutathione reductase gene from *Brassica campestris* improved resistance to oxidative stress in *Escherichia coli*. *Biochemical and Biophysical Research Communications* 326, 618–623.

Zhao, F.J., Wang, J.R., Barker, J.H.A., Schat, H., Bleeker, P.M. and McGrath, S.P. (2003) The role of phytochelatins in arsenic tolerance in the hyperaccumulator *Pteris vittata*. *New Phytologist* 159, 403–410.

Zhao, F.J., Ma, J.F., Meharg, A.A. and McGrath, S.P. (2009) Arsenic uptake and metabolism in plants. *New Phytologist* 181, 777–794.

12 Plant Responses to Abiotic Stresses in Sustainable Agriculture

Zlatko Stoyanov Zlatev*
Department of Plant Physiology and Biochemistry, Agricultural University-Plovdiv, Bulgaria

Abstract

In recent decades the significance of sustainable agriculture has risen to become one of the most important directions in agriculture. In both conventional and sustainable agriculture, plants are exposed to abiotic and biotic factors of the environment. These include factors such as drought, flooding, low temperature, heat, high light intensity, UV-B radiation and soil salinization. Environmental factors that limit plant growth and development are considered as stress factors. Maintaining growth and crop productivity under adverse environmental stresses is presumably the major challenge facing sustainable agriculture. Better understanding physiological and biochemical mechanisms of plant acclimation to stress conditions, and the relationship between plants and environment is the first step to meet this challenge.

In this chapter recent information about the plant physiological reaction to different abiotic stress factors, as well as physiological and biochemical bases of acclimation are analysed. It is now known that tolerance to abiotic stress is complex and many authors suggest that the plasticity of cell metabolism and its fast acclimation to changes in environmental conditions is a main essential step in stress tolerance.

12.1 Introduction

Plants, either growing naturally or under cultivation, are often exposed to various environmental conditions (Zlatev and Lidon, 2012). These include factors such as drought, low temperature, heat, high light intensity, UV-B radiation and high salinity. Abiotic factors that limit growth and development of plants are considered as stress factors. Maintaining growth and crop productivity under adverse environmental stress is presumably the major challenge facing sustainable agriculture. To meet this challenge, it is necessary to have a full knowledge of the physiological and biochemical mechanisms of plant acclimation to growth in stressed conditions, as well as the interaction between plants and environment (Zlatev, 2013).

Global climatic changes and low input of mineral nutrients will probably emerge as the most significant factors for limiting growth and productivity of plants in sustainable agriculture. Considering the high level of organization of living organisms, including the plants, it is obvious that there are complex and multiple relations with the environment. The strength and duration of the environmental factor, the

*E-mail: zl_zlatev@abv.bg

genetic make-up of the particular plant, as well as their interactions will determine the degree of influence on the plant. For each of the physiological processes in the live system, there always exists the so-called 'stability limit'.

Any deviation of the environment from the stability limit of the system results in stress, causing disturbances in structure and functional activity of the system. The degree to which plants will be affected by the various stresses will be determined by the plant's capacity and sensitivity, which are genetically determined, as well as the duration, intensity and nature of the stress (Bhadula *et al.*, 1998; Chaves *et al.*, 2009). Plants have built-in buffering capacity, i.e. a given norm of reaction towards concrete external conditions, which helps them overcome or tolerate the various stresses. Hence, plants that can better maintain their physiological processes within the reaction norm under different environmental conditions will have greater acclimation capacity (Valladares *et al.*, 2007).

A decrease in photosynthetic carbon assimilation and growth is usually seen as the effect of stress factors at the whole plant level. Considering the importance of this, this chapter is focused mainly on recent information about the effects of abiotic stresses on plant growth, water relations and photosynthesis, as well as mechanisms of acclimation.

12.2 Effects of Abiotic Factors on Photosynthesis

Water deficit occurring during drought, singly or combined with high air temperature and extremely high light intensity, is the most important factor that causes physiological disturbances in plants. Thus, undoubtedly, drought is a complex stress disturbing plants at various levels of their organization (Yordanov *et al.*, 2000; Wentworth *et al.*, 2006). During drought the dehydration process is developed, and there are specific, fundamental changes in water relations, physiological and biochemical processes, structure of membrane, as well as ultrastructure of subcellular organelles (Tuba *et al.*, 1996; Sarafis, 1998; Yordanov *et al.*, 2003). The complexity of plant responses to drought at whole-plant level is due to the integration of effects of stress and responses at all levels of organization over space and time (Zlatev and Yordanov, 2004).

At whole organism level, the observed reduction in growth of plants during limited water availability is associated mainly with alterations in carbon and nitrogen metabolism, which are dependent to a great extent on leaf photosynthesis. Considering the importance of photosynthetic responses to low water availability it has been the subject of study and debate for decades, specifically focused on determining the factors that are most limiting for photosynthesis under this stress (Zlatev and Yordanov, 2004). Soil drought and growing-leaf water deficit in this case lead to a permanent depression of photosynthesis (Yordanov *et al.*, 2003; Chaves *et al.*, 2009). Decreased photosynthesis is the result of stomatal and mesophyllic (non-stomatal) limitations (Yordanov *et al.*, 2000; Zlatev and Lidon, 2012).

Tolerance to drought therefore requires the steady maintenance of the activity of the photosynthetic apparatus. Rapid stomatal closure is generally perceived as a first response in plants to water stress to reduce or minimize transpirational water loss (Cornic, 1994; Lawlor, 1995), which in turn restricts CO_2 diffusion into the leaves (Chaves, 1991; Flexas and Medrano, 2002). This has been confirmed in many experiments where a decrease in net photosynthetic rate (A_n) was observed under water stress conditions. Mechanisms responsible for this decrease in photosynthesis could be a lowered internal CO_2 concentration (C_i) at the acceptor site of ribulose-1,5-bisphosphate carboxylase/oxygenase (Rubisco) (Cornic *et al.*, 1992), or direct inhibition of enzymes involved in photosynthesis such as Rubisco (Haupt-Herting and Fock, 2000) or ATP synthase (Tezara *et al.*, 1999; Nogués and Baker, 2000). It is well known that photosystem II (PSII) is highly drought resistant (Yordanov *et al.*, 2003), but under conditions of water deficit photosynthetic electron transport through PSII also decreases (Chen and Hsu, 1995; Chakir and Jensen, 1999). Many *in vivo* studies showed that water stress resulted in damage to the oxygen-evolving complex of PSII (Lu and Zhang, 1999; Skotnica *et al.*, 2000). Water stress also leads to degradation of D1 protein, an essential component of the reaction centre of PSII (Cornic, 1994; He *et al.*, 1995).

However, the exact mechanism by which the water deficit decreases electron transport remains still unclear.

It has also been demonstrated by previous workers that the water deficit-induced photosynthetic inhibition could be linked to the changes in many of the biochemical processes (Lauer and Boyer, 1992). There are many reports suggesting that stomatal limitations of photosynthesis are the primary reaction, which are then followed by the other changes in photosynthetic reactions (Chaves, 1991; Zlatev and Yordanov, 2004; Zlatev and Lidon, 2012). At present it is agreed that the observed inhibition in photosynthesis is due to both stomatal and non-stomatal factors (Shangguan et al., 1999). Non-stomatal limitation of photosynthesis has been linked to a decrease in efficiency of carboxylation (Jia and Gray, 2004), inhibited ribulose-1,5-bisphospate (RuBP) regeneration (Tezara et al., 1999), reduced availability of functional Rubisco (Kanechi et al., 1995), or to disturbances in functional activity of PSII.

Determination of maximal CO_2 assimilation (A_{max}) under severe water stress allows evaluation of non-stomatal limitations of photosynthesis, because it reflects all of those mesophyllic changes. Drought stress inhibits leaf gas exchange, reduces the maximal carboxylation efficiency and increases the CO_2 compensation point of young bean plants (Table 12.1). This treatment also changes the shape of CO_2 curves of photosynthesis (Zlatev and Yordanov, 2004).

It is demonstrated that both stomatal and mesophyllic factors have a role to play in decreased photosynthesis, but their relative ratio changes significantly. The drought-tolerant species control stomatal function to allow some carbon fixation at stress, thus improving water use efficiency, or open stomata rapidly when water deficit is stopped through watering (Lawlor, 2002). Stomatal resistance is more closely connected to soil moisture than to leaf water parameters (Davies and Zhang, 1991). At the end of the stress period stomatal limitation values of stressed plants are higher than in the control plants, suggesting enhanced stomatal limitation.

Water stress also led to an inhibition of both activity of Rubisco and capacity for RuBP regeneration, as detailed in Table 12.1. Lawlor and Cornic (2002) suggested that the main reasons for decreased A_{max} when relative water content (RWC) is low are impaired metabolism (synthesis and storage of ATP, RuBP synthesis inhibition, but inhibition of photosynthetic enzymes including Rubisco was of lesser importance). This has been corroborated in studies on leaves of bean plants cv. Dobrudjanski ran (Table 12.1). Thus, photosynthetic carbon assimilation may be regulated through a balance between carboxylation ability of Rubisco, RuBP utilization and its regeneration. It has also been suggested that some of the steps of Calvin cycle taking part in RuBP regeneration are inhibited. Regeneration of RuBP may be inhibited either by a decreased supply of reductants and ATP from

Table 12.1. Effect of soil drought on leaf gas exchange in first trifoliate leaves of control and water-stressed bean plants (Zlatev and Yordanov, 2004).

	α ($\mu mol\ m^{-2}\ s^{-1}\ mol^{-1}$)	Γ ($\mu mol\ mol^{-1}$)	A_{max} ($\mu mol\ m^{-2}\ s^{-1}$)	$Ac_a=350$ ($\mu mol\ m^{-2}\ s^{-1}$)	$C_{i(Ca=350)}$ ($\mu mol\ mol^{-1}$)	SL (%)
Control						
Plovdiv 10	0.160	39.3	22.3	14.93	204	23.9
Dobrudjanski ran	0.110	45.6	23.1	13.80	245	20.4
Prelom	0.124	37.1	22.9	14.30	228	22.3
Drought-stressed						
Plovdiv 10	0.064	122.6	6.9	3.31	223	40.0
Dobrudjanski ran	0.033	133.8	3.4	2.28	286	19.4
Prelom	0.059	117.1	7.6	3.90	248	30.5

α, maximal carboxylation efficiency; Γ, CO_2 compensation point; A_{max}, maximal CO_2 assimilation at saturating CO_2; $Ac_a=350$, net CO_2 assimilation at 350 $\mu mol\ mol^{-1}$ ambient CO_2 concentration; $C_{i(Ca=350)}$, intercellular CO_2 concentration at 350 $\mu mol\ mol^{-1}$ ambient CO_2 concentration; SL, stomatal limitation of photosynthesis electron transport or by an inhibition of enzymes of this cycle other than Rubisco (Baker et al., 1997; Nogués and Baker, 2000). The large depression in A_{max} observed in bean plants at the end of stress period was accompanied by large changes in the photochemical efficiency of electron transport through PSII (Y) (Table 12.2). These results suggest that inhibited regeneration of RuBP can be connected to an inhibition in non-cyclic electron transport and the ability to synthesize ATP and reductants. The same results are reported for sunflower where water deficit-induced reduction of RuBP regeneration has been correlated to a decrease in supply of ATP due to inhibition in activity of ATP synthase (Tezara et al., 1999). Inactivation of Rubisco is a reason for decrease in maximal carboxylation efficiency (α) (Allen et al., 1997).

Despite significant limitation of photosynthetic assimilation evaluated, this was not accompanied with reduction of C_i (Table 12.1). On the other hand, an increase (10–14%) in C_i at $C_a = 350$ µmol mol^{-1} was observed in the studied bean genotypes. The observed increase in C_i could be due to increased mesophyllic resistance for transport of CO_2 or increased respiration that is confirmed by the increased CO_2 compensation point. Decreased CO_2 diffusion into the mesophyll of leaf might not be the only reason for inhibited A_n during water stress, because A_n could not be restored to normal values even with high external CO_2 concentrations (1500 µmol mol^{-1}). Observed decrease in A_n under drought may also be due to direct inhibition of Calvin cycle reactions by changes in ionic or osmotic conditions, which affect activities of enzymes such as ATP synthase and Rubisco (Tezara et al., 1999; Haupt-Herting and Fock, 2000; Zlatev and Yordanov, 2004). The presumption that not only stomatal but also biochemical factors are included in the response of photosynthesis to water stress is supported by the inhibited rate of A_{max}, increased CO_2 compensation points and inhibited α.

Though both stomatal and non-stomatal factors contribute to the decreased photosynthetic rate during drought, their proportion (stomatal limitation of photosynthesis; SL) changes significantly. The drought-tolerant species control stomatal function to allow some carbon fixation at stress, thus improving water use efficiency, or open stomata rapidly when water deficit is relieved (Lawlor, 2002). Stomatal conductance is more closely linked to soil moisture content than to leaf water status (Davies and Zhang, 1991). At the end of stress treatment the values of SL in the bean leaves increased significantly in cv. Plovdiv 10 and to a lesser extent in cv. Prelom. In cv. Dobrudjanski ran SL is the same as the control, thus indicating that mainly non-stomatal factors determine the photosynthesis inhibition.

Involvement of at least two main phenomena in chlorophyll fluorescence changes under environmental stress conditions was shown by Baker and Horton (1987). The first phenomenon, which results in an increase in minimal fluorescence (F_0), was thought to be due to reduced availability of primary acceptor (Q_A^-). This reduction was due to the fact that inhibition of the electron transport through PSII led to incomplete oxidation of the acceptor (Krause and Weis, 1991; Velikova et al., 1999). Reduction could also be due to the

Table 12.2. Influence of drought on chlorophyll fluorescence parameters in first trifoliate leaves of bean plants (Zlatev and Yordanov, 2004).

Genotype	F_0	F_m	F_v/F_m	Y	qP	qN
Control						
Plovdiv 10	361±13	1900±77	0.810±0.031	0.514±0.026	0.811±0.039	0.569±0.027
Dobrudjanski ran	385±13	2047±70	0.812±0.033	0.497±0.023	0.801±0.041	0.681±0.036
Prelom	382±13	1900±66	0.799±0.029	0.534±0.031	0.816±0.043	0.546±0.027
Drought-stressed						
Plovdiv 10	398±15	1780±74	0.776±0.027	0.324±0.017***	0.584±0.037**	0.745±0.038**
Dobrudjanski ran	433±15*	1721±58*	0.748±0.024	0.204±0.014***	0.457±0.028***	0.984±0.053***
Prelom	403±14	1850±67	0.782±0.028	0.465±0.024*	0.668±0.039*	0.607±0.033

*$P<0.05$; **$P<0.01$; ***$P<0.001$

fragmentation of light-harvesting Chl a/b protein complexes of PSII from the main core complex (Cona et al., 1995). The second mechanism is connected to the quenching of variable fluorescence (F_v) and maximal fluorescence (F_m). Quenching of F_v would show more extensive disturbances to the reaction centres, which prevents charge recombination. The drop of F_m is associated with processes leading to a decrease in the reaction intensities of the water-splitting enzyme complex and with a concomitant electron transport within or around PSII (Aro et al., 1993). Gilmore and Björkman (1995) have established that a quenching of maximal fluorescence (F_m) accompanies increased non-radiative energy dissipation.

Zlatev and Yordanov (2004) established that the increase of F_0/F_m ratio under drought was accompanied by a smaller decrease in F_v/F_m in all the studied genotypes (Table 12.2). These results indicated, to some extent, that the chronic photoinhibition due to inactivation of PSII centres may possibly be related to D1 protein disturbances (Rintamäki et al., 1994; Campos, 1998). Photoinhibitory damages of PSII might have occurred in bean stressed leaves since, as noted earlier by Verhoeven et al. (1997), even a light at low intensity is potentially damaging under stress conditions, and may inhibit photosynthetic activity. Saccardy et al. (1998) established that during illumination of *Zea mays* wilted leaves, a significant inhibition of PSII fluorescence efficiency occurs even under low light intensity. Photoinhibitory damage to leaves during drought is facilitated by a decrease in relative water content (RWC; Björkman and Powles, 1984), and the photosynthetic inhibition could practically reflect an inhibition of PSII activity with a simultaneous uncoupling of non-cyclic photophosphorylation, as has been shown in soybean (Younis et al., 1979) and *Nerium oleander* (Björkman and Powles, 1984). The occurrence of photoinhibition was further confirmed in the studied genotypes by the significant decrease of quantum yield of electron transport (Y), since it is an indicator of photochemical efficiency of PSII under steady-state conditions (Table 12.2).

The maximal efficiency of excitation energy capture by 'open' PSII reaction centres is reflected by F_v/F_m, and a decrease in this fluorescence parameter is an indication of photosynthetic inhibition or photoinhibition (Öquist et al., 1992). In the studies on bean, a slight decrease in this ratio was observed in the first trifoliate leaves (Table 12.2). The reason for this decrease may be because a large proportion of absorbed light is not used in the photosynthesis, as shown by the increase in non-photochemical quenching (qN). Photochemical quenching (qP) presented a similar behaviour to Y. Under drought conditions, it is probable that Y is mainly dependent on the proportion of reaction centres which are photochemically 'open' (expressed by qP), rather than on the efficiency with which an absorbed photon can reach a reaction centre. High values of qP are connected to the presence of Q_A in the oxidized state. In this situation, non-photochemical quenching (qN) is low and, when light intensity increases to values close to light saturation, qN increases rapidly corresponding to high rates of energy dissipation (Plesnicar and Pancovic, 1991).

12.3 Effects of Abiotic Factors on Water Relations

It is well known that the major constraints in plant productivity are the abiotic stress factors, but development of tolerance is very difficult due to different mechanisms involved. Among these mechanisms, osmotic adjustment is one of the most important, and could play a primary function (Ludlow and Muchow, 1990).

Plants have developed different mechanisms to cope with water stress, which is developed during drought, and these include escape, tolerance, and avoidance of cell and tissue dehydration (Turner, 1986). Stress avoidance mechanisms include rapid phenological development, increased stomatal and cuticular resistance to water vapour, decrease in leaf area, and changes in orientation and anatomy among others (Jones and Corlett, 1992). Maintenance of cell turgor to allow uninterrupted metabolic reactions under increasing water deficits is one of the mechanisms of tolerance. Both osmotic adjustment and changes in the elastic properties of tissues are involved in stress tolerance (Savé et al., 1993).

In general, osmotic adjustment is very important for maintaining cell turgor during decreased water potential, since it enables water uptake leading to continuation of metabolic reactions and finally growth and productivity of the plant (Zlatev, 2005). In cells where solutes accumulate, a lowering of the osmotic potential is generally considered to be due to osmotic adjustment provided the accumulation of solutes is not merely a result of tissue dehydration (Bray, 1993).

It has been suggested by many authors that plant biochemical processes are more sensitive to turgor and cell volume than to water potential (Jones and Corlett, 1992). Among the biochemical mechanisms involved in maintaining leaf turgor pressure, a decrease in osmotic potential either due to decreasing osmotic water fraction or from an osmotic adjustment (net accumulation of solutes in the symplast) may be important (Zlatev, 2005). Besides, water stress-induced changes in cell wall elasticity, which change the relationship between cell volume and turgor pressure, might contribute to tolerance, as observed in black spruce (Blake et al., 1991) and sunflower (Maury et al., 2000). Parameters of leaf water relations (RWC; Ψ_w, water potential; Ψ_{tlp}, water potential at turgor loss point (MPa); RWC_{tlp}, relative water content at turgor loss point (%); Ψ_o^{100}, osmotic potential at full hydration (MPa); OA, osmotic adjustment (MPa); ε_{vmax}, bulk elastic modulus at full hydration (MPa); α_{pmax}, structure coefficient at full hydration (MPa) may be a useful indication of the capability of the particular species for maintaining functional activity under water deficit (White et al., 2000; Zlatev, 2005).

While studying water relations in young bean plants subjected to water stress, Zlatev (2005) investigated changes in cell wall elasticity, water potential, turgor pressure and wall structural coefficient in the different cultivars using pressure–volume curves (Table 12.3).

Data in Table 12.3 reveal that RWC was significantly reduced in all cultivars under drought, along with changes in Ψ_w and proline content in leaves. RWC in the first trifoliate leaves decreased by 19% for cv. Plovdiv 10 and 32% for cv. Dobrudjanski ran with an intermediate value for cv. Prelom. Water potential, the thermodynamic parameter of plant water status, was reduced to a greater extent, about 78% in the first trifoliate leaf in cv. Dobrudjanski ran. Changes in absolute values of Ψ_w varied between 1.4 and 1.8 MPa, indicating severe water stress, as these values suggest (Hale and Orcutt, 1987).

There were significant differences in proline content among the cultivars during drought (Table 12.3). In the first trifoliate leaf of cv. Prelom, the highest accumulation and increase in relation to control is registered (with 282% rise) and the lowest accumulation and increase in cv. Dobrudjanski ran (with 203% rise), with cv. Plovdiv 10 showing an intermediate value. Analysis of Ψ_w versus RWC^{-1} curves (Richter diagram) allow determination of plant water relations precisely, as well as the rate of osmotic adjustment (OA) under conditions of water deficit (Teulat et al., 1997). A great variation was observed among cultivars for OA (Table 12.4). The highest calculated values were found in stressed plants of cv. Plovdiv 10 and cv. Prelom for the first trifoliate leaf.

Table 12.3. Influence of water stress on relative water content, water potential and proline accumulation in the first trifoliate leaves in young bean plants (Zlatev, 2005).

Genotype	Variants	RWC (%)	Ψ_w (MPa)	Proline accumulation (μg g^{-1}DM)
Plovdiv 10	control	93.2±2.8	−0.5±0.02	18.3±0.9
	water stress	75.6±2.5** (81)[a]	−2.1±0.09*** (24)	50.1±1.6*** (274)
Dobrudjanski ran	control	91.7±2.7	−0.4±0.02	17.7±0.9
	water stress	62.4±2.1*** (68)	−1.8±0.08*** (22)	35.9±1.4*** (203)
Prelom	control	93.8±2.9	−0.6±0.02	38.9±0.9
	water stress	68.6±2.3*** (73)	−2.4±0.09*** (25)	109.7±1.6*** (282)

RWC, relative water content; Ψ_w, water potential
[a]Percentage increase
*P<0.05; **P<0.01;***P<0.001

Table 12.4. Water relation parameters for first trifoliate leaf of bean plants derived from pressure–volume curves (Zlatev, 2005).

Variants	Ψ_{tlp}	RWC_{tlp}	Ψ_o^{100}	OA	ε_{vmax}	α_{pmax}
Plovdiv 10 – control	−1.02	69.8	−0.94		8.54	9.04
Plovdiv 10 – water-stressed	−2.10	60.4	−1.58	0.64	7.48	5.13
Dobrudjanski ran – control	−0.80	76.4	−0.64		8.13	12.60
Dobrudjanski ran – water-stressed	−1.24	71.5	−1.06	0.42	10.54	9.97
Prelom – control	−1.31	67.6	−1.26		11.80	9.34
Prelom – water-stressed	−2.29	58.8	−2.01	0.75	12.99	6.47

Ψ_{tlp}, water potential at turgor loss point (MPa); RWC_{tlp}, relative water content at turgor loss point (%); Ψ_o^{100}, osmotic potential at full hydration (MPa); OA, osmotic adjustment (MPa); ε_{vmax}, bulk elastic modulus at full hydration (MPa); α_{pmax}, structure coefficient at full hydration (MPa)

In contrast, low OA was found in plants of cv. Dobrudjanski ran (0.42 MPa).

The Ψ_o^{100} derived from the PV-curves was significantly decreased by the water deficit (Table 12.4). Ψ_o^{100} was highest in leaves of cv. Dobrudjanski ran for both control and stressed plants, followed by cv. Plovdiv 10. Plants of cv. Prelom had lowest values. Under drought, turgor loss point (TLP) declined. TLP in cv. Dobrudjanski ran had highest Ψ_w and RWC in both control and water-stressed plants while cv. Prelom had the lowest values.

In cv. Prelom, the mean maximum leaf bulk elastic modulus (ε_{vmax}) of the control plants was 11.80 MPa. Both in this cultivar and in cv. Dobrudjanski ran water stress slightly increased the ε_{vmax}, whereas in cv. Plovdiv 10 ε_{vmax} was lower in the stressed plants as compared to control ones. These results suggest that water deficit caused changes in cell wall properties of leaves, making them less elastic in cv. Plovdiv 10. The relation of ε_{vmax} to the Ψ_p (the structure coefficient α_{pmax}) was also affected, but indifferently. α_{pmax} decreased significantly in all the studied cultivars. The lowest values for OA are found in stressed plants of cv. Dobrudjanski ran (0.42 MPa) and higher values in cvs Plovdiv 10 and Prelom.

In this study, the selected bean cultivars showed significant differences in their adaptive response to water stress. The analysis of PV-curves demonstrated an active OA in bean leaves when water deficit was imposed slowly, at a rate of about 0.15 MPa day^{-1}. The degree of OA was related to the decrease in osmotic potential at full turgor in stressed plants. Maintenance of higher RWC under water deficit was observed in cvs Plovdiv 10 and Prelom and this may be due to their ability to accumulate greater amounts of proline and other osmotically active compounds. Such compounds play important roles in the reduction of Ψ_o and in OA. Earlier studies on grapevine (Rodrigues et al., 1993) and wheat and barley (Teulat et al., 1997) have also revealed the relation between decrease of Ψ_o and OA.

It is clear from the above results that leaf water relations in young bean plants were influenced by water deficit. In cv. Dobrudjanski ran Ψ_w was highest, but RWC was reduced to a greater extent, whereas in cv. Prelom changes in RWC were intermediate but changes in Ψ_w were highest, which could be indicative of tolerance to drought. The main difference among cultivars appears to be due to turgor maintenance, which may be more important as an indicator of the physiological status of the cell in these cultivars (Zlatev, 2005).

In comparison to plants receiving normal watering, those under drought showed higher maximum leaf bulk elastic modulus, except cv. Plovdiv 10, where a slight decrease in ε_{vmax} was observed. However, these changes are not due to alteration in cell wall structure but could be a consequence of the lower osmotic potential at full turgor which lead to a greater maximum turgor potential. Stressed leaves reached turgor loss point at lower Ψ_w in comparison to well-watered leaves, indicating that they have an increased capacity to maintain turgor at lower water potentials. This parameter was higher in control plants than in the stressed plants, in spite of the higher ε_{vmax} of the latter (Zlatev, 2005). Similar results were also obtained earlier (Wilson et al., 1980; Rodrigues et al., 1993).

Zlatev (2005) also established differences in accumulation of proline during water stress among bean cultivars. While variability in proline metabolism is quite common in different crop species, it is still not clear whether this amino acid has any role to play in conferring susceptibility or tolerance in the genotypes (Iannucci et al., 2000). Results of Zlatev (2005), along with some other previous workers (Naidu et al., 1992; Iannucci et al., 2000) were of the opinion that proline levels were more closely related to the decrease in RWC than in Ψ_w. It is also suggested that metabolic differences among cultivars may reflect differences in water relations achieved, and not the metabolic differences as such (Navari-Izzo et al., 1990; Zlatev, 2005). According to some authors, it is quite likely that proline accumulation may be a symptom of the development of severe plant water stress, since high proline levels were observed when Ψ_w was lower than –1.5 MPa and leaf turgor was very close to zero. Such a relationship between turgor and proline accumulation could be useful as a possible drought-injury sensor (Iannucci et al., 2000) rather than tolerance.

12.4 Effects of Abiotic Factors on Plant Growth

For crops in sustainable agriculture, both a high growth rate and high water use efficiency along with use of available mineral nutrients are desirable traits. Both growth rate and water use are influenced by the physiological and morphological properties of the different organs and allocation of biomass to these organs. It is evident that there may be conflict among plant traits involved in increasing water use efficiency and those that are involved in promotion of growth rate. Water use efficiency of plants is influenced mainly by the relative growth rate (RGR_{pl}) and the transpiration intensity at plant level (T_{pl}). The RGR_{pl} consist of two components: a morphological one, the leaf area ratio (LAR) and a physiological one, the net assimilation rate (NAR) (Van den Boogaard et al., 1997).

Availability of water determines the allocation pattern maximizing growth or water use efficiency. Under mild water deficit there is an increase in allocation of biomass to roots (Hamblin et al., 1991; Gorai et al., 2010) leading to increased capacity for water uptake by them. However, greater allocation of biomass means a larger root system and this comes with a cost: that of construction, possibly at the expense of construction of photosynthetic tissue along with increased respiratory losses associated with its maintenance. So it can be postulated that while increase in root biomass is associated with more efficient water uptake capacity, it has additional cost in terms of carbon use.

Higher leaf area ratio during favourable conditions is advantageous for the plant since it provides more photosynthesizing area leading to better growth. Differences in leaf area ratio are mainly linked to interspecific variation in relative growth (Poorter and Remkes, 1990). Along with higher photosynthesis, a larger leaf area also helps in more transpiration, which in turn is a drawback during drought. Thus, higher biomass for leaf, though beneficial for growth, is not so favourable for tolerance against water deficit as more water is lost by transpiration. Variation in growth and water use efficiency depends on several factors such as differences in pattern of biomass allocation, rates of water uptake as well as loss of water and carbon from different plant organs. In a plant, assimilation is a result of a combination of leaf area and rate of photosynthesis, and similarly, water use efficiency is dependent on both rate of water uptake and transpiration. Plant growth and development is inhibited by drought.

Ahmadi and Joudi (2007) obtained reduction in dry matter in wheat under water deficiency. It was also reported in previous studies that mild water deficit inhibited RGR_{pl} by 25% (Boutraa and Sanders, 2001). Changes in photosynthetic rate and NAR may be the main reasons for these, since changes in LAR are insignificant. Similar observations were also reported by Poorter and Remkes (1990) for 24 wild species, Van den Boogard et al. (1997)

for ten wheat cultivars, Lutts et al. (2004) for durum wheat and Berova and Zlatev (2002) for young bean plants. Biomass accumulation is inhibited to a greater extent in fresh plant mass than in dry mass (Ramos et al., 1999). This lower influence on dry mass is indicative of disturbances in water relations. This was confirmed by work of other authors such as Konings (1989) in cowpea and Augé et al. (2001) in common bean plants. Young bean plants when exposed to drought conditions showed a decrease in dry biomass (14–17%) with a concomitant increase in dry mass/fresh mass (DM/FM) ratio (Lazcano-Ferrat and Lovatt, 1999). At the plant level, increased DM/FM ratio can be considered as a stress parameter (Augé et al., 2001).

12.5 Oxidative Stress

Drought disturbs not only plant water relations, but also other physiological processes such as stomatal conductance, photosynthetic rate and, ultimately, growth. Diffusion of CO_2 into the mesophyll of leaves is reduced by enhanced stomatal closing which in turn leads to accumulation of NADPH. Electrons produced by the electron transport occurring in the thylakoid membrane having no available electron acceptor, are passed on to oxygen, which, after accepting the electrons, becomes converted to superoxide radical ($O_2^{\bullet}$) (Cadenas, 1989).

Superoxide radical gets reduced to H_2O_2 and both of these are toxic to cells at high concentrations. These also react to produce hydroxyl radical ($OH^{\bullet}$), which is highly toxic, by Haber-Weiss reaction (Sairam et al., 1998). Increased production of reactive oxygen species (ROS) under different abiotic stresses, and their deleterious effects, have been the subject of studies by several workers (Malenčić et al., 2000; Blokhina et al., 2003). Among the toxic effects caused by ROS the most immediate and important are lipid peroxidation leading to membrane injuries, degradation of protein and inactivation of enzymes (Sairam et al., 2005), inducing oxidative stress. It is therefore necessary that tolerant genotypes should not only be able to retain physiological processes at sufficient level but should also have a high ROS scavenging activity. Antioxidative enzymes such as superoxide dismutase (SOD), catalase (CAT) and ascorbate peroxidase (APOX), which are present in the tissues, are over-expressed during stresses and provide tolerance, along with reduced lipid peroxidation (Bowler et al., 1992). Up-regulation of these antioxidative enzymes will no doubt be regulated by gene expression and protein synthesis variations (Foyer et al., 1994; Scandalios et al., 1997).

While studying oxidative stress responses of bean cultivars varying in drought tolerance, Zlatev et al. (2006) observed no significant differences among cultivars in levels of lipid peroxidation (LPO) under normal well-watered conditions, which was measured as MDA content (Table 12.5). However, changes were observed during drought in MDA content, as well as injury index (I%), which was higher in cv. Dobrudjanski ran than in cvs Plovdiv 10 and Prelom.

Table 12.5. H_2O_2 content, $OH^{\bullet}$, lipid peroxidation and electrolyte leakage, expressed as injury index (I; %), in the leaves of three bean cultivars subjected to water stress (Zlatev et al., 2006).

Cultivar	Treatment	H_2O_2 ($\mu mol\ g^{-1}$ (f.m.))	$OH^{\bullet}$ ($mmol\ g^{-1}$ (f.m.))	MDA ($nmol$ (MDA) g^{-1} (d.m.))	I (%)
Plovdiv 10	Control	4.23±0.21 a/r	0.135±0.011 b/r	114±8.5 b/r	
	Drought	4.65±0.24 a/s	0.208±0.013 a/t	169±9.4 a/s	28±1.8 s
Dobrudjanski ran	Control	4.46±0.19 b/r	0.143±0.009 b/r	147±9.6 b/r	
	Drought	5.91±0.27 a/r	0.483±0.022 a/r	284±12.7 a/r	48±3.1 r
Prelom	Control	3.41±0.17 b/s	0.158±0.012 b/r	124±8.6 b/r	
	Drought	4.53±0.19 a/s	0.301±0.017 a/s	189±10.4 a/s	35±2.3 s

Means ± SE, n = 5. Different letters express significantly different results between control and drought-stressed plants in the same genotype (a, b) or between cultivars within each treatment (r, s, t)

Both H_2O_2 and $OH^•$ production in leaves of all genotypes under water stress increased. The lowest H_2O_2 content was observed in cv. Prelom and the highest in cv. Dobrudjanski ran, both under control and stress conditions. Differences between cvs Prelom and Plovdiv 10 under water stress were not significant, with the latter exhibiting the smallest increase (c. 10 %). A very high increase (238%) in $OH^•$ content under water deficit was shown in cv. Dobrudjanski ran while cv. Plovdiv 10 showed the minimum increase (54%).

In case of antioxidative enzymes, cv. Plovdiv 10 had significantly higher APOX activity in all conditions in comparison to the other two cultivars (Table 12.6). APOX activity increased significantly during water stress in all cultivars: 38% and 115% in cvs Dobrudjanski ran and Prelom, respectively. Plodiv 10 revealed a 196% and a 63% increase in comparison to cvs Dobrudjanski ran and Prelom, respectively. In the case of SOD, cv. Plovdiv 10 exhibited the highest activity in control plants but no significant differences were obtained between cultivars under drought conditions. Chloroplastic SOD activity increased significantly under water stress in all genotypes, with an increase of more than two-fold in cv. Dobrudjanski ran. Catalase (CAT) activity showed significant variation among the cultivars and following drought treatment. Under both control and stress conditions cv. Plovdiv 10 maintained highest CAT activity, followed by cvs Prelom and Dobrudjanski ran. While CAT activity increased 225% in cv. Plovdiv 10 and 265% in cv. Prelom (Table 12.6) under drought, a significant decrease (c. 27%) was observed in cv. Dobrudjanski ran.

The results obtained from this study showed that, at the end of the period of drought, a state of oxidative stress was imposed in all genotypes, as evidenced by an increased I%, lipid peroxidation, H_2O_2 and $OH^•$ production, leading to membrane damage. H_2O_2, a strong oxidant produced mainly by scavenging of superoxide radical, is injurious to cells at higher concentration, causing oxidative damage, lipid peroxidation, and disruption of metabolic function and losses of cellular integrity (Foyer et al., 1997; Velikova et al., 2000). But it is also known that at low concentrations in the early stages of stress it plays multifunctional roles. It diffuses over relatively long distances, and hence causes changes in redox status of surrounding cells and tissues. In such cases, it would be present at relatively lower concentrations and may act in signalling of antioxidative responses (Foyer et al., 1997). It is thus clear that in order to achieve an efficient control, other than scavenging capacity, cells need to have the ability to fine-tune the H_2O_2 levels. This is because H_2O_2, though toxic to some cellular components, at low concentrations, is essential to plants for various biosynthetic reactions and also in signal transduction pathways, contributing to plant defence (Schreck and Baeuerle, 1991). Further, it is also evident that the stress-induced production of H_2O_2 in the mesophyll cells is probably associated with changes in the cell wall structure (Scandalios et al., 1997). H_2O_2 also plays a role in lignin biosynthesis involving peroxidase-mediated

Table 12.6. Changes in the antioxidant enzyme activities in the leaves of three bean (*Phaseolus vulgaris* L.) cultivars (Plovdiv 10, Dobrudjanski ran and Prelom) submitted to drought (Zlatev et al., 2006).

Cultivar	Treatment	APOX (μmol Asc mg^{-1} Chl min^{-1})	SOD (U mg^{-1} Chl min^{-1})	CAT (μmol H_2O_2 mg^{-1} Chl min^{-1})
Plovdiv 10	Control	917±56 a/r	442.6±24.9 b/r	241.2±19.8 b/r
	Drought	1037±79 a/r	593.1±29.5 a/r	784.9±25.9 a/r
Dobrudjanski ran	Control	254±16 b/s	341.7±22.7 b/s	114.7± 8.7 a/s
	Drought	350±21 a/t	689.8±29.4 a/r	83.6± 4.2 b/t
Prelom	Control	296±11 b/s	438.6±21.8 b/r	138.4±10.2 b/s
	Drought	635±30 a/s	620.5±24.1 a/r	504.6±14.1 a/s

APOX, ascorbate peroxidase; SOD, superoxide dismutase; CAT, catalase. Means ± SE, n = 5. Different letters express significantly different results between control and drought-stressed plants in the same genotype (a, b) or between cultivars within each treatment (r, s, t)

oxidative polymerization of cinnamyl alcohols, and several authors have suggested that different enzyme systems may be responsible for hydrogen peroxide production on the surface of plant cells (Lütje et al., 2000). Therefore it is quite evident that drought-induced increment in level of H_2O_2, which causes oxidative damage, may eventually also have a signal function.

Lipid peroxidation is enhanced by H_2O_2, OH• and other ROS (Sairam et al., 2005) under drought, indicating an overall enhancement of the total oxidative lipid metabolism in leaves, which indicates a relationship between drought and oxidative stress (Munné-Bosch et al., 2001). Lipid peroxidation causes a decrease in membrane stability and CMS (cell membrane stability), which is generally used as an indicator of drought tolerance (Premachandra et al., 1990). Lower LPO (lipid peroxidation) and higher membrane stability (lower electrolyte leakage) has been reported in drought-tolerant genotypes of maize (Pastori and Trippi, 1992) and wheat (Sairam et al., 1998). It was reported by Zlatev et al. (2006) that under drought conditions, bean cv. Plovdiv 10, which had comparatively lower I% and LPO, also showed higher APOX and CAT activity, as compared to the other two cultivars. There are several previous reports that indicate the involvement of APOX and CAT in H_2O_2 scavenging (Jagtap and Bhargava, 1995) and thus have a role in the antioxidant defence against oxidative stress. The higher activities of APOX and CAT would have contributed to a lesser (ca. 10%) increase of H_2O_2 when exposed to drought.

On the other hand, in cv. Dobrudjanski ran (susceptible cultivar) activity of CAT under water stress decreased significantly (Table 12.2), which might be responsible for the observed maximum accumulation of H_2O_2 in this cultivar. Though it was proposed by Du and Klessig (1997) that binding to salicylic acid or other cellular components may inactivate CAT, it is difficult to confirm these data under physiological conditions.

Enhanced activities of catalase and ascorbate peroxidase in drought-tolerant genotypes of different crops such as pea, tomato, sorghum and wheat have been reported (Gillham and Dodge, 1987; Walker and McKersie, 1993; Jagtap and Bhargava, 1995; Sairam et al., 1998, respectively). Such patterns of CAT and APOX activities have been observed not only during water stress but also other stresses such as excess iron (Hendry and Brocklebank, 1985), arsenic (Stoeva et al., 2003) and acid rain (Velikova et al., 2000). It has been suggested that enhancement of activities of enzymes which scavenge ROS during stress could be either due to an adaptive change in catalytic properties or to the transcription of the corresponding silent genes (Sgherri and Navari-Izzo, 1995).

According to Zlatev et al. (2006), the increased activities of SOD, APOX and CAT when water stress was imposed in relatively tolerant bean cvs. Plovdiv 10 and Prelom, may be due to increased levels of free radicals or other ROS in plant cells and correlate with a temporal coordination of the production of H_2O_2 via SOD and destruction of this peroxide by APOX and CAT. It is believed that such coordinated responses are responsible for plant tolerance to oxidative stress (Foyer et al., 1994). Increased SOD activity could also alter the expression of other metabolic processes associated with water stress. It has been reported that enhanced activity of Cu, Zn SOD in transgenic plants was associated with increased activity of APOX, while several authors have reported increases in SOD activity in plants under oxidative stress (Gupta et al., 1993; Kang and Saltveit, 2002).

The results presented in the investigation by Zlatev et al. (2006) indicate that drought stress-induced oxidative stress led to membrane disturbances. Lower lipid peroxidation and higher membrane stability, as observed in relatively tolerant cultivars, are related to the activity of antioxidant enzymes such as APOX, SOD and CAT, which provide protection against oxidative stress. On the basis of all the data on stress tolerance evaluation of bean plants using several parameters, authors suggested that cvs Plovdiv 10 and Prelom can be considered as tolerant, and cv. Dobrudjanski ran as sensitive to drought.

On the basis of results of the author and literature analysed in this article, it can be assumed that mechanisms of plant tolerance to abiotic stresses are very complex. Results further support the observations of previous authors that the flexibility of plant cell metabolism and its acclimation to changes in environmental conditions is a first and very important step in stress avoidance. The broader the extent of acclimation capacity of plants, the better they are protected against different abiotic stresses. The changes of plant development are always connected to changes in their metabolic activity.

In spite of intensive analyses of the problem of stress tolerance, many of its aspects remain to be explored. Abiotic stresses induce expression of particular genes and this is associated in most cases with adaptive responses of plants. The functions of many of these responses are still not established. One valuable approach to understanding stress resistance mechanisms is to identify the key metabolic steps that are most sensitive to a given stress factor.

References

Ahmadi, A. and Joudi, M. (2007) Effects of timming and defoliation intensity on grouth, yield and gas exchange rate of wheat grown under well-watered and drought condititions. *Pakistan Journal of Biological Sciences* 10(21), 3794–3800.

Allen, D.J., McKee, I.F., Farage, P.K. and Baker, N.R. (1997) Analysis of the limitation to CO_2 assimilation to exposure of leaves of two *Brassica napus* cultivars to UV-B. *Plant Cell Environment* 20, 633–640.

Aro, E.-M., Virgin, I. and Andersson, B. (1993) Photoinhibition of photosystem II. Inactivation, protein damage and turnover. *Biochemical and Biophysical Acta* 1143, 113–134.

Augé, R.M., Kubikova, E. and Moore, J.L. (2001) Foliar dehydration tolerance of mycorrhizal cowpea, soybean and bush bean. *New Phytologist* 151, 535–541.

Baker, N.R. and Horton, P. (1987) Chlorophyll fluorescence quenching during photoinhibition. In: Kyle, D.J., Osmond, C.B. and Arntzen, C.J. (eds) *Photoinhibition*. Elsevier Scientific Publisher, Amsterdam, the Netherlands, pp. 85–94.

Baker, N.R., Nogues, S. and Allen, D.J. (1997) Photosynthesis and photoinhibition. In: Lumsden, P.J. (ed.) *Plants and UV-B: responses to environmental change*. Cambridge University Press, Cambridge, UK, pp. 95–111.

Berova, M. and Zlatev, Z. (2002) Influence of soil drought on growth and biomass partitioning in young bean (*Phaseolus vulgaris* L.) plants. *Annual Report of the Bean Improvement Cooperative* 45, 190–191.

Bhadula, S.K., Yang, G.P., Sterzinger, A. and Ristic, Z. (1998) Synthesis of a family of 45 K heat shock proteins in a drought and heat resistant line of maize under controlled and field conditions. *Journal of Plant Physiology* 152, 104–111.

Björkman, O. and Powles, S.B. (1984) Inhibition of photosynthetic reactions under water stress: interaction with light level. *Planta* 161, 490–504.

Blake, T.J., Bevilacqua, E. and Zwiazek, J.J. (1991) Effects of repeated stress on turgor pressure and cell elasticity changes in black spruce seedlings, *Canadian Journal of Forest Research* 21, 1329–1333.

Blokhina, O., Virolainen, E. and Fagerstedt, K.V. (2003) Antioxidants, oxidative damage and oxygen deprivation stress: a review. *Annals of Botany* 91, 179–194.

Boutraa, T. and Sanders, F.E. (2001) Effects of interactions of moisture regime and nutrient addition on nodulation and carbon partitioning in two cultivars of bean (*Phaseolus vulgaris* L.). *Journal of Agronomy and Crop Science* 186, 229–237.

Bowler, C., Van Montagu, M. and Inze, D. (1992) Superoxide dismutase and stress tolerance. *Annual Review of Plant Physiology and Plant Molecular Biology* 43, 83–116.

Bray, E.A. (1993) Molecular responses to water deficits. *Plant Physiology* 103, 1035–1040.

Cadenas, S.E. (1989) Biochemistry of oxygen toxicity. *Annual Review of Biochemistry* 58, 79–110.

Campos, P.S. (1998) Effects of water stress on photosynthetic performance and membrane integrity in *Vigna* spp. The role of membrane lipids in drought tolerance. PhD thesis, Universidade Nova de Lisboa, Lisboa.

Chakir, S. and Jensen, M. (1999) How does *Lobaria pulmoria* regulate Photosystem II during progressive desiccation and osmotic water stress? A chlorophyll fluorescence study at room temperature and at 77 K. *Physiologia Plantarum* 105, 257–265.

Chaves, M. (1991) Effects of water deficits on carbon assimilation. *Journal of Experimental Botany* 42, 1–16.

Chaves, M.M., Flexas, J. and Pinheiro, C. (2009) Photosynthesis under drought and salt stress: regulation mechanisms from whole plant to cell. *Annals of Botany* 103, 551–560.

Chen, Y.H. and Hsu, B.D. (1995) Effect of dehydration on the electron transport of *Chlorella*. An *in vivo* fluorescence study. *Photosynthesis Research* 46, 295–299.

Cona, A., Kučera, T., Masojídek, J., Geiken, B., Mattoo, A.K. and Giardi, M.T. (1995) Long-term drought stress symptom: structural and functional reorganization of photosystem II. In: Mathis, M. (ed.) *Photosynthesis: From Light to Biosphere*. Vol. IV. Kluwer Academic Publisher, Dordrecht, the Netherlands, pp. 521–524.

Cornic, G. (1994) Drought stress and high light effects on leaf photosynthesis. In: Baker, N.R. and Boyer, J.R. (eds) *Photoinhibition of Photosynthesis: from molecular mechanisms to the field*. Bios Scientific Publishers, Oxford, UK, pp. 297–313.

Cornic, G., Ghashghaie, J., Genty, B. and Briantais, J.-M. (1992) Leaf photosynthesis is resistant to a mild drought stress. *Photosynthetica* 27, 295–309.

Davies, W.J. and Zhang, J.H. (1991) Root signals and the regulation of growth and development of plants in drying soil. *Annual Review of Plant Physiology and Plant Molecular Biology* 42, 55–76.

Du, H. and Klessig, D.F. (1997) Identification of a soluble, high-affinity salicylic acid-binding protein in tobacco. *Plant Physiology* 113, 1319–1327.

Flexas, J. and Medrano, H. (2002) Energy dissipation in C3 plants under drought. *Functional Plant Biology* 29, 1209–1215.

Foyer, C.H., Descourvières, P. and Kunert, K.J. (1994) Protection against oxygen radicals: an important defense mechanism studied in transgenic plants. *Plant Cell and Environment* 17, 507–523.

Foyer, C.H., Lopez-Delgado, H., Dat, J.F. and Scott, I.M. (1997) Hydrogen peroxide- and glutathione-associated mechanisms of acclimatory stress tolerance and signaling. *Physiologia Plantarum* 100, 241–254.

Gillham, D.J. and Dodge, A.D. (1987) Chloroplast superoxide and hydrogen peroxide scavenging systems from pea leaves: seasonal variations. *Plant Science* 50, 105–109.

Gilmore, A.M. and Björkman, O. (1995) Temperature sensitive coupling and uncoupling of ATPase-mediated, nonradiative energy dissipation: similarities between chloroplasts and leaves. *Planta* 197, 646–654.

Gorai, M., Hachef, A. and Neffati, M. (2010) Differential responses in growth and water relationship of *Medicago sativa* (L.) cv. Gabès and *Astragalus gombiformis* (Pom.) under water-limited conditions. *Emirates Journal of Food and Agriculture* 22(1), 1–12.

Gupta, A.S., Webb, R.P., Holaday, A.S. and Allen, D. (1993) Overexpression of SOD protects plants from oxidative stress. Induction of ascorbate peroxidase in superoxide dismutase-overexpressing plants. *Plant Physiology* 103, 1067–1073.

Hale, M.G. and Orcutt, D.M. (1987) *The Physiology of Plants Under Stress*. John Wiley & Sons, New York.

Hamblin, A., Tennant, D. and Perry, W. (1991) The cost of stress: dry matter partitioning changes with seasonal supply of water and nitrogen to dryland wheat. *Plant and Soil* 122, 47–58.

Haupt-Herting, S. and Fock, H.P. (2000) Exchange of oxygen and its role in energy dissipation during drought stress in tomato plants. *Physiologia Plantarum* 110, 489–495.

He, J.X., Wang, J. and Liang, H.G. (1995) Effects of water stress on photochemical function and protein metabolism of photosystem II in wheat leaves. *Physiologia Plantarum* 93, 771–777.

Hendry, G.A.F. and Brocklebank, K.J. (1985) Iron-induced oxygen radical metabolism in waterlogged plants. *New Phytologist* 101, 199–206.

Iannucci, A., Rascio, A., Russo, M., Di Fonzo, N. and Martiniello, P. (2000) Physiological responses to water stress following a conditioning period in berseem clover. *Plant and Soil* 223, 217–227.

Jagtap, V. and Bhargava, S. (1995) Variation in antioxidant metabolism of drought tolerant and drought susceptible varieties of *Sorghum bicolor* (L.) Moench, exposed to high light, low water and high temperature stress. *Journal of Plant Physiology* 145, 195–197.

Jia, Y. and Gray, V.M. (2004) Interrelationships between nitrogen supply and photosynthetic parameters in *Vicia faba* L. *Photosynthetica* 41, 605–610.

Jones, H.G. and Corlett, J.E. (1992) Current topics in drought physiology. *Journal of Agricultural Science* 119, 291–296.

Kanechi, M., Kunitomo, E., Inagaki, N. and Maekawa, S. (1995) Water stress effects on ribulose-1,5-bisphosphate carboxylase and its relationship to photosynthesis in sunflower leaves. In: Mathis, M. (ed.) *Photosynthesis: from light to biosphere*, Vol. IV. Kluwer Academic Publisher, Dordrecht, the Netherlands, pp. 597–600.

Kang, H.M. and Saltveit, E. (2002) Effect of chilling on antioxidant enzymes and DPPH-radical scavenging activity of high- and low-vigour cucumber seedling radicles. *Plant Cell and Environment* 25, 1233–1238.

Konings, H. (1989) Physiological and morphological differences between plants with a high NAR or a high LAR as related to environmental conditions. In: Lambers, H., Cambridge, M., Konings, H. and Pons, T.L. (eds) *Causes and Consequences of Variation in Growth Rate and Productivity of Higher Plants*. SPB Academic Publishing, The Hague, pp. 101–123.

Krause, G.H. and Weis, E. (1991) Chlorophyll fluorescence and photosynthesis: the basics. *Annual Review of Plant Physiology and Plant Molecular Biology* 42, 313–349.

Lauer, K.J. and Boyer, J.S. (1992) Internal CO_2 measured directly in leaves: abscisic acid and low leaf water potential cause opposing effects. *Plant Physiology* 98, 1310–1316.

Lawlor, D.W. (1995) Effects of water deficit on photosynthesis. In: Smirnoff, N. (ed.) *Environment and Plant Metabolism*. Bios Scientific Publishers, Oxford, UK, pp. 129–160.

Lawlor, D.W. (2002) Limitation of photosynthesis in water-stressed leaves. Stomatal metabolism and the role of ATP. *Annals of Botany* 89, 871–885.

Lawlor, D.W. and Cornic, C. (2002) Photosynthetic carbon assimilation and associated metabolism in relation to water deficits in higher plants. *Plant Cell Environment* 25, 275–291.

Lazcano-Ferrat, I. and Lovatt, C.J. (1999) Relationship between relative water content, nitrogen pools, and growth of *Phaseolus vulgaris* L. and *P. acutifolius* Gray during water deficit. *Crop Science* 39, 467–475.

Lu, C. and Zhang, J. (1999) Effects of water stress on photosystem II photochemistry and its thermostability in wheat plants. *Journal of Experimental Botany* 50, 1199–1206.

Ludlow, M.M. and Muchow, R.C. (1990) A critical evaluation of traits for improving crop yields in water-limited environment. *Advances in Agronomy* 43, 107–153.

Lütje, S., Böttger, M. and Döring, O. (2000) Are plants stacked neutrophyles? Comparison of pathogen induced oxidative burst in plants and mammals. In: Esser, K., Kadereit, J.W., Lütje, U. and Runge, M. (eds) *Progress in Botany*, Vol. II. Springer, Berlin, pp. 187–222.

Lutts, S., Almansouri, M. and Kinet, J.-M. (2004) Salinity and water stress have contrasting effects on the relationship between growth and cell viability during and after stress exposure in durum wheat. *Plant Science* 167, 9–18.

Malenčić, D., Gašić, O., Popović, M. and Boza, P. (2000) Screening for antioxidant properties of *Salvia reflexa* Hornem. *Phytotherapy Research* 14, 546–548.

Maury, P., Berger, M., Mojayad, F. and Planchon, C. (2000) Leaf water characteristics and drought acclimation in sunflower genotypes. *Plant and Soil* 223, 153–160.

Munné-Bosch, S., Jubany-Marí, T. and Alegre, L. (2001) Drought-induced senescence is characterized by a loss of antioxidant defenses in chloroplasts. *Plant Cell Environment* 24, 1319–1327.

Naidu, B.P., Aspinall, D. and Paleg, L.G. (1992) Variability in proline-accumulating ability of barley (*Hordeum vulgare* L.) cultivars induced by vapor pressure deficit. *Plant Physiology* 98, 716–722.

Navari-Izzo, F., Quartacci, M.F. and Izzo, R. (1990) Water-stress induced changes in protein and free amino acids in field-grown maize and sunflower. *Plant Physiology and Biochemistry* 28, 531–537.

Nogués, S. and Baker, N.R. (2000) Effects of drought on photosynthesis in Mediterranean plants grown under enhanced UV-B radiation. *Journal of Experimental Botany* 51, 1309–1317.

Öquist, G., Anderson, J.M., Mc Caffery, S. and Show, W.S. (1992) Mechanistic differences of photoinhibition of sun and shade plants. *Planta* 188, 422–431.

Pastori, G.M. and Trippi, V.S. (1992) Oxidative stress induces high rate of glutathione reductase synthesis in a drought resistant maize strain. *Plant Cell Physiology* 33, 957–961.

Plesnicar, M. and Pancovic, D. (1991) Relationship between chlorophyll fluorescence and photosynthetic O_2 evolution in several *Helianthus* species. *Plant Physiology and Biochemistry* 29, 681–688.

Poorter, H. and Remkes, C. (1990) Leaf area ratio and net assimilation rate of 24 wild species differing in relative growth rate. *Oecologia* 83, 553–559.

Premachandra, G.S., Saneoka, H. and Ogata, S. (1990) Cell membrane stability, an indicator of drought tolerance, as affected by applied nitrogen in soybean. *Journal of Agricultural Science* 115, 63–66.

Ramos, M.L., Gordon, A.J., Minchin, F.R., Sprent, J.I. and Parsons, R. (1999) Effect of water stress on nodule physiology and biochemistry of a drought tolerant cultivar of common bean (*Phaseolus vulgaris* L.). *Annals of Botany* 83, 57–63.

Rintamäki, E., Salo, R. and Aro, E.M. (1994) Rapid turnover of the D1 reaction-center protein of photosystem II as a protection mechanism against photoinhibition in a moss *Ceratodon purpureus* (Hedw.) Brid. *Planta* 193, 520–529.

Rodrigues, M.L., Chaves, M.M., Wendler, R., David, M.M., Quick, W.P., Leegood, R.C., Stitt, M. and Pereira, J.S. (1993) Osmotic adjustment in water stressed grapevine leaves in relation to carbon assimilation. *Australian Journal of Plant Physiology* 20, 309–321.

Saccardy, K., Pineau, B., Roche, O. and Cornic, G. (1998) Photochemical efficiency of photosystem II and xanthophylls cycle components in *Zea mays* leaves exposed to water stress and high light. *Photosynthesis Research* 56, 57–66.

Sairam, R.K., Deshmukh, P.S. and Saxena, D.C. (1998) Role of antioxidant systems in wheat genotypes tolerance to water stress. *Biologia Plantarum* 41, 387–394.

Sairam, R.K., Srivastava, G.C., Agarwal, S. and Meena R.C. (2005) Differences in antioxidant activity in response to salinity stress in tolerant and susceptible wheat genotypes. *Biologia Plantarum* 49, 85–91.

Sarafis, V. (1998) Chloroplasts: a structural approach. *Journal of Plant Physiology* 152, 248–264.

Savé, R., Peñuelas, J., Marfá, O. and Serrano, L. (1993) Changes in leaf osmotic and elastic properties and canopy structure of strawberries under mild water stress. *Horticultural Science* 28, 925–927.

Scandalios, J.G., Guan, L. and Polidoros, A.N. (1997) Catalases in plants: gene structure, properties, regulation and expression. In: Scandalios, J.G. (ed.) *Oxidative Stress and the Molecular Biology of Antioxidant Defenses*. Cold Spring Harbor Laboratory Press, New York, pp. 343–398.

Schreck, R. and Baeuerle, P.A. (1991) The role of oxygen radicals as second messengers. *Trends in Cell Biology* 1, 39–42.

Sgherri, C.L.M. and Navari-Izzo, F. (1995) Sunflower seedlings subjected to increasing water deficit stress: oxidative stress and defence mechanisms. *Physiologia Plantarum* 93, 25–30.

Shangguan, Z., Shao, M. and Dyckmans, J. (1999) Interaction of osmotic adjustment and photosynthesis in winter wheat under soil drought. *Journal of Plant Physiology* 154, 753–758.

Skotnica, J., Matouškova, M., Nauš, J., Lazár, D. and Dvořák, L. (2000) Thermoluminescence and fluorescence study of changes in Photosystem II photochemistry in desiccating barley leaves. *Photosynthesis Research* 65, 29–40.

Stoeva, N., Berova, M. and Zlatev, Z. (2003) Effect of arsenic on some physiological indices in maize (Zea mays L.). *Journal of Environmental Protection and Ecology* 4, 496–501.

Teulat, B., Rekika, D., Nachit, M.M. and Monneveux, P. (1997) Comparative osmotic adjustments in barley and tetraploid wheats. *Plant Breeding* 116, 519–523.

Tezara, W., Mitchell, V.J., Driscoll, S D. and Lawlor, D.W. (1999) Water stress inhibits plant photosynthesis by decreasing coupling factor and ATP. *Nature* 401, 914–917.

Tuba, Z., Lichtenthaler, H.K., Csintalan, Z., Nagy, Z. and Szente, K. (1996) Loss of chlorophylls, cessation of photosynthetic CO_2 assimilation and respiration in the poikilochlorophyllous plant *Xerophyta scabrida* during desiccation. *Physiologia Plantarum* 96, 383–388.

Turner, N.C. (1986) Adaptation to water deficits: a changing perspective. *Australian Journal of Plant Physiology* 13, 175–189.

Valladares, F., Gianoli, E. and Gómez, J.M. (2007) Ecological limits to plant phenotypic plasticity. *New Phytologist* 176, 749–763.

Van den Boogaard, R., Alewijnse, D., Veneklaas, E.J. and Lambers, H. (1997) Growth and water-use efficiency of ten *Triticum aestivum* cultivars at different water availability in relation to allocation of biomass. *Plant Cell Environment* 20, 200–210.

Velikova, V., Tsonev, T. and Yordanov, I. (1999) Light and CO_2 responses of photosynthesis and chlorophyll fluorescence characteristics in bean plants after simulated acid rain. *Physiologia Plantarum* 107, 77–83.

Velikova, V., Yordanov, I. and Edreva, A. (2000) Oxidative stress and some antioxidant systems in acid rain-treated bean plants. Protective role of exogenous polyamines. *Plant Sciences* 151, 59–66.

Verhoeven, A.S., Demmig-Adams, B. and Adams, B.B. (1997) Enhanced employment of the xanthophylls cycle and thermal energy dissipation in spinach exposed to high light and N stress. *Plant Physiology* 113, 817–824.

Walker, M.A. and McKersie, B.D. (1993) Role of ascorbate-glutathione antioxidant system in chilling resistance of tomato. *Journal of Plant Physiology* 141, 234–239.

Wentworth, M., Murchie, M.H., Gray, J.E., Villegas, D., Pastenes, C., Pinto, M. and Horton, P. (2006) Differential adaptation of two varieties of common bean to abiotic stress. II. Acclimation of photosynthesis. *Journal of Experimental Botany* 57(3), 699–709.

White, D.A., Turner, N.C. and Galbraith, J.H. (2000) Leaf water relations and stomatal behavior of four allopatric *Eucalyptus* species planted in Mediterranean southwestern Australia. *Tree Physiology* 20, 1157–1165.

Wilson, J.R., Ludlow, M.M., Fisher, M.J. and Schulze, E.D. (1980) Adaptation to water stress of the leaf water relations of four tropical forage species. *Australian Journal of Plant Physiology* 7, 207–220.

Yordanov, I., Velikova, V. and Tsonev, T. (2000) Plant responses to drought, acclimation, and stress tolerance. *Photosynthetica* 38, 171–186.

Yordanov, I., Velikova, V. and Tsonev, T. (2003) Plant responses to drought and stress tolerance. *Bulgarian Journal of Plant Physiology* Special issue, 187–206.

Younis, H.M., Boyer, J.S. and Govindjee (1979) Conformation and activity of chloroplast coupling factor exposed to low chemical potential of water in cells. *Biochimica Biophysica Acta* 548, 328–340.

Zlatev, Z.S. (2005) Effects of water stress on leaf water relations of young bean plants. *Journal of Central European Agriculture* 6(1), 5–14.

Zlatev, Z.S. (2013) Drought-induced changes and recovery of photosynthesis in two bean cultivars (*Phaseolus vulgaris* L.). *Emirates Journal of Food and Agriculture* 25(12), 1014–1023.

Zlatev, Z. and Lidon, F.C. (2012) An overview on drought induced changes in plant growth, water relations and photosynthesis. *Emirates Journal of Food and Agriculture* 24(1), 57–72.

Zlatev, Z. and Yordanov, I. (2004) Effects of soil drought on photosynthesis and chlorophyll fluorescence in common bean plants. *Bulgarian Journal of Plant Physiology* 30(3–4), 3–18.

Zlatev, Z., Lidon, F.C., Ramalho, J.C. and Yordanov, I.T. (2006) Comparison of resistance to drought of three bean cultivars. *Biologia Plantarum* 50(3), 389–394.

13 Interactive Role of Polyamines and Reactive Oxygen Species in Stress Tolerance of Plants

Rup Kumar Kar*

Plant Physiology & Biochemistry Laboratory, Department of Botany, Visva-Bharati University, Santiniketan, West Bengal, India

---

Abstract

Plants encounter abiotic stresses regularly under natural conditions apart from abrupt natural calamities for which the plant may not be prepared. Accordingly, plants are also armoured with protective or adaptive mechanisms that combat the potential stress-induced injuries. Some small molecules including polyamines may play a definite role in such mechanisms, an overall idea of which has been revealed so far. Polyamines are aliphatic amines, polycationic in nature and are known to play a role in a wide range of plant processes including growth and development. Although their role in stress tolerance has been established for last two decades, the exact nature of their involvement is not clear. Recently, besides protective roles, their involvement in signalling for stress tolerance has been revealed through molecular and genetic approaches. Also, polyamines may mediate signalling via reactive oxygen species; polyamine metabolism in apoplast assumes importance in this regard. Moreover, it is understood that PA-ROS-mediated signalling under stress may have a cross-talk with the phytohormones, figuring a further complex network of signalling for stress tolerance, analysis of which will be a challenging task in near future.

13.1 Introduction

Since the development of the outer layer of this planet physicochemical changes in soil and atmosphere have been going on with time. Along with these, life that originated in the meantime has also been subjected to change or alterations in form and functionalities in order to cope. So evolution becomes meaningful in making the life variable according to the changes in nature. Plants are evolved while passing through radical changes in their immediate environment giving rise to diversity in the form of species. Underlying these events that have been taking place across a long time span, plants are always exposed to short-term changes (with reference to geological time scale) due to fluctuations in their ambient environment, which may be considered as abiotic stress. For survival plants called for certain alterations in their morphological forms as well as their metabolism. Such alterations are genetically stable and become useful for the changed environment and are considered as adaptation. On the other hand, for short term (temporary) fluctuations in the habitat plants adopted certain mechanisms at the metabolic and molecular level

*E-mail: rk_kar@rediffmail.com

that only become operative under occurrence of stress. An understanding of such conditional mechanisms for combating stress is of paramount importance particularly for crop improvement as well as for sustaining biodiversity.

Most of the wild species that are exposed to oscillating environments bear the potential tolerance mechanism against stress and express the relevant genes at the time of stress imposition. However, in the case of crop plants, repeated attempts for the selection of yield-related parameters might have resulted in the loss of such traits for tolerance. Present-day research is turning towards the retrieval of such characters from the gene pool of wild counterparts of respective crop species. For doing this one has to acquire a thorough understanding of metabolism and related gene expression behind the plant response to stress as a part of tolerance mechanism. Stress tolerance may be accrued in a plant at physiological and metabolic level through different lines. Examples are alteration in membrane behaviour as well as channel activities, associated osmoregulatory processes, accumulation of proteins (e.g. LEA, osmotin etc.) and small molecules (e.g. proline, sugars, polyamines etc.) conferring osmoregulation and also osmoprotection, activation of antioxidant system to reduce the oxidative load. No one of these is mutually exclusive and more than one mechanism may operate at the same time in a plant challenged by environmental stress. In the recent past most of these lines of defence against stress have been characterized at molecular level but still lacking is the knowledge on exact signalling event(s) that triggers such defence processes upon exposure to stress.

Polyamines (PAs) are unique among the low molecular weight molecules that accumulate in response to stress because of their role in signalling for stress tolerance besides giving protection. This chapter is mainly focused on the role of PAs in inducing stress tolerance and their interaction with reactive oxygen species (ROS), another group of molecules that also gained importance recently as signalling components for stress tolerance.

13.2 Polyamines: Ubiquitous Polycations Having Multiple Tasks in Cell Physiology

For a long time PAs have been chemically well known with a biological origin (Galston, 1991) and these take the form of polycationic hydrocarbon chains with two or more amino groups that occur widely among all kinds of living organisms (Takahashi and Kakehi, 2010). The most common PAs are spermidine (triamine) and spermine (tetra-amine), but also include diamines such as putrescine and cadavarine, which are typically known for their involvement in fundamental cellular processes like cell division, DNA replication, transcription and translation, enzyme activity modulation (Tabor and Tabor, 1999) and also several plant growth and developmental processes (Galston and Kaur-Sawhney, 1995; Galston et al., 1997; Kusano et al., 2007). In bacteria and mammals PAs are considered for their role in modulating RNA secondary structure (Igarashi and Kashiwagi, 2000). Initially PAs were assigned a role of stabilizing membrane and macromolecules because of their polycationic nature so as to bind with anionic macromolecules (Bachrach, 2005). Now it appears that this group of natural compounds has also a regulatory role in fundamental cell functions (Alcázar et al., 2006a; Kusano et al., 2008). In fact PAs are vitally needed, as organisms such as yeast and even plants do not survive when putrescine and spermidine are genetically depleted (Imai et al., 2004; Urano et al., 2005). Several physiological processes in plants are reported to be controlled by PAs. Major processes are embryogenesis, root formation, flowering, leaf senescence and stress tolerance (Kar and Choudhuri, 1986; Galston and Kaur-Sawhney, 1990; Kumar et al., 1997; Alcázar et al., 2006a; Kusano et al., 2008). Among all these processes association of PAs with stress in particular has gained importance in recent times and is the focal point of the present chapter. Thus a role of PAs has been envisaged in case of metal stress, chilling stress, drought and salinity (Chattopadhayay et al., 2002; Groppa et al., 2003; Yamaguchi et al., 2007; Groppa and Benavides, 2008; Duan et al., 2008). Other roles, such as involvement in protein phosphorylation, conformational change of DNA and regulation of gene expression by influencing transcription factors, are also suggested for PAs (Panagiotidis et al., 1995; Martin-Tanguy, 2001).

13.3 Polyamines: Titre, Biosynthesis and Regulation

Polyamine titre at cellular level depends on the biosynthesis from its precursors (ornithine and arginine), metabolism to different endproducts and conjugation with phenolic compounds (soluble conjugates) or binding with proteins and nucleic acids (insoluble bound forms). For a long time observations have been made that under stresses PA titre increases in plants (Richards and Coleman, 1952; Smith, 1973; Flores, 1983; McDonald and Kushad, 1986; Kuehn et al., 1990). In most of these cases it was due to accumulation of diamine or putrescine. The biosynthesis of PAs starts with the formation of the simplest polyamine, putrescine. Putrescine originates either directly from ornithine or indirectly from arginine via agmatine. In the case of former the reaction is catalysed by ornithine decarboxylase (ODC), which is mostly found in mammals and fungi. On the other hand, plants and bacteria mostly rely upon a second pathway where arginine is converted to putrescine via successive steps catalysed by ODC followed by agmatine iminohydrolase and N-carbamoylputrescine amidohydrolase (reviewed in Gill and Tuteja, 2010; Alcázar et al., 2010; Takahashi and Kakehi, 2010). In fact, no gene encoding ODC has been reported in the genome of Arabidopsis thaliana (Hanfrey et al., 2001). Moreover, overexpression of arginine decarboxylase (ADC) augments putrescine accumulation in plants (Alcázar et al., 2005). However, ODC activity has been reported in other plants and relevant genes were characterized (Michael et al., 1996; Imanishi et al., 1998). Regarding localization, ADC activity has been mainly found in chloroplasts (in leaves) and nuclei (roots), indicating subcellular compartmentalization of ADC pathway depending on specific functions (Alcázar et al., 2010). For example, ADC is associated with photosystem II in chloroplasts producing PAs, probably to stabilize thylakoid membranes under stress. On the other hand, ODC has been reported to occur in the nucleus (Hanfrey et al., 2001). It is speculated that the ADC pathway might have arisen from a cyanobacterial ancestor, while ODC could have a bacterial origin (Illingworth et al., 2003).

Higher PAs such as spermidine and spermine are synthesized by addition of aminopropyl moieties successively to putrescine, which is catalysed by the enzymes spermidine synthase (SPDS) and spermine synthase (SPMS), respectively. Such aminopropyl moieties are provided by decarboxylated S-adenosyl methionine that arises from S-adenosyl methionine (SAM) by the activity of S-adenosyl methionine decarboxylase (SAMDC), a rate-limiting cytosolic enzyme. The metabolic pathway for PAs along with its connection with the other pathways has recently been elucidated in detail in *Arabidopsis* (Bitrián et al., 2012).

Polyamines are present in all kinds of tissues but still their levels in specific tissues or organs are controlled by critical regulation of the biosynthesis pathway. Thus the enzymes of this pathway are regulated mostly by different environmental cues including stresses. In *Arabidopsis*, ADC is encoded by two genes *ADC1* and *ADC2*. While *ADC1* is constitutively expressed in all tissues *ADC2* expression is induced by stresses. Spermidine synthase is also encoded by two genes, *SPDS1* and *SPDS2*, but spermine synthase is encoded by a single gene, *SPMS* (Alcázar et al., 2010). Interestingly, assemblage of SPDS and SPMS forming a metabolon, as proved by protein–protein interaction studies in *Arabidopsis*, suggests metabolic canalization for production of spermine from putrescine (Alcázar et al., 2011).

Level of PAs in tissues may also be regulated by their catabolism that occurs by the activity of two groups of oxidases: diamine oxidases (DAOs) and polyamine oxidases (PAOs). DAOs are copper-containing enzymes reported from both monocots and dicots and act preferably on putrescine and cadaverine producing 4-aminobutanol, hydrogen peroxide and ammonia as end-products. On the other hand, PAOs are non-covalently bound with flavin adenine dinucleotide (FAD) occurring at high level in monocots. PAOs oxidize the secondary amino group of spermidine and spermine producing 1,3-diaminopropane along with 4-aminobutanol and H_2O_2 (Cona et al., 2006). Another group of PAOs is reported from plants that resembles mammalian enzyme and is responsible for back conversion of spermine to spermidine (Moschou et al., 2008a). Both DAOs and PAOs are usually localized in cytoplasm and cell wall, supplying H_2O_2 necessary for lignification in the latter location. However, H_2O_2 produced as a result of PA oxidation may participate in signalling for defence against stress, which will be discussed in later section.

13.4 Polyamines: Role in Stress Tolerance

13.4.1 Increased biosynthesis as a stress response

Association of PAs with stress responses in plants perhaps has been realized with the earlier observation of enhanced putrescine biosynthesis under K^+ deficiency (Richards and Coleman, 1952) and subsequently establishment of relevance of ADC pathway for putrescine synthesis in response to abiotic stress (Flores and Galston, 1982). Such stress-induced accumulation of PAs has been ascribed for stress tolerance (Bouchereau et al., 1999). PAs are now considered for acquisition of tolerance against a range of stresses like salt stress, chilling stress, osmotic stress, oxidative stress and also for pathogenic defence (Shen et al., 2000; Chattopadhyay et al., 2002; Capell et al., 2004; Takahashi et al., 2004; Liu and Moriguchi, 2006).

Involvement of PAs in stress tolerance has been proven physiologically by applying PAs exogenously or pharmacologically inhibiting PA biosynthesis by using specific inhibitors (Bouchereau et al., 1999; Alcázar et al., 2006a). The role of PAs in stress tolerance has been demonstrated by exogenous application of PAs that resulted in reduction of cold-induced electrolyte leakage in tomato (Kim et al., 2002) and alleviation of salt stress effects in apple (Liu and Moriguchi, 2006). In case of blocked PA synthesis, exogenous application of PA can restore tolerance. As the putrescine biosynthesis is the basic pathway leading to other PAs, its regulation is pivotal for stress responses. In this regard, the ADC pathway is preferably activated by stress since high putrescine accumulation is effectively inhibited by treatment with α-difluoromethylarginine (DFMA), a suicidal inhibitor of ADC (Galston et al., 1997). Indeed, the involvement of ADC activity in stress-induced PA accumulation may be correlated with expression of ADC. Among the two paralogues, *ADC1* and *ADC2*, the latter is responsive to drought, oxidative stress, salinity and biotic stress (Alcázar et al., 2006a). This was also corroborated by gene expression analysis in a recent study on drought tolerance of rice cultivars (Do et al., 2013). However, the same study showed higher accumulation of spermine as the predominant PA. Similarly, Zapata et al. (2004) also found a decrease in putrescine level with an increase in spermidine and/or spermine level in response to salinity and they correlated an increased ratio of (spermidine + spermine)/putrescine with higher salt tolerance. Though ADC activity is primarily required for PA biosynthesis, accumulation of higher PAs, spermidine and spermine, depends on regulation of SAMDC, which is probably suppressed by stress. So plant defence against stress relies more on the enhanced expression of *SAMDC* which results in accumulation of spermidine and spermine (Kuznetsov and Shevyakova, 2007). Besides, spermidine synthase (SPDS) and spermine synthase (SPMS) are also equally important for drought tolerance as the expression of *SPDS1* and *SPMS* increased many-fold under drought (Alcázar et al., 2006b).

Molecular and genetic approaches extend further support to the role of PAs in plant stress tolerance. Recent studies with either loss-of-function mutants or over-expression transgenics clearly defined such a role (Alcázar et al., 2006a; Kusano et al., 2008; Gill and Tuteja, 2010). Thus mutants for decreased activity of ADC or SPMS showed more stress-sensitivity (Kasinathan and Wingler, 2004; Yamaguchi et al., 2006). Similarly, insertion mutants of ADC (*adc2-1*) have less putrescine increase under salt stress and are more sensitive to stress, which can be recovered by exogenous application of putrescine (Urano et al., 2004). Other mutants of *ADC1* and *ADC2* are found to be more sensitive to freezing and can be rescued by application of exogenous putrescine (Cuevas et al., 2008). Similarly, a mutant of spermine synthase (*spms*) showed more sensitivity to drought and saline stress (Yamaguchi et al., 2007). On the other hand, over-expression of *ADC2* and *SAMDC1* resulting in increase in the levels of putrescine and spermine, respectively, led to enhanced stress tolerance (Alcázar et al., 2006a, 2010). Even heterologous over-expression of *ODC*, *ADC*, *SAMDC* and *SPDS* from plant and animal sources can result in tolerance against broad spectrum stress conditions (Alcázar et al., 2010).

13.4.2 Interaction with reactive oxygen species

Most of the stresses have an intimate connection with oxidative metabolism and ROS. Oxidative metabolism involves generation of superoxide ($O_2^{\bullet-}$), hydrogen peroxide (H_2O_2) and hydroxyl radical ($OH^{\bullet}$), which can damage important macromolecules like DNA, lipids, proteins and other molecules (Apel and Hirt, 2004). Plants have antioxidant machinery that can combat oxidative load depending on the balance between the rate of ROS production and scavenging. ROS-scavenging enzymes are superoxide dismutases (SODs), catalase, peroxidases, glutathione reductases and ascorbate peroxidase. Different antioxidant molecules like ascorbate, reduced glutathione, carotenoids and tocopherol also play an important role in ROS management under stress. Polyamines may interact with ROS in connection with stress tolerance. Thus PAs are reported to have direct ROS scavenging activity (Drolet et al., 1986), though PAs conjugated with different phenolic compounds are found to be more potent in this regard (Langebartels et al., 1991). Besides, PAs may be involved in activating the cellular antioxidant system in response to stress. Thus genes for peroxidase in tobacco plants (Hiraga et al., 2000) and that for SOD in case of roots of a halophyte (Aronova et al., 2005) have been found to be induced by spermidine and cadaverine, respectively. He et al. (2008) demonstrated enhanced enzymatic and non-enzymatic antioxidant activity following increased spermidine content in a *SPDS*-over-expressing transgenic *Pyrus communis* in response to salt stress. Exogenous application of PAs also may confer stress tolerance by up-regulating the antioxidant system (Yiu et al., 2009). Treatment with putrescine and spermidine prevented decline in activities of antioxidant enzymes due to chilling in case of a chilling-sensitive cucumber cultivar (Zhang et al., 2009).

Apparently PAs, as well as their protective effect (directly by binding with the macromolecules and/or by scavenging ROS as an antioxidant), participate in a signalling process to activate the antioxidant system. It has already been discussed that PA catabolism is linked with ROS generation, particularly H_2O_2, which may be associated with plant defence and abiotic stress responses (Cona et al., 2006). Among ROS, H_2O_2 is relatively more stable and capable of diffusion intracellularly and intercellularly for signalling (Neil et al., 2002). Therefore, PA itself being synthesized due to stress undergoes oxidative degradation to produce H_2O_2 that may participate in the signalling process instead of in a damaging action. Polyamine oxidizing enzymes (PAO), a flavoprotein enzyme and DAO, a copper enzyme located in the apoplast, can generate H_2O_2 that can induce signalling for activation of the antioxidant system. Over-expressing a maize *PAO* gene in tobacco resulted in production of H_2O_2 and subsequent induction of antioxidant genes required for stress adaptation (Rea et al., 2004; Moschou et al., 2008b). In the case of programmed cell death (PCD) as a hypersensitive response during defence against plant pathogens, a 'spermine signalling pathway' operates. Thus accumulation of spermine in apoplast followed by generation of H_2O_2 due to action of PAO up-regulates defence-related genes like PR proteins and MAPK (Cona et al., 2006; Kusano et al., 2008; Moschou et al., 2008a). However, the amount of H_2O_2 is crucial and depending on endogenous polyamine level it results in either tolerance or PCD. Moschou et al. (2008c) proposed that spermidine exodus followed by oxidation in the apoplast creates a ROS signature to determine tolerance response. Antioxidant potential is low in the apoplastic space, a compartment where ROS can act as a signalling component. However, inside the cellular compartment the fate of ROS depends on the relative antioxidant activity. A ROS-signalling cascade may also be used for signalling amplification by PA recycling loop where PA synthesis and back-conversion pathway induced by stress caused recurrent generation of H_2O_2 (Alcázar et al., 2012).

Regarding further downstream signalling of H_2O_2 to bring about stress tolerance, various signalling cascades and components are reported for different plant species and kinds of stress. Downstream signalling occurs via calcium, reversible phosphorylation or MAPK cascade (reviewed in Kar, 2011). Final targets are the genes involved in cellular protection and DNA and protein repair processes. Indeed, H_2O_2 can up-regulate a number of genes that can orchestrate to build up stress tolerance.

13.4.3 Integration with phytohormone signalling

Plant hormones are involved in a comprehensive manner in regulating plant physiology and metabolism including plant responses to stresses. It is already established that ROS can act in coordination with phytohormones in several instances (Kwak et al., 2006). Therefore, an integration of PAs in this network having an interactive role for stress tolerance may easily be understood. Indeed, recent researches point towards an intricate cross-talk of PAs with hormonal signalling in regulating stress responses (Alcázar et al., 2010). Recently, cross-talks of PAs with two stress-related plant hormones, abscisic acid and ethylene, have been reviewed in detail (Bitrián et al., 2012).

Abscisic acid (ABA) appears to be the most important as regards its role in tolerance against most of the stresses. Thus a number of drought-inducible genes are ABA-responsive. It was found that ABA can induce PA biosynthesis under water and salt stresses (Urano et al., 2004; Alcázar et al., 2006b). In Arabidopsis, exogenous application of ABA up-regulated the expression of ADC2 and SPMS (Urano et al., 2003). Alcázar et al. (2006b), using two ABA mutants, aba2-3 (ABA biosynthesis mutant) and abi1-1 (ABA action mutant), showed that polyamine biosynthesis genes ADC2, SPDS1 and SPMS are highly responsive to drought stress. These genes have ABA-responsive elements (ABRE) or ABRE-related motifs in their promoters indicating ABA-regulation of PA biosynthesis during stress. Interestingly, during dehydration ABA-up-regulated ADC2 expression resulted in putrescine accumulation but progressive reduction in spermine level. This was explained as putrescine to spermine canalization linked with back-conversion (spermine to putrescine) leading to ROS signal amplification through a PA recycling loop. Besides, ROS can interact with ABA signalling during stress responses as was reported in the case of involvement of copper amine oxidase-generated H_2O_2 in ABA-induced stomatal closure in response to drought stress (An et al., 2008). Such a signalling pathway may also involve Ca^{2+} ions (Alcázar et al., 2010). Another observation based on complementation of adc1 mutants with ABA and reciprocal complementation of aba2-3 with putrescine (Cuevas et al., 2008, 2009) reveals a positive feed-back loop and raises a possibility of regulation of ABA biosynthesis by PAs (Alcázar et al., 2010). Thus, nced3 (mutant defective of an important step in ABA biosynthesis) did not show PA responses to dehydration (Urano et al., 2009). On the other hand, low expression of NCED3 was noted in the adc1 mutant at low temperature (Cuevas et al., 2008).

Ethylene is another hormone associated with stress responses, and the most important regulatory enzyme of ethylene biosynthesis pathway is ACC synthase, which is induced in response to stress. Though ethylene synthesis is reported to be stimulated by ROS during osmotic stress (Ke and Sun, 2004) and H_2O_2 may share the ethylene signalling pathway as was found during drought-induced stomatal closing (Desikan et al., 2005), a mutual inhibition of biosynthesis of ethylene and PAs because of competition for common precursor, SAM, made these two opposing each other in respect of stress responses. Thus SAM has been shown to be diverted to PAs in the case of silencing ACC synthase and ACC oxidase by antisense technology (Wi and Park, 2002), which is known to increase tolerance to abiotic stresses. However, an absence of such competition for SAM has also been recorded in the case of tomato plant transformed with yeast *SAMDC* that produced ethylene and PAs simultaneously during fruit ripening (Mehta *et al.*, 2002). But in the case of defence against a fungal species, tomato plant overexpressing yeast *spermidine synthase* showed susceptibility to the pathogen but repressed ethylene biosynthesis (Nambeesan *et al.*, 2012). A very recent work on effects of soil drying on rice cultivars revealed that mild stress has favoured PA synthesis while suppressing ethylene evolution, resulting in better grain filling of inferior spikelets (Chen *et al.*, 2013). All these observations indicate a complex interaction of PA biosynthesis and ethylene biosynthesis pathways having a common metabolite, SAM.

13.5 Concluding Remarks

Apparently, a complex interacting network of ROS and hormone signalling exists, integrating PA metabolism as a component for general plant responses to stresses. Although PAs may confer defence against stress by direct protection of cellular macromolecules and metabolites, a signalling role of PAs is becoming realized following recent works using molecular tools. A PA recycling loop involving PA canalization and subsequent back-conversion resulting in recurrent generation of ROS (H_2O_2) in the apolast suggests the possibility of further amplification of ROS signalling, a process that has gained importance in case of different plant responses, particularly host defence against pathogens. Similar ROS signal propagation involving NADPH oxidase, instead of PAs, that creates an auto-propagative wave in response to stimuli has already been indicated (Miller *et al.*, 2009; Mittler *et al.*, 2011). A clear-cut picture will emerge only after thorough study of the PA profile at subcellular level and its temporal regulation under the control of hormones imparting a specific ROS signature that is translated into the ultimate fate of the cellular system under stress. Complete understanding of such mechanisms will definitely help for better crop management under marginal habitats. This could be a challenging task for plant biologists in the coming years when global warming and food security are becoming the major issues to the nations worldwide.

References

Alcázar, R., García-Martínez, J.L., Cuevas, J.C., Tiburcio, A.F. and Altabella, T. (2005) Overexpression of ADC2 in *Arabidopsis* induces dwarfism and late-flowering through GA deficiency. *The Plant Journal* 43, 425–436.

Alcázar, R., Marco, F., Cuevas, J.C., Patrón, M., Ferrando, A., Carrasco, P., Tiburcio, A.F. and Altabella, T. (2006a) Involvement of polyamines in plant response to abiotic stress. *Biotechnology Letters* 28, 1867–1876.

Alcázar, R., Cuevas, J.C., Patrón, M., Altabella, T. and Tiburcio, A.F. (2006b) Abscisic acid modulates polyamine metabolism under water stress in *Arabidopsis thaliana*. *Physiologia Plantarum* 128, 448–455.

Alcázar, R., Altabella, T., Marco, F., Bortolotti, C., Reymond, M., Koncz, C., Carrasco, P. and Tiburcio, A.F. (2010) Polyamines: molecules with regulatory functions in plant abiotic stress tolerance. *Planta* 231, 1237–1249.

Alcázar, R., Bitrian, M., Bartels, D., Koncz, C., Altabella, T. and Tiburcio, A.F. (2011) Polyamine metabolic canalization in response to drought stress in *Arabidopsis* and the resurrection plant *Craterostigma plantagineum*. *Plant Signaling & Behavior* 6, 243–250.

Alcázar, R., Bitrián, M., Zarza, X. and Tiburcio, A.F. (2012) Polyamine metabolism and signaling in plant abiotic stress protection. In: Muñoz-Torrero, D., Haro, D. and Vallès, J. (eds) *Recent Advances in Pharmaceutical Sciences II*. Transworld Research Network, Trivandrum, Kerala, India, pp. 29–47.

An, Z.F., Jing, W., Liu, Y.L. and Zhang, W.H. (2008) Hydrogen peroxide generated by copper amine oxidase is involved in abscisic acid-induced stomatal closure in *Vicia faba*. *Journal of Experimental Botany* 59, 815–825.

Apel, K. and Hirt, H. (2004) Reactive oxygen species: metabolism, oxidative stress and signal transduction. *Annual Review of Plant Biology* 55, 373–379.

Aronava, E.E., Shevyakova, N.I., Sretsenko, L.A. and Kuznetsov, V.I.V. (2005) Cadaverine-induced induction of superoxide dismutase gene expression in *Mesembryanthemum crystallinum* L. *Doklady Biological Sciences* 403, 1–3.

Bachrach, U. (2005) Naturally occurring polyamines: interaction with macromolecules. *Current Protein Peptic Science* 6, 559–566.

Bitrián, M., Zarza, X., Altabella, T., Tiburcio, A.F. and Alcázar, R. (2012) Polyamines under abiotic stress: metabolic crossroads and hormonal crosstalks in plants. *Metabolites* 2, 516–528.

Bouchereau, A., Aziz, A., Larher, F. and Martin-Tanguy, J. (1999) Polyamines and environmental challenges: recent development. *Plant Science* 140, 103–125.

Capell, T., Bassie, L. and Christou, P. (2004) Modulation of polyamine biosynthetic pathway in transgenic rice confers tolerance to drought stress. *Proceedings of the National Academy of Sciences USA* 101, 9909–9914.

Chattopadhayay, M.K., Tiwari, B.S., Chattopadhyay, G., Bose, A., Sengupta, D.N. and Ghosh, B. (2002) Protective role of exogenous polyamines on salinity-stressed rice (*Oryza sativa*) plants. *Physiologia Plantarum* 116, 192–199.

Chen, T., Xu, Y., Wang, J., Wang, Z., Yang, J. and Zhang, J. (2013) Polyamines and ethylene interact in rice grains in response to soil drying during grain filling. *Journal of Experimental Botany* doi: 10.1093/jxb/ert115.

Cona, A., Rea, G., Angelini, R., Federico, R. and Tavladoraki, P. (2006) Functions of amine oxidases in plant development and defence. *Trends in Plant Science* 11, 80–88.

Cuevas, J.C., Lopez-Cobollo, R., Alcázar, R., Zarza, X., Koncz, C., Altabella, T., Salinas, J., Tiburcio, A.F. and Ferrando, A. (2008) Putrescine is involved in *Arabidopsis* freezing tolerance and cold acclimation by regulating abscisic acid levels in response to low temperature. *Plant Physiology* 148, 1094–1105.

Cuevas, J.C., Lopez-Cobollo, R., Alcazar, R., Zarza, X., Koncz, C., Altabella, T., Salinas, J., Tiburcio, A.F. and Ferrando, A. (2009) Putrescine as a signal to modulate the indispensable ABA increase under cold stress. *Plant Signaling & Behavior* 4, 219–220.

Desikan, R., Hancock, J.T., Bright, J., Harrison, J., Weir, I., Hooley, R. and Neill, S.J. (2005) A role for ETR1 in hydrogen peroxide signaling in stomatal guard cells. *Plant Physiology* 137, 831–834.

Do, P.T., Degenkolbe, T., Erban, A., Heyer, A.G., Kopka, J., Köhl, K.I., Hincha, D.K. and Zuther, E. (2013) Dissecting rice polyamine metabolism under controlled long-term drought stress. *PLoS ONE* 8(4), e60325. doi:10.1371/journal.pone.0060325.

Drolet, G., Dumbroff, E.B., Leggee, R.L. and Thompson, J.E. (1986) Radical scavenging properties of polyamines. *Phytochemistry* 25, 367–371.

Duan, J.J., Li, J., Guo, S.R. and Kang, Y.Y. (2008) Exogenous spermidine affects polyamine metabolism in salinity-stressed *Cucumis sativus* roots and enhances short-term salinity tolerance. *Journal of Plant Physiology* 165, 1620–1635.

Flores, H. (1983) Studies on the Physiology and Biochemistry of Polyamines in Higher Plants. PhD Thesis, Yale University, New Haven, Connecticut.

Flores, H.E. and Galston, A.W. (1982) Polyamines and plant stress – activation of putrescine biosynthesis by osmotic shock. *Science* 217, 1259–1261.

Galston, A.W. (1991) On the trail of a new regulatory system in plants. *The New Biologist* 3, 450–453.

Galston, A.W. and Kaur-Sawhney, R. (1990) Polyamines in plant physiology. *Plant Physiology* 94, 406–410.

Galston, A.W. and Kaur-Sawhney, R. (1995) Polyamines as endogenous growth regulators. In: Davies, P.J. (ed.) *Plant Hormones. Physiology, Biochemistry and Molecular Biology*. Kluwer Academic Publishers, Dordrecht, the Netherlands, pp. 158–178.

Galston, A.W., Kaur-Sawhney, R., Atabella, T. and Tibutcio, A.F. (1997) Plant polyamines in reproductive activity and response to abiotic stress. *Botanica Acta* 110, 197–207.

Gill, S.S. and Tuteja, N. (2010) Polyamines and abiotic stress tolerance in plants. *Plant Signaling & Behavior* 5, 26–33.

Groppa, M.D. and Benavides, M.P. (2008) Polyamines and abiotic stress: recent advances. *Amino Acids* 34, 35–45.

Groppa, M.D., Benavides, M.P. and Tomaro, M.L. (2003) Polyamine metabolism in sunflower and wheat leaf discs under cadmium or copper stress. *Plant Science* 161, 481–488.

Hanfrey, C., Sommer, S., Mayer, M.J., Burtin, D. and Michael, A.J. (2001) Arabidopsis polyamine biosynthesis: absence of ornithine decarboxylase and the mechanism of arginine decarboxylase activity. *The Plant Journal* 27, 551–560.

He, L., Ban, Y., Inoue, H., Matsuda, N., Liu, J. and Moriguchi, T. (2008) Enhancement of spermidine content and antioxidant capacity in transgenic pear shoots overexpressing apple spermidine synthase in response to salinity and hyperosmosis. *Phytochemistry* 69, 2133–2141.

Hiraga, S., Ito, H., Yamakawa, H., Ohtsubo, N., Seo, S., Mitsuhara, I., Matsui, H., Honma, M. and Ohashi, Y. (2000) An HR-induced tobacco peroxidase gene is responsive to spermine, but not to salicylate, methyl-jasmonate and ethaphon. *Molecular Plant-Microbe Interactions* 13, 210–216.

Igarashi, K. and Kashiwagi, K. (2000) Polyamines: mysterious modulators of cellular functions. *Biochemical and Biophysical Research Communications* 271, 559–564.

Illingworth, C., Mayer, M.J., Elliott, K., Hanfrey, C., Walton, N.J. and Michael, A.J. (2003) The diverse bacterial origins of the *Arabidopsis* polyamine biosynthetic pathway. *FEBS Letters* 549, 26–30.

Imai, A., Matsuyama, T., Hanzawa, Y., Akiyama, T., Tamaoki, M., Saji, H., Shirano, Y., Kato, T., Hayashi, H., Shibata, D., Tabata, S., Komeda, Y. and Takahashi, T. (2004) Spermidine synthase genes are essential for survival of *Arabidopsis*. *Plant Physiology* 135, 1565–1573.

Imanishi, S., Hashizume, K., Nakakita, M., Kojima, H., Matsubayashi, Y., Hashimoto, T., Sakagami, Y., Yamada, Y. and Nakamura, K. (1998) Differential induction by methyl jasmonate of genes encoding ornithine decarboxylase and other enzymes involved in nicotine biosynthesis in tobacco cell cultures. *Plant Molecular Biology* 38, 1101–1111.

Kar, R.K. (2011) Plant responses to water stress: role of reactive oxygen species. *Plant Signaling & Behavior* 6, 1741–1745.

Kar, R.K. and Choudhuri, M.A. (1986) Effects of light and spermine on senescence of Hydrilla and spinach leaves. *Plant Physiology* 80, 1030–1033.

Kasinathan, V. and Wingler, A. (2004) Effect of reduced arginine decarboxylase activity on salt tolerance and on polyamine formation during salt stress in *Arabidopsis thaliana*. *Physiologia Plantarum* 121, 101–107.

Ke, D. and Sun, G. (2004) The effect of reactive oxygen species on ethylene production induced by osmotic stress in etiolated mungbean seedling. *Plant Growth Regulation* 44, 199–206.

Kim, T.E., Kim, S.-K., Han, T.J., Lee, J.S. and Chang, S.C. (2002) ABA and polyamines act independently in primary leaves of clod-stressed tomato (*Lycopersicon esculentum*). *Physiologia Plantarum* 115, 370–376.

Kuehn, G.D., Rodriguez-Garay, B., Bagga, S. and Phillips, G.C. (1990) Novel occurrence of common polyamines in higher plants. *Plant Physiology* 94, 855–857.

Kumar, A., Altabella, T., Taylor, M.A. and Tiburcio, A.F. (1997) Recent advances in polyamine research. *Trends in Plant Science* 2, 124–130.

Kusano, T., Yamaguchi, K., Berberich, T. and Takahashi, Y. (2007) Advances in polyamine research in 2007. *Journal of Plant Research* 120, 345–350.

Kusano, T., Berberich, T., Tateda, C. and Takahashi, Y. (2008) Polyamines: essential factors for growth and survival. *Planta* 228, 367–381.

Kuznetsov, V. and Shevyakova, N. (2007) Polyamines and stress tolerance in plants. *Plant Stress* 1, 50–71.

Kwak, J.M., Nguyen, V. and Schroeder, J.I. (2006) The role of reactive oxygen species in hormonal responses. *Plant Physiology* 141, 323–329.

Langebartels, C., Kerner, K.J., Leonardi, S., Schraudner, M., Trost, M., Heiller, W. and Sanderman, H. (1991) Biochemical plant response to ozone. Differential induction of polyamine and ethylene biosynthesis in tobacco. *Plant Physiology* 95, 882–887.

Liu, J.H. and Moriguchi, T. (2006) ADC pathway plays an important role in salt stress response of apple *in vitro* callus. *Plant Genomics in China* 124, 1315–1325.

Martin-Tanguy, J. (2001) Metabolism and function of polyamines in plants: recent development (new approaches). *Plant Growth Regulation* 34, 135–148.

McDonald, R.E. and Kushad, M.M. (1986) Accumulation of putrescine during chilling injury of fruits. *Plant Physiology* 82, 324–326.

Mehta, R.A., Cassol, T., Li, N., Ali, N., Handa, A.K. and Mattoo, A.K. (2002) Engineered polyamine accumulation in tomato enhances phytonutrient content, juice quality, and vine life. *Nature Biotechnology* 20, 613–618.

Michael, A.J., Furze, J.M., Rhodes, M.J.C. and Burtin, D. (1996) Molecular cloning and functional identification of a plant ornithine decarboxylase cDNA. *Biochemical Journal* 314, 241–248.

Miller, G., Schlauch, K., Tam, R., Cortes, D., Torres, M.A., Shulaev, V., Dangl, J.L. and Mittler, R. (2009) The plant NADPH oxidase RBOHD mediates rapid systemic signaling in response to diverse stimuli. *Science Signaling* 2, ra45.

Mittler, R., Vanderauwera, S., Suzuki, N., Miller, G., Tognetti, V.B., Vandepoele, K., Gollery, M., Shulaev, V. and Van Breusegem, F. (2011) ROS signaling: the new wave? *Trends in Plant Science* 16, 300–309.

Moschou, P.N., Paschalidis, K.A. and Roubelakis-Angelakis, K.A. (2008a) Plant polyamine catabolism: the state of the art. *Plant Signaling & Behavior* 3, 1061–1066.

Moschou, P.N., Delis, I.D., Paschalidis, K.A. and Roubelakis-Angelakis, K.A. (2008b) Transgenic tobacco plants overexpressing polyamine oxidase are not able to cope with oxidative burst generated by abiotic factors. *Physiologia Plantarum* 133, 140–156.

Moschou, P.N., Paschalidis, K.A., Delis, I.D., Andriopoulou, A.H., Lagiotis, G.D., Yakoumakis, D.I. and Roubelakis-Angelakis, K.A. (2008c) Spermidine exodus and oxidation in the apoplast induced by abiotic stress is responsible for H_2O_2 signatures that direct tolerance responses in tobacco. *The Plant Cell* 20, 1708–1724.

Nambeesan, S., AbuQamar, S., Laluk, K., Mattoo, A.K., Mickelbart, M.V., Ferruzzi, M.G., Mengiste, T. and Handa, A.K. (2012) Polyamines attenuate ethylene-mediated defense responses to abrogate resistance to *Botrytis cinerea* in tomato. *Plant Physiology* 158, 1034–1045.

Neil, S., Desikan, R. and Hancock, J. (2002) Hydrogen peroxide signaling. *Current Opinion in Plant Biology* 5, 388–395.

Panagiotidis, C.A., Artandi, S., Calame, K. and Silverstein, S.J. (1995) Polyamines alter sequence-specific DNA-protein interactions. *Nucleic Acids Research* 23, 1800–1809.

Rea, G., Concetta de Pinto, M., Tavazza, R., Biondi, S., Gobbi, V., Ferrante, P., de Gara, L., Federico, R., Angelini, R. and Tavladoraki, P. (2004) Ectopic expression of maize polyamine oxidase and pea copper amine oxidase in the cell wall of tobacco plants. *Plant Physiology* 134, 1414–1426.

Richards, F.J. and Coleman, R.G. (1952) Occurrence of putrescine in potassium-deficient barley. *Nature* 170, 460–464.

Shen, W.Y., Nada, K. and Tachibana, S. (2000) Involvement of polyamines in the chilling tolerance of cucumber cultivars. *Plant Physiology* 124, 431–439.

Smith, T.A. (1973) Amine levels in mineral-deficient *Hordeum vulgare* leaves. *Phytochemistry* 12, 2093–2095.

Tabor, C.W. and Tabor, H. (1999) It all started on a streetcar in Boston. *Annual Review of Biochemistry* 68, 1–32.

Takahashi, T. and Kakehi, J. (2010) Polyamines: ubiquitous polycations with unique roles in growth and stress responses. *Annals of Botany* 105, 1–6.

Takahashi, Y., Uehara, Y., Berberich, T., Ito, A., Saitoh, H. and Miyazaki, A. (2004) A subset of hypersensitive response marker genes, including HSR203J, is the downstream target of a spermine signal transduction pathway in tobacco. *The Plant Journal* 40, 586–595.

Urano, K., Yoshiba, Y.M., Nanjo, T., Igarashi, Y., Seki, M., Sekiguchi, F., Yamaguchi-Shinozaki, K. and Shinozaki, K. (2003) Characterization of *Arabidopsis* genes involved in biosynthesis of polyamines in abiotic stress responses and developmental stages. *Plant, Cell & Environment* 26, 1917–1926.

Urano, K., Yoshiba, Y., Nanjo, T., Ito, T., Yamaguchi-Shinozaki, K. and Shinozaki, K. (2004) *Arabidopsis* stress-inducible gene for arginine decarboxylase AtADC2 is required for accumulation of putrescine in salt tolerance. *Biochemical and Biophysical Research Communications* 313, 369–375.

Urano, K., Hobo, T. and Shinozaki, K. (2005) *Arabidopsis* ADC genes involved in polyamine biosynthesis are essential for seed development. *FEBS Letters* 579, 1557–1564.

Urano, K., Maruyama, K., Ogata, Y., Morishita, Y., Takeda, M., Sakurai, N., Suzuki, H., Saito, K., Shibata, D., Kobayashi, M., Yamaguchi-Shinozaki, K. and Shinozaki, K. (2009) Characterization of the ABA-regulated global responses to dehydration in *Arabidopsis* by metabolomics. *The Plant Journal* 57, 1065–1078.

Wi, S.J. and Park, K.Y. (2002) Antisense expression of carnation cDNA encoding ACC synthase or ACC oxidase enhances polyamine content and abiotic stress tolerance in transgenic tobacco plants. *Molecules and Cells* 13, 209–220.

Yamaguchi, K., Takahashi, Y., Berberich, T., Imai, A., Miyazaki, A., Takahashi, T., Michael, A. and Kusano, T. (2006) The polyamine spermine protects against high salt stress in *Arabidopsis thaliana*. *FEBS Letters* 22, 6783–6788.

Yamaguchi, K., Takahashi, Y., Berberich, T., Imai, A., Takahashi, T., Michael, A.J. and Kusano, T. (2007) A protective role for the polyamine spermine against drought stress in Arabidopsis. *Biochemical and Biophysical Research Communications* 352, 486–490.

Yiu, J.C., Juang, L.D., Fang, D.Y.T., Liu, C.W. and Wu, S.J. (2009) Exogenous putrescine reduces flooding-induced oxidative damage by increasing the antioxidant properties of Welsh onion. *Scientia Horticulturae* 120, 306–314.

Zapata, P.J., Serrano, M., Pretel, M.T., Amoros, A. and Botella, M.A. (2004) Polyamines and ethylene changes during germination of different plant species under salinity. *Plant Science* 167, 781–788.

Zhang, W., Jiang, B., Li, W., Song, H., Yu, Y. and Chen, J. (2009) Polyamines enhance chilling tolerance of cucumber (*Cucumis sativus* L.) through modulating antioxidative system. *Scientia Horticulturae* 122, 200–208.

14 Indirect and Direct Benefits of the Use of *Trichoderma harzianum* Strain T-22 in Agronomic Plants Subjected to Abiotic and Biotic Stresses

Antonella Vitti, Maria Nuzzaci, Antonio Scopa and Adriano Sofo*
School of Agricultural, Forestry, Food and Environmental Sciences, University of Basilicata, Potenza, Italy

Abstract

Biological control of several plant diseases has been successfully achieved by the use of *Trichoderma harzianum* strain T-22, which acts through chemiotropic mycoparasitic interactions with the target fungal or bacterial organism. Since this strain can colonize the roots of most plant species across a wide range of soil types, it is particularly important for agronomic purposes. On the other hand, the study on the effect of T-22 or its derived substances against plant viruses (e.g. Cucumber mosaic virus – CMV) and the pathogenic and molecular aspects involved in this kind of three-way cross-talk between the plant, virus and antagonist are very little known. Besides the use of T-22 as a biocontrol agent, it has been reported that this fungus can also directly improve root growth and plant development in the absence of pathogens. Several mechanisms have been proposed for this, such as production of some unidentified growth-regulating compounds by the fungus, the increased availability of nutrients for plants and induction of certain root morphological changes. All these findings indicate the versatility through which T-22 can directly increase plant tolerance against abiotic stresses, such as drought, salinity and soils with low fertility. In spite of their theoretical and practical importance, the mechanisms responsible for the growth response due to the direct (growth-promoting) and indirect (antipathogenic) actions of T-22 in agronomic plants have not been investigated extensively. This chapter, based on the most significant and updated studies published in the last years by our research group, aims to contribute to a better understanding of the fundamental biochemical and physiological aspects of the antipathogenic and plant growth-promoting activities of T-22 on some economically important crops. This could promote a rational and non-empirical inclusion of this important fungal species into modern agricultural sustainable practices.

14.1 Introduction

Plant life on emerged land has made been possible by the symbiosis between plants and related microorganisms. Mycorrhization is the demonstration of the importance in establishing symbiosis between root system and some microorganisms, which makes possible an enduring protection of cultivated plants and a better use of nutrients, so improving plant tolerance to diseases. It is a symbiotic relationship between the mycelium of a fungus

*E-mail: adriano.sofo@unibas.it

and the roots of a plant (Lynch, 1990). In soils, numerous microorganisms co-exist in association with plant roots, inducing morphological and physiological changes in the roots in order to promote the adaptability and survival of both symbionts (Rigamonte et al., 2010). Some microorganisms live specifically in the rhizosphere or on plant root surfaces, and these can have many effects on plant performance and may also affect plant community structure. The plant root surface is surrounded by a specific microflora, and the microorganisms distributed there have specific roles in the decomposition of organic matter. Diverse substances are secreted and deleterious microorganisms, which could inhibit plant growth, may be suppressed (Hyakumachi and Kubota, 2004). Mycorrhizated plants are generally able to tolerate pathogens and compensate for root damage and photosynthate drain by pathogens because mycorrhiza are able to enhance host nutrition and the overall plant growth. Arbuscular fungi (e.g. *Glomus* spp.) are known to enhance plant tolerance to pathogens but also to abiotic stresses (Hrynkiewicz and Baum, 2012), enhancing photosynthetic capacity and delaying senescence.

The microorganisms that populate soils, as mycorrhizae, endophytes, saprophytes, but also phytopathogens and entomopathogens, represent a good resource in transformation of organic matter, offering products of enormous potential, such as secondary metabolites, antibiotics and catabolic enzymes (Arora, 2003). Among them, some species of bacteria and fungi are effective also as biocontrol agents (BCAs). These fungal antagonists reduce the growth of plant pathogens by antibiosis, competition and parasitism (Mathivanan et al., 2008). They also induce various defence responses in host plants, such as systemic acquired resistance (SAR) and/or induced systemic resistance (ISR). For this purpose, many scientists proposed the use of mycorrhizae associated to biocontrol microorganisms as a solution for increasing plant tolerance/resistance against both biotic and abiotic stresses, for increasing plant productivity in degraded soils and for reducing agricultural environmental impact. The use of microorganisms for biocontrolling plant pathogens has been shown to be very efficacious for some fungi of the genus *Trichoderma*, *Glomus*, *Streptomyces* and some species of bacteria (e.g. *Bacillus subtilis* and *Agrobacterium radiobacter*). In fact, some of these fungi interact with other fungi in a mechanism called mycoparasitism, wherein one fungus directly kills and obtains nutrients from other fungi. Mycoparasitism is one of the most important biocontrol mechanisms (Mukherjee, 2011).

Besides the use of *Trichoderma* as a biocontrol agent, this fungus can directly stimulate root and shoot growth without the presence of pathogens (Sofo et al., 2012). This direct effect could be due to some growth-regulating compounds produced by the fungus, the increased availability of nutrients for plants, and some induced change in root morphology. All these findings indicate the versatility through which *Trichoderma* can directly increase plant tolerance to different kinds of abiotic stresses. In the context of plant defence by biotic stresses, understanding biochemical and molecular mechanisms deriving from the host–pathogen–*Trichoderma* interaction is without doubt essential for investigating the dynamics of infectious processes. This knowledge could be also useful for the development of new strategies for controlling phytopathogens, particularly viruses, against which chemical treatments have no effect.

Thanks to recent studies, it is now possible to develop new strategies based on the use of peptaibols, a class of linear peptides biosynthesized by many species of *Trichoderma* (Daniel and Filho, 2007). Trichokonins, which are antimicrobial peptaibols isolated from *Trichoderma pseudokoningii* SMF2, have been reported to induce tobacco systemic resistance against tobacco mosaic virus (TMV) through activation of multiple plant defence pathways. This is based on an elicitor-like cellular response, i.e. enhancement of production of superoxide anion radical and peroxide in tobacco plants and also enhancement of enzymes such as peroxidase (POD), which are involved in resistance, up-regulation of antioxidative enzyme genes known to be associated with the reactive oxygen species (ROS) intermediate-mediated signalling pathway, and of salicylic acid (SA)-, ethylene (ET)- and jasmonic acid (JA)-mediated defence pathway marker genes (Luo et al., 2010). This finding implies the antiviral potential of peptaibols, supporting the hypothesis of using them as biocontrol antiviral agents. Therefore, *Trichoderma* spp., already used as BCAs against bacterial and fungal phytopathogens, could be advantageously used also against viruses. Considering the theoretical and practical importance of the broad range of mechanisms responsible for the growth response due to the direct (growth-promoting) and indirect (antipathogenic) actions of *Trichoderma*, these need to be investigated in more detail.

This chapter, based on the most significant and updated studies published in the last years, aims to contribute to a better understanding of the fundamental biochemical and physiological aspects of the antipathogenic and plant growth-promoting activities of *Trichoderma* on some important economically important crops. In particular, the strain T-22 of *T. harzianum* is of key importance because it is often used as active ingredient in many commercial biocontrol products. This could promote a rational and non-empirical inclusion of this important fungal species in modern agricultural sustainable practices.

14.2 The Genus *Trichoderma harzianum* Strain T-22

The filamentous ascomycetous fungi *Trichoderma* spp. are abundant and present in many soil types. These fungi are able to infect plant roots, invading the first or second layers of cells of the root epidermis (Harman *et al.*, 2004a). *Trichoderma* spp. show a number of different activities between strains (Harman *et al.*, 2004b). They are rarely associated with diseases of living plants (Gams and Bissett, 2002). On the contrary, many *Trichoderma* species (e.g. *T. harzianum*, *T. viride*) are used as BCAs by antagonizing many plant pathogenic fungi. Indeed, approximately 60% of all commercial biocontrol formulations are based on *Trichoderma* (Verma *et al.*, 2007).

By working as a deterrent, T-22 protects the roots from the assault of pathogenic fungi (e.g. *Fusarium*, *Pythium*, *Rhizoctonia* and *Sclerotinia*). Establishing itself in the rhizosphere, T-22 can grow on the root system, along which it establishes a barrier against pathogens. As long as the root system remains active, T-22 continues to grow, feeding on the root exudates and subtracting the nutrients that the pathogens use to feed (Tataranni *et al.*, 2012).

Biocontrol studies have confirmed the effectiveness of *Trichoderma* spp. in plant protection not only against many pathogenic fungi (Akrami *et al.*, 2011), but also bacteria (Segarra *et al.*, 2009) and viruses (Luo *et al.*, 2010), probably due to the induction of hypersensitive response (HR), systematic acquired resistance (SAR) and induced systematic resistance (ISR) (Kaewchai *et al.*, 2009). However, the antiviral effects of *Trichoderma* spp. and the associated biochemical and molecular mechanisms implicated are still scarcely known. Plant resistance induced by *Trichoderma* spp. at a molecular level is due to the release of specific defence metabolites and enzymes, such as: (i) phenyl-alanine ammonia-lyase (PAL) and chalcone synthase (CHS), involved in the biosynthesis of phytoalexins (HR response); (ii) chitinases and glucanases, that include pathogenesis-related proteins (PR) and SAR response; and (iii) other enzymes involved in the response to oxidative stress (Benítez *et al.*, 2004).

It was demonstrated that T-22 improves growth in maize plants, increasing root formation (size and area of main and secondary roots) and, at the same time, enhancing crop yields, tolerance to drought and resistance to compacted soils (Harman, 2000; Harman *et al.*, 2004c). This improved plant growth was probably due to direct effects on plants because of a better solubilization of soil nutrients or by a direct enhancing plant uptake of nutrients linked to the presence of T-22 in the agroecosystem (Yedidia *et al.*, 2001). More recently, it was demonstrated that plant overall morphology and metabolism of plant colonized by T-22 caused enhanced root growth and suberification (Sofo *et al.*, 2011, 2012) and the induction of the synthesis of antimicrobial phenolic compounds (Mathivanan *et al.*, 2008). The enhanced plant growth due to T-22 was confirmed also in terms of total biomass and root development, not only in herbaceous plants but also in tree species (Sofo *et al.*, 2010). Furthermore, the beneficial effects of T-22 application depend on the treated plant genotype, as recently demonstrated by Tucci *et al.* (2011) on tomato plants.

Although the capability of *Trichoderma* spp. to alleviate the effects of various abiotic stresses on plants is recognized, an understanding of the mechanisms that control the factors implied in the specific plant stress are still missing. Using T-22 in organic management systems can surely improve plant physiological status using a holistic approach that adopts specific practices for promoting plant defence mechanisms, such as tolerance and/or resistance to pathogens (Woo *et al.*, 2006).

14.3 Abiotic and Biotic Stresses in Plants

Plant growth and productivity are affected by various environmental stresses to which plants are subjected during their lifespan. Due to their sessile conditions, plants cannot avoid these stresses and must have strong defences to face them. Indeed, molecular, biochemical, physiological and morphological characteristics of plants are markedly affected by the exposure to abiotic and biotic stresses. The activation of induced defence in plants is mediated through the synthesis of molecules with signal functions acting as hormones or stimulators of plant growth and development (Vitti *et al.*, 2013). Among phytohormones, a prevailing role in biotic stress signalling is played by SA, JA and ET, while abscisic acid (ABA) plays a role in the response to some abiotic stresses such as drought, low temperature and osmotic stress (Fraire-Velázquez *et al.*, 2011).

For example, it is known that SA-induced resistance to viruses in tobacco and *Arabidopsis thaliana* is partly mediated by a pathway involving signals transduced through changes in reactive oxygen species (ROS) in the mitochondria (Singh *et al.*, 2004). In fact, SA impedes electron flow through the respiratory electron transport chain and enhances ROS levels in the mitochondria (Mayers *et al.*, 2005). Resistance to TMV is altered in transgenic tobacco plants with altered levels of alternative oxidase (AOX), an enzyme that negatively regulates mitochondrial ROS levels (Gilliland *et al.*, 2003).

In *A. thaliana*, as in tobacco, SA treatments inhibited the systemic movement of another virus, cucumber mosaic virus (CMV). At the same time, in squash, SA induced resistance to CMV and this was most likely due to inhibition of viral cell-to-cell movement. This means that the mechanisms of SA-induced resistance may differ markedly between host species (Mayers *et al.*, 2005). ROS are important second messengers in the responses of plants to various other biotic and abiotic stresses (Kwon *et al.*, 2007; Wahid *et al.*, 2007; Miller *et al.*, 2010; Torres, 2010).

Recently, Vitti and co-workers (2013) demonstrated that changes in root morphology observed in *A. thaliana* seedlings subjected to both biotic (CMV) and abiotic (excess cadmium) stressors are probably due to modifications in hormonal balances. As shown in Fig. 14.1, in our experience, evident variations

Fig. 14.1. *Arabidopsis thaliana* Columbia ecotype control plants (left), inoculated with CMV (centre) and treated with cadmium (right) observed 12 days after the viral infection or the exposure to cadmium.

occurred in plant growth, in terms of both shoot and root development and also in leaf colour (from green of the control plants to brownish violet of inoculated and, overall, treated plants). Molecular, biochemical, physiological and morphological characteristics of plants are markedly affected also by the exposure to some heavy metals. Indeed, treatments of plants with some metals induce changes in root morphology, caused by a hormonal inbalance, mainly governed by the auxin/cytokinin ratio (Sofo et al., 2013).

14.4 Benefits of the Use of *Trichoderma harzianum* Strain T-22 in Stressed Crops

A broad range of genetic traits and environmental conditions are able to affect the complex phenotype of mycorrhizal fungi, as in *Trichoderma* spp., as well as their ecological performance (Buée et al., 2009). The interactions between microbes and plant roots are known to have significant effects on plant nutrient condition and tolerance to pathogens (Altomare et al., 1999). Many studies included the use of proteomic (Grinyer et al., 2005) and functional genomic analysis in the attempt to obtain more information on the changes that occur in the *Trichoderma* spp., plant and pathogen expressomes when they interact with each other, especially when an increase in disease resistance is generated (Woo et al., 2006). In a recent study, the dynamics of gene expression in the roots of *Arabidopsis* colonized by *Trichoderma* were investigated, demonstrating that this colonization has induced deep changes in plant transcripts, through plant gene modulation, together with resistance to both biotic and abiotic stresses (Brotman et al., 2013).

The mechanism of the interaction *Trichoderma*–plant–pathogen is very complex and includes not only the colonization of rhizosphere and phyllosphere and mycoparasitism, but also antagonism against nematodes, production of extracellular hydrolytic enzymes and secondary metabolites that could be toxic to plant pathogens, as well as induction of systemic resistance against different pathogens' promotion. These *Trichoderma*–plant interactions can also result in better plant growth and root development (Harman et al., 2004c; Mathivanan et al., 2008). In particular, T-22 is adapted for facing many fungal or bacterial pathogens in a broad range of plant species (Sofo et al., 2010; Tataranni et al., 2012). Therefore, T-22 is considered a very efficacious BCA for the control of plant diseases.

14.4.1 Benefits of T-22 against abiotic stresses

It has been established recently that the change in phytohormone levels, particularly auxins and cytokinins, is one of the direct mechanisms by which T-22 acts for promotion of plant growth in fruit rootstocks (Sofo et al., 2011). Thus, T-22 seems to promote plant growth and development, so acting as a plant growth-promoting microorganism, that in turn determines a higher tolerance of the plants against abiotic stresses (Sofo et al., 2011). It was also discovered that soil colonization by T-22 enhances plant growth in terms of total biomass and root development by about 20% and 30%, respectively (Sofo et al., 2010).

The cross-talk between the different plant hormones, whose levels change after plant inoculation with T-22, results in synergetic or antagonistic interactions that play crucial roles in the response of plants to abiotic stresses, such as drought, salinity and toxic metals (Baroni et al., 2004; Peleg and Blumwald, 2011). An example of this is depicted in Fig. 14.2, where cherry seedlings inoculated with T-22 and subjected to water deficit appear to be more developed than control un-inoculated plants. Thus, it is possible that plant hormones play central roles in the ability of plants to adapt to changing environments by mediating growth, development, nutrient allocation and source/sink transitions.

In a study by Mastouri et al. (2010) it was shown that under either biotic stress caused by *Pythium ultimum* or different abiotic stresses

Fig. 14.2. Cherry seedlings (*Prunus cerasus* x *P. canescens*) inoculated with T-22, grown in sterile perlite and subjected to drought stress (right) and control un-inoculated plants subjected to the same degree of drought (left).

such as drought, salinity, elevated or low temperature, treatment of tomato seeds with T-22 led to more rapid and uniform germination in comparison to no treatment.

More recently, the same authors demonstrated that the application of T-22 to tomato seedlings enhanced the tolerance to water deficit by improving the antioxidant defence mechanism (e.g. higher activity of ascorbate and glutathione-recycling enzymes) (Mastouri *et al.*, 2012). It is proposed that in addition to the hormonal factors, T-22 allows plants to tolerate abiotic stresses more efficiently by increased root suberification and hardening, as well as acidification of the soil, which would favour the diffusion of cations from the soil to the root against the concentration gradient and an increased availability of some inorganic compounds indispensable for plants (Sofo *et al.*, 2012).

14.4.2 Benefits of T-22 against biotic stresses

Biocontrol by T-22 is related to its ability to compete with soil pathogens rather than to its control of plant diseases. Therefore, T-22 does not act by producing compounds that are toxic to the pathogens, but rather by inducing change in the physiology and metabolism of the plants leading to development of resistance to the disease (Harman *et al.*, 2008). For that reason, various mechanisms are involved, foremost mycoparasitism and antibiosis (Howell, 2003; Vitale *et al.*, 2012). In the case of mycoparasitism, recognition, binding and enzymatic disruption of the target cell wall take place (Woo and Lorito, 2007). On the other hand, inhibition or destruction of the microorganism target by metabolites or by the production of antibiotics able to inhibit their growth (antibiosis) were also observed.

Fig. 14.3. (a) *Nicotiana tabacum* cv. Xanthi infected with CMV (left); (b) the same plant infected with CMV and also treated with *Trichoderma harzianum* T-22 (right).

In such case, antibiotics can stop spore germination (action known as fungistasis) or alternatively destroy the cells (veritable antibiosis) (Benítez *et al.*, 2004).

The range of pathogens controlled by T-22 is broad and includes fungi, bacteria and viruses (Harman *et al.*, 2004a). Among plant-pathogenic fungi, the following are the most represented: *Botrytis cinerea*, *Fusarium*, *Pythium* and *Rhizoctonia* (Kaewchai *et al.*, 2009). The efficacy of *Trichoderma* spp. action is obviously related to the specific interaction between plant–pathogen–antagonist. For example, Vitale and co-authors (2012) demonstrated that T-22 was able to act as a BCA of collar and root rot caused by different *Calonectria pauciramosa* isolates on red clover (*Trifolium pratense*) and, specifically, that the degree of virulence and T-22 effects in controlling infections were highly variable among the isolates tested. In our experience, preliminary results of current studies conducted in our laboratories seem to indicate a potential antiviral activity of T-22 against the infection of CMV, strain Fny, on tobacco plants, as shown in Fig. 14.3, where the plant treated with the fungus does not show the symptoms induced by the virus.

14.5 Conclusion

In T-22-inoculated plants subjected to different types of adverse environmental conditions, comparative proteomics experiments should be carried out to identify specific proteins involved in plant resistance against specific stresses. For this kind of analysis, 2D-electrophoretic cells, protein fractionation and isoelectrofocusing techniques and MALDI-TOF MS are commonly used. Accurate microscopic analyses should be carried out through electron (SEM and ESEM), epifluorescence and light microscopes in order to ascertain T-22 persistence and evaluate their colonization. Finally, comparative proteomics experiments could be of primary importance to identify specific proteins involved in the common response of T-22-inoculated plants to face abiotic stresses.

Plant stresses contribute significantly to crop damage and yield loss. In agriculture, annual crop losses by phytopathogenic microorganisms in the field and also during post-harvest exceed €500 billion (Tataranni *et al.*, 2012). The balance of beneficial and detrimental effects is reflected in many other areas of agriculture and horticulture. In such a scenario, in modern agro-industry, fungi such as *T. harzianum* strain T-22 offer many established beneficial roles, particularly as biofertilizers, mycorrhizae, and BCAs of pathogens, pests and weeds. In addition to their biocontrol characteristics, T-22 also exhibits plant growth-promoting activity, acting as powerful biostimulants. The utilization of T-22 or other microorganisms as biostimulants can cause a reduction in the use of fertilizers and fungicides in agricultural production, with consequent benefits for the environment. This is necessary to help maintain ecosystems and to develop sustainable agriculture.

References

Akrami, M., Golzary, H. and Ahmadzadeh, M. (2011) Evaluation of different combinations of *Trichoderma* species for controlling *Fusarium* rot of lentil. *African Journal of Biotechnology* 10(14), 2653–2658.

Altomare, C., Norvell, W.A., Björkman, T. and Harman, G.E. (1999) Solubilization of phosphates and micronutrients by the plant-growth-promoting and biocontrol fungus *Trichoderma harzianum* Rifai 1295-22. *Applied and Environmental Microbiology* 65(7), 2926–2933.

Arora, D.K. (2003) *Fungal Biotechnology in Agricultural, Food, and Environmental Applications*. Marcel Dekker, New York.

Baroni, F., Boscaglia, A., Di Lella, L.A., Protano, G. and Riccobono, F. (2004) Arsenic in soil and vegetation of contaminated areas in southern Tuscany (Italy). *Journal of Geochemical Exploration* 81, 1–14.

Benítez, T., Rincón, A.M., Limón, M.C. and Codón, A.C. (2004) Biocontrol mechanisms of *Trichoderma* strains. *International Microbiology* 7, 249–260.

Brotman, Y., Landau, U., Cuadros-Inostroza, A., Takayuki, T., Fernie, A.R., Chet, I., Viterbo, A. and Willmitzer, L. (2013) *Trichoderma*-plant root colonization: escaping early plant defense responses and activation of the antioxidant machinery for saline stress tolerance. *PLoS Pathogens* 9(3), 1–15.

Buée, M., De Boer, W., Martin, F., van Overbeek, L. and Jurkevitch, E. (2009) The rhizosphere zoo: an overview of plant-associated communities of microorganisms, including phages, bacteria, archaea, and fungi, and of some of their structuring factors. *Plant Soil* 321, 189–212.

Daniel, J.F.S. and Filho, E.R. (2007) Peptaibols of *Trichoderma*. *Natural Product Reports* 24, 1128–1141.

Fraire-Velázquez, S., Rodríguez-Guerra, R. and Sánchez-Calderón L. (2011) Abiotic and biotic stress response crosstalk in plants. In: Shanker, A. (ed.) *Abiotic Stress Response in Plants - Physiological, Biochemical and Genetic Perspectives*. InTech Europe, Croatia, pp. 3–26.

Gams, W. and Bissett, J. (2002) Morphology and identification of *Trichoderma*. In: Kubicek, P. and Harman E. (eds) Trichoderma *and* Gliocladium *- Basic biology, taxonomy and genetics*, Vol. 1. Taylor & Francis, London, pp. 3–34.

Gilliland, A., Singh, D.P., Hayward, J.M., Moore, C.A., Murphy, A.M., York, C.J., Slator, J. and Carr, J.P. (2003) Genetic modification of alternative respiration has differential effects on antimycin A-induced *versus* salicylic acid-induced resistance to *Tobacco mosaic virus*. *Plant Physiology* 132, 1518–1528.

Grinyer, J., Hunt, S., McKay, M., Herbert, B.R. and Nevalainen, H. (2005) Proteomic response of the biological control fungus *Trichoderma atroviride* to growth on the cell walls of *Rhizoctonia solani*. *Current Genetics* 47, 381–383.

Harman, G.E. (2000) Myths and dogmas of biocontrol. Changes in perceptions derived from research on *Trichoderma harzianum* T22. *Plant Disease* 84, 377–393.

Harman, G.E., Howell, C.R., Viterbo, A., Chet, I. and Lorito, M. (2004a) *Trichoderma* species – opportunistic, avirulent plant symbionts. *Nature Reviews Microbiology* 2, 43–56.

Harman, G.E., Lorito, M. and Lynch, J.M. (2004b) Uses of *Trichoderma* spp. to alleviate or remediate soil and water pollution. In: Laskin, A.I., Bennett, J.W. and Gadd, G.M. (eds) *Advances in Applied Microbiology*, Vol. 56. Elsevier Academic Press, San Diego, California, pp. 313–330.

Harman, G.E., Petzoldt, R., Comis, A. and Chen, J. (2004c) Interactions between *Trichoderma harzianum* strain T22 and maize inbred line Mo17 and effects of this interaction on diseases caused by *Pythium ultimum* and *Colletotrichum graminicola*. *Phytopathology* 94(2), 147–153.

Harman, G.E., Björkman, T., Ondik, K. and Shoresh, M. (2008) Changing paradigms on the mode of action and uses of *Trichoderma* spp. for biocontrol. *Outlooks on Pest Management* 19, 24–29.

Howell, R.C. (2003) Mechanisms employed by *Trichoderma* species in the biological control of plant diseases: the history and evolution of current concepts. *Plant Disease* 87, 4–10.

Hrynkiewicz, K. and Baum, C. (2012) The potential of rhizosphere microorganisms to promote the plant growth in disturbed soils. In: Malik, A. and Grohmann, E. (eds) *Environmental Protection Strategies for Sustainable Development*, Vol. 14, *Strategies for Sustainability*. Springer Science+Business Media BV, the Netherlands, pp. 35–64.

Hyakumachi, M. and Kubota, M. (2004) Fungi as plant growth promoter and disease suppressor. In: Arora, D.K. (ed.) *Mycology Series. Fungal Biotechnology in Agricultural, Food, and Environmental Applications*, Vol. 21. Marcel Dekker, New York, pp. 101–110.

Kaewchai, S., Soytong, K. and Hyde, K.D. (2009) Mycofungicides and fungal biofertilizers. *Fungal Diversity* 38, 25–50.

Kwon, S.J., Kwon, S.I., Bae, M.S., Cho, E.J. and Park, O.K. (2007) Role of the methionine sulfoxide reductase MsrB3 in cold acclimation in *Arabidopsis*. *Plant and Cell Physiology* 48, 1713–1723.

Luo, Y., Zhang, D.D., Dong, X.W., Zhao, P.B., Chen, L.L., Song, X.Y., Wang, X.J., Chen, X.L., Shi, M. and Zhang, Y.Z. (2010) Antimicrobial peptaibols induce defense responses and systemic resistance in tobacco against *tobacco mosaic virus*. FEMS *Microbiology Letters* 313, 120–126.

Lynch, J.M. (1990) Beneficial interactions between micro-organisms and roots. *Biotechnology Advances* 8(2), 335–346.

Mastouri, F., Bjorkman, T. and Harman, G.E. (2010) Seed treatment with *Trichoderma harzianum* alleviates biotic, abiotic, and physiological stresses in germinating seeds and seedlings. *Phytopathology* 100, 1213–1221.

Mastouri, F., Bjorkman, T. and Harman, G.E. (2012) *Trichoderma harzianum* enhances antioxidant defense of tomato seedlings and resistance to water deficit. *Molecular Plant-Microbe Interactions* 25, 1264–1271.

Mathivanan, N., Prabavathy, V.R. and Vijayanandraj, V.R. (2008) The effect of fungal secondary metabolites on bacterial and fungal pathogens. In: Karlovsky, P. (ed.) *Secondary Metabolites in Soil Ecology*, Vol. 14. Springer, Berlin, pp. 129–140.

Mayers, C.N., Lee, K.C., Moore, C.A., Wong, S.M. and Carr, J.P. (2005) Salicylic acid-induced resistance to *Cucumber mosaic virus* in squash and *Arabidopsis thaliana*: contrasting mechanisms of induction and antiviral action. *Molecular Plant-Microbe Interactions* 18(5), 428–434.

Miller, G., Suzuki, N., Ciftci-Yilmaz, S. and Mittler, R. (2010) Reactive oxygen species homeostasis and signalling during drought and salinity stresses. *Plant, Cell & Environment* 33, 453–467.

Mukherjee, P.K. (2011) Genomics of biological control – whole genome sequencing of two mycoparasitic *Trichoderma* spp. *Current Science* 101(3), 268.

Peleg, Z. and Blumwald, E. (2011) Hormone balance and abiotic stress tolerance in crop plants. *Current Opinion in Plant Biology* 14, 290–295.

Rigamonte, T.A., Pylro, V.S. and Duarte, G.F. (2010) The role of mycorrhization helper bacteria in the establishment and action of ectomycorrhizae associations. *Brazilian Journal of Microbiology* 41, 832–840.

Segarra, G., Van der Ent, S., Trillas, I. and Pieterse, C.M.J. (2009) MYB72, a node of convergence in induced systemic resistance triggered by a fungal and a bacterial beneficial microbe. *Plant Biology* 11, 90–96.

Singh, D.P., Moore, C.A., Gilliland, A. and Carr, J.P. (2004) Activation of multiple anti-viral defence mechanisms by salicylic acid. *Molecular Plant Pathology* 5, 57–63.

Sofo, A., Milella, L. and Tataranni, G. (2010) Effects of *Trichoderma harzianum* strain T-22 on the growth of two *Prunus* rootstocks during the rooting phase. *Journal of Horticultural Science & Biotechnology* 85(6), 497–502.

Sofo, A., Scopa, A., Manfra, M., De Nisco, M., Tenore, G., Trisi, J., Di Fiori, R. and Novellino, E. (2011) *Trichoderma harzianum* strain T-22 induces changes in phytohormone levels in cherry rootstocks (*Prunus cerasus* × *P. canescens*). *Plant Growth Regulation* 65, 421–425.

Sofo, A., Tataranni, G., Xiloyannis, C., Dichio, B. and Scopa, A. (2012) Direct effects of *Trichoderma harzianum* strain T-22 on micropropagated shoots of GiSeLa6® (*Prunus cerasus* × *Prunus canescens*) rootstock. *Environmental and Experimental Botany* 76, 33–38.

Sofo, A., Vitti, A., Nuzzaci, M., Tataranni, G., Scopa, A., Vangronsveld, J., Remans, T., Falasca, G., Altamura, M.M., Degola, F. and Sanità di Toppi, L. (2013) Correlation between hormonal homeostasis and morphogenic responses in *Arabidopsis thaliana* seedlings growing in a Cd/Cu/Zn multi-pollution context. *Physiologia Plantarum* doi:10.1111/ppl.12050.

Tataranni, G., Dichio, B. and Xiloyannis, C. (2012) Soil fungi-plant interaction. In: Montanaro, G. (ed.) *Advances in Selected Plant Physiology Aspects*. InTech Europe, Croatia, pp. 161–188.

Torres, M.A. (2010) ROS in biotic interactions. *Physiologia Plantarum* 138, 414–429.

Tucci, M., Ruocco, M., De Masi, L., De Palma, M. and Lorito, M. (2011) The beneficial effect of *Trichoderma* spp. on tomato is modulated by the plant genotype. *Molecular Plant Pathology* 12(4), 341–354.

Verma, M., Brar, S.K., Tyagi, R.D., Surampalli, R.Y. and Valéro, J.R. (2007) Antagonistic fungi, *Trichoderma* spp.: Panoply of biological control. *Biochemical Engineering Journal* 37, 1–20.

Vitale, A., Cirvilleri, G., Castello, I., Aiello, D. and Polizzi, G. (2012) Evaluation of *Trichoderma harzianum* strain T22 as biological control agent of *Calonectria pauciramosa*. *BioControl* 57, 687–696.

Vitti, A., Nuzzaci, M., Scopa, A., Tataranni, G., Remans, T., Vangronsveld, J. and Sofo, A. (2013) Auxin and cytokinin metabolism and root morphological modifications in *Arabidopsis thaliana* seedlings infected with *Cucumber mosaic virus* (CMV) or exposed to cadmium. *International Journal of Molecular Sciences* 14, 6889–6902.

Wahid, A., Gelani, S., Ashraf, M. and Foolad, M. (2007) Heat tolerance in plants: an overview. *Environmental and Experimental Botany* 61, 199–223.

Woo, S.L. and Lorito, M. (2007) Exploiting the interactions between fungal antagonists, pathogens and the plant for biocontrol. In: Vurro, M. and Gressel J. (eds) *Novel Biotechnologies for Biocontrol Agent Enhancement and Management.* Springer, the Netherlands, pp. 107–130.

Woo, S.L., Scala, F., Ruocco, M. and Lorito, M. (2006) The molecular biology of the interactions between *Trichoderma* spp., phytopathogenic fungi, and plants. *The American Phytopathological Society* 96(2), 181–185.

Yedidia, I., Srivastva, A.K., Kapulnik, Y. and Chet, I. (2001) Effect of *Trichoderma harzianum* on microelement concentrations and increased growth of cucumber plants. *Plant Soil* 235, 235–242.

ID# 15 Role of Microorganisms in Alleviation of Abiotic Stresses for Sustainable Agriculture

Usha Chakraborty,[1]* Bishwanath Chakraborty,[2] Pannalal Dey[1] and Arka Pratim Chakraborty[2]

Plant Biochemistry Laboratory[1] and Immuno-Phytopathology Laboratory[2], Department of Botany, University of North Bengal, Siliguri, India

Abstract

Abiotic stresses affect plants in different ways and are causes of reduction in crop productivity. In order to increase crop productivity it becomes necessary to evolve efficient low-cost technologies for abiotic stress management. Soil microorganisms, surviving in the soil under extreme conditions, have shown great properties, which, if exploited can serve agriculture for increasing and maintaining crop productivity. While it is well established that beneficial soil microorganisms can promote growth and increase productivity through mechanisms such as nutrient mobilization, hormone secretion and disease suppression, it is also becoming increasingly clear that their effects may be more far-reaching. Several studies have reported that soil microorganisms may have mechanisms for alleviation of abiotic stresses in plants such as water and temperature stress, salinity, heavy metals etc. Some of these include tolerance to salinity, drought (*Azospirillum* sp., *Pseudomonas syringae*, *P. fluorescens*, *Bacillus* sp.) and nutrient deficiency (*Bacillus polymyxa*, *Pseudomonas alacaligenes*). Other than bacteria, salinity- and drought-tolerant isolates of *Trichoderma harzianum* and the effect of other strains of *Trichoderma* in amelioration of such abiotic stresses have also been reported. Arbuscular mycorrhizal fungi (*Glomus mosseae*, *G. etunicatum*, *G. intraradices*, *G. fasciculatum*, *G. macrocarpum*, *G. coronatum* etc.) help in alleviating abiotic stresses in different crops by enhancing nutrient uptake (phosphorus, nitrogen, magnesium and calcium), biochemical (accumulation of proline, betaines, polyamines, carbohydrates and antioxidants), physiological, molecular and ultra-structural changes. In the present chapter, we attempt an overview of current knowledge on how plant–rhizobacteria, plant–*Trichoderma* as well as plant–mycorrhiza interactions help in alleviating abiotic stress conditions in different crop systems, which can be used for sustainable agriculture.

15.1 Introduction

Plants, which remain rooted to the soil, are subjected to varying types of abiotic stresses throughout the course of their lifespan and they have to develop mechanisms for coping with such stresses in order to survive. Agriculture is extremely sensitive to environmental changes such as high and low temperatures, drought, flooding, salinity, freezing, change in pH, strong light, UV and heavy metals. Such adverse environmental conditions have a negative impact on crop production, which has the potential to become a major problem for food security, particularly in tropical regions. Abiotic stress management in plants

*E-mail: ucnbu2012@gmail.com

needs to be taken up on a priority basis, keeping in mind that the technology adopted should be ecofriendly as well as cost-effective. This is a major challenge for agriculture. To this end, extensive research is being carried out worldwide to develop strategies to cope with abiotic stresses, through development of heat- and drought-tolerant varieties, shifting the crop calendars, resource management practices etc. (Venkateswarlu and Shanker, 2009). However, while many of these technologies, because of their high cost, may not reach the farmers, there is another strategy that has high potential to help plants withstand abiotic stresses, is highly ecofriendly and cost-effective. This strategy involves the utilization of multi-faceted traits of several microorganisms with an established role in plant growth promotion, nutrient management and disease control. The last two decades have witnessed some reports on the utilization of such microbes for induction of tolerance against abiotic stresses. Yang et al. (2009) proposed the term 'induced systemic tolerance' (IST) for changes in plants induced by plant growth-promoting rhizobacteria (PGPR), which in turn leads to enhanced tolerance to abiotic stress. Plants live in close symbiotic association with a group of fungi, known as arbuscular mycorrhizal fungi (AMF), which confer on the plants an improved ability for nutrient uptake as well as increased tolerance to abiotic and biotic stresses while the fungi themselves acquire a protected ecological niche and plant photosynthates (Ruiz-Lozano, 2003). This interaction of plant–AMF can be successfully exploited for development of abiotic stress management strategies.

Here we review the recent work done in the area of abiotic stress alleviation through microbes and the mechanisms of the observed alleviation.

15.2 Role of Plant Growth-Promoting Rhizobacteria in Alleviation of Abiotic Stresses

Soil, consisting of both inorganic and organic matter, is specialized because of the metabolic activities of the millions of microbes present therein. In spite of the high metabolic activity in the soil, the living microbes occupy less than 5% of the total space. These microorganisms are involved in the decomposition of organic matter as well as solubilization of nutrients, which become available to the roots. However, microbial activity is not uniform throughout the soil, but is concentrated in the region of the root, known as the rhizosphere (Lynch, 1990; Pinton et al., 2001). This is mainly because of the fact that plants, through exudates and sloughed off root tissues and cells provide more than 85% of total organic carbon to the soil (Barber and Martin, 1976). Proper root health, capacity for nutrient acquisition as also enhanced ability to withstand abiotic stresses depends, to a great extent, on the rhizosphere microorganisms (Bowen and Rovira, 1999). It is thus natural that attempts are being made to exploit these microorganisms to develop strategies for achieving maximum crop yield, which is limited by various environmental factors as well as the genetic potential of the crop (Cook, 2002). These soil microorganisms are thus important resources in agriculture and play significant roles in maintenance of life. Many of these bacteria are beneficial, and have the ability to promote growth of plants. These are generally termed as plant growth promoting rhizobacteria (PGPR). These bacteria are known to remain associated with plant roots and act in the soil for growth promotion, either directly, or indirectly, through a number of mechanisms (Glick, 1995). Among rhizobacteria there is a gradient of root proximity and intimacy as follows: (i) bacteria that live in close proximity to the roots, utilizing metabolites leaked from roots as carbon and nitrogen sources; (ii) bacteria colonizing the rhizoplane (root surface); (iii) bacteria residing in root tissue, inhabiting spaces between cortical cells; and (iv) bacteria living inside cells in specialized root structures, or nodules, which generally fall into two groups, the legume-associated rhizobia and the woody plant-associated *Frankia* species (Gray and Smith, 2005). Rhizobacteria that establish inside plant roots, forming more intimate associations, are endophytes. Interestingly, quite a number of these beneficial bacteria, both free living rhizospheric as well as endophytic, have shown the ability of inducing tolerance against abiotic stresses.

Currently not much information is available about the plant–bacteria interactions at the molecular level in rhizosphere as it is not a typical 'gene-for-gene' interaction (Nautiyal et al., 2008). Many gram-positive and -negative PGPR colonize the plant rhizosphere and confer beneficial effects. The beneficial effects are generally brought about by direct and indirect mechanisms, which in turn can be correlated with their ability to form biofilm, chemotaxis and production of exopolysaccharide, IAA and ACC deaminase (Glick, 1995, 2004). An indirect mechanism of increasing plant growth by such bacteria is through suppression of diseases caused by fungi, bacteria, viruses and nematodes (Nautiyal et al., 2008). Rhizosphere bacteria that have been found to have beneficial effects on various plants include species of the genera *Bacillus, Pseudomonas, Erwinia, Caulobacter, Serratia, Arthrobacter, Micrococcus, Flavobacterium, Chromobacterium, Agrobacterium, Hyphomycrobium, Ochrobactrum* and free-living nitrogen-fixing bacteria (Foster et al., 1983; Prithiviraj et al., 2003; Chakraborty et al., 2004). Interestingly, researches in several laboratories have now confirmed that several of these PGPR can also help the plant in withstanding abiotic stresses (Bashan and Levanony, 1990; Bashan and de-Basahan, 2010). Reports are now accumulating on application of PGPR as elicitors for tolerance to abiotic stresses in plants and raising the possibility for incorporation of microbial genes into plant and diverse microbial species (Apse et al., 1999; Yang et al., 2009). The beneficial plant–microbial interactions are very frequent in nature, where PGPR help the plants to overcome various stresses. Microbial communities offer a potentially powerful opportunity for understanding these beneficial interactions. Consequently, changes in the structure or function of microbial communities may have a major impact on ecosystem activities (Khan et al., 2011).

Undoubtedly, water stress is perhaps the most alarming condition faced by a plant, as it affects the water relations of a plant at cellular and whole plant level, decreasing productivity and causing economic losses in agriculture. Inoculation of plants with beneficial micro-organisms which are adapted to adverse conditions promotes plant growth and protects the plants against the deleterious effects of some environmental stresses (Glick et al., 1997; Timmusk and Wagner, 1999; Marulanda et al., 2007, 2008). Bacterial cells accumulate compatible solutes such as amino acids, quaternary amines and sugars that prevent degenerative processes and improve cell growth under adverse osmotic conditions (Potts, 1994). One of the earliest reports on induction of drought tolerance by PGPR was that by Timmusk and Wagner (1999) in *Arabidopsis thaliana*. They showed that *Paenibacillus polymyxa* induced tolerance against drought by over-expression of drought-responsive gene *ERD15 (EARLY RESPONSE TO DEHYDRATION 15)*. It was further shown in wheat that inoculation of *Azospirillum brasilense* Sp245 under low water regime resulted in a better water status and an additional 'elastic adjustment' leading to better grain yield and mineral quality (Creus et al., 2004). Plants treated with exopolysaccharide (EPS)-producing bacteria display increased resistance to water stress (Bensalim et al., 1998). The EPS-producing strain *Pseudomonas putida* strain GAP-P45 forms biofilm on the root surface of sunflower seedlings and imparts tolerance to plants against drought stress. The inoculated seedlings showed improved soil aggregation and root-adhering soil and higher relative water content (RWC) in the leaves (Sandhya et al., 2009, 2011).

Soil salinization occurs due to accumulation of dissolved salts in the soil leading to plant growth inhibition (Carmen and Roberto, 2012). Increasing salinity is a major environmental stress and is perceived as a substantial constraint to crop production. It is expected that with increasing salinization of cultivable land, within the next 25 years there would be a land loss of 30%, which could increase up to 50% by the middle of this century linked with devastating global effects (Wang et al., 2003). Salt tolerance in plants is a very complex character, linked to stress and other developmental responses. Use of microbes provides a useful technology for improving stress-tolerance capacity in plants (Lynch and Thompson, 1982; Cossins, 1994). Stimulation of root growth and effective root area for enhanced water and nutrient uptake is the most important stress-management tool because a healthy, strong and proliferated root system plays a major role in helping the plant to maintain optimal growth and development under stress conditions (Lynch and Thompson, 1982; Gibson et al., 1994).

With global warming and other related phenomena, earth is now witnessing several extreme temperature conditions – from high to very high. This constitutes temperature stress and is now a major concern in agriculture. Water scarcity also leads to temperature increase. Similar to other stresses, microbes, which themselves are thermotolerant, can also be used for conferring temperature tolerance to crops. The ability of a thermotolerant strain of *Pseudomonas* AKM-P6 was used to alleviate the heat stress in sorghum seedlings (Ali et al., 2009). Several workers have also reported the ability of cold-tolerant bacteria to induce cold tolerance in plants (Barka et al., 2006; Chang et al., 2007; Selvakumar et al., 2008a, b; Mishra et al., 2009a, b). Since ice nucleation has been recognized as a cause of frost damage of plants, attempts are now being made to identify bacteria from the phyllosphere which have low ice nucleating activity and use them as foliar sprays with a view to overcome frost damage (Selvakumar et al., 2012).

Table 15.1 presents a list of plant growth promoting bacteria reported for alleviation of abiotic stresses.

15.3 Plant Growth-Promoting Fungi in Abiotic Stress Amelioration

15.3.1 Mycorrhizal fungi

Other than beneficial bacteria, either free living or symbiotic, most terrestrial plants have a symbiotic association in their root system with a group of fungi known as arbuscular mycorrhizal fungi (AMF). Such plants have an improved ability for nutrient acquisition and exhibit enhanced tolerance to different stresses while the fungus acquires a protected ecological niche and plant photosynthates (Smith and Read, 2008). Several studies have revealed that AMF spore population in the soil increases during both water and salt stress, resulting in enhanced tolerance (Augé, 2001; Ruíz-Lozano, 2003; Ruíz-Lozano et al., 2006; Jahromi et al., 2008; Estrada et al., 2012). Although it is clear that AMF mitigate growth reduction caused by osmotic stress, the mechanism involved remains unresolved (Ruíz-Lozano et al., 2012). While studying the influence of salinity on mycorrhizal association, it was revealed that the fungal development was affected, reducing fungal mycelia formation and host root colonization (Giri et al., 2007; Sheng, M. et al., 2008). On the other hand, contrary to these reports, a few other studies reported no reduction or even increasing fungal development (Aliasgharzadeh et al., 2001; Yamato et al., 2008). Probably, during the evolutionary process mechanisms evolved that allow specific AMF to develop a higher tolerance to salinity. In fact, it has been reported by several workers that mycorrhizal fungi occur naturally in saline environments (Juniper and Abbott, 1993) and Copeman et al. (1996) suggested that the observed differences in fungal behaviour and efficiency can be due to the origin of the AMF (Ruíz-Lozano and Azcón, 2000; Estrada et al., 2013). Recent reports on abiotic stress alleviation in different plants by AMF have been listed in Table 15.2.

15.3.2 Trichoderma

Trichoderma, traditionally known as a biocontrol agent, has in recent times been used for plant growth promotion and alleviation against abiotic stresses, bringing into focus its multifunctional traits. Several reports are now available indicating the ability of different species of *Trichoderma* to confer tolerance to abiotic stresses in plants, though the mechanisms involved still remain unclear (Donoso et al., 2008; Bae et al., 2009). Treatment of tomato seeds with *T. harzianum* was reported by Mastouri et al. (2010) to have several beneficial effects: from accelerating seed germination and enhancing seedling vigour to amelioration of abiotic stresses such as water, salinity and heat. They suggested that *T. harzianum*

Table 15.1. Bacteria listed as PGPR as well as conferring abiotic stress tolerance.

Bacterial isolates	Plant	Type of stress	References
Azospirillum brasilense	Hordeum vulgare	Salt	Hamaoui et al. (2001)
	Lactuca sativa		Barassi et al. (2006)
	Pisum sativum		Dardanelli et al. (2008)
	Cicer arietinum		Omar et al. (2009)
Bacillus amylolequifaciens, B. insolitus Microbacterium sp. Pseudomonas syringae	Triticum aestivum	Salt	Ashraf et al. (2004)
P. fluorescens	Arachis hypogaea	Salt	Saravanakumar and Samiyappan (2007)
B. subtilis	Arabidopsis thaliana	Salt	Zhang, H. et al. (2008)
Achromobacter piechaudii	Lycopersicon esculentum	Salt, drought	Mayak et al. (2004a, b)
Azospirillum brasilense	Zea mays	Drought	Casanovas et al. (2002)
Pseudomonas spp.			Sandhya et al. (2010)
Rhizobium sp.	Helianthus annuus	Drought	Alami et al. (2000)
P. putida P45			Sandhya et al. (2009)
Bacillus	Lactuca sativa	Drought	Arkhipova et al. (2007)
P. mendocina			Kohler (2008)
B. megaterium	Trifolium	Drought	Marulanda et al. (2007)
Pseudomonas sp.	Pisum sativum	Drought	Arshad et al. (2008)
Variovorax paradoxus			Belimov et al. (2009)
Paenibacillus polymyxa Rhizobium tropici	Vigna radiata	Drought	Figueiredo et al. (2008)
Pseudomonas spp.	Asparagus	Drought	Liddycoat et al. (2009)
Azospirillum sp.	Triticum aestivum	Drought	Creus et al. (2004)
B. safensis Ochrobactrum pseudogregnonense			Chakraborty et al. (2013)
Enterobacter cloacae P. putida	Lycopersicon esculentum	Flooding	Grichko and Glick (2001)
Pseudomonas sp. AMK-P6	Sorghum bicolor	Heat	Ali et al. (2009)
P. putida	Brassica napus	Cold	Chang et al. (2007)
Burkholderia phytofirmans	Vitis vinifera	Cold and heat	Bakra et al. (2006)
Sanguibacter sp. Pseudomonas sp.	Nicotiana tabacum	Heavy metals	Mastretta et al. (2009)
B. subtilis Pantoea agglomerans	Avena sativa	Heavy metals	Pishchik et al. (2009)
P. fluorescens Microbacterium sp.	Brassica napus	Heavy metals	Sheng, X. et al. (2008)
Methylobacterium oryzae Burkholderia sp.	Lycopersicon esculentum	Ni and Cd	Madhaiyan et al. (2007)
B. subtilis, Bacillus sp., B. megaterium	Oryza sativa	Iron toxicity	Terré et al. (2007)

probably acts by inducing protection in the host against oxidative stress. Delayed wilt response to drought was achieved in rice seedlings by treatment with *Trichoderma*, along with a concomitant delay in water stress-induced physiological changes such as leaf greenness, stomatal conductance and photosynthesis (Shukla *et al.*, 2012a). Among all their isolates Th 56 was most effective in inducing tolerance against drought, as well as direct promotion of seedling and root growth in rice plants.

Table 15.2. List of AMF along with the plants conferring abiotic stress tolerance.

Arbuscular mycorrhizal fungi	Plant	Type of stress	References
Glomus etunicatum	Carthamus tinctorius	Salt	Abbaspour (2010)
G. intraradices	Zea mays	Salt	Estrada et al. (2013)
	Trigonella foenum-graecum		Evelin et al. (2013)
	Fragaria ananassa		Fan et al. (2010)
G. viscosum	Medicago sativa L.	Salt	Campanelli et al. (2013)
G. etunicatum	Brachiaria humidicola	Salt	Mergulhão et al. (2002)
G. mosseae	Piper nigrum	Heavy metals	Abdel Latef (2011)
G. mosseae	Zea mays	Heavy metals	Abdelmoneim and Almagrabi (2013)
Aculaospora laevis			
G. intraradices	Thlaspi sp.	Heavy metals	Hildebrandt et al. (2007)
G. etunicatum	–	Heavy metals	Pawlowska and Charvat (2004)
G. intraradices			
G. macrocarpum	Zea mays	Heavy metals	de Andrade and da Silveira (2008)
G. mosseae	Dalbergia sissoo, Acacia nilotica	Nutrient deficiency	Kaushik and Mandal (2005)
		Nutrient	Wu et al. (2011)
	Poncirus trifoliata	Heat	Wu (2011)
G. mosseae, G. sp. R10	Fragaria ananassa	Heat and cold	Matsubara et al. (2004)
G. aggregatum			
G. fasciculatum			
Gigaspora margarita			
G. mosseae	Poncirus trifoliata	Drought	Fan and Liu (2011)
G. deserticola	Pepper	Drought	Garmendia et al. (2005)
G. intraradices	Rosa hybrida L.	Drought	Pinior et al. (2005)
	Lactuca sativa		Alguacil et al. (2009)
	Cicer arietinum		Sohrabi et al. (2012)
G. etunicatum, G. versiform	Cicer arietinum	Drought	Sohrabi et al. (2012)

15.4 Mechanisms of Stress Alleviation by Microbes

Abiotic stresses caused due to water deficit, excess salt, extreme temperature variations and other environmental conditions affect plants at various levels. Production of reactive oxygen species (ROS) during such stresses leads to cellular damage involving metabolic toxicity, membrane damage and also inhibition of photosynthesis, changes in hormone levels etc. However, plants are able to overcome these stresses to a great extent, mainly due to the array of defence mechanisms that can become activated under such conditions. Some of the defence mechanisms include regulation of plant hormones, ROS scavenging mechanisms, compartmentation or exclusion of excess ions which cause osmotic disbalance as well as accumulation of metabolites, which protect the cells against osmotic damage (Mahajan and Tuteja, 2005; Parida and Das, 2005; Santner et al., 2009; Shao et al., 2009; Des Marais and Juenger, 2010).

Several mechanisms have been proposed for the observed alleviation of abiotic stresses in plants by different microbes. Some of these have been discussed below.

15.4.1 Hormones

Since lateral root formation is associated with a number of abiotic stresses, and auxin induces lateral root formation, it is believed that many of the responses induced by the stresses may be mediated through the action of auxins. Changes in auxin metabolism induced by abiotic stresses have been mostly shown to be through changes in its transport and catabolism (Carmen and Roberto, 2012). It has been reported that expression of PIN genes are altered during drought or salinity affecting auxin transport and inhibiting polar auxin transport (Potters et al., 2009). It is also probable that the hydrolysis of auxin conjugates leads to increases in free auxin, which in turn inhibits root elongation and provides protection against stresses as has been reported in *Arabidopsis* by Junghans et al. (2006).

Inoculation with *Azospirillum*, a plant-growth promoter, consistently led to changes in root morphology, which has been linked to the production of growth hormones, with auxin being the most important (Spaepen et al., 2008). The involvement of auxin in enhancement of lateral growth was further confirmed by comparing IAA-attenuate mutants with their parental wild types. Enhanced secretion of growth hormones in maize inoculated with *Pseudomonas fluorescens* was correlated to elevated tolerance to drought by Ansary et al. (2012). IAA-mediated improvement in root growth may be direct, or again may be through the reduction in levels of ethylene as a relationship exists between IAA and ACC, which is the precursor of ethylene (Lugtenberg and Kamilova, 2009).

Since ethylene is involved in several regulatory processes, its biosynthesis is under both transcriptional and post-transcriptional regulation, being affected by a number of environmental factors (Hardoim et al., 2008). It has been reported that under stress, as an immediate response, a small amount of ethylene is produced, which promotes the activity of stress-related genes. This is followed by the second phase, within a few days of the stressful stimulus, where a larger amount of ethylene is produced, leading to the specific growth-inhibitory activities of ethylene such as chlorosis, senescence and abscission. The beneficial effects of several plant growth-promoting bacteria are due to their ability to produce ACC deaminase, which degrades ACC to nitrogen with release of energy. Thus ACC availability for ethylene production becomes lessened, which in turn leads to a lessening of the deleterious effects of ethylene (Glick et al., 2007). Such bacteria that inhibit ethylene biosynthesis induce better roots, which in turn would be of help to the plants in increasing their water uptake capacity under drought.

Since the ACC deaminase-producing bacteria are soil inhabiting, living in close proximity to the root system, it is quite probable that the ACC is exuded into the soil by the roots and the bacteria utilize it for their growth purposes after degrading it initially. Thus both the bacteria and the plant would benefit by this continuous ACC deaminase function (Glick et al., 1998).

Such PGPR strains that promote growth as well as alleviate abiotic stresses have been used in several cases. Growth promotion of tomato seedling under salinity stress by *Achromobacter piechaudii*, which is an ACC deaminase producer, was reported by Mayak et al. (2004a). Similar results have also been reported with *P. fluorescens* in maize (Kausar and Shahzad, 2006), as well as *P. putida* UW4 and 14 other halotolerant bacteria in canola (Siddikee et al., 2010). The ability of ACC deaminase-producing *Pseudomonas* spp. in drought alleviation leading to better growth and yield in pea was recorded by Arshad et al. (2008).

Comparison of nodulation of *Medicago trunculata* by IAA over-producing strain (Mt-RD64) of *Sinorhizobium meliloti* with wild-type strain under salt stress revealed that in the former, IAA accumulation in nodules and roots was much higher than in shoots (Bianco and Defez, 2009). It was shown by transcriptional analysis of ethylene signalling genes that in the Mt-RD64 plants, under salt stress, this pathway was not induced, and hence less damage was obtained. In another study, it was also shown that *Bacillus* sp. TW4, an ACC deaminase-producing bacterium, could protect pepper plants from certain abiotic stresses. This protection was correlated with down-regulation of genes linked with ethylene metabolism such as *caACCO* (encoding ACC oxidase) and *caLTPI* (Sziderics et al., 2007). Nautiyal et al. (2013) reported 0.37- and 0.80-fold less expression of ethylene responsive element binding proteins (EREBP) in salt-stressed (S) and 1.5-fold more expression in B+S in hydroponic condition and they emphasized the role of *NBRISN13* in salt stress management similar to over-expression of

EREBP in salt-tolerant transgenic *Arabidopsis* plants (Zhang *et al.*, 2012). EREBP are reported to have a role in hormone metabolism, ethylene signal transduction, disease resistance response and abiotic stress conditions (Zhang *et al.*, 2010).

Increased ABA content in leaves and decreased endogenous cytokinin levels in water-stressed plants point to their antagonism resulting from metabolic interactions due, in part, to their sharing a common biosynthetic origin (Cowan *et al.*, 1999; Figueiredo *et al.*, 2008). Interaction of hormone-signalling pathways occurs during a plant's response to different stresses with cross-communication between SA, JA and ethylene involved in defence responses, ABA associated with plant development and stresses and IAA and gibberellins involved in root development and growth. Gibberellin signalling occurs by degradation of DELLA proteins, which are growth-repressing proteins. Other than ACC deaminase-producing bacteria, it has also been reported that certain *Trichoderma* species also promote plant growth through the activity of ACC deaminase, reducing ethylene biosynthesis, which is also linked to degradation of DELLA proteins. Reciprocal regulation of ethylene and IAA biosynthesis also occurs in the roots and, probably, IAA from *Trichoderma* contributes to exogenous IAA-stimulated ethylene synthesis through the activation of ACC synthase, which in turn triggers an increase in ABA biosynthesis (Stepanova *et al.*, 2007).

15.4.2 Protective metabolites

Certain specific metabolites such as specific proteins, glycine betaines, certain amino acids, amides, imino acids and polyamines generally accumulate during drought and salt tolerance in plants (Parida and Das, 2005; Shukla *et al.*, 2012a). When plants face salt stress, proline accumulates in the cell and helps substantially in cytoplasmic osmotic adjustment (Leigh *et al.*, 1981). Under salt stress, proline also helps the plant cell by stabilizing subcellular structures such as membranes and proteins, scavenging free radicals and buffering cellular redox potential (Ashraf and Foolad, 2007).

High accumulation of proline during stress documented in several plant species might be either due to increased biosynthesis or decreased degradation. Shukla *et al.* (2012b) showed that PGPR-treated plants showed low proline content under 100 mM NaCl stress compared to uninoculated plants, but the level was higher than the basal level of proline in uninoculated plants under no salt stress. They suggested that PGPR-treated plants do not face much salt stress, therefore the proline accumulation is less in the presence of PGPR.

It is clear from several studies using mutants or transgenics, that proline metabolism during stress is very complex and it may play multiple roles to help plants survive under stress conditions. It may provide a carbon and nitrogen store, act as osmolyte or exhibit an antioxidant property for scavenging ROS. Another proposed function of proline is to act as molecular chaperone and stabilize protein structures, as well as enhancing certain enzyme activities (Kavi Kishor *et al.*, 2005; Verbruggen and Hermans, 2008).

In cases where inoculation of plants subjected to abiotic stresses with PGPR led to alleviation of these stresses, increased proline biosynthesis was observed by several authors (Kohler *et al.*, 2009; Jha *et al.*, 2010; Sandhya *et al.*, 2010; Vardharajula *et al.*, 2011). However, increased synthesis of proline and/or other compatible solutes which provide protection to plants against osmotic stress requires additional energy and may occur at the expense of growth (Munns and Tester, 2008). Increased proline content was also correlated with reduced salt-stress symptoms such as chlorosis, necrosis and drying in *Medicago trunclata* inoculated by IAA over-producing strain (Mt-RD64) of *Sinorhizobium meliloti* in comparison with wild-type strain (Mt-1021) under salt stress of 150 mM NaCl (Bianco and Defez, 2009). Transgenic *A. thaliana* containing *proBA* genes from *Bacillus subtilis* showed increased tolerance to osmotic stresses as well as enhanced levels of proline (Chen *et al.*, 2007). Increased production of proline along with decreased electrolyte leakage, maintenance of relative water content of leaves and selective uptake of K$^+$ ions resulted in salt tolerance in *Zea mays* co-inoculated with *Rhizobium* and *Pseudomonas* (Bano and Fatima, 2009).

Other than bacteria which induce higher proline accumulation in plants under abiotic stresses, AMF colonization has also been reported to increase production of proline. At different salt concentrations ranging from 50 to 200mM NaCl, higher proline concentration was reported in soybean plants inoculated with AMF in comparison to those that were not; in AMF plants again, higher accumulation of proline was obtained in roots than in shoots. Sharifi et al. (2007) opined that higher proline in roots may be due to the necessity of maintaining osmotic balance at the primary site of water absorption. Contrary to the above, it was suggested in an earlier study that under different levels of salinity, inoculation with AMF in *Vicia faba* did not enhance proline accumulation (Rabie and Almadini, 2005). It has further been suggested that higher accumulation of proline is not correlated to tolerance to salinity but rather to stressed condition, as potassium ions were more involved in maintaining osmotic balance (Wang et al., 2004). Results from studies on various crops involving abiotic stresses, specifically osmotic stresses such as salinity or drought, accumulation of proline and role of AMF in inducing proline accumulation have not yielded any conclusive evidence regarding induction of proline by AMF. The inconsistency of the different reports points to a complex mechanism operating in plants either during direct tolerance to stresses or tolerance induced by microorganisms, specially AMF, which needs to be elucidated further for a thorough understanding. Besides proline, betaines, which are quaternary ammonium compounds and N-methylated derivatives of amino acids, are also osmolytes which generally accumulate under salt stress. Besides being non-toxic cellular osmolytes, they can also stabilize the structures and activities of enzymes and protein complexes and maintain the integrity of membranes against the damaging effects of excessive salt (Gorham, 1995). Interestingly, it was observed that accumulation of these glycine betaines increased about two-fold in AMF-inoculated plants under salt stress as against non-AMF plants (Al-Garni, 2006).

While determining the role of free amino acids in tolerance to drought, it was reported that when cacao seedlings were subjected to drought, contents of amino acids such as proline, γ-aminobutyric acid (GABA), arginine, histidine, leucine and valine increased, but when colonized by *T. hamatum* DIS 219b (drought tolerant) there was no increase in accumulation (Bae et al., 2009). Similar results were also reported in rice by Shukla et al. (2012a), where accumulation of drought-induced metabolites was less when tolerance to drought was induced by treatment with *T. harzianum*. They suggested that regulation of different physiological pathways would be one of the mechanisms of inducing tolerance to drought by *T. harzianum*.

Another group of small cationic molecules reported to be involved in a plant's response to different environmental stresses such as salinity and other osmotic stresses are the polyamines, consisting mainly of three compounds: putrescine, spermidine and spermine (Besford et al., 1993; Delauney and Verma, 1993). The probable role of polyamines in regulation of root development under saline conditions has also been established (Couée et al., 2004). An increase in total free polyamine pools induced by inoculation with AMF in plants under salinity has been reported in several plants. In *Lotus glaber* plants, differences in polyamine concentrations were apparent between mycorrhizal salt-tolerant and salt-sensitive genotypes, between mycorrhizal and non-mycorrhizal plants as well as between roots and shoots (Sannazzaro et al., 2007). Authors suggested that this modulation of polyamine pools may be one of the mechanisms in the host–AMF interaction leading to tolerance to salt stress.

Soluble sugars such as trehalose and sugar alcohol-mannitol, play important roles in maintaining the osmotic potential of plants under osmotic stresses such as drought and salinity, thereby affording protection against these stresses. Trehalose is also the main storage carbohydrate in AMF and is present in the extra radical mycelium as well as in spores (Becard et al., 1991). Following inoculation with AMF, trehalose accumulation becomes enhanced in plants and may be involved in protection from abiotic stresses. Trehalose is perceived to protect biological structures from damage due to desiccation (Hoekstra et al., 1992; Schubert et al., 1992). This is very important in rhizobial interaction with legumes during signalling for plant growth as well as adaptation against stresses (Lopez et al., 2008; Suarez et al., 2008). Inoculation of *Phaseolus vulgaris* plants with *Rhizobium etli* over-expressing trehalose-6-phosphate synthase gene had more nodules, increased nitrogenase activity and tolerance to drought in comparison with those inoculated with wild-type *R. etli*. In such cases where trehalose-6-phosphate synthase gene was over-expressed, microarray analysis of 7200 expressed sequence tags from nodules revealed up-regulation of genes involved in stress tolerance, which indicates a probable signalling mechanism of trehalose (Shukla et al., 2012b). While studying the effect of salt-stress on trehalose content in AMF, it was observed that salt stress of 0.5 M NaCl did not induce any additional accumulation of trehalose in the extra-radical hyphae of *Glomus intraradices*. However, moderate transient activation of enzymes involved in trehalose metabolism such as trehalose-6-phosphate phosphatase and neutral trehalase were observed (Ocon et al., 2007). Thus, reports from different studies strengthen the view that trehalose can be exploited as an osmoprotectant, which can provide protection to different plants against abiotic stresses. Since microorganisms have the ability to accumulate trehalose, they may have important roles in protection to abiotic stresses provided by different groups of microorganisms such as AMF, symbiotic bacteria or free-living PGPR.

15.4.3 Maintaining ion homoeostasis

Salinity causes an imbalance in the ratio of ion homoeostasis in the plant system. With excess NaCl in the soil, it is quite natural that the uptake of Na^+ is enhanced while that of K^+ is reduced. Potassium is essential for several metabolic processes, such as stomatal movements and protein synthesis, where it is required for the binding of tRNA to ribosomes (Blaha et al., 2000). With Na^+ competing with K^+, and a subsequent higher $Na^+:K^+$ ratio than normal, these processes are disrupted (Giri et al., 2007).

Plants try to maintain low salt composition in the cytosol by extrusion through the plasma membrane using the SOS pathway or by scavenging in the vacuole through NHX1 antiporters. Salinity impedes the ratio of Ca^{2+} and K^+ in the cell. However, an increase in K^+ concentration can alleviate the deleterious effect of salinity on growth and yield (Giri et al., 2007). In a study conducted by Shukla et al. (2012b), authors obtained a low Na^+ content and a higher K^+ content in the presence of PGPR under salinity leading to a higher $K^+:Na^+$ ratio in plants under salt stress due to the restricted Na^+ uptake and enhanced K^+ uptake. Similarly, several PGPR are reported to reduce the salt toxicity in various other plants by lowering the Na^+ concentration and increasing the K^+ concentration (Hamdia et al., 2004; Nadeem et al., 2006; Yildirim et al., 2006; Bano and Fatima, 2009; Kohler et al., 2009).

Expression of salt-responsive genes *NHX1*, *SOS1*, *BZ8*, *SAPK4* and *SNRK2* has been shown to be enhanced in the presence of salt. *NHX1* and *SOS1* were reported to be involved in Na^+/H^+ exchange, and reduced cellular Na^+ (Apse et al., 1999; Shi et al., 2000, 2003; Diedhiou et al., 2008; Hussain et al., 2008; Ying et al., 2011). Repression of *NHX1* and *SOS1* to 0.56-fold under B+S as compared to S in hydroponics and similar expression in S and C under soil conditions implies the supportive role of bacterial strain SN13 interaction with rice plant. The *SAPK4* gene acts as regulatory factor in salt stress acclimatization, ion homoeostasis, growth and development, therefore its 1.3-fold up-regulation in SN13 alone implies SN13 mediated low $Na^+ Cl^-$ intake under the hydroponic system (Diedhiou et al., 2008).

Ability of volatile organic compounds (VOCs) secreted by certain bacteria have been known to inhibit pathogens and thus help in the biocontrol mechanism as well as plant growth promotion. This trait of such bacteria has long been considered as a characteristic of PGPR. It is also now established that besides direct inhibition of pathogens, the VOCs may also act as signals for inducing systemic resistance in the plants. Moreover, under salt stress conditions, in *Arabidopsis*, the expression of HKT1, which is involved in entry of Na^+ to roots, is down-regulated by the VOCs secreted by *Bacillus subtilis* GB03, which are responsible for stress alleviation. Down-regulation of HKT1 leads to a balanced distribution of Na⁺ in the cells (Zhang, H. et al., 2008). However, HKT1 adjusts Na⁺ and K⁺ levels differentially in roots and shoots. This has been corroborated by studies using mutants and it is now established clearly that bacterial VOCs control Na⁺ homoeostasis under salinity.

Under saline conditions, enhancing the uptake of K⁺ and inhibition of Na⁺ to shoot tissues may be one of the mechanisms for ion homoeostasis achieved in plants by mycorrhizal inoculation (Giri et al., 2007; Sharifi et al., 2007; Zuccarini and Okurowska, 2008). Enhancing absorption of K leads to increased K⁺:Na⁺ ratio in roots and shoots of plants inoculated with AMF (Giri et al., 2007). In such cases, activities of K⁺ and Na⁺ transporters, as well as H⁺ pumps, all of which contribute to the transport of various cations, are regulated (Parida and Das, 2005). Increased activity of Na⁺/H⁺ antiporter ensures rapid compartmentalization of Na⁺ in the vacuoles or apoplasm (Ouziad et al., 2006). Increased Na⁺ and decreased K⁺, which results from excessive salinity, disrupts different enzymatic processes requiring K, as also protein synthesis. This is generally overcome by the higher K⁺:Na⁺ ratio (Colla et al., 2008).

15.4.4 Nutrient up-take enhancement

Stress tolerance of plants depends to a great degree on the plant's health, and plants with enhanced nutrient uptake capacities have been shown to have greater tolerance. It is thus not surprising that several bacteria and fungi, which have the ability to improve growth through different mechanisms, including enhanced nutrient uptake, also have the ability to induce tolerance against abiotic stresses. One of the causes for the adverse effects of abiotic stresses on plant growth and development being the imbalance of nutrient uptake and metabolism, it was reported in several cases that addition of macronutrients exogenously leads to a certain degree of stress alleviation (Endris and Mohammed, 2007; Heidari and Jamshid, 2010). Phosphorus nutrition is one of the most important in plants, second only to nitrogen, as it is not only involved in metabolic processes but is also a part of the structural make-up of plants, being components of membrane phospholipids, phosphoproteins, as well as nucleotides. Salinity tends to decrease phosphorus uptake and accumulation in plants, leading to deficiency symptoms (Navarro et al., 2001; Rogers et al., 2003; Parida and Das, 2004). In the soil, phosphorus can exist either as inorganic salts or as part of the organic composition, but has limited mobility and solubility (Hayat et al., 2010). Hence, solubilization and mobilization of insoluble phosphates into soluble forms generally improves plant growth as well as ability to withstand abiotic stresses. Soil microorganisms, mainly the group of bacteria known as PGPR, have the ability to convert insoluble phosphates into soluble, available forms and thus increase phosphorus uptake by plants, leading to better growth. Some of the insoluble phosphorous forms which are solubilized by PGPR are rock phosphate, tri- and dicalcium phosphate (Chakraborty et al., 2009, 2010; Khan et al., 2009; Richardson et al., 2009). The organic forms of phosphate are initially converted to the inorganic forms, which are further solubilized for uptake. Synthesis of acid and alkaline phosphatases by the bacteria and their secretion into soil, where they hydrolyse the insoluble phosphates into soluble and available forms, is considered a major mechanism of phosphorous mobilization by PGPR. This was shown in the case of IAA- over-producing *Sinorhizobium meliloti* RD64, which had better phosphate solubilizing activity linked to high phosphatase activity and improved plant growth-promoting ability in comparison to *Mt*-1021 (Bianco and Defez, 2010). The IAA-over-producing strain could also protect the plants against abiotic stresses such as salinity.

Phosphorus mobilization is also one of the characteristic traits of mycorrhizal fungi which improve nutrient uptake by the plant. In a study conducted on *Trifolium alexandrium*, it was observed that the decrease in phosphorus concentration in the plants due to salinity could be overcome by mycorrhization (Shokri and Maadi, 2009). It has been suggested that improved growth rate, enhanced nitrogen fixation and nodulation, as well as higher antioxidant activity obtained in AM-inoculated legumes may be correlated to improved phosphorus nutrition (Feng et al., 2002; Alguacil et al., 2003; Garg and Manchanda, 2008). It is also probable that the negative effects of excess Na⁺ and Cl⁻ ions during salinity stress may be reduced by AMF due to increased phosphate uptake, which helps to maintain vacuolar membrane integrity facilitating vacuolar compartmentalization of such ions (Cantrell and Linderman, 2001).

15.4.5 Antioxidant mechanisms

Normal cellular metabolism such as respiration and photosynthesis release ROS in very low quantities as by-products which have certain signalling roles during growth and development. However, the concentration of such ROS increase during various abiotic stresses and at high concentrations these become toxic to cellular metabolism and cause cell and tissue damage. Plants have evolved an array of mechanisms to counteract these ROS, which help in scavenging the ROS and minimize their damage. Chakraborty et al. (2013), from their studies on wheat, showed that in susceptible varieties GY and MW, both superoxide dismutase (SOD) and catalase (CAT) declined from the onset of drought; application of either *Bacillus safensis* or *Ochrobactrum pseudogregnonense* helped to maintain higher levels of the two enzymes and thus helped alleviate drought. Besides, even in those varieties where there was an initial increase in enzyme activities followed by a decline, bacterial treatments helped maintain higher levels of activities of these enzymes. One of the mechanisms of alleviation of drought may be the ability to tilt the balance from oxidatively stressed condition to a more antioxidative state, thereby providing tolerance against stress.

In lettuce plants subjected to salinity and inoculated with PGPR strains, alleviation of stress was obtained along with a concomitant induction of two antioxidant enzymes, catalase and peroxidase. Kohler et al. (2010) suggested that these enzymes could have a role in the observed stress alleviation by the PGPR. Inoculation with *Pseudomonas mendocina* and fertilization led to increases of about 30% in plant growth, whereas a decrease in growth was obtained under salinity stress. However, better plant growth as evidenced by greater shoot biomass was obtained in bacterized plants in comparison to non-bacterized ones under both medium and high salt levels. It was reported by Bianco and Defez (2009) that under high salt treatments, *S. meliloti* Mt-RD64 plants showed much less oxidative damage (reduced chlorosis, necrosis, and drying) compared with salt-stressed *S. meliloti* Mt-1021 treated plants. They correlated this effect to enhanced activities of antioxidant enzymes such as peroxidase, ascorbate peroxidase, glutathione reductase and superoxide dismutase. Under salt-stressed conditions 1.2-fold higher expression of MAPK5 in comparison to control and its repression to 0.28-fold in presence of *B. amyloliquefaciens* SN13, state its (SN13) ability to ameliorate salt stress in soil (Nautiyal et al., 2013). This is also in accordance to the prior report of Lee et al. (2011). Increased transcript level of NADP-Me2 (1.35-fold) in hydroponic system under salt condition (S) and 0.57-fold repression in soil system and expression of NADP-Me2 in *B. amyloliquefaciens*-treated plants under normal and saline-stressed conditions emphasize the supportive role of bacteria in ameliorating salt stress (Nautiyal et al., 2013) as NADP-malic enzyme is involved in different metabolic pathways and provides osmotic tolerance and plant defence through malate degradation and stomatal conductance (Liu et al., 2007). It was further reported by Nautiyal et al. (2013) that inhibitory effects of over-produced ROS by salt and osmotic stress are suppressed by up-regulation of GIG and CAT in B+S treatment as compared to control. The observed up-regulation of GIG (2.95-fold) in salt-stressed condition in hydroponics was in accordance with a previous report (Nadeem et al., 2007). Induction of antioxidant level, a prerequisite for resistance mechanism (Chakraborty et al., 2011), is well demonstrated by the up-regulation of catalase (1.6-fold) under hydroponic conditions thereby emphasizing the role of SN13 as an elicitor which enhances defence enzyme activities and confers resistance against salt stress.

Nutrients and antioxidants have been demonstrated to act together in synergy to reduce ROS level more effectively than antioxidant alone (Hussain et al., 2008). Since PGPR enables plants to increase the uptake of nutrients, authors hypothesized that treatment of SN13 may help the rice plants in maintaining adequate nutrition as evident with up-regulation of BADH in B+S in hydroponics (Shirasawa et al., 2006). Up-regulation of SERK1 (a leucine-rich receptor-like kinase) in salt (S) and down-regulation in B (in hydroponics) and more or less equal expression in B+S as compared to C emphasizes its role in salt stress alleviation.

However, in a study by Sandhya et al. (2010), it was observed that alleviation of drought in maize plants inoculated with species of *Pseudomonas*, namely *P. entomophila*, *P. stutzeri*, *P. putida*, *P. syringae* and *P. montelli*, could not be correlated with increased activities of antioxidative enzymes; rather, there was a lowering of activities. In the same vein, Omar et al. (2009) also reported that in barley salt stress led to antioxidative enzymes CAT and POX.

Plants inoculated with *Azospirillum brasilense* under salinity stress improved crop growth and ameliorated the deleterious effects of salt stress. However, this amelioration could not be correlated with increases in antioxidative enzymes since bacterization did not increase activities of these enzymes. Authors who obtained such results explained that since bacterial inoculation alleviated salinity stress, plants were under less stressed conditions, which was reflected in lower antioxidative enzyme activities.

It has been proposed that alleviation of abiotic stress by *T. harzianum* T-22 could be by controlling the damage caused by the ROS (Mastouri et al., 2010). An increase in level of lipid peroxide content in young seedlings was found under osmotic stress, but T-22-treated seedlings had significantly lower levels of lipid peroxides. Authors proposed a model wherein they suggested that *T. harzianum* strain T-22 increases seedling vigour and ameliorates stress by inducing plant physiological protection against oxidative damage. Interestingly, an increase in the level of several families of protective proteins, including GSH-dependent enzymes such as glutathione reductase and glutathione S-transferase, in *Trichoderma*-treated maize and other seedlings has been reported previously (Alfano et al., 2006, 2007; Bae et al., 2006; Bailey et al., 2006; Shoresh and Harman, 2008). The mechanisms whereby *Trichoderma* spp. induce such changes are not known; however, enhanced ROS level could act as a signal to regulate expression of some of the related genes. A transient increase in intracellular ROS has been detected 5–10 min after treating soybean cell culture with culture filtrate of *T. atroviride* (Navazio et al., 2007). Such signals, along with Ca^{++} signalling (Navazio et al., 2007), can induce plant ROS-scavenging mechanisms (Mittler, 2002), resulting in elevated protection against the oxidative damage.

15.5 Conclusions

It is quite apparent that a gamut of environmental conditions such as extremely variable climatic conditions, water scarcity, urbanization, over-population, salinization, global warming, etc. to name a few, has been putting enormous pressure on survival and productivity of plants in general, and crop plants in particular. Thus with increasing abiotic stress conditions, appropriate techniques for management would be needed to ensure sufficient crop productivity for feeding the millions of hungry mouths. Among the potentially useful management systems, those based on cost-effective, low-cost technologies, utilization of microorganisms with multi-faceted traits for improvement of crop growth and yield, as well as abiotic/biotic stress alleviation offer a tempting prospect. There have been innumerable studies which have brought out the efficacies of certain bacteria and fungi in crop protection and growth improvement. Their mechanisms of action, both in the soil, and within the plant, have been worked out in several crop systems. It can thus be hoped that, in coming years, such microorganisms could be routinely used in the field for sustainable agriculture and these will be available to farmers as low-input technologies.

References

Abbaspour, H. (2010) Investigation of the effects of vesicular arbuscular mycorrhiza on mineral nutrition and growth of *Carthamus tinctorius* under salt stress conditions. *Russian Journal of Plant Physiology* 57, 526–531.

Abdel Latef, A.A.H. (2011) Influence of arbuscular mycorrhizal fungi and copper on growth, accumulation of osmolyte, mineral nutrition and antioxidant enzyme activity of pepper (*Capsicum annuum* L.). *Mycorrhiza* 21, 495–503.

Abdelmoneim, T.S. and Almagrabi, O.A. (2013) Improved tolerance of maize plants to heavy metals stress by inoculation with arbuscular mycorrhizal fungi. *Archives des Sciences* 66, 155–167.

Alami, Y., Achouak, W., Marol, C. and Heulin, T. (2000) Rhizosphere soil aggregation and plant growth promotion of sunflowers by exopolysaccharide producing *Rhizobium* sp. strain isolated from sunflower roots. *Applied Environmental Microbioliogy* 66, 3393–3398.

Alfano, G., Bos, J., Cakir, C., Horst, L., Ivey, M., Madden, L.V., Kamoun, S. and Hoitink, H. (2006) Modulation of gene expression in tomato by *Trichoderma hamatum* 382. *Phytopathology (Abstr.)* 96, S4.

Alfano, G., Ivey, M.L.L., Cakir, C., Bos, J.I.B., Miller, S.A., Madden, L.V., Kamoun, S. and Hoitink, H.A.J. (2007) Systemic modulation of gene expression in tomato by *Trichoderma hamatum* 382. *Phytopathology* 97, 429–437.

Al-Garni, S.M.S. (2006) Increasing NaCl-salt tolerance of a halophytic plant *Phragmites australis* by mycorrhizal symbiosis. *American-Eurasian Journal of Agricultural and Environmental Science* 1, 119–126.

Alguacil, M.M., Hernandez, J.A., Caravaca, F., Portillo, B. and Roldan, A. (2003) Antioxidant enzyme activities in shoots from three mycorrhizal shrub species afforested in a degraded semi-arid soil. *Physiologia Plantarum* 118, 562–570.

Alguacil, M.M., Kohler, J., Caravaca, F. and Roldán, A. (2009) Differential effects of *Pseudomonas mendocina* and *Glomus intraradices* on lettuce plants physiological response and aquaporin PIP2 gene expression under elevated atmospheric CO_2 and drought. *Microbial Ecology* 58, 942–951.

Ali, S.Z., Sandhya, V., Grover, M., Kishore, N., Rao, L.V. and Venkateswarlu, B. (2009) *Pseudomonas* sp. strain AKM-P6 enhances tolerance of sorghum seedlings to elevated temperatures. *Biology and Fertility of Soils* 46, 45–55.

Aliasgharzadeh, N., Saleh Rastin, N., Towfighi, H. and Alizadeh, A. (2001) Occurrence of arbuscular mycorrhizal fungi in saline soils of the Tabriz Plain of Iran in relation to some physical and chemical properties of soil. *Mycorrhiza* 11, 119–122.

Ansary, M.H., Rahmani, H.A., Ardakani, M.R., Paknejad, F., Habibi, D. and Mafakheri, S. (2012) Effect of *Pseudomonas fluorescens* on proline and phytohormonal status of maize (*Zea mays* L.) under water deficit stress. *Annals of Biological Research* 3, 1054–1062.

Apse, M.P., Ahron, G.S., Snedden, W.A. and Blumwald, E. (1999) Salt tolerance conferred by overexpression of a vacuolar Na^+/H^+ antiport in *Arabidopsis*. *Science* 285, 1256–1258.

Arkhipova, T.N., Prinsen, E., Veselov, S.U., Martinenko, E.V., Melentiev, A.I. and Kudoyarova, G.R. (2007) Cytokinin producing bacteria enhance plant growth in drying soil. *Plant and Soil* 292, 305–315.

Arshad, M., Sharoona, B. and Mahmood, T. (2008) Inoculation with *Pseudomonas* spp. containing ACC deaminase partially eliminate the effects of drought stress on growth, yield and ripening of pea (*Pisum sativum* L.). *Pedosphere* 18, 611–620.

Ashraf, M. and Foolad, M.R. (2007) Roles of glycine betaine and proline in improving plant abiotic stress resistance. *Environmental and Experimental Botany* 59, 207–216.

Ashraf, M., Hasnain, S., Berge, O. and Mahmood, T. (2004) Inoculating wheat seedlings with exopolysaccharide-producing bacteria restricts sodium uptake and stimulates plant growth under salt stress. *Biology and Fertility of Soils* 40, 157–162.

Augé, R.M. (2001) Water relations, drought and vesicular mycorrhizal fungi symbiosis. *Mycorrhiza* 11, 3–42.

Bae, H., Sicher, R.C., Kim, S.H., Kim, M.S., Strem, M.D. and Bailey, B.A. (2006) The response of *Theobroma cacao* (cacao) to abiotic and biotic stresses and the role of beneficial endophytes. *Plant Biology* 2006, 186.

Bae, H., Sicher, R.C., Kim, M.S., Kim, S.H., Strem, M.D., Melnick, R.L. and Bailey, B.A. (2009) The beneficial endophyte *Trichoderma hamatum* isolate DIS 219b promotes growth and delays the onset of the drought response in *Theobroma cacao*. *Journal of Experimental Botany* 60, 3279–3295.

Bailey, B.A., Bae, H., Strem, M.D., Roberts, D.P., Thomas, S.E., Crozier, J., Samuels, G.J., Choi, I.Y. and Holmes, K.A. (2006) Fungal and plant gene expression during the colonization of cacao seedlings by endophytic isolates of four *Trichoderma* species. *Planta* 224, 1449–1464.

Bano, A. and Fatima, M. (2009) Salt tolerance in *Zea mays* (L). following inoculation with *Rhizobium* and *Pseudomonas*. *Biology and Fertility of Soils* 45, 405–413.

Barassi, C.A., Ayrault, G., Creus, C.M., Sueldo, R.J. and Sobrero, M.T. (2006) Seed inoculation with *Azospirillum* mitigates NaCl effects on lettuce. *Scientia Horticulturae* 109, 8–14.

Barber, D.A. and Martin, J.K. (1976) The release of organic substances by cereal roots into soil. *New Phytologist* 76, 69–80.

Barka, E., Nowak, J. and Clement, C. (2006) Enhancement of chilling resistance of inoculated grapevine plantlets with a plant growth-promoting rhizobacterium, *Burkholderia phytofirmans* strain PsJN. *Applied Environmental Microbiology* 72, 7246–7252.

Bashan, Y. and de-Bashan, L.E. (2010) How the plant growth-promoting bacterium *Azospirillum* promotes plant growth – a critical assessment. *Advances in Agronomy* 108, 77–136.

Bashan, Y. and Levanony, H. (1990) Current status of *Azospirillum* inoculation technology: *Azospirillum* as a challenge for agriculture. *Canadian Journal of Microbiology* 36, 591–608.

Becard, G., Doner, L.W., Rolin, D.B., Douds, D.D. and Pfeffer, P.E. (1991) Identification and quantification of trehalose in vesicular arbuscular mycorrhizal fungi *in vivo* 13C NMR and HPLC analyses. *New Phytologist* 108, 547–552.

Belimov, A.A., Dodd, I.C., Hontzeas, N., Theobald, J.C., Safronova, V.I. and Davies, W.J. (2009) Rhizosphere bacteria containing 1-aminocyclopropane-1- carboxylate deaminase increase yield of plants grown in drying soil via both local and systemic hormone signalling. *New Phytologist* 181, 413–423.

Bensalim, S., Nowak, J. and Asiedu, S.K. (1998) A plant growth promoting rhizobacterium and temperature effects on performance of 18 clones of potato. *American Journal of Potato Research* 75, 145–152.

Besford, R.T., Richardson, C.M., Campos, J.L. and Tiburico, A.F. (1993) Effect of polyamines on stabilization of molecular complexes in thyllakoid membranes of osmotically stressed oat leaves. *Planta* 189, 201–206.

Bianco, C. and Defez, R. (2009) *Medicago truncatula* improves salt tolerance when nodulated by an indole-3-acetic acid-overproducing *Sinorhizobium meliloti* strain. *Journal of Experimental Botany* doi:10.1093/jxb/erp140.

Bianco, C. and Defez, R. (2010) Improvement of phosphate solubilization and *Medicago* plant yield by an indole-3-acetic acid-overproducing strain of *Sinorhizobium meliloti*. *Applied and Environmental Microbiology* 76, 4626–4632.

Blaha, G., Stelzl, U., Spahn, C.M.T., Agrawal, R.K., Frank, J. and Nierhaus, K.H. (2000) Preparation of functional ribosomal complexes and effect of buffer conditions on tRNA positions observed by cryoelectron microscopy. *Methods in Enzymology* 317, 292–309.

Bowen, G.D. and Rovira, A.D. (1999) The rhizosphere and its management to improve plant growth. *Advances in Agronomy* 66, 1–102.

Campanelli, A., Ruta, C., Mastro, G.D. and Morone-Fortunato, I. (2013) The role of arbuscular mycorrhizal fungi in alleviating salt stress in *Medicago sativa* L. var. icon. *Symbiosis* 59, 65–76.

Cantrell, I.C. and Linderman, R.G. (2001) Preinoculation of lettuce and onion with VA mycorrhizal fungi reduces deleterious effects of soil salinity. *Plant and Soil* 233, 269–281.

Carmen, B. and Roberto, D. (2012) Soil bacteria support and protect plants against abiotic stresses. *Abiotic Stress in Plants – Mechanisms and Adaptations* 7, 144–170.

Casanovas, E.M., Barassi, C.A. and Sueldo, R.J. (2002) *Azospirillum* inoculation mitigates water stress effects in maize seedlings. *Cereal Research Communications* 30, 343–350.

Chakraborty, U., Chakraborty, B.N., Tongden, C., RoyChowdhury, P. and Basnet, M. (2004) Assessment of tea rhizosphere microorganisms as plant growth promoters. *Tea* 25, 38–47.

Chakraborty, U., Chakraborty, B.N., Basnet, M. and Chakraborty, A.P. (2009) Evaluation of *Ochrobactrum anthropi* TRS-2 and its talc based formulation for enhancement of growth of tea plants and management of brown root rot disease. *Journal of Applied Microbiology* 107, 625–634.

Chakraborty, U., Chakraborty, B.N. and Chakraborty, A.P. (2010) Influence of *Serratia marcescens* TRS-1 on growth promotion and induction of resistance in *Camellia sinensis* against *Fomes lamaoensis*. *Journal of Plant Interactions* 5, 261–272.

Chakraborty, U., Roy, S., Chakraborty, A.P., Dey, P. and Chakraborty, B. (2011) Plant growth promotion and amelioration of salinity stress in crop plants by a salt-tolerant bacterium. *Recent Research in Science and Technology* 3, 61–70.

Chakraborty, U., Chakraborty, B.N., Chakraborty, A.P. and Dey, P.L. (2013) Water stress amelioration and plant growth promotion in wheat plants by osmotic stress tolerant bacteria. *World Journal of Microbiology and Biotechnology* 29, 789–803.

Chang, W.S., van de Mortel, M., Nielsen, L., de Guzman, G.N., Li, X. and Halverson, L.J. (2007) Alginate production by *Pseudomonas putida* creates a hydrated microenvironment and contributes to biofilm architecture and stress tolerance under water-limiting conditions. *Journal of Bacteriology* 189, 8290–8299.

Chen, M., Wei, H., Cao, J., Liu, R., Wang, Y. and Zheng, C. (2007) Expression of *Bacillus subtilis* proAB genes and reduction of feedback inhibition of proline synthesis increases proline production and confers osmotolerance in transgenic *Arabidopsis*. *Journal of Biochemistry and Molecular Biology* 40, 396–403.

Colla, G., Rouphael, Y., Cardarelli, M., Tullio, M., Rivera, C.M. and Rea, E. (2008) Alleviation of salt stress by arbuscular mycorrhizal in zucchini plants grown at low and high phosphorus concentration. *Biology and Fertility of Soils* 44, 501–509.

Cook, R.J. (2002) Advances in plant health management in the twentieth century. *Annual Review of Phytopathology* 38, 95–116.

Copeman, R.H., Martin, C.A. and Stutz, J.C. (1996) Tomato growth in response to salinity and mycorrhizal fungi from saline or nonsaline soils. *Horticultural Science* 31, 341–344.

Cossins, A.R. (1994) Homeoviscous adaptation of biological membranes and its functional significance. In: Cossins, A.R. (ed.) *Temperature Adaptation of Biological Membranes*. Portland Press, London, pp. 63–76.

Couée, I., Hummel, I., Sulmon, C., Gowsbet, G. and El Armani, A. (2004) Involvement of polyamines in root development. *Plant Cell, Tissue and Organ Culture* 76, 1–10.

Cowan, A.K., Cairns, A.L.P. and Bartels-Rahm, B. (1999) Regulation of abscisic acid metabolism: towards a metabolic basis for abscisic acid-cytokinin antagonism. *The Journal of Experimental Botany* 50, 595–603.

Creus, C.M., Sueldo, R.J. and Barassi, C.A. (2004) Water relations and yield in *Azospirillum*-inoculated wheat exposed to drought in the field. *Canadian Journal of Botany* 82, 273–281.

Dardanelli, M.S., Fernández de Córdoba, F.J., Rosario Espuny, M., Rodríguez Carvajal, M.A., Soria Díaz, M.E., Gil Serrano, A.M., Okon, Y. and Megías, M. (2008) Effect of *Azospirillum brasilense* coinoculated with *Rhizobium* on *Phaseolus vulgaris* flavonoids and Nod factor production under salt stress. *Soil Biology & Biochemistry* 40, 2713–2721.

de Andrade, S.A.L. and da Silveira, A.P.D. (2008) Mycorrhiza influence on maize development under Cd stress and P supply. *Brazilian Journal of Plant Physiology* 20, 39–50.

Delauney, A.J. and Verma, D.P.S. (1993) Proline biosynthesis and osmoregulation in plants. *The Plant Journal* 4, 215–223.

Des Marais, D.L. and Juenger, T.E. (2010) Pleiotropy, plasticity, and the evolution of plant abiotic stress tolerance. *Annals of the New York Academy of Sciences* 1206, 56–79.

Diedhiou, C.J., Popova, O.V., Dietz, K., Arl, J. and Golldack, D. (2008) The SNF1-type serinethreonine protein kinase SAPK4 regulates stress-responsive gene expression in rice. *BMC Plant Biology* 8, 49–62.

Donoso, E.P., Bustamante, R.O., Carú, M. and Niemeyer, H.M. (2008) Water deficit as a driver of the mutualistic relationship between the fungus *Trichoderma harzianum* and two wheat genotypes. *Applied and Environmental Microbiology* 74, 1412–1417.

Endris, S. and Mohammed, M.J. (2007) Nutrient acquisition and yield response of barley exposed to salt stress under different levels of potassium nutrition. *International Journal of Environmental Science and Technology* 4, 323–330.

Estrada, B., Barea, J.M., Aroca, R. and Ruiz-Lozano, J.M. (2012) A native *Glomus intraradices* strain from a Mediterranean saline area exhibits salt tolerance and enhanced symbiotic efficiency with maize plants under salt stress conditions. *Plant Soil* DOI 10.1007/s11104-012-1409-y.

Estrada, B.E., Aroca, R., Barea, J.M. and Ruiz-Lozano, J.M. (2013) Native arbuscular mycorrhizal fungi isolated from a saline habitat improved maize antioxidant systems and plant tolerance to salinity. *Plant Science* 201, 43–51.

Evelin, H., Giri, B. and Kapoor, R. (2013) Ultrastructural evidence for AMF mediated salt stress mitigation in *Trigonella foenum-graecum*. *Mycorrhiza* 23, 71–86.

Fan, L., Dalpé, Y., Fang, C., Dubé, C. and Khanizadeh, S. (2010) Influence of arbuscular mycorrhizae on biomass and root morphology of selected strawberry cultivars under salt stress. *Botany* 89, 1–7.

Fan, Q.J. and Liu, J.H. (2011) Colonization with arbuscular mycorrhizal fungus affects growth, drought tolerance and expression of stress-responsive genes in *Poncirus trifoliata*. *Acta Physiologiae Plantarum* 33, 1533–1542.

Feng, G., Zhang, F.S., Li, X.L., Tian, C.Y., Tang, C. and Rengel, Z. (2002) Improved tolerance of maize plants to salt stress by arbuscular mycorrhiza is related to higher accumulation of soluble sugars in roots. *Mycorrhiza* 12, 185–190.

Figueiredo, M.V.B., Burity, H.A., Martinez, C.R. and Chanway, C.P. (2008) Alleviation of drought stress in the common bean (*Phaseolus vulgaris* L.) by co-inoculation with *Paenibacillus polymyxa* and *Rhizobium tropici*. *Applied Soil Ecology* 40, 182–188.

Foster, R.C., Rovira, A.D. and Cock, T.W. (1983) *Ultrastructure of the Root–Soil Interface*. The American Phytopathological Society, St Paul, Minnesota, 157 pp.

Garg, N. and Manchanda, G. (2008) Effect of arbuscular mycorrhizal inoculation of salt-induced nodule senescence in *Cajanus cajan* (pigeonpea). *Journal of Plant Growth Regulators* 27, 115–124.

Garmendia, I., Goicoechea, N. and Aguirreolea, J. (2005) Moderate drought influences the effect of arbuscular mycorrhizal fungi as biocontrol agents against Verticillium-induced wilt in pepper. *Mycorrhiza* 15, 345–356.

Gibson, S., Arondel, V., Iba, K. and Somerville, C. (1994) Cloning of a temperature-regulated gene encoding a chloroplast Omega-3 desaturase from *Arabidopsis thaliana*. *Plant Physiology* 106, 1615–1621.

Giri, B., Kapoor, R. and Mukerji, K.G. (2007) Improved tolerance of *Acacia nilotica* to salt stress by arbuscular mycorrhiza, *Glomus fasciculatum*, may be partly related to elevated K^+/Na^+ ratios in root and shoot tissues. *Microbial Ecology* 54, 753–760.

Glick, B.R. (1995) The enhancement of plant growth by free-living bacteria. *Canadian Journal of Microbiology* 41, 109–117.

Glick, B.R. (2004) Bacterial ACC deaminase and the alleviation of plant stress. *Advances in Applied Microbiology* 56, 291–312.

Glick, B.R., Liu, C., Ghosh, S. and Dumbroff, E.B. (1997) The effect of the plant growth promoting rhizobacterium *Pseudomonas putida* GR 12-2 on the development of canola seedlings subjected to various stresses. *Soil Biology & Biochemistry* 29, 1233–1239.

Glick, B.R., Penrose, D.M. and Li, J. (1998) A model for the lowering of plant ethylene concentrations by plant growth-promoting bacteria. *Journal of Theoretical Biology* 190, 63–68.

Glick, B.R., Cheng, Z., Czarny, J. and Duan, J. (2007) Promotion of plant growth by ACC deaminase-producing soil bacteria. *European Journal of Plant Pathology* 119, 329–339.

Gorham, J. (1995) Betaines in higher plants – biosynthesis and role in stress metabolism. In: Wallgrove, R.M. (ed.) *Amino Acids and their Derivatives in Higher Plants*. Cambridge University Press, Cambridge, pp. 171–203.

Gray, E.J. and Smith, D.L. (2005) Intracellular and extracellular PGPR: commonalities and distinctions in the plant–bacterium signaling processes. *Soil Biology & Biochemistry* 37, 395–412.

Grichko, V.P. and Glick, B.R. (2001) Amelioration of flooding stress by ACC deaminase-containing plant growth-promoting bacteria. *Plant Physiology and Biochemistry* 39, 11–17.

Hamaoui, B., Abbadi, J.M., Burdman, S., Rashid, A., Sarig, S. and Okon, Y. (2001) Effects of inoculation with *Azospirillum brasilense* on chickpeas (*Cicer arietinum*) and faba beans (*Vicia faba*) under different growth conditions. *Agronomie* 21, 553–560.

Hamdia, M.A., Shaddad, M.A.K. and Doaa, M.M. (2004) Mechanism of salt tolerance and interactive effect of *Azospirillum brasilense* inoculation on maize cultivars grown under salt stress conditions. *Plant Growth Regulation* 44, 165–174.

Hardoim, P.R., van Overbeek, S.V. and van Elsas, J.D. (2008) Properties of bacterial endophytes and their proposed role in plant growth. *Trends in Microbiology* 16, 463–471.

Harman, G.E., Howell, C.R., Viterbo, A., Chet, I. and Lorito, M. (2004) *Trichoderma* species – opportunistic avirulent plant symbionts. *Nature Reviews Microbiology* 2, 43–56.

Hayat, R., Ali, S., Amara, U., Khalid, R. and Ahmed, I. (2010) Soil beneficial bacteria and their role in plant growth promotion: a review. *Annals of Microbiology* 60, 579–598.

Heidari, M. and Jamshid, P. (2010) Interaction between salinity and potassium on grain yield, carbohydrate content and nutrient uptake in pearl millet. *Journal of Agricultural and Biological Science* 5, 39–46.

Hildebrandt, U., Regvar, M. and Bothe, H. (2007) Arbuscular mycorrhiza and heavy metal tolerance. *Phytochemistry* 68, 139–146.

Hoekstra, F.A., Crow, J.H., Crowe, L.M., Van Roekel, T. and Vermeer, T. (1992) Do phospholipids and sucrose determine membrane phase transitions in dehydrating pollen species? *Plant, Cell and Environment* 15, 601–606.

Hussain, T.M., Chandrasekhar, T., Hazara, M., Sultan, Z., Saleh, B.K. and Gopal, G.R. (2008) Recent advances in salt stress biology – a review. *Biotechnology and Molecular Biology Reviews* 3, 8–13.

Jahromi, F., Aroca, R., Porcel, R. and Ruiz-Lozano, J.M. (2008) Influence of salinity on the *in vitro* development of *Glomus intraradices* and on the *in vivo* physiological and molecular responses of mycorrhizal lettuce plants. *Microbial Ecology* 55, 45–53.

Jha, Y., Subramanian, R.B. and Patel, S. (2010) Combination of endophytic and rhizospheric plant growth promoting rhizobacteria in *Oryza sativa* shows higher accumulation of osmoprotectant against saline stress. *Acta Physiologiae Plantarum* DOI 10.1007/s11738-010-0604-9.

Junghans, U., Polee, A., Duchting, P., Weiler, E., Kuhlman, B., Grubber, F. and Teichmann, T. (2006) Adaptation to high salinity in poplar involves changes in xylem anatomy and auxin physiology. *Plant, Cell and Environment* 29, 1519–1531.

Juniper, S. and Abbott, L.K. (1993) Vesicular-arbuscular mycorrhizas and soil salinity. *Mycorrhiza* 4, 45–57.

Kausar, R. and Shahzad, S.M. (2006) Effect of ACC-deaminase containing rhizobacteria on growth promotion of maize under salinity stress. *Journal of Agriculture & Social Sciences* 2, 216–218.

Kaushik, J.C. and Mandal, B.S. (2005) The role of mycorrhiza in stree management for seedling growth of *Dalbergia sissoo* and *Acacia nilotica*. *Bulletin of the NIE* 15, 133–137.

Kavi Kishor, P.B., Sangam, S., Amrutha, R.N., Sri Laxmi, P., Naidu, K.R., Rao, K.R.S.S., Rao, S., Reddy, K.J., Theriappan, P. and Sreenivasulu, N. (2005) Regulation of proline biosynthesis, degradation, uptake and transport in higher plants: its implications in plant growth and abiotic stress tolerance. *Current Science* 88, 424–438.

Khan, A.A., Jilani, G., Akhtar, M.S., Naqvi, S.M.S. and Rasheed, M. (2009) Phosphorus solubilizing bacteria: occurrence, mechanisms and their role in crop production. *Research Journal of Agriculture and Biological Sciences* 1, 48–58.

Khan, N., Mishra, A., Chauhan, P.S. and Nautiyal, C.S. (2011) Induction of *Paenibacillus lentimorbus* biofilm by sodium alginate and $CaCl_2$ alleviates drought stress in chickpea. *Annals of Applied Biology* 159, 372–386.

Kohler, J. (2008) Plant growth promoting rhizobacteria and arbuscular mycorrhizal fungi modify alleviation biochemical mechanisms in water stressed plants. *Functional Plant Biology* 35, 141–151.

Kohler, J., Hernandez, J.A., Caravaca, F. and Roldàn, A. (2009) Induction of antioxidant enzymes is involved in the greater effectiveness of a PGPR versus AM fungi with respect to increasing the tolerance of lettuce to severe salt stress. *Environmental and Experimental Botany* 65, 245–252.

Kohler, J., Caravaca, F. and Roldàn, A. (2010) An AM fungus and a PGPR intensify the adverse effects of salinity on the stability of rhizosphere soil aggregates of *Lactuca sativa*. *Soil Biology & Biochemistry* 42, 429–434.

Lee, S.K., Kim, B.G., Kwon, T.R., Jeong, M.J., Park, S.R., Lee, J.W., Byun, M.O., Kwon, H.B., Matthews, B.F., Hong, C.B. and Park, S.C. (2011) Overexpression of the mitogen-activated protein kinase gene OsMA-PK33 enhances sensitivity to salt stress in rice (*Oryza sativa* L.). *Journal of Biosciences* 36, 139–151.

Leigh, R.A., Ahmad, N. and Wyn-Jones, R.G. (1981) Assessment of glycine betaine and proline compartmentation by analysis isolated beet vacuoles. *Planta* 153, 34–41.

Liddycoat, S.M., Greenberg, B.M. and Wolyn, D.J. (2009) The effect of plant growth-promoting rhizobacteria on asparagus seedling and germinating seeds subjected to water stress under greenhouse conditions. *Canadian Journal of Microbiology* 55, 388–394.

Liu, S., Cheng, Y., Zhang, X., Guan, Q., Nishiuchi, S., Hase, K. and Takano, T. (2007) Expression of an NADP-malic enzyme gene in rice (*Oryza sativa* L.) is induced by environmental stresses; over-expression of the gene in *Arabidopsis* confers salt and osmotic stress tolerance. *Plant Molecular Biology* 64, 49–58.

Lopez, M., Tejera, N.A., Iribarne, C., Lluch, C. and Herrera-Cervera, J. (2008) Trehalose and trehalase in root nodulae of *Medicago truncatula* and *Phaseolus vulgaris* in response to salt stress. *Physiologia Plantarum* 134, 575–582.

Lugtenberg, B. and Kamilova, F. (2009) Plant-growth-promoting rhizobacteria. *Annual Review of Microbiology* 63, 541–556.

Lynch, D.V. and Thompson Jr, G.A. (1982) Low temperature-induced alterations in the chloroplast and microsomal membrane of *Dunaliella salina*. *Plant Physiology* 69, 1369–1375.

Lynch, J.M. (1990) *The Rhizosphere*. John Wiley & Sons, New York.

Madhaiyan, M., Poonguzhali, S. and Sa, T. (2007) Metal tolerating methylotrophic bacteria reduces nickel and cadmium toxicity and promotes plant growth of tomato (*Lycopersicon esculentum* L.). *Chemosphere* 69, 220–228.

Mahajan, S. and Tuteja, N. (2005) Cold, salinity and drought stresses: an overview. *Archives of Biochemistry and Biophysics* 444, 139–158.

Marulanda, A., Porcel, R., Barea, J.M. and Azcon, R. (2007) Drought tolerance and antioxidant activities in lavender plants colonized by native drought tolerant or drought sensitive *Glomus* species. *Microbial Ecology* 54(3), 543–552.

Marulanda, A., Azcón, R., Ruiz-Lozano, J.M. and Aroca, R. (2008) Differential effects of a *Bacillus megaterium* strain on *Lactuca sativa* plant growth depending on the origin of the arbuscular mycorrhizal fungus coinoculated: physiologic and biochemical traits. *Journal of Plant Growth Regulation* 27, 10–18.

Mastouri, F., Thomas Björkman, T. and Harman, G.E. (2010) Seed treatment with *Trichoderma harzianum* alleviates biotic, abiotic, and physiological stresses in germinating seeds and seedlings. *Phytopathology* 100, 1213–1221.

Mastretta, C., Taghavi, S., van der Lelie, D., Mengoni, A., Galardi, F., Gonnelli, C., Barac, T., Boulet, J., Weyens, N. and Vangronsveld, J. (2009) Endophytic bacteria from seeds of *Nicotiana tabacum* can reduce cadmium phytotoxicity. *International Journal of Phytoremediation* 11, 251–267.

Matsubara, Y., Hirano, I., Sassa, D. and Koshikawa, K. (2004) Alleviation of high temperature stress in strawberry (*Fragaria ananassa*) plants infected with arbuscular mycorrhizal fungi. *Environment Control in Biology* 42, 105–111.

Mayak, S., Tirosh, T. and Glick, B.R. (2004a) Plant growth-promoting bacteria confer resistance in tomato plants to salt stress. *Plant Physiology and Biochemistry* 42, 565–572.

Mayak, S., Tirosh, T. and Glick, B.R. (2004b) Plant growth-promoting bacteria that confer resistance to water stress in tomatoes and peppers. *Plant Science* 166, 525–530.

Mergulhão, A.C.E.S., Burity, H.A., Tabosa, J.N., Figueiredo, M.V.B. and Maia, L.C. (2002) Influence of NaCl on *Brachiaria humidicola* inoculated or not with *Glomus etunicatum*. *Investigacion Agraria, Produccion y Proteccion Vegetales* 17, 220–227.

Mishra, P.K., Bisht, S.C., Pooja, R., Joshi, P., Suyal, P., Bisht, J.K. and Srivastva, A.K. (2009a) Enhancement of chilling tolerance and productivity of inoculated wheat with cold tolerant plant growth promoting *Pseudomonas* spp. PPERs23. In: *Abstracts of 4th USSTC* 10–12 November.

Mishra, P.K., Bisht, S.C., Ruwari, P., Selvakumar, G. and Bisht, J.K. (2009b) Enhancement of chilling tolerance of inoculated wheat seedlings with cold tolerant plant growth promoting Pseudomonads from NW Himalayas. In: *Abstracts of first Asian PGPR conference*, Angrau, Hyderabad, 22–24 June.

Mittler, R. (2002) Oxidative stress, antioxidants and stress tolerance. *Trends in Plant Science* 7, 405–410.

Munns, R. and Tester, M. (2008) Mechanisms of salinity tolerance. *Annual Review of Plant Biology* 59, 651–681.

Nadeem, S.M., Zahir, Z.A., Naveed, M., Arshad, M. and Shahzad, S.M. (2006) Variation in growth and ion uptake of maize due to inoculation with plant growth promoting rhizobacteria under salt stress. *Soil Environment* 25, 78–84.

Nadeem, S.M., Zahair, Z.A., Naveed, M. and Arshad, M. (2007) Preliminary investigation on inducing salt tolerance in maize through inoculation with rhizobacteria containing ACC deaminase activity. *Canadian Journal of Microbiology* 53, 1141–1149.

Nautiyal, C.S., Srivastava, S. and Chauhan, P.S. (2008) Rhizosphere colonization: molecular determinants from plant-microbe coexistence perspective. In: Nautiyal, C.S. and Dion, P. (eds) *Molecular Mechanisms of Plant, Microbe Coexistence*. Soil Biology Series, Springer, Berlin, pp. 99–124.

Nautiyal, C.S., Srivastava, S., Chauhan, P.S., Seem, K., Mishra, A. and Sopory, S.K. (2013) Plant growth-promoting bacteria *Bacillus amyloliquefaciens* NBRISN13 modulates gene expression profile of leaf and rhizosphere community in rice during salt stress. *Plant Physiology and Biochemistry* 66, 1–9.

Navarro, J.M., Botella, M.A., Cerda, A. and Martinez, V. (2001) Phosphorus uptake and translocation in salt-stressed melon plants. *Journal of Plant Physiology* 158, 375–381.

Navazio, L., Baldan, B., Moscatiello, R., Zuppini, A., Woo, S.L., Mariani, P. and Lorito, M. (2007) Calcium-mediated perception and defense responses activated in plant cells by metabolite mixtures secreted by the biocontrol fungus *Trichoderma atroviride*. *BMC Plant Biology* 7, 41–49.

Ocon, A., Hampp, R. and Requena, N. (2007) Trehalose turnover during abiotic stress in arbuscular mycorrhizal fungi. *New Phytologist* 174, 879–891.

Omar, M.N.A., Osman, M.E.H., Kasim, W.A. and Abd El-Daim, I.A. (2009) Improvement of salt tolerance mechanisms of barley cultivated under salt stress using *Azospirillum brasiliense*. *Tasks for Vegetation Science* 44, 133–147.

Ouziad, F., Wilde, P., Schmelzer, E., Hilderbrandt, U. and Bothe, H. (2006) Analysis of expression of aquaporins and Na^+/H^+ transporters in tomato colonized by arbuscular mycorrhizal fungi and affected by salt stress. *Environment and Experimental Botany* 57, 177–186.

Parida, A.K. and Das, A.B. (2004) Effects of NaCl stress on nitrogen and phosphorous metabolism in a true mangrove *Bruguiera parviflora* grown under hydroponic culture. *Journal of Plant Physiology* 161, 921–928.

Parida, A.K. and Das, A.B. (2005) Salt tolerance and salinity effects on plants. *Ecotoxicology and Environmental Safety* 60, 324–349.

Pawlowska, T.E. and Charvat, I. (2004) Heavy metal stress and developmental patterns in arbuscular mycorrhizal fungi. *Applied and Environmental Microbiology* 70, 6643–6649.

Pinior, A., Grunewaldt-Stöcker, G., von Alten, H. and Strasser, R.J. (2005) Mycorrhizal impact on drought stress tolerance of rose plants probed by chlorophyll a fluorescence, proline content and visual scoring. *Mycorrhiza* 15, 596–605.

Pinton, R., Varanini, Z. and Nannipieri, P. (2001) The rhizosphere as a site of biochemical interactions among soil components, plants and microorganisms. In: Pinton, R., Varanini, Z. and Nannipieri, P. (eds) *The Rhizosphere*. Marcel Dekker, New York, pp. 1–18.

Pishchik, V.N., Provorov, N.A., Vorobyov, N.I., Chizevskaya, E.P., Safronova, V.I., Tuev, A.N. and Kozhemyakov, A.P. (2009) Interactions between plants and associated bacteria in soils contaminated with heavy metals. *Microbiology* 78, 785–793.

Potters, G., Pasternak, T.P., Guisez, Y. and Jansen, M.A.K. (2009) Different stresses, similar morphogenic responses: integrating a plethora of pathways. *Plant, Cell and Environment* 32, 158–169.

Potts, M. (1994) Desiccation tolerance of prokaryotes. *Microbiological Reviews* 58, 755–805.

Prithiviraj, B., Zhou, X., Souleimanov, A., Kahn, W. and Smith, D.L. (2003) A host-specific bacteria-to-plant signal molecule (Nod factor) enhances germination and early growth of diverse crop plants. *Planta* 216, 437–445.

Rabie, G.H. and Almadini, A.M. (2005) Role of bioinoculants in development of salt-tolerance of *Vicia faba* plants under salinity stress. *African Journal of Biotechnology* 4, 210–222.

Richardson, A.E., Barea, J.M., McNeill, A.M. and Prigent-Cobaret, C. (2009) Acquisition of phosphorus and nitrogen in the rhizosphere and plant growth promotion by microorganisms. *Plant and Soil* 321, 305–339.

Rogers, M.C., Grieve, C. and Shannon, M. (2003) Plant growth and ion relations in lucerne (*Medicago sativa* L.) in response to the combined effects of NaCl and P. *Plant and Soil* 253, 187–194.

Ruíz-Lozano, J.M. (2003) Arbuscular mycorrhizal symbiosis and alleviation of osmotic stress, new perspectives for molecular studies. *Mycorrhiza* 13, 309–317.

Ruíz-Lozano, J.M. and Azcón, R. (2000) Symbiotic efficiency and infectivity of an autochthonous arbuscular mycorrhizal *Glomus* sp. from saline soils and *Glomus deserticola* under salinity. *Mycorrhiza* 10, 137–143.

Ruíz-Lozano, J.M., Porcel, R. and Aroca, R. (2006) Does the enhanced tolerance of arbuscular mycorrhizal plants to water deficit involve modulation of drought-induced plant genes? *New Phytologist* 171, 693–698.

Ruíz-Lozano, J.M., Porcel, R., Azcón, R. and Aroca, R. (2012) Regulation by arbuscular mycorrhizae of the integrated physiological response to salinity in plants. New challenges in physiological and molecular studies. *The Journal of Experimental Botany* 63, 4033–4044.

Sandhya, V., Ali, Sk.Z., Grover, M., Reddy, G. and Venkateswarlu, B. (2009) Alleviation of drought stress effects in sunflower seedlings by exopolysaccharides producing *Pseudomonas putida* strain P45. *Biology and Fertility of Soils* 46, 17–26.

Sandhya, V., Ali, Sk.Z., Grover, M., Reddy, G. and Venkateswarlu, B. (2010) Effect of plant growth promoting *Pseudomonas* spp. on compatible solutes, antioxidant status and plant growth of maize under drought stress. *Plant Growth Regulation* 62, 21–30.

Sandhya, V., Ali, S.Z., Grover, M., Reddy, G. and Bandi, V. (2011) Drought tolerant plant growth promoting *Bacillus* spp.: effect on growth, osmolytes, and antioxidant status of maize under drought stress. *Journal of Plant Interactions* 6, 1–14.

Sannazzaro, A.I., Echeverria, M., Alberto, E.O., Ruiz, O.A. and Menéndez, A.B. (2007) Modulation of polyamine balance in *Lotus glaber* by salinity and arbuscular mycorrhiza. *Plant Physiology and Biochemistry* 45, 39–46.

Santner, A., Calderon-Villalobos, L.I.A. and Estelle, M. (2009) Plant hormones are versatile chemical regulators of plant growth. *Nature Chemical Biology* 5, 301–307.

Saravanakumar, D. and Samiyappan, R. (2007) ACC deaminase from *Pseudomonas fluorescens* mediated saline resistance in groundnut (*Arachis hypogaea*) plants. *Journal of Applied Microbiology* 102, 1283–1292.

Schubert, A., Wyss, P. and Wiekman, A. (1992) Occurrence of trehalose in vesicular-arbuscular mycorrhizal fungi and in mycorrhizal roots. *Journal of Plant Physiology* 140, 41–45.

Selvakumar, G., Kundu, S., Joshi, P., Nazim, S., Gupta, A.D., Mishra, P.K. and Gupta, H.S. (2008a) Characterization of a cold-tolerant plant growth-promoting bacterium *Pantoea dispersa*1A isolated from a subalpine soil in the North Western Indian Himalayas. *World Journal of Microbiology and Biotechnology* 24, 955–960.

Selvakumar, G., Mohan, M., Kundu, S., Gupta, A.D., Joshi, P., Nazim, S. and Gupta, H.S. (2008b) Cold tolerance and plant growth promotion potential of *Serratia marcescens* strain SRM(MTCC8708) isolated from flowers of summer squash (*Cucurbita pepo*). *Letters in Applied Microbiology* 46, 171–175.

Selvakumar, G., Panneerselvam, P. and Ganeshamurthy, A.N. (2012) Bacterial mediated alleviation of abiotic stress in crops. In: Maheswari, D.K. (ed.) *Bacteria in Agrobiology: Stress Management*. Springer-Verlag, Berlin, pp. 205–224.

Shao, H.B., Chu, L.Y., Jaleel, C.A., Manivannan, P., Panneerselvam, R. and Shao, M.A. (2009) Understanding water deficit stress-induced changes in the basic metabolism of higher plants – biotechnologically and sustainably improving agriculture and the ecoenvironment in arid regions of the globe. *Critical Reviews in Biotechnology* 29, 131–151.

Sharifi, M., Ghorbanli, M. and Ebrahimzadeh, H. (2007) Improved growth of salinity-stressed soybean after inoculation with salt pre-treated mycorrhizal fungi. *Journal of Plant Physiology* 164, 1144–1151.

Sheng, M., Tang, M., Chan, H., Yang, B., Zhang, F. and Huang, Y. (2008) Influence of arbuscular mycorrhizae on photosynthesis and water status of maize plants under salt stress. *Mycorrhiza* 18, 287–296.

Sheng, X.F., Xia, J.J., Jiang, C.Y., He, L.Y. and Qian, M. (2008) Characterization of heavy metal-resistant endophytic bacteria from rape (*Brassica napus*) roots and their potential in promoting the growth and lead accumulation of rape. *Environmental Pollution* 15, 1164–1170.

Shi, H., Ishitani, M., Kim, C. and Zhu, J.K. (2000) The *Arabidopsis thaliana* salt tolerance gene SOS1 encodes a putative Na^+H^+ antiporter. *Proceedings of National Academy of Sciences* 97, 6896–6901.

Shi, H., Lee, B.H., Wu, S.J. and Zhu, J.K. (2003) Overexpression of plasma membrane Na^+/H^+ antiporter gene improves salt tolerance in *Arabidopsis thaliana*. *Nature Biotechnology* 21, 81–85.

Shirasawa, K., Takabe, T., Takabe, T. and Kishitani, S. (2006) Accumulation of glycine betaine in rice plants that overexpress choline monooxygenase from spinach and evaluation of their tolerance to abiotic stress. *Annals of Botany* 98, 565–571.

Shokri, S. and Maadi, B. (2009) Effects of arbuscular mycorrhizal fungus on the mineral nutrition and yield of *Trifolium alexandrium* plants under salinity stress. *Journal of Agronomy* 8, 79–83.

Shoresh, M. and Harman, G.E. (2008) The molecular basis of shoot responses of maize seedlings to *Trichoderma harzianum* T22 inoculation of the root: a proteomic approach. *Plant Physiology* 147, 2147–2163.

Shukla, N., Awasthi, R.P. and Rawat, L.J.K. (2012a) Biochemical and physiological responses of rice (*Oryza sativa* L.) as influenced by *Trichoderma harzianum* under drought stress. *Plant Physiology and Biochemistry* 54, 78–88.

Shukla, P.S., Agarwal, P.K. and Jha, B. (2012b) Improved salinity tolerance of *Arachis hypogaea* (L.) by the interaction of halotolerant plant growth promoting rhizobacteria. *Journal of Plant Growth Regulation* 31, 195–206.

Siddikee, M.A., Chauhan, P.S., Anandham, R., Han, G.H. and Sa, T. (2010) Isolation, characterization, and use for plant growth promotion under salt stress, of ACC deaminase-producing halotolerant bacteria derived from coastal soil. *Journal of Microbiology and Biotechnology* 20, 1577–1584.

Smith, S.E. and Read, D.J. (2008) *Mycorrhizal Symbiosis*, 3rd edn. Academic Press, Elsevier, London.

Sohrabi, Y., Heidari, G., Weisany, W., Golezani, K.G. and Mohammadi, K. (2012) Changes of antioxidative enzymes, lipid peroxidation and chlorophyll content in chickpea types colonized by different *Glomus* species under drought stress. *Symbiosis* 56, 5–18.

Spaepen, S., Boddelaere, S., Croonenborghs, A. and Vanderleyden, J. (2008) Effect of *Azospirillum brasiliense* indole-3-acetic acid production on inoculated wheat plants. *Plant and Soil* 312, 15–23.

Stepanova, A.N., Yun, J., Likhacheva, A.V. and Alonso, J.M. (2007) Multilevel interactions between ethylene and auxin in *Arabidopsis* roots. *Plant Cell* 19, 2169–2185.

Suarez, R., Wong, A., Ramirez, M., Barraza, A., Orozco, M. del C., Cevallos, M.A., Lara, M., Hernandez, G. and Iturriaga, G. (2008) Improvement of drought tolerance and grain yield in common bean by overexpressing trehalose-6-phosphate synthase in rhizobia. *Molecular Plant-Microbe Interactions* 21, 958–966.

Sziderics, A.H., Rasche, F., Trognitz, F., Sessitsch, A. and Wilhelm, E. (2007) Bacterial endophytes contribute to abiotic stress adaptation in pepper plants (*Capsicum annuum* L.). *Canadian Journal of Microbiology* 53, 1195–1202.

Terré, S., Asch, F., Padham, J., Sikora, R.A. and Becker, M. (2007) Influence of root zone bacteria on root iron plaque formation in rice subjected to iron toxicity. In: Tielkes, E. (ed.) *Utilisation of Diversity in Land Use Systems: Sustainable and Organic Approaches to Meet Human Needs*. Tropentag, Witzenhausen, Germany, p. 446.

Timmusk, S. and Wagner, E.G.H. (1999) The plant growth promoting rhizobacterium *Paenibacillus polymyxa* induces changes in *Arabidopsis thaliana* gene expression: a possible connection between biotic and abiotic stress responses. *Molecular Plant-Microbe Interactions* 12, 951–959.

Vardharajula, S., Ali, S.Z., Grover, M., Reddy, G. and Bandi, V. (2011) Drought-tolerant plant growth promoting *Bacillus* spp.: effect on growth, osmolytes, and antioxidant status of maize under drought stress. *Journal of Plant Interactions* 6, 1–14.

Venkateswarlu, B. and Shanker, A.K. (2009) Climate change and agriculture: adaptation and mitigation strategies. *Indian Journal of Agronomy* 54, 226–230.

Verbruggen, N. and Hermans, C. (2008) Proline accumulation in plants: a review. *Amino Acids* 35, 753–759.

Wang, S., Wan, C. and Wang, Y. (2004) The characteristics of Na^+, K^+ and free proline distribution in several drought-resistance plants of the Alxa Desert, China. *Journal of Arid Environment* 56, 525–539.

Wang, W., Vinocur, B. and Altman, A. (2003) Plant responses to drought, salinity and extreme temperatures: towards genetic engineering for stress tolerance. *Planta* 218, 1–14.

Wu, Q.S. (2011) Mycorrhizal efficacy of trifoliate orange seedlings on alleviating temperature stress. *Plant, Soil and Environment* 57, 459–464.

Wu, Q.S., Li, G.H. and Zou, Y.N. (2011) Roles of arbuscular mycorrhizal fungi on growth and nutrient acquisition of peach (*Prunus persica* L. Batsch) seedlings. *Journal of Animal & Plant Sciences* 21, 746–750.

Yamato, M., Ikeda, S. and Iwase, K. (2008) Community of arbuscular mycorrhizal fungi in coastal vegetation on Okinawa Island and effect of the isolated fungi on growth of sorghum under salt-treated conditions. *Mycorrhiza* 18, 241–249.

Yang, J., Kloepper, J.W. and Ryu, C.M. (2009) Rhizosphere bacteria help plants tolerate abiotic stress. *Trends in Plant Science* 14, 1–4.

Yildirim, E., Taylor, A.G. and Spittler, T.D. (2006) Ameliorative effects of biological treatments on growth of squash plants under salt stress. *Scientia Horticulturae* 111, 1–6.

Ying, S., Zhang, D.F., Li, H.Y., Liu, Y.H., Shi, Y.S., Song, Y.C., Wang, T.Y. and Li, Y. (2011) Cloning and characterization of a maize SnRK2 protein kinase gene confers enhanced salt tolerance in transgenic *Arabidopsis*. *Plant Cell Reports* 30, 1683–1699.

Zhang, H., Kim, M.S., Sun, Y., Dowd, S.E., Shi, H. and Pare, P.W. (2008) Soil bacteria confer plant salt tolerance by tissue-specific regulation of the sodium transporter HKT1. *Molecular Plant-Microbe Interactions* 21, 737–744.

Zhang, H., Liu, W., Wan, L., Li, F., Dai, L., Li, D., Zhang, Z.R. and Huang, Y. (2010) Functional analyses of ethylene response factor JERF3 with the aim of improving tolerance to drought and osmotic stress in transgenic rice. *Transgenic Research* 19, 809–818.

Zhang, Y., Zhao, L., Wang, Y., Yang, B. and Chen, S. (2008) Enhancement of heavy metal accumulation by tissue specific co-expression of iaaM and ACC deaminase genes in plants. *Chemosphere* 72, 564–571.

Zhang, Z., Wang, J., Zhang, R. and Huang, R. (2012) The ethylene response factor AtERF98 enhances tolerance to salt through the transcriptional activation of ascorbic acid synthesis in *Arabidopsis*. *The Plant Journal* 71, 273–287.

Zuccarini, P. and Okurowska, P. (2008) Effects of mycorrhizal colonization and fertilization on growth and photosynthesis of sweet basil under salt stress. *Journal of Plant Nutrition* 31, 497–513.

Index

abiotic stress factors see osmotic adjustment (OA); oxidative stress; plant growth and abiotic factors
abscisic acid (ABA)
    ABA-responsive element (ABRE) binding protein 66, 75–6
    with drought and salt stresses 60
    increase with drought stress 71
    signalling 66
ACC deaminase-producing bacteria 238–9
ACC synthase 217–18
alkaloids 117
allene oxide cyclase (AOC) 62
aluminium (Al) tolerance 169–71
antibiotics and anticancer drugs 16
antioxidant defence genes/sources 30
antioxidant mechanisms, with bacterial treatments 243–4
antioxidants, work of 37
antioxidative enzymes 144, 204–5, 205
antioxidative systems 90–3
    enzymatic/non-enzymatic antioxidant defences 90, 91
    generation of reactive oxygen species in cells 91
    leaf CAT studies 91–3
    oxidative stress 90
    photosynthesis and respiration limitation 93
    root and leaf SOD studies 93
    ROS production issues 93
    salt-sensitive rice studies 92
    salt-sensitive sorghum genotypes 93
    salt-stressed tomatoes 92
    salt-tolerant cotton 92
    salt-tolerant cultivars 92
    stomatal closure during salt stress 90
    time course issues for enzyme activity 91
apoplastic ascorbate 124–5
Arabidopsis transcription activation factor (ATAF) 72
arbuscular mycorrhizal fungi (AMF) 233, 235, 237, 240
arginine decarboxylase (ADC) 214
arsenic 178–89
    arsenic and phosphate 178–9, 188–9
    groundwater contamination 179
    toxicity 179
    uptake and transport in plants 179–80
        arsenic hyper-accumulator plants 179
        in tomato plants 179–80
        toxicity issues 179
arsenic, influencing effects on 180–8
    antioxidant enzymes 181–3
        ascorbic acid oxidase (AOX) activity 182
        H_2O_2 activity 182–3
        ROS generation 182
    arsenic binding proteins 187–8
        with phytochelatin (PC) 187–8
    glutathione (GSH) 185–7
        GSH-S-transferases (GST) 186–7
    nitrogen metabolism 184–5
    oxidative stress markers 181
        ROS production 181
    photosynthetic phosphorylation 184
    respiratory cycle 183–4
        inhibition of succinic oxidase system 183
        SH groups 183
        wheat 183–4
    seedling growth 180–1
        shoot and root growth reduction 180
        stimulation at low concentrations 181

ascorbate (AsA) 114–15
ascorbic acid oxidase (AOX) 182
ATAF proteins 74–5
AtHK1 as osmosensor 84
ATP binding 13
ATP-binding cassettes (ABC) transporters 139
ATP-bound DnaK 12–13
ATPases
    and Hsp90 16, 17
    see also dynamic hexameric ATPases
auxin 237–8
Azospirillum 238

bacterial J-proteins 13
breeding strategies 66–7
    marker-assisted selection (MAS) 66

cadmium (Cd) stress in plants 136–43, 165
    cadmium as a pollutant 136–7, 152–3
    availability in soil 137
    ethylene in sulfur-mediated alleviation 152
    genes involved in Cd tolerance 140
    metal cation homoeostasis 138
    reactive oxygen species (ROS) generation/signalling 142–3
    rhizosphere, effects on 137
    sulfur uptake/assimilation regulation 143
    transportation and accumulation 138–9
        ATP-binding cassettes (ABC) transporters 139
    uptake process and rate 137–8
    see also ethylene
cadmium (Cd), toxicity effects in plants 139–42
    molecular aspects 142
    morphological aspects 139–40
    physiological and biochemical 140–2
        effect on nitrogen based processes 141
        inhibition of electron transport 141
        nitrogen, effects on 141
cadmium tolerance, detoxification mechanisms 143–50
    antioxidants, role of 144–6
        catalase 145
        cysteine synthesis 146
        glutathione (GSH) 145–6
        superoxide dismutase 144–5
    epibrassinolide (EBR) 148
    indole acetic acid (IAA) 149
    lipid peroxidation (LPO) 148
    nitric oxide (NO) signalling 148
    phytochelates (PCs), role of 146–7
    phytohormones 147–8
    sulfur involvement 149–50
    synthesis of protein 147
        heat-shock proteins (Hsps) 147
    tocopherol cyclase 146
    uptake of Cu, Fe, Zn, manganese (Mn) and phosphorus (P) 148
Calvin cycle 198–9
carbon metabolism and utilization 34
catalase (CAT) activity 25–6, 25, 145, 206
catalytic activity, Hsp90 16
cation diffusion family (CDF) 168
CCT oligomer 10
    operation 10
    structure 10
cell turgor adjustment 200–3
    turgor loss point 201–3
cellular homoeostasis 93
chaperonins
    Hsp60 family 9–10
    see also E. coli GroE chaperonin ensemble/system
chickpea roots and abiotic stress 60–3
    ABA increase during drought 63
    ABA signalling 66
    allene oxide cyclase (AOC) issues 62
    enzyme activities 63
    epigenetic control 66
    histone acetyltransferases (HATs) 66
    hormonal response to salt stress 62–3
    JA synthesis data 61
    OPDA, JA and JA-Ile concentrations 64
    root systems 61
    SNP associated alternative tags (SAAT) 62
    superSAGE analysis 61–2, 62
    Taqman probe studies 63
    transcription factor TCP4 66
chlorophyll fluorescence with bean plants 199
chloroplastic sHsps 33
chromium (Cr), plant stress from 165
cis-acting elements 73
cis-binding factor (CBF) 72, 73–4
Clean Air Act (USA) 113
cobalt (Co), plant stress from 165
copper (Cu), plant stress from 165
coronatine 56
cross-tolerance 93–5
    benefits 94
    cellular homoeostasis 93
    hydrogen peroxide as a signalling molecule 94–5
        maize seedlings 94
        rice seedlings 94
        with secondary oxidative stress 95
    pathway disruption 93

cysteine synthesis 146
cytokinin, plant/root regulation properties 60

dehydration responsive element binding
       regulon (DREB) 72, 73–4
   control of 73
DNA methylation modifications 35
DnaJ proteins 13–14
DnaK-DnaJ binary complex 12
   see also Hsp70 (DnaK) family
drought see salt stresses and drought
drought tolerance
   engineering functional proteins 76–7
   signalling factors 76
   see also genetic engineering crop plants for
          drought tolerance; photosynthesis,
          plant responses to drought; salt
          stresses and drought
drought/water deficient stress 71–2
   abscisic acid (ABA) increase 71
dynamic hexameric ATPases 10–12
   DnaK-DnaJ binary complex 12
   Hsp70 family chaperones 11–12
   Hsp100 family chaperones 10, 11
   removal/dissolution of stress 10
   yeast Hsp104 11
      structure 11

E. coli GroE chaperonin ensemble/system 9–10
   operational process 9–10
   structure 9
endoplasmic reticulum (ER) 2
engineering functional proteins for drought
       tolerance 76–7
epibrassinolide (EBR) 148
epigenetic regulation of HS in plants 34–5
   DNA methylation modifications 35
   ONSEN insertions 35
   small interfering RNAs (siRNAs) 35
   small RNAs (smRNAs) 35
   stress memory issues 34–5
ERF transcription factors 46
ethylene 150–2
   ethylene biosynthesis 150–1
      sulfur/ethylene relationship 150–1
   signalling highway 151–2
      ethylene insensitive (EIN) 151
      ethylene response factor (ERF) 151
      ethylene response sensor (ERS) 151
      GAF domain 151
   in sulfur-mediated alleviation
      of cadmium stress 152
ethylene, and flooding 45–7
   ERF transcription factors 46
   ethylene biosynthesis 45–6

hyponasty 46
O_2 dependence 46
O_2 sensing mechanisms 47
   with PCD 46
SUB1A 47
in submergence tolerance 45, 45
ethylene responsive element binding proteins
       (EREBP) 238–9
eukaryotes 2
eukaryotic cells 3
eukaryotic chaperonin see CCT oligomer
eukaryotic Hsp40 (J protein) 13
expressed sequence tag analysis 168

flavonoids 117
flooding, plant tolerance to 43–9
   Hsp generation 44
   low oxygen 'escape' strategies (LOES) 44
   low oxygen quiescence strategies (LOQS) 44
   reduced O_2 studies 44
   ROS generation 44
   soil microbe issues 44
   submergence periods 44
   waterlogging 43–4
   see also ethylene, and flooding

geldanamycin 17
genetic engineering crop plants for drought
       tolerance 71–7
   engineering functional proteins 76–7
   strategies for stress tolerance 72
   stress tolerant genes 72
   see also transcription factors
global warming and water shortage 35–7
   reproductive process vulnerability 36–7
   thermotolerance 36
glucosinolates 125
glutathione (GSH) 29, 114–15, 137, 145–6, 185–7
glycine-betaine (GB) 89–90
glycoalkaloids 125
groundwater contamination, arsenic 179

heat stress (HS) response 23–4
   protective gene expression 24
   reactive oxygen species (ROS) 24
   soluble sugars, sensitivity of 24
   thermotolerance 24
heat tolerance mechanisms 28–35
   antioxidants and enzymes 28
   HS transcription factors (HSFs) 28
   see also heat-shock proteins (Hsps) and
          transcription factors; reactive
          oxygen species (ROS) defence
          from HS

heat-shock proteins (Hsps) 2, 6, 147
    groupings 6
    Hsp70 families 6
    Hsp90 families 6
    see also Hsp....; small heat-shock
        proteins (sHsps)
heat-shock proteins (Hsps) and transcription
        factors 30–1, 32–4
    chloroplastic sHsps 33
    heat-shock granules (HSG) 32–3
    HS transcription factors (HSFs) 33
    Hsp synthesis 32
    Hsp101 32
    low molecular weight (LMW) Hsps 33
    transcription factors (TFs) 33
heavy metal toxicity 164–72
    cadmium, plant stress from 165
    chromium (Cr), plant stress from 165
    cobalt (Co), plant stress from 165
    copper (Cu), plant stress from 165
    effects and sources 166
    miRNA stress regulation 171–2
    nickel (Ni), plant stress from 165
    zinc (Zn), plant stress from 165
heavy metal toxicity, molecular physiology and
        metalloid stress tolerance 168–71
    aluminium (Al) tolerance 169–71
    cation diffusion family (CDF) 168
    influx-type transporters 169
    metal hyper-accumulation 169
    organic acid (OA) excretion 170
    Zn transporters 168–9
heavy metal toxicity, and reactive oxygen
        species (ROS) 165–8
    Al-induced oxidative stress 167–8
    mechanism 166
high-affinity potassium transporters (HKT) 85–6
histidine kinases 84
histone acetyltransferases (HATs) 66
hormones, stress alleviation
        mechanisms 237–9
    ACC deaminase-producing bacteria 238–9
    auxin properties 237–8
    Azospirillum 238
    EREBP 238–9
    indole acetic acid (IAA) 238
    PGPR strains 238
HPD motif 14
HS transcription factors (HSFs) 28, 33
Hsp60 family of chaperones 9–10
        see also E. coli GroE chaperonin
            ensemble/system
Hsp70 (DnaK) family 12–13
    ATP-bound DnaK 12–13
    properties 12
    substrate binding 12–13
    substrate binding domain (SBD) 12

Hsp90 and co-shaperones in plant and animal
        pathology 16–17
    and ATPase activity 17
    MEEVD motif 17
    plant NLR proteins 17
    resistance (R) genes 17
Hsp90 (multifaceted chaperone) 14–16
    antibiotics and anticancer drugs 16
    ATP binding 15
    catalytic activity 16
    critical process regulation 14
    with fungi and higher eukaryotes 14
    with geldanamycin 17
    human orthologue 16
    N-terminal domain (NTD) 15–16
    structure 15
Hsp101 32
hydrogen peroxide as a signalling
        molecule 94–5
    with secondary oxidative stress 95
hydroxide radical (OH*) 166
hyperosmotic and oxidative stress 60
hyperthermia 2
    in eukaryotic cells 3
hyponasty 46
hypoxia-inducible transcriptional
        factors (HIFs) 5

indole acetic acid (IAA) 149, 238
induced systematic tolerance (IST) 233
influx-type transporters 169
inorganic ions 88
Intergovernmental Panel on Climate Change
        (IPCC) 113
ion homoeostasis 85–7, 241–2
    salt uptake issues 85–6, 241
        high-affinity potassium transporters
            (HKT) 85–6
        non-selective cation (NSCC)
            channels 86
    sodium extrusion and compartmentation 86–7
        Na^+/H^+ exchangers (antiporters) 86
        Salt Overly Sensitive (SOS) mutants 86–7
ionic- and osmotic-disequilibrium components 59

jasmonate biosynthesis pathway 55–6, 56
    allene oxide cyclase (AOC) 56–7
    coronatine 56
    JA responses to pathogens 57
    JA-responsive transcription factors 57
    OPDA import step 55
    oxylipin branches 56
        synthesis pathways 57
    plastids and peroxisomes steps 55
    wounding and JA 57

jasmonate signalling 55
    see also nitric oxide (NO) regulation
        of jasmonic acid signalling; roots,
        stress response; salt stresses and
        drought
jasmonic acid (JA) 54–5

leaf area ratio 203
leaf senescence 24–6
leaf water relations 200–3, 202
lipids
    effects of ozone 120–1
    lipid peroxidation (LPO) 148, 204, 206
Long-Range Transboundary Air Pollution
    Convention 113
low molecular weight (LMW) Hsps 33
low oxygen 'escape' strategies (LOES) 44
low oxygen quiescence strategies (LOQS) 44

maize
    reproduction under HS 27
        relative water content (RWC) issues 27
marker-assisted selection (MAS) 66
MEEVD motif 15, 17
metabolites see protective metabolites
metal cation homoeostasis 138
metal hyper-accumulation 169
microorganisms, in alleviation of abiotic
        stress 232–44
    use of microorganisms 232–3
    arbuscular mycorrhizal fungi (AMF) 233
    induced systematic tolerance (IST) 233
    ion homoeostasis 241–2
    nutrient up-take enhancement 242–3
    phosphorus mobilization 242
    see also hormones, stress alleviation
        mechanisms; plant growth-promoting
        rhizobacteria (PGPR); protective
        metabolites
microRNAs (miRNA) and heavy metal
        stress 171–2
minerals, effects of ozone 125
mitochondrial ROS 5
Model for Ozone and Related Chemical
        Traces (MOZART) 113
molecular chaperones 1–18
    co-chaperones 4
    heat stress effects limitation 3–4
    with linear polypeptide chain 3
    mediation of macromolecular trafficking 4
    role in DNA replication/repair 3
    see also CCT oligomer; E. coli GroE
        chaperonin ensemble/system;
        heat-shock proteins (Hsps)
mycorrhizal fungi 235

mycorrhization 222–4
    arbuscular mycorrhizal fungi (AMF) 233,
        235, 237
    mycorrhizated plants 223
    organic matter transformation
        by microorganisms 223
    peptaibols 223
    ROS intermediate-mediated signalling
        pathway 223
    Trichoderma, beneficial properties 223–4
    see also trichoderma harzianum strain T-22
myeloblastosis oncogene (MYB) 72, 76
myelocytomatosis oncogene (MYC) 72, 76

N-terminal domain (NTD) 15–16, 17
Na+/H+ exchangers (antiporters) 86
NAC proteins/family 74–5
nickel (Ni), plant stress from 165
nitrate reductase (NR) 47
nitric oxide (NO) regulation of jasmonic acid
        signalling 63–6
    ABA and NO cooperation 65
    long-distance signalling 65
    miRNA response role 65–6
    NO-mediated promoters 65
    NO-response promoters 65
nitric oxide (NO) and submergence tolerance 47–9
    alternative oxidase (AOX) induction 49
    ethylene, role of 48
    ethylene signalling influence 48–9
    generation during flooding 47
    haemoglobins, role of 48
    Hb/NO cycle 48
    nitrate reductase (NR) action 47
    production during hypoxia 48
nitrogen metabolism 184–5
non-selective cation (NSCC) channels 86
North American Crop Loss Assessment Network
        (NCLAN) 113
nuclear factor (NF-Y) TFs 74
nutrient up-take enhancement 242–3

ONSEN insertions 35
organic acid (OA) excretion 170
organic matter transformation by microorganisms 223
organic solutes 88–9
ornithine decarboxylase (ODC) 214
osmotic adjustment (OA) 200–3
    cell turgor adjustment 200–3
    leaf water relations 200–3, 202
    pressure-volume (PV) curves 202, 202
    proline content/accumulation 201, 201, 203
    relative water content (RWC) 201, 201
    turgor loss point (TLP) 201–3
    water potential 200, 201

osmotic adjustment (OA) for salt stressed
        plants 87–90
    glycine-betaine (GB) 89–90
    high proline concentration in leaves 89
    inorganic ions 88
    organic compound accumulation 88
    organic solutes 88–9
    osmoregulation 87
    osmotic adaptation of halophytes 88
    regulation by organic compounds 87
    subcellular compartmentalization 87
    vacuolated cells 87
oxidative stress 90, 204–7
    antioxidative enzymes 204, 205
    catalase (CAT) activity 206
    H_2O_2, importance of 204, 205–6
    lipid peroxidation (LPO) 204, 206
    membrane disturbances 206
    see also heavy metal toxicity, and reactive
            oxygen species (ROS)
ozone 112–28
    ozone problems worldwide 112–14, 128
    formation from NOx, VOCs and sunlight 112–13
ozone, detrimental effects mechanisms 114–16
    ascorbate (AsA) 114–15
    carbon translocation problems 116
    glutathione (GSH) 114–15
    PEP-case 115
    photosynthetic process 115–16
    protective mechanisms 114–15
    reactive oxygen species (ROS) 114
ozone, negative impacts on quality/yield 118–28
    quality/yield changes 118
    carbohydrates 121–3
        carbohydrate effects 121
        effects on major crops, forages and
                trees 122
        fruit crop value 123
        potato 121
        sucrose content 121–3
    forage for ruminant animals 127–8
        forage and ozone 127
        clovers under oxygen exposure 127
        digestibility and protein content 127
        grassland species 127–8
        straw 128
    lipids 120–1
        seed lipid content effects 121
    minerals 125–6
        about effects on minerals 125
        content effects for major crops, 126
    physical and sensory aspects 125–7
        flavour modifications 126
        visible injury on leaves 125–6
    proteins 119–20
        effects of ozone stress 119
        legumes 120

        rice 120
        wheat 119–20
    secondary compounds elicited by ozone 123–5
        apoplastic ascorbate 124–5
        effects on major crops, forages and
                trees 124
        glucosinolates 125
        glycoalkaloids 125
ozone, plant defence metabolism changes 116–18
    alkaloids 117
    chemical composition changes 116
    flavonoids 117
    isoprenoid biosynthetic pathway 117
    phenylpropanoid metabolism/
            pathways 116–17
    secondary metabolites 116, 118
    terpenoids 117

peptaibols 223
persistent oxidative stress effects 5
phenylpropanoid metabolism/pethways 116–17
phosphate/phosphorus 178–9, 180
    energy transfer properties 180
    protein metabolism properties 180
    similarities to arsenic 179
phosphorus mobilization/nutrition 242
photochemical quenching 200
photosynthesis
    limitation by osmotic/ionic effects 93
    process problems 115–16
photosynthesis, plant responses to drought 197–200
    whole organism level 197
    whole plant level 197
    Calvin cycle 198–9
    chlorophyll fluorescence changes 199–200, 199
    leaf gas exchange 198, 198
    mesophyllic factors 198
    need to maintain photosynthesis 197–8
    photochemical quenching 200
    photoinhibitory damage of PSII 200
    Rubisco activity inhibition 198
    RuBP regeneration inhibition 198–9
    stomatal/non-stomatal factors 198–9
photosynthetic phosphorylation 184
phytochelates (PCs) 137, 146–7, 187
phytohormones 147–8
    phytohormone signalling 217–18
plant growth and abiotic factors 203, 203–4
    biomasss issues 203, 204
    leaf area ratio 203
    water availability 203
    water use efficiency/deficiency 203
plant growth-promoting fungi 235–7
    arbuscular mycorrhizal fungi (AMF) 235, 237
    mycorrhizal fungi 235
    Trichoderma 235–6

plant growth-promoting rhizobacteria
    (PGPR) 233, 238
    about PGPRs 233–4
    beneficial effect mechanisms 234
    global warming issues 235
    induction of drought tolerance 234
    plant growth promoting bacteria 236
    salt tolerance/soil salinization 234–5
    water stress performance 234
plant NLR proteins and Hsp90 17
plant roots see roots...
plant stress communication 54
    jasmonate signalling 55
polyamine oxidising enzymes (PAOs) 216
polyamines (PAs) 212–18, 240
    stress and polyamines 212–13, 218
    arginine decarboxylase (ADC) 214
    biosynthesis and regulation 214–15
    levels in tissues 214–15
    ornithine decarboxylase (ODC) 214
    polyamine titre 214
    tasks/roles/processes in cell physiology 213–14
polyamines (PAs), role in stress tolerance 215–18
    increased biosynthesis as stress
        response 215–16
        ADC involvement 215
        molecular and genetic approaches 215–16
    integration with phytohormone
        signalling 217–18
        abscisic acid (ABA) role 217
        ACC synthase role 217–18
    interaction with ROS species 216–17
        polyamine oxidising enzymes
            (PAOs) 216
        signalling processes 216–17
potato, effects of ozone 121
pressure-volume (PV) curves 202, 202
programmed cell death (PCD) process see
    senescence
proline accumulation/metabolism 239–40
proline concentration in leaves 89
proline content/accumulation 201, 201, 203
protective gene expression 24
protective metabolites 239–41
    AMF colonization 240
    polyamines 240
    proline accumulation/metabolism 239–40
    soluble sugars 240
    trehalose 241
protein mis-folding and aggregation 4

reactive oxygen species (ROS) 4–5, 24, 114
    with antioxidative systems 93
    from animal exposure to low oxygen 5
    from ozone 114
    generation 4, 5

H_2O_2 issues 5
hypoxia-inducible transcriptional factors (HIFs) 5
mitochondrial ROS 5
persistent oxidative stress effects 5
ROS intermediate-mediated signalling
    pathway 223
see also heavy metal toxicity, and reactive
    oxygen species (ROS)
reactive oxygen species (ROS) defence from
    HS 29–32
    oxidative damage from HS 29
    antioxidant defence genes/sources 30–1
    antioxidant enzymes role evaluations 29
    antioxidant metabolites 29–32
    Arabidopsis thaliana response studies 29
    enzyme stability 29
    glutathione (GSH) 29
    heat-shock proteins and transcription
        factors 30–1
    tocopherols and carotenoids 32
relative water content (RWC) 201, 201
reproductive developments under HS 26–8
    reproduction under HS 26
    maize 27
    rice 27–8
    wheat 27
reproductive process vulnerability 36–7
resistance (R) genes (plants) 17
respiration, limitation by osmotic/ionic effects 93
rhizobacteria see plant growth-promoting
    rhizobacteria (PGPR)
rhizosphere 137
rice
    effects of ozone 120
    reproduction under HS 27–8
RNA polymerase 72
roots, stress response 58–9
    ABA regulation 58
    jasmonates involvement 58
    local triggering 59
    nitric oxide (NO) issues 58–9
    signalling compounds 58
        9-HPOT 58
        JA and Me-JA 58
    symbiotic microorganisms 58
ROS see reactive oxygen species (ROS)
Rubisco activity/content 25
RuBP regeneration 198–9

salinity, effect on growth and development 82–3
    carbon flux issues 83
    factors determining plant response 82
    hormonal signals from roots 83
    osmotic and ionic components 82–3, 83
    see also stress perception for salt and
        osmotic stress

Salt Overly Sensitive (SOS) mutants 84, 85, 86
salt stress tolerance, physiology and
    biochemistry 81–95
  salinity tolerance 81–2
  see also antioxidative systems; cross-tolerance;
    osmotic adjustment (OA) for salt
    stressed plants; salinity, effect on
    growth and development; stress
    perception, salt and osmotic stress;
    sugarcane salt tolerance study
salt stresses and drought 59–60
  ABA involvement 60
  abiotic stress signalling 59
  crop production problems 59
  cytokinin regulation properties 60
  defence mechanisms 60
  DELLA proteins 60
  hyperosmotic and oxidative stress 60
  ionic and osmotic responses to drought/
    salinity 60
  ionic- and osmotic-disequilibrium
    components 59
  jasmonates role 59–60
  stress-dependent physiological/biochemical
    changes in plants 59
senescence 24–6
  catalase (CAT) activity 25–6, 25
  leaf senescence 24–6
  Rubisco activity/content 25
  sugars as signalling molcules 26
  superoxide dismutase (SOD) 25–6, 25
signal transduction, osmotic and ionic
    stress 84–5
  Ca^{2+}-dependent signalling proteins 85
  Na^+ sensor control of Ca^{2+} 85
  salt-dependent protein-nucleic acid
    interactions 85
  salt-induced osmotic stress triggers 84
signalling factors in drought tolerance 76
small heat-shock proteins (sHsps) 6–9
  at non-permissive temperatures 7
  in biomedical field 7
  heat/tolerance enhancement properties 7
  see also Yeast Hsp12
small heat-shock proteins (sHsps), structural
    characteristics 7–8
  M. jannaschii Hsp16.5 8
  structural relationship to amino acids 7–8
  wheat Hsp16.9 8
  X-ray crystallographic analysis 8
small interfering RNAs (siRNAs) 35
small RNAs (smRNAs) 35
SNP associated alternative tags (SAAT) 62
soluble sugars, sensitivity to HS 24
spikelet formation in wheat 27
stability limit, sustainable agriculture 196–7
stomatal closure during salt stress 90

stress
  about stress 1–2
  cytotoxic hazard 2
  defence by Hsps 2
  endoplasmic reticulum (ER) 2
  in the eukaryotes 2
  hyperthermia 2
  prevalent forms 1–2
stress memory issues 34–5
stress perception, salt and osmotic stress 83–5
  AtHK1 as osmosensor 84
  direct mechanisms 83–4
  extracellular Na^+ sensing 84
  histidine kinases 84
  indirect mechanisms 84
  see also signal transduction, osmotic and
    ionic stress
stress response see roots, stress response
SUB1A 47
subcellular compartmentalization 87
substrate binding domain (SBD) 12
sugar metabolism issues 34
  carbohydrate availability 34
  soluble sugar HS sensitivity 34
  tomato studies 34
sugarcane salt tolerance study 102–9
  cellular-level 103
  germination stage 103
  plant material 103
  salt concentrations 103
  statistical analysis 103–4
  young plants-level 103
sugarcane salt tolerance study, results 104–8
  at germination stage 104, 104
  cellular level 106–8
    calli necrosis changes 106–7
    relative fresh weight growth
      (RFWC) 107
  whole plant growth 105–6
    some relative growths 106
  whole plant survival 104–6
    germination percentages 105
    mortality percentages 105
sugars, as signalling molcules 26
sulfur uptake/assimilation
    regulation 143–4, 144
  superoxide dismutase 144–5
  superoxide dismutase (SOD) 25–6, 25, 93, 182
  superSAGE analysis of chickpea
    roots 61–2
sustainable agriculture 196–207
  stability limit 196–7

terpenoids 117
thermotolerance 24, 36, 36
tocopherol cyclase 146

transcription factors (TFs) 33, 72–6
    ABRE-binding protein 75–6
    Arabidopsis transcription activation factor (ATAF) 72
    ATAF proteins 74–5
    cis-acting elements 73
    cis-binding factor (CBF) 72, 73–4
    dehydration responsive element binding regulon (DREB) 72, 73–4
    drought responsible TFs 73
    ERD1 74–5
    myelo-blastosis oncogene (MYB) 72, 76
    myelo-cytomatosis oncogene (MYC) 72, 76
    NAC proteins/family 74–5
    nuclear factor (NF-Y) TFs 74
    RNA polymerase 72
    transgenic approach 74
    WRKY TFs 76
    zinc-finger homoeodomain (ZFHD) 72, 75
Trichoderma, beneficial properties 223, 235–6
Trichoderma harzianum strain T-22 224–8
    relevent abiotic and biotic stresses 225–6
    benefits of use 226–8
        against abiotic stresses 226–7
        against biotic stresses 227–8
    for growth improvement 224
    protection against fungi, bacteria and viruses 224
    root protective properties 224

tropospheric ozone see ozone
turgor loss point (TLP) 201–3

vacuolated cells 87
volatile organic compounds (VOCs) 241–2

water potential 201, 201
water relations 200–3, 202
    see also osmotic adjustment (OA)
water use efficiency/deficiency 203
water-deficient stress see genetic engineering crop plants for drought tolerance
waterlogging 43–4
wheat
    effects of ozone 119–20
    reproduction under HS 27
        spikelet formation 27
WRKY TFs 76

Yeast Hsp12 8–9
    chaperoning activities 9
    unusual characteristics 8–9

zinc (Zn), plant stress from 165
zinc-finger homoeo-domain (ZF-HD) 72, 75